P9-AZV-734

Homo Deus

ALSO BY YUVAL NOAH HARARI

Sapiens: A Brief History of Humankind

Yuval Noah Harari

Homo Deus

A Brief History of Tomorrow

HARPER

An Imprint of HarperCollins*Publishers*

Translated by the author.

First published as *The History of Tomorrow* in Hebrew in Israel in 2015 by Kinneret Zmora-Bitan Dvir.

Previously published in Great Britain in 2016 by Harvill Secker, a division of Penguin Random House Group Ltd.

FIRST U.S. EDITION

Library of Congress Cataloging-in-Publication Data has been applied for.

ISBN: 978-0-06-246431-6

17 18 19 20 21 OFF/RRD 10 9 8 7 6 5 4 3 2 1

To my teacher, S. N. Goenka (1924–2013),
who lovingly taught me important things.

Contents

1. *In vitro* fertilisation: mastering creation.

I

The New Human Agenda

At the dawn of the third millennium, humanity wakes up, stretching its limbs and rubbing its eyes. Remnants of some awful nightmare are still drifting across its mind. 'There was something with barbed wire, and huge mushroom clouds. Oh well, it was just a bad dream.' Going to the bathroom, humanity washes its face, examines its wrinkles in the mirror, makes a cup of coffee and opens the diary. 'Let's see what's on the agenda today.'

For thousands of years the answer to this question remained unchanged. The same three problems preoccupied the people of twentieth-century China, of medieval India and of ancient Egypt. Famine, plague and war were always at the top of the list. For generation after generation humans have prayed to every god, angel and saint, and have invented countless tools, institutions and social systems – but they continued to die in their millions from starvation, epidemics and violence. Many thinkers and prophets concluded that famine, plague and war must be an integral part of God's cosmic plan or of our imperfect nature, and nothing short of the end of time would free us from them.

Yet at the dawn of the third millennium, humanity wakes up to an amazing realisation. Most people rarely think about it, but in the last few decades we have managed to rein in famine, plague and war. Of course, these problems have not been completely solved, but they have been transformed from incomprehensible and uncontrollable forces of nature into manageable challenges. We don't need to pray to any god or saint to rescue us from them. We know quite well what needs to be done in order to prevent famine, plague and war – and we usually succeed in doing it.

True, there are still notable failures; but when faced with such failures we no longer shrug our shoulders and say, 'Well, that's the way things work in our imperfect world' or 'God's will be done'. Rather, when famine, plague or war break out of our control, we feel that somebody must have screwed up, we set up a commission of inquiry, and promise ourselves that next time we'll do better. And it actually works. Such calamities indeed happen less and less often. For the first time in history, more people die today from eating too much than from eating too little; more people die from old age than from infectious diseases; and more people commit suicide than are killed by soldiers, terrorists and criminals combined. In the early twenty-first century, the average human is far more likely to die from bingeing at McDonald's than from drought, Ebola or an al-Qaeda attack.

Hence even though presidents, CEOs and generals still have their daily schedules full of economic crises and military conflicts, on the cosmic scale of history humankind can lift its eyes up and start looking towards new horizons. If we are indeed bringing famine, plague and war under control, what will replace them at the top of the human agenda? Like firefighters in a world without fire, so humankind in the twenty-first century needs to ask itself an unprecedented question: what are we going to do with ourselves? In a healthy, prosperous and harmonious world, what will demand our attention and ingenuity? This question becomes doubly urgent given the immense new powers that biotechnology and information technology are providing us with. What will we do with all that power?

Before answering this question, we need to say a few more words about famine, plague and war. The claim that we are bringing them under control may strike many as outrageous, extremely naïve, or perhaps callous. What about the billions of people scraping a living on less than $2 a day? What about the ongoing AIDS crisis in Africa, or the wars raging in Syria and Iraq? To address these concerns, let us take a closer look at the world of the early twenty-first century, before exploring the human agenda for the coming decades.

The Biological Poverty Line

Let's start with famine, which for thousands of years has been humanity's worst enemy. Until recently most humans lived on the very edge of the biological poverty line, below which people succumb to malnutrition and hunger. A small mistake or a bit of bad luck could easily be a death sentence for an entire family or village. If heavy rains destroyed your wheat crop, or robbers carried off your goat herd, you and your loved ones may well have starved to death. Misfortune or stupidity on the collective level resulted in mass famines. When severe drought hit ancient Egypt or medieval India, it was not uncommon that 5 or 10 per cent of the population perished. Provisions became scarce; transport was too slow and expensive to import sufficient food; and governments were far too weak to save the day.

Open any history book and you are likely to come across horrific accounts of famished populations, driven mad by hunger. In April 1694 a French official in the town of Beauvais described the impact of famine and of soaring food prices, saying that his entire district was now filled with 'an infinite number of poor souls, weak from hunger and wretchedness and dying from want, because, having no work or occupation, they lack the money to buy bread. Seeking to prolong their lives a little and somewhat to appease their hunger, these poor folk eat such unclean things as cats and the flesh of horses flayed and cast onto dung heaps. [Others consume] the blood that flows when cows and oxen are slaughtered, and the offal that cooks throw into the streets. Other poor wretches eat nettles and weeds, or roots and herbs which they boil in water.'[1]

Similar scenes took place all over France. Bad weather had ruined the harvests throughout the kingdom in the previous two years, so that by the spring of 1694 the granaries were completely empty. The rich charged exorbitant prices for whatever food they managed to hoard, and the poor died in droves. About 2.8 million French – 15 per cent of the population – starved to death between 1692 and 1694, while the Sun King, Louis XIV, was dallying with his mistresses in Versailles. The following year, 1695, famine struck Estonia, killing a fifth of the population. In 1696 it was the turn of Finland, where a quarter to a third of people died. Scotland suffered from severe famine between 1695 and 1698, some districts losing up to 20 per cent of their inhabitants.[2]

Most readers probably know how it feels when you miss lunch, when you fast on some religious holiday, or when you live for a few days on vegetable shakes as part of a new wonder diet. But how does it feel when you haven't eaten for days on end and you have no clue where to get the next morsel of food? Most people today have never experienced this excruciating torment. Our ancestors, alas, knew it only too well. When they cried to God, 'Deliver us from famine!', this is what they had in mind.

During the last hundred years, technological, economic and political developments have created an increasingly robust safety net separating humankind from the biological poverty line. Mass famines still strike some areas from time to time, but they are exceptional, and they are almost always caused by human politics rather than by natural catastrophes. There are no longer natural famines in the world; there are only political famines. If people in Syria, Sudan or Somalia starve to death, it is because some politician wants them to.

In most parts of the planet, even if a person has lost his job and all of his possessions, he is unlikely to die from hunger. Private insurance schemes, government agencies and international NGOs may not rescue him from poverty, but they will provide him with enough daily calories to survive. On the collective level, the global trade network turns droughts and floods into business opportunities, and makes it possible to overcome food shortages quickly and cheaply. Even when wars, earthquakes or tsunamis devastate entire countries, international efforts usually succeed in preventing famine. Though hundreds of millions still go hungry almost every day, in most countries very few people actually starve to death.

Poverty certainly causes many other health problems, and malnutrition shortens life expectancy even in the richest countries on earth. In France, for example, 6 million people (about 10 per cent of the population) suffer from nutritional insecurity. They wake up in the morning not knowing whether they will have anything to eat for lunch; they often go to sleep hungry; and the nutrition they do obtain is unbalanced and unhealthy – lots of starch, sugar and salt, and not enough protein and vitamins.[3] Yet nutritional insecurity isn't famine, and France of the early twenty-first century isn't France of 1694. Even in the worst slums around Beauvais or Paris, people don't die because they have not eaten for weeks on end.

The same transformation has occurred in numerous other countries, most notably China. For millennia, famine stalked every Chinese regime from the Yellow Emperor to the Red communists. A few decades ago China was a byword for food shortages. Tens of millions of Chinese starved to death during the disastrous Great Leap Forward, and experts routinely predicted that the problem would only get worse. In 1974 the first World Food Conference was convened in Rome, and delegates were treated to apocalyptic scenarios. They were told that there was no way for China to feed its billion people, and that the world's most populous country was heading towards catastrophe. In fact, it was heading towards the greatest economic miracle in history. Since 1974 hundreds of millions of Chinese have been lifted out of poverty, and though hundreds of millions more still suffer greatly from privation and malnutrition, for the first time in its recorded history China is now free from famine.

Indeed, in most countries today overeating has become a far worse problem than famine. In the eighteenth century Marie Antoinette allegedly advised the starving masses that if they ran out of bread, they should just eat cake instead. Today, the poor are following this advice to the letter. Whereas the rich residents of Beverly Hills eat lettuce salad and steamed tofu with quinoa, in the slums and ghettos the poor gorge on Twinkie cakes, Cheetos, hamburgers and pizza. In 2014 more than 2.1 billion people were overweight, compared to 850 million who suffered from malnutrition. Half of humankind is expected to be overweight by 2030.[4] In 2010 famine and malnutrition combined killed about 1 million people, whereas obesity killed 3 million.[5]

Invisible Armadas

After famine, humanity's second great enemy was plagues and infectious diseases. Bustling cities linked by a ceaseless stream of merchants, officials and pilgrims were both the bedrock of human civilisation and an ideal breeding ground for pathogens. People consequently lived their lives in ancient Athens or medieval Florence knowing that they might fall ill and die next week, or that

2. Medieval people personified the Black Death as a horrific demonic force beyond human control or comprehension.

an epidemic might suddenly erupt and destroy their entire family in one swoop.

The most famous such outbreak, the so-called Black Death, began in the 1330s, somewhere in east or central Asia, when the flea-dwelling bacterium *Yersinia pestis* started infecting humans bitten by the fleas. From there, riding on an army of rats and fleas, the plague quickly spread all over Asia, Europe and North Africa, taking less than twenty years to reach the shores of the Atlantic Ocean. Between 75 million and 200 million people died – more than a quarter of the population of Eurasia. In England, four out of ten people died, and the population dropped from a pre-plague high of 3.7 million people to a post-plague low of 2.2 million. The city of Florence lost 50,000 of its 100,000 inhabitants.[6]

The authorities were completely helpless in the face of the calamity. Except for organising mass prayers and processions, they had no idea how to stop the spread of the epidemic – let alone cure it. Until the modern era, humans blamed diseases on bad air, malicious demons and angry gods, and did not suspect

3. The real culprit was the minuscule *Yersinia pestis* bacterium.

the existence of bacteria and viruses. People readily believed in angels and fairies, but they could not imagine that a tiny flea or a single drop of water might contain an entire armada of deadly predators.

The Black Death was not a singular event, nor even the worst plague in history. More disastrous epidemics struck America, Australia and the Pacific Islands following the arrival of the first Europeans. Unbeknown to the explorers and settlers, they brought with them new infectious diseases against which the natives had no immunity. Up to 90 per cent of the local populations died as a result.[7]

On 5 March 1520 a small Spanish flotilla left the island of Cuba on its way to Mexico. The ships carried 900 Spanish soldiers along with horses, firearms and a few African slaves. One of the slaves, Francisco de Eguía, carried on his person a far deadlier cargo. Francisco didn't know it, but somewhere among his trillions of cells a biological time bomb was ticking: the smallpox virus. After Francisco landed in Mexico the virus began to multiply exponentially within his body, eventually bursting out all over his skin in a terrible rash. The feverish Francisco was taken to bed in the house of a Native American family in the town of Cempoallan. He infected the family members, who infected the neighbours. Within ten days Cempoallan became a graveyard. Refugees spread the disease from Cempoallan to the nearby towns. As town after town succumbed to the plague, new waves of terrified refugees carried the disease throughout Mexico and beyond.

The Mayas in the Yucatán Peninsula believed that three evil gods – Ekpetz, Uzannkak and Sojakak – were flying from village to village at night, infecting people with the disease. The Aztecs blamed it on the gods Tezcatlipoca and Xipetotec, or perhaps on the black magic of the white people. Priests and doctors were consulted. They advised prayers, cold baths, rubbing the body with bitumen and smearing squashed black beetles on the sores. Nothing helped. Tens of thousands of corpses lay rotting in the streets, without anyone daring to approach and bury them. Entire families perished within a few days, and the authorities ordered that the houses were to be collapsed on top of the bodies. In some settlements half the population died.

In September 1520 the plague had reached the Valley of Mexico, and in October it entered the gates of the Aztec capital, Tenochtitlan – a magnificent metropolis of 250,000 people. Within two months at least a third of the population perished, including the Aztec emperor Cuitláhuac. Whereas in March 1520, when the Spanish fleet arrived, Mexico was home to 22 million people, by December only 14 million were still alive. Smallpox was only the first blow. While the new Spanish masters were busy enriching themselves and exploiting the natives, deadly waves of flu, measles and other infectious diseases struck Mexico one after the other, until in 1580 its population was down to less than 2 million.[8]

Two centuries later, on 18 January 1778, the British explorer Captain James Cook reached Hawaii. The Hawaiian islands were densely populated by half a million people, who lived in complete isolation from both Europe and America, and consequently had never been exposed to European and American diseases. Captain Cook and his men introduced the first flu, tuberculosis and syphilis pathogens to Hawaii. Subsequent European visitors added typhoid and smallpox. By 1853, only 70,000 survivors remained in Hawaii.[9]

Epidemics continued to kill tens of millions of people well into the twentieth century. In January 1918 soldiers in the trenches of northern France began dying in the thousands from a particularly virulent strain of flu, nicknamed 'the Spanish Flu'. The front line was the end point of the most efficient global supply network the world had hitherto seen. Men and munitions were pouring in from Britain, the USA, India and Australia. Oil was sent from the Middle East, grain and beef from Argentina, rubber from Malaya and copper from Congo. In exchange, they all got Spanish Flu. Within a few months, about half a billion people – a third of the global population – came down with the virus. In India it killed 5 per cent of the population (15 million people). On the island of Tahiti, 14 per cent died. On Samoa, 20 per cent. In the copper mines of the Congo one out of five labourers perished. Altogether the pandemic killed between 50 million and 100 million people in less than a year. The First World War killed 40 million from 1914 to 1918.[10]

Alongside such epidemical tsunamis that struck humankind every few decades, people also faced smaller but more regular waves of infectious diseases, which killed millions every year. Children who lacked immunity were particularly susceptible to them, hence they are often called 'childhood diseases'. Until the early twentieth century, about a third of children died before reaching adulthood from a combination of malnutrition and disease.

During the last century humankind became ever more vulnerable to epidemics, due to a combination of growing populations and better transport. A modern metropolis such as Tokyo or Kinshasa offers pathogens far richer hunting grounds than medieval Florence or 1520 Tenochtitlan, and the global transport network is today even more efficient than in 1918. A Spanish virus can make its way to Congo or Tahiti in less than twenty-four hours. We should therefore have expected to live in an epidemiological hell, with one deadly plague after another.

However, both the incidence and impact of epidemics have gone down dramatically in the last few decades. In particular, global child mortality is at an all-time low: less than 5 per cent of children die before reaching adulthood. In the developed world the rate is less than 1 per cent.[11] This miracle is due to the unprecedented achievements of twentieth-century medicine, which has provided us with vaccinations, antibiotics, improved hygiene and a much better medical infrastructure.

For example, a global campaign of smallpox vaccination was so successful that in 1979 the World Health Organization declared that humanity had won, and that smallpox had been completely eradicated. It was the first epidemic humans had ever managed to wipe off the face of the earth. In 1967 smallpox had still infected 15 million people and killed 2 million of them, but in 2014 not a single person was either infected or killed by smallpox. The victory has been so complete that today the WHO has stopped vaccinating humans against smallpox.[12]

Every few years we are alarmed by the outbreak of some potential new plague, such as SARS in 2002/3, bird flu in 2005, swine flu in 2009/10 and Ebola in 2014. Yet thanks to efficient counter-measures these incidents have so far resulted in a comparatively small number of victims. SARS, for example, initially raised fears of a new Black Death, but eventually ended with the death of less than 1,000 people worldwide.[13] The Ebola outbreak in West Africa seemed at first to spiral out of control, and on 26 September 2014 the WHO described it as 'the most severe public health emergency seen in modern times'.[14] Nevertheless, by early 2015 the epidemic had been reined in, and in January 2016 the WHO declared it over. It infected 30,000 people (killing 11,000 of them), caused massive economic damage throughout West Africa, and sent shockwaves of anxiety across the world; but it did not spread beyond West Africa, and its death toll was nowhere near the scale of the Spanish Flu or the Mexican smallpox epidemic.

Even the tragedy of AIDS, seemingly the greatest medical failure of the last few decades, can be seen as a sign of progress. Since its first major outbreak in the early 1980s, more than 30 million people have died of AIDS, and tens of millions more have suffered debilitating physical and psychological damage. It was hard to understand and treat the new epidemic, because AIDS is a uniquely devious disease. Whereas a human infected with the smallpox virus dies within a few days, an HIV-positive patient may seem perfectly healthy for weeks and months, yet go on infecting others unknowingly. In addition, the HIV virus itself does not kill. Rather, it destroys the immune system, thereby exposing the patient to numerous other diseases. It is these secondary diseases that actually kill AIDS victims. Consequently, when AIDS began to spread, it was especially difficult to understand what was happening. When two patients were admitted to a New York hospital in 1981, one ostensibly dying from pneumonia and the other from cancer, it was not at all evident that both were in fact victims of the HIV virus, which may have infected them months or even years previously.[15]

However, despite these difficulties, after the medical community became aware of the mysterious new plague, it took scientists just two years to identify it, understand how the virus spreads and suggest effective ways to slow down the epidemic. Within another ten years new medicines turned AIDS from a death sentence into a chronic condition (at least for those wealthy enough to afford the treatment).[16] Just think what would have happened if AIDS had erupted in 1581 rather than 1981. In all likelihood, nobody back then would have figured out what caused the epidemic, how it moved from person to person, or how it could be halted (let alone cured). Under such conditions, AIDS might have killed a much larger proportion of the human race, equalling and perhaps even surpassing the Black Death.

Despite the horrendous toll AIDS has taken, and despite the millions killed each year by long-established infectious diseases such as malaria, epidemics are a far smaller threat to human health today than in previous millennia. The vast majority of people die from non-infectious illnesses such as cancer and heart disease, or simply from old age.[17] (Incidentally cancer and heart disease are of course not new illnesses – they go back to antiquity. In previous eras, however, relatively few people lived long enough to die from them.)

Many fear that this is only a temporary victory, and that some unknown cousin of the Black Death is waiting just around the corner. No one can guarantee that plagues won't make a comeback, but there are good reasons to think that in the arms race between doctors and germs, doctors run faster. New infectious diseases appear mainly as a result of chance mutations in pathogen genomes. These mutations allow the pathogens to jump from animals to humans, to overcome the human immune system, or to resist medicines such as antibiotics. Today such mutations probably occur and disseminate faster than in the past, due to human impact on the environment.[18] Yet in the race against medicine, pathogens ultimately depend on the blind hand of fortune.

Doctors, in contrast, count on more than mere luck. Though science owes a huge debt to serendipity, doctors don't just throw different chemicals into test tubes, hoping to chance upon some new medicine. With each passing year doctors accumulate more and better knowledge, which they use in order to design more effective medicines and treatments. Consequently, though in 2050 we will undoubtedly face much more resilient germs, medicine in 2050 will likely be able to deal with them more efficiently than today.[19]

In 2015 doctors announced the discovery of a completely new type of antibiotic – teixobactin – to which bacteria have no resistance as yet. Some scholars believe teixobactin may prove to be a game-changer in the fight against highly resistant germs.[20] Scientists are also developing revolutionary new treatments that work in radically different ways to any previous medicine. For example, some research labs are already home to nano-robots, which may one day navigate through our bloodstream, identify illnesses and kill pathogens and cancerous cells.[21] Microorganisms may have 4 billion years of cumulative experience fighting organic enemies, but they have exactly zero experience fighting bionic predators, and would therefore find it doubly difficult to evolve effective defences.

So while we cannot be certain that some new Ebola outbreak or an unknown flu strain won't sweep across the globe and kill millions, we will not regard it as an inevitable natural calamity. Rather, we will see it as an inexcusable human failure and demand the heads of those responsible. When in late summer 2014 it seemed for a few terrifying weeks that Ebola was gaining the upper hand over the global health authorities, investigative committees were hastily set up. An initial report published on 18 October 2014 criticised the World Health Organization for its unsatisfactory reaction to the outbreak, blaming the epidemic on corruption and inefficiency in the WHO's African branch. Further criticism was levelled at the international community as a whole for not responding quickly and forcefully enough. Such criticism assumes that humankind has the knowledge and tools to prevent plagues, and if an epidemic nevertheless gets out of control, it is due to human incompetence rather than divine anger. Similarly, the fact that AIDS continued to infect and kill millions in sub-Saharan Africa years after doctors had understood its mechanisms is rightly seen as the result of human failings rather than of cruel fortune.

So in the struggle against natural calamities such as AIDS and Ebola, the scales are tipping in humanity's favour. But what about the dangers inherent in human nature itself? Biotechnology enables us to defeat bacteria and viruses, but it simultaneously turns humans themselves into an unprecedented threat. The same tools that enable doctors to quickly identify and cure new illnesses may also enable armies and terrorists to engineer even more terrible diseases and doomsday pathogens. It is therefore likely that major epidemics will continue to endanger humankind in the future only if humankind itself creates them, in the service of some ruthless ideology. The era when humankind stood helpless before natural epidemics is probably over. But we may come to miss it.

Breaking the Law of the Jungle

The third piece of good news is that wars too are disappearing. Throughout history most humans took war for granted, whereas peace was a temporary and precarious state. International relations were governed by the Law of the Jungle, according to which even if two polities lived in peace, war always remained an option. For example, even though Germany and France were at peace in 1913, everybody knew that they might be at each other's throats in 1914. Whenever politicians, generals, business people and ordinary citizens made plans for the future, they always left room for war. From the Stone Age to the age of steam, and from the Arctic to the Sahara, every person on earth knew that at any moment the neighbours might invade their territory, defeat their army, slaughter their people and occupy their land.

During the second half of the twentieth century this Law of the Jungle has finally been broken, if not rescinded. In most areas wars became rarer than ever. Whereas in ancient agricultural societies human violence caused about 15 per cent of all deaths, during the twentieth century violence caused only 5 per cent of deaths, and in the early twenty-first century it is responsible for about 1 per cent of global mortality.[22] In 2012 about 56 million people died throughout the world; 620,000 of them died due to human violence (war killed 120,000 people, and crime killed another 500,000). In contrast, 800,000 committed suicide, and 1.5 million died of diabetes.[23] Sugar is now more dangerous than gunpowder.

Even more importantly, a growing segment of humankind has come to see war as simply inconceivable. For the first time in history, when governments, corporations and private individuals consider their immediate future, many of them don't think about war as a likely event. Nuclear weapons have turned war between superpowers into a mad act of collective suicide, and therefore forced the most powerful nations on earth to find alternative and peaceful ways to resolve conflicts. Simultaneously, the global economy has been transformed from a material-based economy into a knowledge-based economy. Previously the main sources of wealth were material assets such as gold mines, wheat fields and oil wells. Today the main source of wealth is knowledge. And whereas you can conquer oil fields through war, you cannot acquire knowledge that way. Hence as knowledge became the most important economic resource, the profitability of war declined and wars became increasingly restricted to those parts of the world – such as the Middle East and Central Africa – where the economies are still old-fashioned material-based economies.

In 1998 it made sense for Rwanda to seize and loot the rich coltan mines of neighbouring Congo, because this ore was in high demand for the manufacture of mobile phones and laptops, and Congo held 80 per cent of the world's coltan reserves. Rwanda earned $240 million annually from the looted coltan. For poor Rwanda that was a lot of money.[24] In contrast, it would have made no sense for China to invade California and seize Silicon Valley, for even if the Chinese could somehow prevail on the battlefield, there were no silicon mines to loot in Silicon Valley. Instead, the Chinese have earned billions of dollars from cooperating with hi-tech giants such as Apple and Microsoft, buying their software and manufacturing their products. What Rwanda earned from an entire year of looting Congolese coltan, the Chinese earn in a single day of peaceful commerce.

In consequence, the word 'peace' has acquired a new meaning. Previous generations thought about peace as the temporary absence of war. Today we think about peace as the implausibility of war. When in 1913 people said that there was peace between France and Germany, they meant that 'there is no war going on at present between France and Germany, but who knows what next year will bring'. When today we say that there is peace between France and Germany, we mean that it is inconceivable under any foreseeable circumstances that war might break out between them. Such peace prevails not only between France and Germany, but between most (though not all) countries. There is no scenario for a serious war breaking out next year between Germany and Poland, between Indonesia and the Philippines, or between Brazil and Uruguay.

This New Peace is not just a hippie fantasy. Power-hungry governments and greedy corporations also count on it. When Mercedes plans its sales strategy in eastern Europe, it discounts the possibility that Germany might conquer Poland. A corporation importing cheap labourers from the Philippines is not worried that Indonesia might invade the Philippines next year. When the Brazilian government convenes to discuss next year's budget, it's unimaginable that the Brazilian defence minister will rise from his seat, bang his fist on the table and shout, 'Just a minute! What if we want to invade and conquer Uruguay? You didn't take that into account. We have to put aside $5 billion to finance this conquest.' Of course, there are a few places where defence ministers still say such things, and there are regions where the New Peace has failed to take root. I know this very well because I live in one of these regions. But these are exceptions.

There is no guarantee, of course, that the New Peace will hold indefinitely. Just as nuclear weapons made the New Peace possible in the first place, so future technological developments might set the stage for new kinds of war. In particular, cyber warfare may destabilise the world by giving even small countries and non-state actors the ability to fight superpowers effectively. When the USA fought Iraq in 2003 it brought havoc to Baghdad and Mosul, but not a single bomb was dropped on Los Angeles or Chicago. In the future, though, a country such as North Korea or Iran could use logic bombs to shut down the power in California, blow up refineries in Texas and cause trains to collide in Michigan ('logic bombs' are malicious software codes planted in peacetime and operated at a distance. It is highly likely that networks controlling vital infrastructure facilities in the USA and many other countries are already crammed with such codes).

However, we should not confuse ability with motivation. Though cyber warfare introduces new means of destruction, it doesn't necessarily add new incentives to use them. Over the last seventy years humankind has broken not only the Law of the Jungle, but also the Chekhov Law. Anton Chekhov famously said that a gun appearing in the first act of a play will inevitably be fired

4. Nuclear missiles on parade in Moscow. The gun that was always on display but never fired.

in the third. Throughout history, if kings and emperors acquired some new weapon, sooner or later they were tempted to use it. Since 1945, however, humankind has learned to resist this temptation. The gun that appeared in the first act of the Cold War was never fired. By now we are accustomed to living in a world full of undropped bombs and unlaunched missiles, and have become experts in breaking both the Law of the Jungle and the Chekhov Law. If these laws ever do catch up with us, it will be our own fault – not our inescapable destiny.

What about terrorism, then? Even if central governments and powerful states have learned restraint, terrorists might have no such qualms about using new and destructive weapons. That is certainly a worrying possibility. However, terrorism is a strategy of weakness adopted by those who lack access to real power. At least in the past, terrorism worked by spreading fear rather than by causing significant material damage. Terrorists usually don't have the strength to defeat an army, occupy a country or destroy entire cities. Whereas in 2010 obesity and related illnesses killed about 3 million people, terrorists killed a total of 7,697 people across the globe, most of them in developing countries.[25] For the average American or European, Coca-Cola poses a far deadlier threat than al-Qaeda.

How, then, do terrorists manage to dominate the headlines and change the political situation throughout the world? By provoking their enemies to overreact. In essence, terrorism is a show. Terrorists stage a terrifying spectacle of violence that captures our imagination and makes us feel as if we are sliding back into medieval chaos. Consequently states often feel obliged to react to the theatre of terrorism with a show of security, orchestrating immense displays of force, such as the persecution of entire populations or the invasion of foreign countries. In most cases, this overreaction to terrorism poses a far greater threat to our security than the terrorists themselves.

Terrorists are like a fly that tries to destroy a china shop. The fly is so weak that it cannot budge even a single teacup. So it finds a bull, gets inside its ear and starts buzzing. The bull goes wild with fear and anger, and destroys the china shop. This is what happened in the Middle East in the last decade. Islamic fundamentalists could never have toppled Saddam Hussein by themselves. Instead they enraged the USA by the 9/11 attacks, and the USA destroyed the Middle Eastern china shop for them. Now they flourish in the wreckage. By themselves, terrorists are too weak to drag us back to the Middle Ages and re-establish the Jungle Law. They may provoke us, but in the end, it all depends on our reactions. If the Jungle Law comes back into force, it will not be the fault of terrorists.

Famine, plague and war will probably continue to claim millions of victims in the coming decades. Yet they are no longer unavoidable tragedies beyond the understanding and control of a helpless humanity. Instead, they have become manageable challenges. This does not belittle the suffering of hundreds of millions of poverty-stricken humans; of the millions felled each year by malaria, AIDS and tuberculosis; or of the millions trapped in violent vicious circles in Syria, the Congo or Afghanistan. The message is not that famine, plague and war have completely disappeared from the face of the earth, and that we should stop worrying about them. Just the opposite. Throughout history people felt these were unsolvable problems, so there was no point trying to put an end to them. People prayed to God for miracles, but they themselves did not seriously attempt to exterminate famine, plague and war. Those arguing that the world of 2016 is as hungry, sick and violent as it was in 1916 perpetuate this age-old defeatist view. They imply that all the huge efforts humans have made during the twentieth century have achieved nothing, and that medical research, economic reforms and peace initiatives have all been in vain. If so, what is the point of investing our time and resources in further medical research, novel economic reforms or new peace initiatives?

Acknowledging our past achievements sends a message of hope and responsibility, encouraging us to make even greater efforts in the future. Given our twentieth-century accomplishments, if people continue to suffer from famine, plague and war, we cannot blame it on nature or on God. It is within our power to make things better and to reduce the incidence of suffering even further.

Yet appreciating the magnitude of our achievements carries another message: history does not tolerate a vacuum. If incidences of famine, plague and war are decreasing, something is bound to take their place on the human agenda. We had better think very carefully about what it is going to be. Otherwise, we might gain complete victory in the old battlefields only to be caught completely unaware on entirely new fronts. What are the projects that will replace famine, plague and war at the top of the human agenda in the twenty-first century?

One central project will be to protect humankind and the planet as a whole from the dangers inherent in our own power. We have managed to bring famine, plague and war under control thanks largely to our phenomenal economic growth, which provides us with abundant food, medicine, energy and raw materials. Yet this same growth destabilises the ecological equilibrium of the planet in myriad ways, which we have only begun to explore. Humankind has been late in acknowledging this danger, and has so far done very little about it. Despite all the talk of pollution, global warming and climate change, most countries have yet to make any serious economic or political sacrifices to improve the situation. When the moment comes to choose between economic growth and ecological stability, politicians, CEOs and voters almost always prefer growth. In the twenty-first century, we shall have to do better if we are to avoid catastrophe.

What else will humanity strive for? Would we be content merely to count our blessings, keep famine, plague and war at bay, and protect the ecological equilibrium? That might indeed be the wisest course of action, but humankind is unlikely to follow it. Humans are rarely satisfied with what they already have. The most common reaction of the human mind to achievement is not satisfaction, but craving for more. Humans are always on the lookout for something better, bigger, tastier. When humankind possesses enormous new powers, and when the threat of famine, plague and war is finally lifted, what will we do with ourselves? What will the scientists, investors, bankers and presidents do all day? Write poetry?

Success breeds ambition, and our recent achievements are now pushing humankind to set itself even more daring goals. Having secured unprecedented levels of prosperity, health and harmony, and given our past record and our current values, humanity's next targets are likely to be immortality, happiness and divinity. Having reduced mortality from starvation, disease and violence, we will now aim to overcome old age and even death itself. Having saved people from abject misery, we will now aim to make them positively happy. And having raised humanity above the beastly level of survival struggles, we will now aim to upgrade humans into gods, and turn *Homo sapiens* into *Homo deus*.

The Last Days of Death

In the twenty-first century humans are likely to make a serious bid for immortality. Struggling against old age and death will merely carry on the time-honoured fight against famine and disease, and manifest the supreme value of contemporary culture: the worth of human life. We are constantly reminded that human life is the most sacred thing in the universe. Everybody says this: teachers in schools, politicians in parliaments, lawyers in courts and actors on theatre stages. The Universal Declaration of Human Rights adopted by the UN after the Second World War – which is perhaps the closest thing we have to a global constitution – categorically states that 'the right to life' is humanity's most fundamental value. Since death clearly violates this right, death is a crime against humanity, and we ought to wage total war against it.

Throughout history, religions and ideologies did not sanctify life itself. They always sanctified something above or beyond earthly existence, and were consequently quite tolerant of death. Indeed, some of them have been downright fond of the Grim Reaper. Because Christianity, Islam and Hinduism insisted that the meaning of our existence depended on our fate in the afterlife, they viewed death as a vital and positive part of the world. Humans died because God decreed it, and their moment of death was a sacred metaphysical experience exploding with meaning. When a human was about to breathe his last, this was the time to call priests, rabbis and shamans, to draw out the balance of life, and to embrace one's true role in the universe. Just try to imagine Christianity, Islam or Hinduism in a world without death – which is also a world without heaven, hell or reincarnation.

Modern science and modern culture have an entirely different take on life and death. They don't think of death as a metaphysical mystery, and they certainly don't view death as the source of life's meaning. Rather, for modern people death is a technical problem that we can and should solve.

How exactly do humans die? Medieval fairy tales depicted Death as a figure in a hooded black cloak, his hand gripping a large scythe. A man lives his life, worrying about this and that, running here and there, when suddenly the Grim Reaper appears before him, taps him on the shoulder with a bony finger and says, 'Come!' And the man implores: 'No, please! Wait just a year, a month, a day!' But the hooded figure hisses: 'No! You must come NOW!' And this is how we die.

In reality, however, humans don't die because a figure in a black cloak taps them on the shoulder, or because God decreed

5. Death personified as the Grim Reaper in medieval art.

it, or because mortality is an essential part of some great cosmic plan. Humans always die due to some technical glitch. The heart stops pumping blood. The main artery is clogged by fatty deposits. Cancerous cells spread in the liver. Germs multiply in the lungs. And what is responsible for all these technical problems? Other technical problems. The heart stops pumping blood because not enough oxygen reaches the heart muscle. Cancerous cells spread because a chance genetic mutation rewrote their instructions. Germs settled in my lungs because somebody sneezed on the subway. Nothing metaphysical about it. It is all technical problems.

And every technical problem has a technical solution. We don't need to wait for the Second Coming in order to overcome death. A couple of geeks in a lab can do it. If traditionally death was the speciality of priests and theologians, now the engineers are taking over. We can kill the cancerous cells with chemotherapy or nanorobots. We can exterminate the germs in the lungs with antibiotics. If the heart stops pumping, we can reinvigorate it with medicines and electric shocks – and if that doesn't work, we can implant a new heart. True, at present we don't have solutions to all technical problems. But this is precisely why we invest so much time and money in researching cancer, germs, genetics and nanotechnology.

Even ordinary people, who are not engaged in scientific research, have become used to thinking about death as a technical problem. When a woman goes to her physician and asks, 'Doctor, what's wrong with me?' the doctor is likely to say, 'Well, you have the flu,' or 'You have tuberculosis,' or 'You have cancer.' But the doctor will never say, 'You have death.' And we are all under the impression that flu, tuberculosis and cancer are technical problems, to which we might someday find a technical solution.

Even when people die in a hurricane, a car accident or a war, we tend to view it as a technical failure that could and should have been prevented. If the government had only adopted a better policy; if the municipality had done its job properly; and if the military commander had taken a wiser decision, death would have been avoided. Death has become an almost automatic reason for law-suits and investigations. 'How could they have died? Somebody somewhere must have screwed up.'

The vast majority of scientists, doctors and scholars still distance themselves from outright dreams of immortality, claiming that they are trying to overcome only this or that particular problem. Yet because old age and death are the outcome of nothing but par-ticular problems, there is no point at which doctors and scientists are going to stop and declare: 'Thus far, and not another step. We have overcome tuberculosis and cancer, but we won't lift a finger to fight Alzheimer's. People can go on dying from that.' The Universal Declaration of Human Rights does not say that humans have 'the right to life until the age of ninety'. It says that every human has a right to life, period. That right isn't limited by any expiry date.

An increasing minority of scientists and thinkers consequently speak more openly these days, and state that the flagship enterprise of modern science is to defeat death and grant humans eternal youth. Notable examples are the gerontologist Aubrey de Grey and the polymath and inventor Ray Kurzweil (winner of the 1999 US National Medal of Technology and Innovation). In 2012 Kurzweil was appointed a director of engineering at Google, and a year later Google launched a sub-company called Calico whose stated mission is 'to solve death'.[26] In 2009 Google appointed another immortality true-believer, Bill Maris, to preside over the Google Ventures investment fund. In a January 2015 interview, Maris said, 'If you ask me today, is it possible to live to be 500, the answer is yes.' Maris backs up his brave words with a lot of hard cash. Google Ventures is investing 36 per cent of its $2 billion portfolio in life sciences start-ups, including several ambitious life-extending projects. Using an American football analogy, Maris explained that in the fight against death, 'We aren't trying to gain a few yards. We are trying to win the game.' Why? Because, says Maris, 'it is better to live than to die'.[27]

Such dreams are shared by other Silicon Valley luminaries. PayPal co-founder Peter Thiel has recently confessed that he aims to live for ever. 'I think there are probably three main modes of approaching [death],' he explained. 'You can accept it, you can deny it or you can fight it. I think our society is dominated by people who are into denial or acceptance, and I prefer to fight it.' Many people are likely to dismiss such statements as teenage fantasies. Yet Thiel is somebody to be taken very seriously. He is one of the most successful and influential entrepreneurs in Silicon Valley with a private fortune estimated at $2.2 billion.[28] The writing is on the wall: equality is out – immortality is in.

The breakneck development of fields such as genetic engineering, regenerative medicine and nanotechnology fosters ever more optimistic prophecies. Some experts believe that humans will overcome death by 2200, others say 2100. Kurzweil and de Grey are even more sanguine. They maintain that anyone possessing a healthy body and a healthy bank account in 2050 will have a serious shot at immortality by cheating death a decade at a time. According to Kurzweil and de Grey, every ten years or so we will march into the clinic and receive a makeover treatment that will not only cure illnesses, but will also regenerate decaying tissues, and upgrade hands, eyes and brains. Before the next treatment is due, doctors will have invented a plethora of new medicines, upgrades and gadgets. If Kurzweil and de Grey are right, there may already be some immortals walking next to you on the street – at least if you happen to be walking down Wall Street or Fifth Avenue.

In truth they will actually be a-mortal, rather than immortal. Unlike God, future superhumans could still die in some war or accident, and nothing could bring them back from the netherworld. However, unlike us mortals, their life would have no expiry date. So long as no bomb shreds them to pieces or no truck runs them over, they could go on living indefinitely. Which will probably make them the most anxious people in history. We mortals daily take chances with our lives, because we know they are going to end anyhow. So we go on treks in the Himalayas, swim in the sea, and do many other dangerous things like crossing the street or eating out. But if you believe you can live for ever, you would be crazy to gamble on infinity like that.

Perhaps, then, we had better start with more modest aims, such as doubling life expectancy? In the twentieth century we have almost doubled life expectancy from forty to seventy, so in the twenty-first century we should at least be able to double it again to 150. Though falling far short of immortality, this would still revolutionise human society. For starters, family structure, marriages and child–parent relationships would be transformed. Today, people still expect to be married 'till death us do part', and much of life revolves around having and raising children. Now try to imagine a person with a lifespan of 150 years. Getting married at forty, she still has 110 years to go. Will it be realistic to expect her marriage to last 110 years? Even Catholic fundamentalists might baulk at that. So the current trend of serial marriages is likely to intensify. Bearing two children in her forties, she will, by the time she is 120, have only a distant memory of the years she spent raising them – a rather minor episode in her long life. It's hard to tell what kind of new parent–child relationship might develop under such circumstances.

Or consider professional careers. Today we assume that you learn a profession in your teens and twenties, and then spend the rest of your life in that line of work. You obviously learn new things even in your forties and fifties, but life is generally divided into a learning period followed by a working period. When you live to be 150 that won't do, especially in a world that is constantly being shaken by new technologies. People will have much longer careers, and will have to reinvent themselves again and again even at the age of ninety.

At the same time, people will not retire at sixty-five and will not make way for the new generation with its novel ideas and aspirations. The physicist Max Planck famously said that science advances one funeral at a time. He meant that only when one generation passes away do new theories have a chance to root out old ones. This is true not only of science. Think for a moment about your own workplace. No matter whether you are a scholar, journalist, cook or football player, how would you feel if your boss were 120, his ideas were formulated when Victoria was still queen, and he was likely to stay your boss for a couple of decades more?

In the political sphere the results might be even more sinister. Would you mind having Putin stick around for another ninety years? On second thought, if people lived to 150, then in 2016 Stalin would still be ruling in Moscow, going strong at 138, Chairman Mao would be a middle-aged 123-year-old, and Princess Elizabeth would be sitting on her hands waiting to inherit from the 121-year-old George VI. Her son Charles would not get his turn until 2076.

Coming back to the realm of reality, it is far from certain whether Kurzweil's and de Grey's prophecies will come true by 2050 or 2100. My own view is that the hopes of eternal youth in the twenty-first century are premature, and whoever takes them too seriously is in for a bitter disappointment. It is not easy to live knowing that you are going to die, but it is even harder to believe in immortality and be proven wrong.

Although average life expectancy has doubled over the last hundred years, it is unwarranted to extrapolate and conclude that we can double it again to 150 in the coming century. In 1900 global life expectancy was no higher than forty because many people died young from malnutrition, infectious diseases and violence. Yet those who escaped famine, plague and war could live well into their seventies and eighties, which is the natural life span of *Homo sapiens*. Contrary to common notions, seventy-year-olds weren't considered rare freaks of nature in previous centuries. Galileo Galilei died at seventy-seven, Isaac Newton at eighty-four, and Michelangelo lived to the ripe age of eighty-eight, without any help from antibiotics, vaccinations or organ transplants. Indeed, even chimpanzees in the jungle sometimes live into their sixties.[29]

In truth, so far modern medicine hasn't extended our natural life span by a single year. Its great achievement has been to save us from *premature* death, and allow us to enjoy the full measure of our years. Even if we now overcome cancer, diabetes and the other major killers, it would mean only that almost everyone will get to live to ninety – but it will not be enough to reach 150, let alone 500. For that, medicine will need to re-engineer the most fundamental structures and processes of the human body, and discover how to regenerate organs and tissues. It is by no means clear that we can do that by 2100.

Nevertheless, every failed attempt to overcome death will get us a step closer to the target, and that will inspire greater hopes and encourage people to make even greater efforts. Though Google's Calico probably won't solve death in time to make Google co-founders Sergey Brin and Larry Page immortal, it will most probably make significant discoveries about cell biology, genetic medicines and human health. The next generation of Googlers could therefore start their attack on death from new and better positions. The scientists who cry immortality are like the boy who cried wolf: sooner or later, the wolf actually comes.

Hence even if we don't achieve immortality in our lifetime, the war against death is still likely to be the flagship project of the coming century. When you take into account our belief in the sanctity of human life, add the dynamics of the scientific establishment, and top it all with the needs of the capitalist economy, a relentless war against death seems to be inevitable. Our ideological commitment to human life will never allow us simply to accept human death. As long as people die of something, we will strive to overcome it.

The scientific establishment and the capitalist economy will be more than happy to underwrite this struggle. Most scientists and bankers don't care what they are working on, provided it gives them an opportunity to make new discoveries and greater profits. Can anyone imagine a more exciting scientific challenge than outsmarting death – or a more promising market than the market of eternal youth? If you are over forty, close your eyes for a minute and try to remember the body you had at twenty-five. Not only how it looked, but above all how it *felt*. If you could have that body back, how much would you be willing to pay for it? No doubt some people would be happy to forgo the opportunity, but enough customers would pay whatever it takes, constituting a well-nigh infinite market.

If all that is not enough, the fear of death ingrained in most humans will give the war against death an irresistible momentum. As long as people assumed that death is inevitable, they trained themselves from an early age to suppress the desire to live for ever, or harnessed it in favour of substitute goals. People want to live for ever, so they compose an 'immortal' symphony, they strive for 'eternal glory' in some war, or even sacrifice their lives so that their souls will 'enjoy everlasting bliss in paradise'. A large part of our artistic creativity, our political commitment and our religious piety is fuelled by the fear of death.

Woody Allen, who has made a fabulous career out of the fear of death, was once asked if he hoped to live on for ever on the silver screen. Allen answered that 'I'd rather live on in my apartment.' He went on to add that 'I don't want to achieve immortality through my work. I want to achieve it by not dying.' Eternal glory, nationalist remembrance ceremonies and dreams of paradise are very poor substitutes for what humans like Allen really want – not to die. Once people think (with or without good reason) that they have a serious chance of escaping death, the desire for life will refuse to go on pulling the rickety wagon of art, ideology and religion, and will sweep forward like an avalanche.

If you think that religious fanatics with burning eyes and flowing beards are ruthless, just wait and see what elderly retail moguls and ageing Hollywood starlets will do when they think the elixir of life is within reach. If and when science makes significant progress in the war against death, the real battle will shift from the laboratories to the parliaments, courthouses and streets. Once the scientific efforts are crowned with success, they will trigger bitter political conflicts. All the wars and conflicts of history might turn out to be but a pale prelude for the real struggle ahead of us: the struggle for eternal youth.

The Right to Happiness

The second big project on the human agenda will probably be to find the key to happiness. Throughout history numerous thinkers, prophets and ordinary people defined happiness rather than life itself as the supreme good. In ancient Greece the philosopher Epicurus explained that worshipping gods is a waste of time, that there is no existence after death, and that happiness is the sole purpose of life. Most people in ancient times rejected Epicureanism, but today it has become the default view. Scepticism about the afterlife drives humankind to seek not only immortality, but also earthly happiness. For who would like to live for ever in eternal misery?

For Epicurus the pursuit of happiness was a personal quest. Modern thinkers, in contrast, tend to see it as a collective project. Without government planning, economic resources and scientific research, individuals will not get far in their quest for happiness. If your country is torn apart by war, if the economy is in crisis and if health care is non-existent, you are likely to be miserable. At the end of the eighteenth century the British philosopher Jeremy Bentham declared that the supreme good is 'the greatest happiness of the greatest number', and concluded that the sole worthy aim of the state, the market and the scientific community is to increase global happiness. Politicians should make peace, business people should foster prosperity and scholars should study nature, not for the greater glory of king, country or God – but so that you and I could enjoy a happier life.

During the nineteenth and twentieth centuries, although many paid lip service to Bentham's vision, governments, corporations and laboratories focused on more immediate and well-defined aims. Countries measured their success by the size of their territory, the increase in their population and the growth of their GDP – not by the happiness of their citizens. Industrialised nations such as Germany, France and Japan established gigantic systems of education, health and welfare, yet these systems were aimed to strengthen the nation rather than ensure individual well-being.

Schools were founded to produce skilful and obedient citizens who would serve the nation loyally. At eighteen, youths needed to be not only patriotic but also literate, so that they could read the brigadier's order of the day and draw up tomorrow's battle plans. They had to know mathematics in order to calculate the shell's trajectory or crack the enemy's secret code. They needed a reasonable command of electrics, mechanics and medicine in order to operate wireless sets, drive tanks and take care of wounded comrades. When they left the army they were expected to serve the nation as clerks, teachers and engineers, building a modern economy and paying lots of taxes.

The same went for the health system. At the end of the nineteenth century countries such as France, Germany and Japan began providing free health care for the masses. They financed vaccinations for infants, balanced diets for children and physical education for teenagers. They drained festering swamps, exterminated mosquitoes and built centralised sewage systems. The aim wasn't to make people happy, but to make the nation stronger. The country needed sturdy soldiers and workers, healthy women who would give birth to more soldiers and workers, and bureaucrats who came to the office punctually at 8 a.m. instead of lying sick at home.

Even the welfare system was originally planned in the interest of the nation rather than of needy individuals. When Otto von Bismarck pioneered state pensions and social security in late nineteenth-century Germany, his chief aim was to ensure the loyalty of the citizens rather than to increase their well-being. You fought for your country when you were eighteen, and paid your taxes when you were forty, because you counted on the state to take care of you when you were seventy.[30]

In 1776 the Founding Fathers of the United States established the right to the pursuit of happiness as one of three unalienable human rights, alongside the right to life and the right to liberty. It's important to note, however, that the American Declaration of Independence guaranteed the right to *the pursuit of* happiness, not the right to happiness itself. Crucially, Thomas Jefferson did not make the state responsible for its citizens' happiness. Rather, he sought only to limit the power of the state. The idea was to reserve for individuals a private sphere of choice, free from state supervision. If I think I'll be happier marrying John rather than Mary, living in San Francisco rather than Salt Lake City, and working as a bartender rather than a dairy farmer, then it's my right to pursue happiness my way, and the state shouldn't intervene even if I make the wrong choice.

Yet over the last few decades the tables have turned, and Bentham's vision has been taken far more seriously. People increasingly believe that the immense systems established more than a century ago to strengthen the nation should actually serve the happiness and well-being of individual citizens. We are not here to serve the state – it is here to serve us. The right to the pursuit of happiness, originally envisaged as a restraint on state power, has imperceptibly morphed into the right to happiness – as if human beings have a natural right to be happy, and anything which makes us dissatisfied is a violation of our basic human rights, so the state should do something about it.

In the twentieth century per capita GDP was perhaps the supreme yardstick for evaluating national success. From this perspective, Singapore, each of whose citizens produces on average $56,000 worth of goods and services a year, is a more successful country than Costa Rica, whose citizens produce only $14,000 a year. But nowadays thinkers, politicians and even economists are calling to supplement or even replace GDP with GDH – gross domestic happiness. After all, what do people want? They don't want to produce. They want to be happy. Production is important because it provides the material basis for happiness. But it is only the means, not the end. In one survey after another Costa Ricans report far higher levels of life satisfaction than Singaporeans. Would you rather be a highly productive but dissatisfied Singaporean, or a less productive but satisfied Costa Rican?

This kind of logic might drive humankind to make happiness its second main goal for the twenty-first century. At first glance this might seem a relatively easy project. If famine, plague and war are disappearing, if humankind experiences unprecedented peace and prosperity, and if life expectancy increases dramatically, surely all that will make humans happy, right?

Wrong. When Epicurus defined happiness as the supreme good, he warned his disciples that it is hard work to be happy. Material achievements alone will not satisfy us for long. Indeed, the blind pursuit of money, fame and pleasure will only make us miserable. Epicurus recommended, for example, to eat and drink in moderation, and to curb one's sexual appetites. In the long run, a deep friendship will make us more content than a frenzied orgy. Epicurus outlined an entire ethic of dos and don'ts to guide people along the treacherous path to happiness.

Epicurus was apparently on to something. Being happy doesn't come easy. Despite our unprecedented achievements in the last few decades, it is far from obvious that contemporary people are significantly more satisfied than their ancestors in bygone years. Indeed, it is an ominous sign that despite higher prosperity, comfort and security, the rate of suicide in the developed world is also much higher than in traditional societies.

In Peru, Guatemala, the Philippines and Albania – developing countries suffering from poverty and political instability – about one person in 100,000 commits suicide each year. In rich and peaceful countries such as Switzerland, France, Japan and New Zealand, twenty-five people per 100,000 take their own lives annually. In 1985 most South Koreans were poor, uneducated and tradition-bound, living under an authoritarian dictatorship. Today South Korea is a leading economic power, its citizens are among the best educated in the world, and it enjoys a stable and comparatively liberal democratic regime. Yet whereas in 1985 about nine South Koreans per 100,000 killed themselves, today the annual rate of suicide has more than tripled to thirty per 100,000.[31]

There are of course opposite and far more encouraging trends. Thus the drastic decrease in child mortality has surely brought an increase in human happiness, and partially compensated people for the stress of modern life. Still, even if we are somewhat happier than our ancestors, the increase in our well-being is far less than we might have expected. In the Stone Age, the average human had at his or her disposal about 4,000 calories of energy per day. This included not only food, but also the energy invested in preparing tools, clothing, art and campfires. Today Americans use on average 228,000 calories of energy per person per day, to feed not only their stomachs but also their cars, computers, refrigerators and televisions.[32] The average American thus uses sixty times more energy than the average Stone Age hunter-gatherer. Is the average American sixty times happier? We may well be sceptical about such rosy views.

And even if we have overcome many of yesterday's miseries, attaining positive happiness may be far more difficult than abolishing downright suffering. It took just a piece of bread to make a starving medieval peasant joyful. How do you bring joy to a bored, overpaid and overweight engineer? The second half of the twentieth century was a golden age for the USA. Victory in the Second World War, followed by an even more decisive victory in the Cold War, turned it into the leading global superpower. Between 1950 and 2000 American GDP grew from $2 trillion to $12 trillion. Real per capita income doubled. The newly invented contraceptive pill made sex freer than ever. Women, gays, African Americans and other minorities finally got a bigger slice of the American pie. A flood of cheap cars, refrigerators, air conditioners, vacuum cleaners, dishwashers, laundry machines, telephones, televisions and computers changed daily life almost beyond recognition. Yet studies have shown that American subjective well-being levels in the 1990s remained roughly the same as they were in the 1950s.[33]

In Japan, average real income rose by a factor of five between 1958 and 1987, in one of the fastest economic booms of history. This avalanche of wealth, coupled with myriad positive and negative changes in Japanese lifestyles and social relations, had surprisingly little impact on Japanese subjective well-being levels. The Japanese in the 1990s were as satisfied – or dissatisfied – as they were in the 1950s.[34]

It appears that our happiness bangs against some mysterious glass ceiling that does not allow it to grow despite all our unprecedented accomplishments. Even if we provide free food for everybody, cure all diseases and ensure world peace, it won't necessarily shatter that glass ceiling. Achieving real happiness is not going to be much easier than overcoming old age and death.

The glass ceiling of happiness is held in place by two stout pillars, one psychological, the other biological. On the psychological level, happiness depends on expectations rather than objective conditions. We don't become satisfied by leading a peaceful and prosperous existence. Rather, we become satisfied when reality matches our expectations. The bad news is that as conditions improve, expectations balloon. Dramatic improvements in conditions, as humankind has experienced in recent decades, translate into greater expectations rather than greater contentment. If we don't do something about this, our future achievements too might leave us as dissatisfied as ever.

On the biological level, both our expectations and our happiness are determined by our biochemistry, rather than by our economic, social or political situation. According to Epicurus, we are happy when we feel pleasant sensations and are free from unpleasant ones. Jeremy Bentham similarly maintained that nature gave dominion over man to two masters – pleasure and pain – and they alone determine everything we do, say and think. Bentham's successor, John Stuart Mill, explained that happiness is nothing but pleasure and freedom from pain, and that beyond pleasure and pain there is no good and no evil. Anyone who tries to deduce good and evil from something else (such as the word of God, or the national interest) is fooling you, and perhaps fooling himself too.[35]

In the days of Epicurus such talk was blasphemous. In the days of Bentham and Mill it was radical subversion. But in the early twenty-first century this is scientific orthodoxy. According to the life sciences, happiness and suffering are nothing but different balances of bodily sensations. We never react to events in the outside world, but only to sensations in our own bodies. Nobody suffers because she lost her job, because she got divorced or because the government went to war. The only thing that makes people miserable is unpleasant sensations in their own bodies. Losing one's job can certainly trigger depression, but depression itself is a kind of unpleasant bodily sensation. A thousand things may make us angry, but anger is never an abstraction. It is always felt as a sensation of heat and tension in the body, which is what makes anger so infuriating. Not for nothing do we say that we 'burn' with anger.

Conversely, science says that nobody is ever made happy by getting a promotion, winning the lottery or even finding true love. People are made happy by one thing and one thing only – pleasant sensations in their bodies. Imagine that you are Mario Götze, the attacking midfielder of the German football team in the 2014 World Cup Final against Argentina; 113 minutes have already elapsed, without a goal being scored. Only seven minutes remain before the dreaded penalty shoot-out. Some 75,000 excited fans fill the Maracanã stadium in Rio, with countless millions anxiously watching all over the world. You are a few yards from the Argentinian goal when André Schürrle sends a magnificent pass in your direction. You stop the ball with your chest, it drops down towards your leg, you give it a kick in mid-air, and you see it fly past the Argentinian goalkeeper and bury itself deep inside the net. Gooooooooal! The stadium erupts like a volcano. Tens of thousands of people roar like mad, your teammates are racing to hug and kiss you, millions of people back home in Berlin and Munich collapse in tears before the television screen. You are ecstatic, but not because of the ball in the Argentinian net or the celebrations going on in crammed Bavarian *Biergartens*. You are actually reacting to the storm of sensations within you. Chills run up and down your spine, waves of electricity wash over your body, and it feels as if you are dissolving into millions of exploding energy balls.

You don't have to score the winning goal in the World Cup Final to feel such sensations. If you receive an unexpected promotion at work, and start jumping for joy, you are reacting to the same kind of sensations. The deeper parts of your mind know nothing about football or about jobs. They know only sensations. If you get a promotion, but for some reason don't feel any pleasant sensations – you will not feel satisfied. The opposite is also true. If you have just been fired (or lost a decisive football match), but you are experiencing very pleasant sensations (perhaps because you popped some pill), you might still feel on top of the world.

The bad news is that pleasant sensations quickly subside and sooner or later turn into unpleasant ones. Even scoring the winning goal in the World Cup Final doesn't guarantee lifelong bliss. In fact, it might all be downhill from there. Similarly, if last year I received an unexpected promotion at work, I might still be occupying that new position, but the very pleasant sensations I experienced on hearing the news disappeared within hours. If I want to feel those wonderful sensations again, I must get another promotion. And another. And if I don't get a promotion, I might end up far more bitter and angry than if I had remained a humble pawn.

This is all the fault of evolution. For countless generations our biochemical system adapted to increasing our chances of survival and reproduction, not our happiness. The biochemical system rewards actions conducive to survival and reproduction with pleasant sensations. But these are only an ephemeral sales gimmick. We struggle to get food and mates in order to avoid unpleasant sensations of hunger and to enjoy pleasing tastes and blissful orgasms. But nice tastes and blissful orgasms don't last very long, and if we want to feel them again we have to go out looking for more food and mates.

What might have happened if a rare mutation had created a squirrel who, after eating a single nut, enjoys an everlasting sensation of bliss? Technically, this could actually be done by rewiring the squirrel's brain. Who knows, perhaps it really happened to some lucky squirrel millions of years ago. But if so, that squirrel enjoyed an extremely happy and extremely short life, and that was the end of the rare mutation. For the blissful squirrel would not have bothered to look for more nuts, let alone mates. The rival squirrels, who felt hungry again five minutes after eating a nut, had much better chances of surviving and passing their genes to the next generation. For exactly the same reason, the nuts we humans seek to gather – lucrative jobs, big houses, good-looking partners – seldom satisfy us for long.

Some may say that this is not so bad, because it isn't the goal that makes us happy – it's the journey. Climbing Mount Everest is more satisfying than standing at the top; flirting and foreplay are more exciting than having an orgasm; and conducting ground-breaking lab experiments is more interesting than receiving praise and prizes. Yet this hardly changes the picture. It just indicates that evolution controls us with a broad range of pleasures. Sometimes it seduces us with sensations of bliss and tranquillity, while on other occasions it goads us forward with thrilling sensations of elation and excitement.

When an animal is looking for something that increases its chances of survival and reproduction (e.g., food, partners or social status), the brain produces sensations of alertness and excitement, which drive the animal to make even greater efforts because they are so very agreeable. In a famous experiment scientists connected electrodes to the brains of several rats, enabling the animals to create sensations of excitement simply by pressing a pedal. When the rats were given a choice between tasty food and pressing the pedal, they preferred the pedal (much like kids preferring to play video games rather than come down to dinner). The rats pressed the pedal again and again, until they collapsed from hunger and exhaustion.[36] Humans too may prefer the excitement of the race to resting on the laurels of success. Yet what makes the race so attractive is the exhilarating sensations that go along with it. Nobody would have wanted to climb mountains, play video games or go on blind dates if such activities were accompanied solely by unpleasant sensations of stress, despair or boredom.[37]

Alas, the exciting sensations of the race are as transient as the blissful sensations of victory. The Don Juan enjoying the thrill of a one-night stand, the businessman enjoying biting his fingernails watching the Dow Jones rise and fall, and the gamer enjoying killing monsters on the computer screen will find no satisfaction remembering yesterday's adventures. Like the rats pressing the pedal again and again, the Don Juans, business tycoons and gamers need a new kick every day. Worse still, here too expectations adapt to conditions, and yesterday's challenges all too quickly become today's tedium. Perhaps the key to happiness is neither the race nor the gold medal, but rather combining the right doses of excitement and tranquillity; but most of us tend to jump all the way from stress to boredom and back, remaining as discontented with one as with the other.

If science is right and our happiness is determined by our biochemical system, then the only way to ensure lasting contentment is by rigging this system. Forget economic growth, social reforms and political revolutions: in order to raise global happiness levels, we need to manipulate human biochemistry. And this is exactly what we have begun doing over the last few decades. Fifty years ago psychiatric drugs carried a severe stigma. Today, that stigma has been broken. For better or worse, a growing percentage of the population is taking psychiatric medicines on a regular basis, not only to cure debilitating mental illnesses, but also to face more mundane depressions and the occasional blues.

For example, increasing numbers of schoolchildren take stimulants such as Ritalin. In 2011, 3.5 million American children were taking medications for ADHD (attention deficit hyperactivity disorder). In the UK the number rose from 92,000 in 1997 to 786,000 in 2012.[38] The original aim had been to treat attention disorders, but today completely healthy kids take such medications to improve their performance and live up to the growing expectations of teachers and parents.[39] Many object to this development and argue that the problem lies with the education system rather than with the children. If pupils suffer from attention disorders, stress and low grades, perhaps we ought to blame outdated teaching methods, overcrowded classrooms and an unnaturally fast tempo of life. Maybe we should modify the schools rather than the kids? It is interesting to see how the arguments have evolved. People have been quarrelling about education methods for thousands of years. Whether in ancient China or Victorian Britain, everybody had his or her pet method, and vehemently opposed all alternatives. Yet hitherto everybody still agreed on one thing: in order to improve education, we need to change the schools. Today, for the first time in history, at least some people think it would be more efficient to change the pupils' biochemistry.[40]

Armies are heading the same way: 12 per cent of American soldiers in Iraq and 17 per cent of American soldiers in Afghanistan took either sleeping pills or antidepressants to help them deal with the pressure and distress of war. Fear, depression and trauma are not caused by shells, booby traps or car bombs. They are caused by hormones, neurotransmitters and neural networks. Two soldiers may find themselves shoulder to shoulder in the same ambush; one will freeze in terror, lose his wits and suffer from nightmares for years after the event; the other will charge forward courageously and win a medal. The difference is in the soldiers' biochemistry, and if we find ways to control it we will at one stroke produce both happier soldiers and more efficient armies.[41]

The biochemical pursuit of happiness is also the number one cause of crime in the world. In 2009 half of the inmates in US federal prisons got there because of drugs; 38 per cent of Italian prisoners were convicted of drug-related offences; 55 per cent of inmates in the UK reported that they committed their crimes in connection with either consuming or trading drugs. A 2001 report found that 62 per cent of Australian convicts were under the influence of drugs when committing the crime for which they were incarcerated.[42] People drink alcohol to forget, they smoke pot to feel peaceful, they take cocaine and methamphetamines to be sharp and confident, whereas Ecstasy provides ecstatic sensations and LSD sends you to meet Lucy in the Sky with Diamonds. What some people hope to get by studying, working or raising a family, others try to obtain far more easily through the right dosage of molecules. This is an existential threat to the social and economic order, which is why countries wage a stubborn, bloody and hopeless war on biochemical crime.

The state hopes to regulate the biochemical pursuit of happiness, separating 'bad' manipulations from 'good' ones. The principle is clear: biochemical manipulations that strengthen political stability, social order and economic growth are allowed and even encouraged (e.g., those that calm hyperactive kids in school, or drive anxious soldiers forward into battle). Manipulations that threaten stability and growth are banned. But each year new drugs are born in the research labs of universities, pharmaceutical companies and criminal organisations, and the needs of the state and the market also keep changing. As the biochemical pursuit of happiness accelerates, so it will reshape politics, society and economics, and it will become ever harder to bring it under control.

And drugs are just the beginning. In research labs experts are already working on more sophisticated ways of manipulating human biochemistry, such as sending direct electrical stimuli to appropriate spots in the brain, or genetically engineering the blueprints of our bodies. No matter the exact method, gaining happiness through biological manipulation won't be easy, for it requires altering the fundamental patterns of life. But then it wasn't easy to overcome famine, plague and war either.

It is far from certain that humankind should invest so much effort in the biochemical pursuit of happiness. Some would argue that happiness simply isn't important enough, and that it is misguided to regard individual satisfaction as the highest aim of human society. Others may agree that happiness is indeed the supreme good, yet would take issue with the biological definition of happiness as the experience of pleasant sensations.

Some 2,300 years ago Epicurus warned his disciples that immoderate pursuit of pleasure is likely to make them miserable rather than happy. A couple of centuries earlier Buddha had made an even more radical claim, teaching that the pursuit of pleasant sensations is in fact the very root of suffering. Such sensations are just ephemeral and meaningless vibrations. Even when we experience them, we don't react to them with contentment; rather, we just crave more. Hence no matter how many blissful or exciting sensations I may experience, they will never satisfy me.

If I identify happiness with fleeting pleasant sensations, and crave to experience more and more of them, I have no choice but to pursue them constantly. When I finally get them, they quickly disappear, and because the mere memory of past pleasures will not satisfy me, I have to start all over again. Even if I continue this pursuit for decades, it will never bring me any lasting achievement; on the contrary, the more I crave these pleasant sensations, the more stressed and dissatisfied I will become. To attain real happiness, humans need to slow down the pursuit of pleasant sensations, not accelerate it.

This Buddhist view of happiness has a lot in common with the biochemical view. Both agree that pleasant sensations disappear as fast as they arise, and that as long as people crave pleasant sensations without actually experiencing them, they remain dissatisfied. However, this problem has two very different solutions. The biochemical solution is to develop products and treatments that will provide humans with an unending stream of pleasant sensations, so we will never be without them. The Buddha's suggestion was to reduce our craving for pleasant sensations, and not allow them to control our lives. According to Buddha, we can train our minds to observe carefully how all sensations constantly arise and pass. When the mind learns to see our sensations for what they are – ephemeral and meaningless vibrations – we lose interest in pursuing them. For what is the point of running after something that disappears as fast as it arises?

At present, humankind has far greater interest in the biochemical solution. No matter what monks in their Himalayan caves or philosophers in their ivory towers say, for the capitalist juggernaut, happiness is pleasure. Period. With each passing year our tolerance for unpleasant sensations decreases, and our craving for pleasant sensations increases. Both scientific research and economic activity are geared to that end, each year producing better painkillers, new ice-cream flavours, more comfortable mattresses, and more addictive games for our smartphones, so that we will not suffer a single boring moment while waiting for the bus.

All this is hardly enough, of course. Since *Homo sapiens* was not adapted by evolution to experience constant pleasure, if that is what humankind nevertheless wants, ice cream and smartphone games will not do. It will be necessary to change our biochemistry and re-engineer our bodies and minds. So we are working on that. You may debate whether it is good or bad, but it seems that the second great project of the twenty-first century – to ensure global happiness – will involve re-engineering *Homo sapiens* so that it can enjoy everlasting pleasure.

The Gods of Planet Earth

In seeking bliss and immortality humans are in fact trying to upgrade themselves into gods. Not just because these are divine qualities, but because in order to overcome old age and misery humans will first have to acquire godlike control of their own biological substratum. If we ever have the power to engineer death and pain out of our system, that same power will probably be sufficient to engineer our system in almost any manner we like, and manipulate our organs, emotions and intelligence in myriad ways. You could buy for yourself the strength of Hercules, the sensuality of Aphrodite, the wisdom of Athena or the madness of Dionysus if that is what you are into. Up till now increasing human power relied mainly on upgrading our external tools. In the future it may rely more on upgrading the human body and mind, or on merging directly with our tools.

The upgrading of humans into gods may follow any of three paths: biological engineering, cyborg engineering and the engineering of non-organic beings.

Biological engineering starts with the insight that we are far from realising the full potential of organic bodies. For 4 billion years natural selection has been tweaking and tinkering with these bodies, so that we have gone from amoeba to reptiles to mammals to Sapiens. Yet there is no reason to think that Sapiens is the last station. Relatively small changes in genes, hormones and neurons were enough to transform *Homo erectus* – who could produce nothing more impressive than flint knives – into *Homo sapiens*, who produce spaceships and computers. Who knows what might be the outcome of a few more changes to our DNA, hormonal system or brain structure. Bioengineering is not going to wait patiently for natural selection to work its magic. Instead, bioengineers will take the old Sapiens body, and intentionally rewrite its genetic code, rewire its brain circuits, alter its biochemical balance, and even grow entirely new limbs. They will thereby create new godlings, who might be as different from us Sapiens as we are different from *Homo erectus*.

Cyborg engineering will go a step further, merging the organic body with non-organic devices such as bionic hands, artificial eyes, or millions of nano-robots that will navigate our bloodstream, diagnose problems and repair damage. Such a cyborg could enjoy abilities far beyond those of any organic body. For example, all parts of an organic body must be in direct contact with one another in order to function. If an elephant's brain is in India, its eyes and ears in China and its feet in Australia, then this elephant is most probably dead, and even if it is in some mysterious sense alive, it cannot see, hear or walk. A cyborg, in contrast, could exist in numerous places at the same time. A cyborg doctor could perform emergency surgeries in Tokyo, in Chicago and in a space station on Mars, without ever leaving her Stockholm office. She will need only a fast Internet connection, and a few pairs of bionic eyes and hands. On second thought, why *pairs*? Why not quartets? Indeed, even those are actually superfluous. Why should a cyborg doctor hold a surgeon's scalpel by hand, when she could connect her mind directly to the instrument?

This may sound like science fiction, but it's already a reality. Monkeys have recently learned to control bionic hands and feet disconnected from their bodies, through electrodes implanted in their brains. Paralysed patients are able to move bionic limbs or operate computers by the power of thought alone. If you wish, you can already remote-control electric devices in your house using an electric 'mind-reading' helmet. The helmet requires no brain implants. It functions by reading the electric signals passing through your scalp. If you want to turn on the light in the kitchen, you just wear the helmet, imagine some preprogrammed mental sign (e.g., imagine your right hand moving), and the switch turns on. You can buy such helmets online for a mere $400.[43]

In early 2015 several hundred workers in the Epicenter high-tech hub in Stockholm had microchips implanted into their hands. The chips are about the size of a grain of rice and store personalised security information that enables workers to open doors and operate photocopiers with a wave of their hand. Soon they hope to make payments in the same way. One of the people behind the initiative, Hannes Sjoblad, explained that 'We already interact with technology all the time. Today it's a bit messy: we need pin codes and passwords. Wouldn't it be easy to just touch with your hand?'[44]

Yet even cyborg engineering is relatively conservative, inasmuch as it assumes that organic brains will go on being the command-and-control centres of life. A bolder approach dispenses with organic parts altogether, and hopes to engineer completely non-organic beings. Neural networks will be replaced by intelligent software, which could surf both the virtual and non-virtual worlds, free from the limitations of organic chemistry. After 4 billion years of wandering inside the kingdom of organic compounds, life will break out into the vastness of the inorganic realm, and will take shapes that we cannot envision even in our wildest dreams. After all, our wildest dreams are still the product of organic chemistry.

Breaking out of the organic realm could also enable life to finally break out of planet earth. For four billion years life remained confined to this tiny speck of a planet because natural selection made all organisms utterly dependent on the unique conditions of this flying rock. Not even the toughest bacteria can survive on Mars. A non-organic artificial intelligence, in contrast, will find it far easier to colonize alien planets. The replacement of organic life by inorganic beings may therefore sow the seed of a future galactic empire, ruled by the likes of Mr. Data rather than Captain Kirk.

We don't know where these paths might lead us, nor what our godlike descendants will look like. Foretelling the future was never easy, and revolutionary biotechnologies make it even harder. For as difficult as it is to predict the impact of new technologies in fields like transportation, communication and energy, technologies for upgrading humans pose a completely different kind of challenge. Since they can be used to transform human minds and desires, people possessing present-day minds and desires by definition cannot fathom their implications.

For thousands of years history was full of technological, economic, social and political upheavals. Yet one thing remained constant: humanity itself. Our tools and institutions are very different from those of biblical times, but the deep structures of the human mind remain the same. This is why we can still find ourselves between the pages of the Bible, in the writings of Confucius or within the tragedies of Sophocles and Euripides. These classics were created by humans just like us, hence we feel that they talk about us. In modern theatre productions, Oedipus, Hamlet and Othello may wear jeans and T-shirts and have Facebook accounts, but their emotional conflicts are the same as in the original play.

However, once technology enables us to re-engineer human minds, *Homo sapiens* will disappear, human history will come to an end and a completely new kind of process will begin, which people like you and me cannot comprehend. Many scholars try to predict how the world will look in the year 2100 or 2200. This is a waste of time. Any worthwhile prediction must take into account the ability to re-engineer human minds, and this is impossible. There are many wise answers to the question, 'What would people with minds like ours do with biotechnology?' Yet there are no good answers to the question, 'What would beings with a *different* kind of mind do with biotechnology?' All we can say is that people similar to us are likely to use biotechnology to re-engineer their own minds, and our present-day minds cannot grasp what might happen next.

Though the details are therefore obscure, we can nevertheless be sure about the general direction of history. In the twenty-first century, the third big project of humankind will be to acquire for us divine powers of creation and destruction, and upgrade *Homo sapiens* into *Homo deus*. This third project obviously subsumes the first two projects, and is fuelled by them. We want the ability to re-engineer our bodies and minds in order, above all, to escape old age, death and misery, but once we have it, who knows what else we might do with such ability? So we may well think of the new human agenda as consisting really of only one project (with many branches): attaining divinity.

If this sounds unscientific or downright eccentric, it is because people often misunderstand the meaning of divinity. Divinity isn't a vague metaphysical quality. And it isn't the same as omnipotence. When speaking of upgrading humans into gods, think more in terms of Greek gods or Hindu devas rather than the omnipotent biblical sky father. Our descendants would still have their foibles, kinks and limitations, just as Zeus and Indra had theirs. But they could love, hate, create and destroy on a much grander scale than us.

Throughout history most gods were believed to enjoy not omnipotence but rather specific super-abilities such as the ability to design and create living beings; to transform their own bodies; to control the environment and the weather; to read minds and to communicate at a distance; to travel at very high speeds; and of course to escape death and live indefinitely. Humans are in the business of acquiring all these abilities, and then some. Certain traditional abilities that were considered divine for many millennia have today become so commonplace that we hardly think about them. The average person now moves and communicates across distances much more easily than the Greek, Hindu or African gods of old.

For example, the Igbo people of Nigeria believe that the creator god Chukwu initially wanted to make people immortal. He sent a dog to tell humans that when someone dies, they should sprinkle ashes on the corpse, and the body will come back to life. Unfortunately, the dog was tired and he dallied on the way. The impatient Chukwu then sent a sheep, telling her to make haste with this important message. Alas, when the breathless sheep reached her destination, she garbled the instructions, and told the humans to bury their dead, thus making death permanent. This is why to this day we humans must die. If only Chukwu had a Twitter account instead of relying on laggard dogs and dim-witted sheep to deliver his messages!

In ancient agricultural societies, many religions displayed surprisingly little interest in metaphysical questions and the afterlife. Instead, they focused on the very mundane issue of increasing agricultural output. Thus the Old Testament God *never* promises any rewards or punishments after death. He instead tells the people of Israel that 'If you carefully observe the commands that I'm giving you [. . .] then I will send rain on the land in its season [. . .] and you'll gather grain, wine, and oil. I will provide grass in the fields for your livestock, and you'll eat and be satisfied. Be careful! Otherwise, your hearts will deceive you and you will turn away to serve other gods and worship them. The wrath of God will burn against you so that he will restrain the heavens and it won't rain. The ground won't yield its produce and you'll be swiftly destroyed from the good land that the Lord is about to give you' (Deuteronomy 11:13–17). Scientists today can do much better than the Old Testament God. Thanks to artificial fertilisers, industrial insecticides and genetically modified crops, agricultural production nowadays outstrips the highest expectations ancient farmers had of their gods. And the parched state of Israel no longer fears that some angry deity will restrain the heavens and stop all rain – for the Israelis have recently built a huge desalination plant on the shores of the Mediterranean, so they can now get all their drinking water from the sea.

So far we have competed with the gods of old by creating better and better tools. In the not too distant future, we might create superhumans who will outstrip the ancient gods not in their tools, but in their bodily and mental faculties. If and when we get there, however, divinity will become as mundane as cyberspace – a wonder of wonders that we just take for granted.

We can be quite certain that humans will make a bid for divinity, because humans have many reasons to desire such an upgrade, and many ways to achieve it. Even if one promising path turns out to be a dead end, alternative routes will remain open. For example, we may discover that the human genome is far too complicated for serious manipulation, but this will not prevent the development of brain–computer interfaces, nano-robots or artificial intelligence.

No need to panic, though. At least not immediately. Upgrading Sapiens will be a gradual historical process rather than a Hollywood apocalypse. *Homo sapiens* is not going to be exterminated by a robot revolt. Rather, *Homo sapiens* is likely to upgrade itself step by step, merging with robots and computers in the process, until our descendants will look back and realise that they are no longer the kind of animal that wrote the Bible, built the Great Wall of China and laughed at Charlie Chaplin's antics. This will not happen in a day, or a year. Indeed, it is already happening right now, through innumerable mundane actions. Every day millions of people decide to grant their smartphone a bit more control over their lives or try a new and more effective antidepressant drug. In pursuit of health, happiness and power, humans will gradually change first one of their features and then another, and another, until they will no longer be human.

Can Someone Please Hit the Brakes?

Calm explanations aside, many people panic when they hear of such possibilities. They are happy to follow the advice of their smartphones or to take whatever drug the doctor prescribes, but when they hear of upgraded superhumans, they say: 'I hope I will be dead before that happens.' A friend once told me that what she fears most about growing old is becoming irrelevant, turning into a nostalgic old woman who cannot understand the world around her, or contribute much to it. This is what we fear collectively, as a species, when we hear of superhumans. We sense that in such a world, our identity, our dreams and even our fears will be irrelevant, and we will have nothing more to contribute. Whatever you are today – be it a devout Hindu cricket player or an aspiring lesbian journalist – in an upgraded world you will feel like a Neanderthal hunter in Wall Street. You won't belong.

The Neanderthals didn't have to worry about the Nasdaq, since they were shielded from it by tens of thousands of years. Nowadays, however, our world of meaning might collapse within decades. You cannot count on death to save you from becoming completely irrelevant. Even if gods don't walk our streets by 2100, the attempt to upgrade *Homo sapiens* is likely to change the world beyond recognition in this century. Scientific research and technological developments are moving at a far faster rate than most of us can grasp.

If you speak with the experts, many of them will tell you that we are still very far away from genetically engineered babies or human-level artificial intelligence. But most experts think on a timescale of academic grants and college jobs. Hence, 'very far away' may mean twenty years, and 'never' may denote no more than fifty.

I still remember the day I first came across the Internet. It was back in 1993, when I was in high school. I went with a couple of buddies to visit our friend Ido (who is now a computer scientist). We wanted to play table tennis. Ido was already a huge computer fan, and before opening the ping-pong table he insisted on showing us the latest wonder. He connected the phone cable to his computer and pressed some keys. For a minute all we could hear were squeaks, shrieks and buzzes, and then silence. It didn't succeed. We mumbled and grumbled, but Ido tried again. And again. And again. At last he gave a whoop and announced that he had managed to connect his computer to the central computer at the nearby university. 'And what's there, on the central computer?' we asked. 'Well,' he admitted, 'there's nothing there yet. But you could put all kinds of things there.' 'Like what?' we questioned. 'I don't know,' he said, 'all kinds of things.' It didn't sound very promising. We went to play ping-pong, and for the following weeks enjoyed a new pastime, making fun of Ido's ridiculous idea. That was less than twenty-five years ago (at the time of writing). Who knows what will come to pass twenty-five years from now?

That's why more and more individuals, organisations, corporations and governments are taking very seriously the quest for immortality, happiness and godlike powers. Insurance companies, pension funds, health systems and finance ministries are already aghast at the jump in life expectancy. People are living much longer than expected, and there is not enough money to pay for their pensions and medical treatment. As seventy threatens to become the new forty, experts are calling to raise the retirement age, and to restructure the entire job market.

When people realise how fast we are rushing towards the great unknown, and that they cannot count even on death to shield them from it, their reaction is to hope that somebody will hit the brakes and slow us down. But we cannot hit the brakes, for several reasons.

Firstly, nobody knows where the brakes are. While some experts are familiar with developments in one field, such as artificial intelligence, nanotechnology, big data or genetics, no one is an expert on everything. No one is therefore capable of connecting all the dots and seeing the full picture. Different fields influence one another in such intricate ways that even the best minds cannot fathom how breakthroughs in artificial intelligence might impact nanotechnology, or vice versa. Nobody can absorb all the latest scientific discoveries, nobody can predict how the global economy will look in ten years, and nobody has a clue where we are heading in such a rush. Since no one understands the system any more, no one can stop it.

Secondly, if we somehow succeed in hitting the brakes, our economy will collapse, along with our society. As explained in a later chapter, the modern economy needs constant and indefinite growth in order to survive. If growth ever stops, the economy won't settle down to some cosy equilibrium; it will fall to pieces. That's why capitalism encourages us to seek immortality, happiness and divinity. There's a limit to how many shoes we can wear, how many cars we can drive and how many skiing holidays we can enjoy. An economy built on everlasting growth needs endless projects – just like the quests for immortality, bliss and divinity.

Well, if we need endless projects, why not settle for bliss and immortality, and at least put aside the frightening quest for superhuman powers? Because it is inextricable from the other two. When you develop bionic legs that enable paraplegics to walk again, you can also use the same technology to upgrade healthy people. When you discover how to stop memory loss among older people, the same treatments might enhance the memory of the young.

No clear line separates healing from upgrading. Medicine almost always begins by saving people from falling below the norm, but the same tools and know-how can then be used to surpass the norm. Viagra began life as a treatment for blood-pressure problems. To the surprise and delight of Pfizer, it transpired that Viagra can also overcome impotence. It enabled millions of men to regain normal sexual abilities; but soon enough men who had no impotence problems in the first place began using the same pill to surpass the norm, and acquire sexual powers they never had before.[45]

What happens to particular drugs can also happen to entire fields of medicine. Modern plastic surgery was born in the First World War, when Harold Gillies began treating facial injuries in the Aldershot military hospital.[46] When the war was over, surgeons discovered that the same techniques could also turn perfectly healthy but ugly noses into more beautiful specimens. Though plastic surgery continued to help the sick and wounded, it devoted increasing attention to upgrading the healthy. Nowadays plastic surgeons make millions in private clinics whose explicit and sole aim is to upgrade the healthy and beautify the wealthy.[47]

The same might happen with genetic engineering. If a billionaire openly stated that he intended to engineer super-smart offspring, imagine the public outcry. But it won't happen like that. We are more likely to slide down a slippery slope. It begins with parents whose genetic profile puts their children at high risk of deadly genetic diseases. So they perform *in vitro* fertilisation, and test the DNA of the fertilised egg. If everything is in order, all well and good. But if the DNA test discovers the dreaded mutations – the embryo is destroyed.

Yet why take a chance by fertilising just one egg? Better fertilise several, so that even if three or four are defective there is at least one good embryo. When this *in vitro* selection procedure becomes acceptable and cheap enough, its usage may spread. Mutations are a ubiquitous risk. All people carry in their DNA some harmful mutations and less-than-optimal alleles. Sexual reproduction is a lottery. (A famous – and probably apocryphal – anecdote tells of a meeting in 1923 between Nobel Prize laureate Anatole France and the beautiful and talented dancer Isadora Duncan. Discussing the then popular eugenics movement, Duncan said, 'Just imagine a child with my beauty and your brains!' France responded, 'Yes, but imagine a child with *my* beauty and *your* brains.') Well then, why not rig the lottery? Fertilise several eggs, and choose the one with the best combination. Once stem-cell research enables us to create an unlimited supply of human embryos on the cheap, you can select your optimal baby from among hundreds of candidates, all carrying your DNA, all perfectly natural, and none requiring any futuristic genetic engineering. Iterate this procedure for a few generations, and you could easily end up with superhumans (or a creepy dystopia).

But what if after fertilising even numerous eggs, you find that all of them contain some deadly mutations? Should you destroy all the embryos? Instead of doing that, why not replace the problematic genes? A breakthrough case involves mitochondrial DNA. Mitochondria are tiny organelles within human cells, which produce the energy used by the cell. They have their own set of genes, which is completely separate from the DNA in the cell's nucleus. Defective mitochondrial DNA leads to various debilitating or even deadly diseases. It is technically feasible with current *in vitro* technology to overcome mitochondrial genetic diseases by creating a 'three-parent baby'. The baby's nuclear DNA comes from two parents, while the mitochondrial DNA comes from a third person. In 2000 Sharon Saarinen from West Bloomfield, Michigan, gave birth to a healthy baby girl, Alana. Alana's nuclear DNA came from her mother, Sharon, and her father, Paul, but her mitochondrial DNA came from another woman. From a purely technical perspective, Alana has three biological parents. A year later, in 2001, the US government banned this treatment, due to safety and ethical concerns.[48]

However, on 3 February 2015 the British Parliament voted in favour of the so-called 'three-parent embryo' law, allowing this treatment – and related research – in the UK.[49] At present it is technically unfeasible, and illegal, to replace nuclear DNA, but if and when the technical difficulties are solved, the same logic that favoured the replacement of defective mitochondrial DNA would seem to warrant doing the same with nuclear DNA.

Following selection and replacement, the next potential step is amendment. Once it becomes possible to amend deadly genes, why go through the hassle of inserting some foreign DNA, when you can just rewrite the code and turn a dangerous mutant gene into its benign version? Then we might start using the same mechanism to fix not just lethal genes, but also those responsible for less deadly illnesses, for autism, for stupidity and for obesity. Who would like his or her child to suffer from any of these? Suppose a genetic test indicates that your would-be daughter will in all likelihood be smart, beautiful and kind – but will suffer from chronic depression. Wouldn't you want to save her from years of misery by a quick and painless intervention in the test tube?

And while you are at it, why not give the child a little push? Life is hard and challenging even for healthy people. So it would surely come in handy if the little girl had a stronger-than-normal immune system, an above-average memory or a particularly sunny disposition. And even if you don't want that for your child – what if the neighbours are doing it for theirs? Would you have your child lag behind? And if the government forbids all citizens from engineering their babies, what if the North Koreans are doing it and producing amazing geniuses, artists and athletes that far outperform ours? And like that, in baby steps, we are on our way to a genetic child catalogue.

Healing is the initial justification for every upgrade. Find some professors experimenting in genetic engineering or brain–computer interfaces, and ask them why they are engaged in such research. In all likelihood they would reply that they are doing it to cure disease. 'With the help of genetic engineering,' they would explain, 'we could defeat cancer. And if we could connect brains and computers directly, we could cure schizophrenia.' Maybe, but it will surely not end there. When we successfully connect brains and computers, will we use this technology only to cure schizophrenia? If anybody really believes this, then they may know a great deal about brains and computers, but far less about the human psyche and human society. Once you achieve a momentous breakthrough, you cannot restrict its use to healing and completely forbid using it for upgrading.

Of course humans can and do limit their use of new technologies. Thus the eugenics movement fell from favour after the Second World War, and though trade in human organs is now both possible and potentially very lucrative, it has so far remained a peripheral activity. Designer babies may one day become as technologically feasible as murdering people to harvest their organs – yet remain as peripheral.

Just as we have escaped the clutches of Chekhov's Law in warfare, we can also escape them in other fields of action. Some guns appear on stage without ever being fired. This is why it is so vital to think about humanity's new agenda. Precisely because we have some choice regarding the use of new technologies, we had better understand what is happening and make up our minds about it before it makes up our minds for us.

The Paradox of Knowledge

The prediction that in the twenty-first century humankind is likely to aim for immortality, bliss and divinity may anger, alienate or frighten any number of people, so a few clarifications are in order.

Firstly, this is not what most individuals will actually do in the twenty-first century. It is what humankind as a collective will do. Most people will probably play only a minor role, if any, in these projects. Even if famine, plague and war become less prevalent, billions of humans in developing countries and seedy neighbourhoods will continue to deal with poverty, illness and violence even as the elites are already reaching for eternal youth and godlike powers. This seems patently unjust. One could argue that as long as there is a single child dying from malnutrition or a single adult killed in drug-lord warfare, humankind should focus all its efforts on combating these woes. Only once the last sword is beaten into a ploughshare should we turn our minds to the next big thing. But history doesn't work like that. Those living in palaces have always had different agendas to those living in shacks, and that is unlikely to change in the twenty-first century.

Secondly, this is a historical prediction, not a political manifesto. Even if we disregard the fate of slum-dwellers, it is far from clear that we should be aiming at immortality, bliss and divinity. Adopting these particular projects might be a big mistake. But history is full of big mistakes. Given our past record and our current values, we are likely to reach out for bliss, divinity and immortality – even if it kills us.

Thirdly, reaching out is not the same as obtaining. History is often shaped by exaggerated hopes. Twentieth-century Russian history was largely shaped by the communist attempt to overcome inequality, but it didn't succeed. My prediction is focused on what humankind will *try* to achieve in the twenty-first century – not what it will *succeed* in achieving. Our future economy, society and politics will be shaped by the attempt to overcome death. It does not follow that in 2100 humans will be immortal.

Fourthly, and most importantly, this prediction is less of a prophecy and more a way of discussing our present choices. If the discussion makes us choose differently, so that the prediction is proven wrong, all the better. What's the point of making predictions if they cannot change anything?

Some complex systems, such as the weather, are oblivious to our predictions. The process of human development, in contrast, reacts to them. Indeed, the better our forecasts, the more reactions they engender. Hence paradoxically, as we accumulate more data and increase our computing power, events become wilder and more unexpected. The more we know, the less we can predict. Imagine, for example, that one day experts decipher the basic laws of the economy. Once this happens, banks, governments, investors and customers will begin to use this new knowledge to act in novel ways, and gain an edge over their competitors. For what is the use of new knowledge if it doesn't lead to novel behaviours? Alas, once people change the way they behave, the economic theories become obsolete. We may know how the economy functioned in the past – but we no longer understand how it functions in the present, not to mention the future.

This is not a hypothetical example. In the middle of the nineteenth century Karl Marx reached brilliant economic insights. Based on these insights he predicted an increasingly violent conflict between the proletariat and the capitalists, ending with the inevitable victory of the former and the collapse of the capitalist system. Marx was certain that the revolution would start in countries that spearheaded the Industrial Revolution – such as Britain, France and the USA – and spread to the rest of the world.

Marx forgot that capitalists know how to read. At first only a handful of disciples took Marx seriously and read his writings. But as these socialist firebrands gained adherents and power, the capitalists became alarmed. They too perused *Das Kapital*, adopting many of the tools and insights of Marxist analysis. In the twentieth century everybody from street urchins to presidents embraced a Marxist approach to economics and history. Even diehard capitalists who vehemently resisted the Marxist prognosis still made use of the Marxist diagnosis. When the CIA analysed the situation in Vietnam or Chile in the 1960s, it divided society into classes. When Nixon or Thatcher looked at the globe, they asked themselves who controls the vital means of production. From 1989 to 1991 George Bush oversaw the demise of the Evil Empire of communism, only to be defeated in the 1992 elections by Bill Clinton. Clinton's winning campaign strategy was summarised in the motto: 'It's the economy, stupid.' Marx could not have said it better.

As people adopted the Marxist diagnosis, they changed their behaviour accordingly. Capitalists in countries such as Britain and France strove to better the lot of the workers, strengthen their national consciousness and integrate them into the political system. Consequently when workers began voting in elections and Labour gained power in one country after another, the capitalists could still sleep soundly in their beds. As a result, Marx's predictions came to naught. Communist revolutions never engulfed the leading industrial powers such as Britain, France and the USA, and the dictatorship of the proletariat was consigned to the dustbin of history.

This is the paradox of historical knowledge. Knowledge that does not change behaviour is useless. But knowledge that changes behaviour quickly loses its relevance. The more data we have and the better we understand history, the faster history alters its course, and the faster our knowledge becomes outdated.

Centuries ago human knowledge increased slowly, so politics and economics changed at a leisurely pace too. Today our knowledge is increasing at breakneck speed, and theoretically we should understand the world better and better. But the very opposite is happening. Our new-found knowledge leads to faster economic, social and political changes; in an attempt to understand what is happening, we accelerate the accumulation of knowledge, which leads only to faster and greater upheavals. Consequently we are less and less able to make sense of the present or forecast the future. In 1016 it was relatively easy to predict how Europe would look in 1050. Sure, dynasties might fall, unknown raiders might invade, and natural disasters might strike; yet it was clear that in 1050 Europe would still be ruled by kings and priests, that it would be an agricultural society, that most of its inhabitants would be peasants, and that it would continue to suffer greatly from famines, plagues and wars. In contrast, in 2016 we have no idea how Europe will look in 2050. We cannot say what kind of political system it will have, how its job market will be structured, or even what kind of bodies its inhabitants will possess.

A Brief History of Lawns

If history doesn't follow any stable rules, and if we cannot predict its future course, why study it? It often seems that the chief aim of science is to predict the future – meteorologists are expected to forecast whether tomorrow will bring rain or sunshine; economists should know whether devaluing the currency will avert or precipitate an economic crisis; good doctors foresee whether chemotherapy or radiation therapy will be more successful in curing lung cancer. Similarly, historians are asked to examine the actions of our ancestors so that we can repeat their wise decisions and avoid their mistakes. But it almost never works like that because the present is just too different from the past. It is a waste of time to study Hannibal's tactics in the Second Punic War so as to copy them in the Third World War. What worked well in cavalry battles will not necessarily be of much benefit in cyber warfare.

Science is not just about predicting the future, though. Scholars in all fields often seek to broaden our horizons, thereby opening before us new and unknown futures. This is especially true of history. Though historians occasionally try their hand at prophecy (without notable success), the study of history aims above all to make us aware of possibilities we don't normally consider. Historians study the past not in order to repeat it, but in order to be liberated from it.

Each and every one of us has been born into a given historical reality, ruled by particular norms and values, and managed by a unique economic and political system. We take this reality for granted, thinking it is natural, inevitable and immutable. We forget that our world was created by an accidental chain of events, and that history shaped not only our technology, politics and society, but also our thoughts, fears and dreams. The cold hand of the past emerges from the grave of our ancestors, grips us by the neck and directs our gaze towards a single future. We have felt that grip from the moment we were born, so we assume that it is a natural and inescapable part of who we are. Therefore we seldom try to shake ourselves free, and envision alternative futures.

Studying history aims to loosen the grip of the past. It enables us to turn our head this way and that, and begin to notice possibilities that our ancestors could not imagine, or didn't want us to imagine. By observing the accidental chain of events that led us here, we realise how our very thoughts and dreams took shape – and we can begin to think and dream differently. Studying history will not tell us what to choose, but at least it gives us more options.

Movements seeking to change the world often begin by rewriting history, thereby enabling people to reimagine the future. Whether you want workers to go on a general strike, women to take possession of their bodies, or oppressed minorities to demand political rights – the first step is to retell their history. The new history will explain that 'our present situation is neither natural nor eternal. Things were different once. Only a string of chance events created the unjust world we know today. If we act wisely, we can change that world, and create a much better one.' This is why Marxists recount the history of capitalism; why feminists study the formation of patriarchal societies; and why African Americans commemorate the horrors of the slave trade. They aim not to perpetuate the past, but rather to be liberated from it.

What's true of grand social revolutions is equally true at the micro level of everyday life. A young couple building a new home for themselves may ask the architect for a nice lawn in the front yard. Why a lawn? 'Because lawns are beautiful,' the couple might explain. But why do they think so? It has a history behind it.

Stone Age hunter-gatherers did not cultivate grass at the entrance to their caves. No green meadow welcomed the visitors to the Athenian Acropolis, the Roman Capitol, the Jewish Temple in Jerusalem or the Forbidden City in Beijing. The idea of nurturing a lawn at the entrance to private residences and public buildings was born in the castles of French and English aristocrats in the late Middle Ages. In the early modern age this habit struck deep roots, and became the trademark of nobility.

Well-kept lawns demanded land and a lot of work, particularly in the days before lawnmowers and automatic water sprinklers. In exchange, they produce nothing of value. You can't even graze animals on them, because they would eat and trample the grass. Poor peasants could not afford wasting precious land or time on lawns. The neat turf at the entrance to chateaux was accordingly a status symbol nobody could fake. It boldly proclaimed to every passerby: 'I am so rich and powerful, and I have so many acres and serfs, that I can afford this green extravaganza.' The bigger and neater the lawn, the more powerful the dynasty. If you came to visit a duke and saw that his lawn was in bad shape, you knew he was in trouble.[50]

The precious lawn was often the setting for important celebrations and social events, and at all other times was strictly off-limits. To this day, in countless palaces, government buildings and public venues a stern sign commands people to 'Keep off the grass'. In my former Oxford college the entire quad was formed of a large, attractive lawn, on which we were allowed to walk or sit on only one day a year. On any other day, woe to the poor student whose foot desecrated the holy turf.

Royal palaces and ducal chateaux turned the lawn into a symbol of authority. When in the late modern period kings were toppled and dukes were guillotined, the new presidents and prime ministers kept the lawns. Parliaments, supreme courts, presidential residences and other public buildings increasingly proclaimed their power in row upon row of neat green blades. Simultaneously, lawns conquered the world of sports. For thousands of years humans played on almost every conceivable kind of ground, from ice to desert. Yet in the last two centuries, the really important games – such as football and tennis – are played

6. The lawns of Château de Chambord, in the Loire Valley. King François I built it in the early sixteenth century. This is where it all began.

7. A welcoming ceremony in honour of Queen Elizabeth II – on the White House lawn.

8. Mario Götze scores the decisive goal, giving Germany the World Cup in 2014 – on the Maracanã lawn.

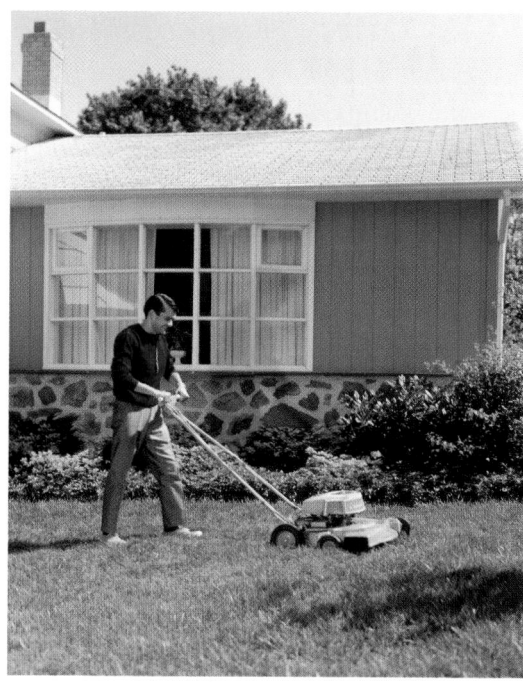

9. Petit-bourgeois paradise.

on lawns. Provided, of course, you have money. In the *favelas* of Rio de Janeiro the future generation of Brazilian football is kicking makeshift balls over sand and dirt. But in the wealthy suburbs, the sons of the rich are enjoying themselves over meticulously kept lawns.

Humans thereby came to identify lawns with political power, social status and economic wealth. No wonder that in the nineteenth century the rising bourgeoisie enthusiastically adopted the lawn. At first only bankers, lawyers and industrialists could afford such luxuries at their private residences. Yet when the Industrial Revolution broadened the middle class and gave rise to the lawnmower and then the automatic sprinkler, millions of families could suddenly afford a home turf. In American suburbia a spick-and-span lawn switched from being a rich person's luxury into a middle-class necessity.

This was when a new rite was added to the suburban liturgy. After Sunday morning service at church, many people devotedly mowed their lawns. Walking along the streets, you could quickly ascertain the wealth and position of every family by the size and quality of their turf. There is no surer sign that something is wrong at the Joneses' than a neglected lawn in the front yard. Grass is nowadays the most widespread crop in the USA after maize and wheat, and the lawn industry (plants, manure, mowers, sprinklers, gardeners) accounts for billions of dollars every year.[51]

The lawn did not remain solely a European or American craze. Even people who have never visited the Loire Valley see US presidents giving speeches on the White House lawn, important football games played out in green stadiums, and Homer and Bart Simpson quarrelling about whose turn it is to mow the grass. People all over the globe associate lawns with power, money and prestige. The lawn has therefore spread far and wide, and is now set to conquer even the heart of the Muslim world. Qatar's newly built Museum of Islamic Art is flanked by magnificent lawns that hark back to Louis XIV's Versailles much more than to Haroun al-Rashid's Baghdad. They were designed and constructed by an American company, and their more than 100,000 square yards of grass – in the midst of the

Arabian desert – require a stupendous amount of fresh water each day to stay green. Meanwhile, in the suburbs of Doha and Dubai, middle-class families pride themselves on their lawns. If it were not for the white robes and black hijabs, you could easily think you were in the Midwest rather than the Middle East.

Having read this short history of the lawn, when you now come to plan your dream house you might think twice about having a lawn in the front yard. You are of course still free to do it. But you are also free to shake off the cultural cargo bequeathed to you by European dukes, capitalist moguls and the Simpsons – and imagine for yourself a Japanese rock garden, or some altogether new creation. This is the best reason to learn history: not in order to predict the future, but to free yourself of the past and imagine alternative destinies. Of course this is not total freedom – we cannot avoid being shaped by the past. But some freedom is better than none.

A Gun in Act I

All the predictions that pepper this book are no more than an attempt to discuss present-day dilemmas, and an invitation to change the future. Predicting that humankind will try to gain immortality, bliss and divinity is much like predicting that people building a house will want a lawn in their front yard. It sounds very likely. But once you say it out loud, you can begin to think about alternatives.

People are taken aback by dreams of immortality and divinity not because they sound so foreign and unlikely, but because it is uncommon to be so blunt. Yet when they start thinking about it, most people realise that it actually makes a lot of sense. Despite the technological hubris of these dreams, ideologically they are old news. For 300 years the world has been dominated by humanism, which sanctifies the life, happiness and power of *Homo sapiens*. The attempt to gain immortality, bliss and divinity merely takes the long-standing humanist ideals to their logical conclusion. It places openly on the table what we have for a long time kept hidden under our napkin.

Yet I would now like to place something else on the table: a gun. A gun that appears in Act I, to fire in Act III. The following chapters discuss how humanism – the worship of humankind – has conquered the world. Yet the rise of humanism also contains the seeds of its downfall. While the attempt to upgrade humans into gods takes humanism to its logical conclusion, it simultaneously exposes humanism's inherent flaws. If you start with a flawed ideal, you often appreciate its defects only when the ideal is close to realisation.

We can already see this process at work in geriatric hospital wards. Due to an uncompromising humanist belief in the sanctity of human life, we keep people alive till they reach such a pitiful state that we are forced to ask, 'What exactly is so sacred here?' Due to similar humanist beliefs, in the twenty-first century we are likely to push humankind as a whole beyond its limits. The same technologies that can upgrade humans into gods might also make humans irrelevant. For example, computers powerful enough to understand and overcome the mechanisms of ageing and death will probably also be powerful enough to replace humans in any and all tasks.

Hence the real agenda in the twenty-first century is going to be far more complicated than what this long opening chapter has suggested. At present it might seem that immortality, bliss and divinity occupy the top slots on our agenda. But once we come nearer to achieving these goals the resulting upheavals are likely to deflect us towards entirely different destinations. The future described in this chapter is merely the future of the past – i.e., a future based on the ideas and hopes that dominated the world for the last 300 years. The real future – i.e., a future born of the new ideas and hopes of the twenty-first century – might be completely different.

To understand all this we need to go back and investigate who *Homo sapiens* really is, how humanism became the dominant world religion and why attempting to fulfil the humanist dream is likely to cause its disintegration. This is the basic plan of the book.

The first part of the book looks at the relationship between *Homo sapiens* and other animals, in an attempt to comprehend what makes our species so special. Some readers may wonder why animals receive so much attention in a book about the future. In my view, you cannot have a serious discussion about the nature and future of humankind without beginning with our fellow animals. *Homo sapiens* does its best to forget the fact, but it is an animal. And it is doubly important to remember our origins at a time when we seek to turn ourselves into gods. No investigation of our divine future can ignore our own animal past, or our relations with other animals – because the relationship between humans and animals is the best model we have for future relations between superhumans and humans. You want to know how super-intelligent cyborgs might treat ordinary flesh-and-blood humans? Better start by investigating how humans treat their less intelligent animal cousins. It's not a perfect analogy, of course, but it is the best archetype we can actually observe rather than just imagine.

Based on the conclusions of this first part, the second part of the book examines the bizarre world *Homo sapiens* has created in the last millennia, and the path that took us to our present crossroads. How did *Homo sapiens* come to believe in the humanist creed, according to which the universe revolves around humankind and humans are the source of all meaning and authority? What are the economic, social and political implications of this creed? How does it shape our daily life, our art and our most secret desires?

The third and last part of the book comes back to the early twenty-first century. Based on a much deeper understanding of humankind and of the humanist creed, it describes our current predicament and our possible futures. Why might attempts to fulfil humanism result in its downfall? How would the search for immortality, bliss and divinity shake the foundations of our belief in humanity? What signs foretell this cataclysm, and how is it reflected in the day-to-day decisions each of us makes? And if humanism is indeed in danger, what might take its place? This part of the book does not consist of mere philosophising or idle future-telling. Rather, it scrutinises our smartphones, dating practices and job market for clues of things to come.

For humanist true-believers, all this may sound very pessimistic and depressing. But it is best not to jump to conclusions. History has witnessed the rise and fall of many religions, empires and cultures. Such upheavals are not necessarily bad. Humanism has dominated the world for 300 years, which is not such a long time. The pharaohs ruled Egypt for 3,000 years, and the popes dominated Europe for a millennium. If you told an Egyptian in the time of Ramses II that one day the pharaohs will be gone, he would probably have been aghast. 'How can we live without a pharaoh? Who will ensure order, peace and justice?' If you told people in the Middle Ages that within a few centuries God will be dead, they would have been horrified. 'How can we live without God? Who will give life meaning and protect us from chaos?'

Looking back, many think that the downfall of the pharaohs and the death of God were both positive developments. Maybe the collapse of humanism will also be beneficial. People are usually afraid of change because they fear the unknown. But the single greatest constant of history is that everything changes.

10. King Ashurbanipal of Assyria slaying a lion: mastering the animal kingdom.

PART I

Homo sapiens Conquers the World

*What is the difference between humans and
all other animals?*

How did our species conquer the world?

Is Homo sapiens *a superior life form,
or just the local bully?*

2

The Anthropocene

With regard to other animals, humans have long since become gods. We don't like to reflect on this too deeply, because we have not been particularly just or merciful gods. If you watch the National Geographic channel, go to a Disney film or read a book of fairy tales, you might easily get the impression that planet Earth is populated mainly by lions, wolves and tigers who are an equal match for us humans. Simba the lion king holds sway over the forest animals; Little Red Riding Hood tries to evade the Big Bad Wolf; and little Mowgli bravely confronts Shere Khan the tiger. But in reality, they are no longer there. Our televisions, books, fantasies and nightmares are still full of them, but the Simbas, Shere Khans and Big Bad Wolves of our planet are disappearing. The world is populated mainly by humans and their domesticated animals.

How many wolves live today in Germany, the land of the Grimm brothers, Little Red Riding Hood and the Big Bad Wolf? Less than a hundred. (And even these are mostly Polish wolves that stole over the border in recent years.) In contrast, Germany is home to 5 million domesticated dogs. Altogether about 200,000 wild wolves still roam the earth, but there are more than 400 million domesticated dogs.[1] The world contains 40,000 lions compared to 600 million house cats; 900,000 African buffalo versus 1.5 billion domesticated cows; 50 million penguins and 20 billion chickens.[2] Since 1970, despite growing ecological awareness, wildlife populations have halved (not that they were prospering in 1970).[3] In 1980 there were 2 billion wild birds in Europe. In 2009 only 1.6 billion were left. In the same year, Europeans raised 1.9 billion chickens for meat and eggs.[4] At present, more than 90 per cent of the large animals of the world (i.e., those weighing more than a few pounds) are either humans or domesticated animals.

Scientists divide the history of our planet into epochs such as the Pleistocene, the Pliocene and the Miocene. Officially, we live in the Holocene epoch. Yet it may be better to call the last 70,000 years the Anthropocene epoch: the epoch of humanity. For during these millennia *Homo sapiens* became the single most important agent of change in the global ecology.[5]

This is an unprecedented phenomenon. Since the appearance of life, about 4 billion years ago, never has a single species changed the global ecology all by itself. Though there had been no lack of ecological revolutions and mass-extinction events, these were not caused by the actions of a particular lizard, bat or fungus. Rather, they were caused by the workings of mighty natural forces such as

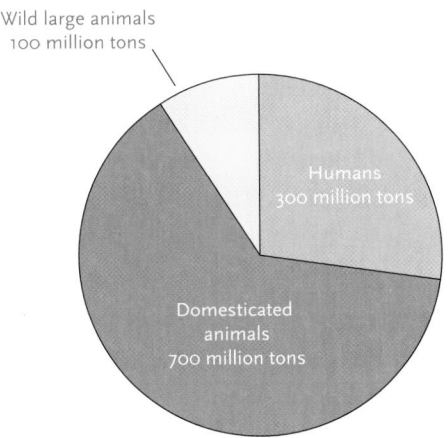

11. Pie chart of global biomass of large animals.

climate change, tectonic plate movement, volcanic eruptions and asteroid collisions.

Some people fear that today we are again in mortal danger of massive volcanic eruptions or colliding asteroids. Hollywood producers make billions out of these anxieties. Yet in reality, the danger is slim. Mass extinctions occur once every many millions of years. Yes, a big asteroid will probably hit our planet sometime in the next 100 million years, but it is very unlikely to happen next Tuesday. Instead of fearing asteroids, we should fear ourselves.

For *Homo sapiens* has rewritten the rules of the game. This single ape species has managed within 70,000 years to change the global ecosystem in radical and unprecedented ways. Our impact is already on a par with that of ice ages and tectonic movements. Within a century, our impact may surpass that of the asteroid that killed off the dinosaurs 65 million years ago.

That asteroid changed the trajectory of terrestrial evolution, but not its fundamental rules, which have remained fixed since the appearance of the first organisms 4 billion years ago. During all those aeons, whether you were a virus or a dinosaur, you evolved according to the unchanging principles of natural selection. In addition, no matter what strange and bizarre shapes life adopted, it remained confined to the organic realm – whether a cactus or a whale, you were made of organic compounds. Now humankind is poised to replace natural selection with intelligent design, and to extend life from the organic realm into the inorganic.

Even if we leave aside these future prospects and only look back on the last 70,000 years, it is evident that the Anthropocene has altered the world in unprecedented ways. Asteroids, plate tectonics and climate change may have impacted organisms all over the globe, but their influence differed from one area to another. The planet never constituted a single ecosystem; rather, it was a collection of many loosely connected ecosystems. When tectonic movements joined North America with South America it led to the extinction of most South American marsupials, but had no detrimental effect on the Australian kangaroo. When the last ice age reached its peak 20,000 years ago, jellyfish in the Persian Gulf and jellyfish in Tokyo Bay both had to adapt to the new climate. Yet since there was no connection between the two populations, each reacted in a different way, evolving in distinct directions.

In contrast, Sapiens broke the barriers that had separated the globe into independent ecological zones. In the Anthropocene, the planet became for the first time a single ecological unit. Australia, Europe and America continued to have different climates and topographies, yet humans caused organisms from throughout the world to mingle on a regular basis, irrespective of distance and geography. What began as a trickle of wooden boats has turned into a torrent of aeroplanes, oil tankers and giant cargo ships that criss-cross every ocean and bind every island and continent. Consequently the ecology of, say, Australia can no longer be understood without taking into account the European mammals or American microorganisms that flood its shores and deserts. Sheep, wheat, rats and flu viruses that humans brought to Australia during the last 300 years are today far more important to its ecology than the native kangaroos and koalas.

But the Anthropocene isn't a novel phenomenon of the last few centuries. Already tens of thousands of years ago, when our Stone Age ancestors spread from East Africa to the four corners of the earth, they changed the flora and fauna of every continent and island on which they settled. They drove to extinction all the other human species of the world, 90 per cent of the large animals of Australia, 75 per cent of the large mammals of America and about 50 per cent of all the large land mammals of the planet – and all before they planted the first wheat field, shaped the first metal tool, wrote the first text or struck the first coin.[6]

Large animals were the main victims because they were relatively few, and they bred slowly. Compare, for example, mammoths (which became extinct) to rabbits (which survived). A troop of mammoths numbered no more than a few dozen individuals, and bred at a rate of perhaps just two youngsters per year. Hence if the local human tribe hunted just three mammoths a year, it would have been enough for deaths to outstrip births, and within a few generations the mammoths disappeared. Rabbits, in contrast, bred like rabbits. Even if humans hunted hundreds of rabbits each year, it was not enough to drive them to extinction.

Not that our ancestors planned on wiping out the mammoths; they were simply unaware of the consequences of their actions. The extinction of the mammoths and other large animals may have been swift on an evolutionary timescale, but slow and gradual in human terms. People lived no more than seventy or eighty years, whereas the extinction process took centuries. The ancient Sapiens probably failed to notice any connection between the annual mammoth hunt – in which no more than two or three mammoths were killed – and the disappearance of these furry giants. At most, a nostalgic elder might have told sceptical youngsters that 'when I was young, mammoths were much more plentiful than these days. And so were mastodons and giant elks. And, of course, the tribal chiefs were honest, and children respected their elders.'

The Serpent's Children

Anthropological and archaeological evidence indicates that archaic hunter-gatherers were probably animists: they believed that there was no essential gap separating humans from other animals. The world – i.e., the local valley and the surrounding mountain chains – belonged to all its inhabitants, and everyone followed a common set of rules. These rules involved ceaseless negotiation between all concerned beings. People talked with animals, trees and stones, as well as with fairies, demons and ghosts. Out of this web of communications emerged the values and norms that were binding on humans, elephants, oak trees and wraiths alike.[7]

The animist world view still guides some hunter-gatherer communities that have survived into the modern age. One of them is the Nayaka people, who live in the tropical forests of south India.

The anthropologist Danny Naveh, who studied the Nayaka for several years, reports that when a Nayaka walking in the jungle encounters a dangerous animal such as a tiger, snake or elephant, he or she might address the animal and say: 'You live in the forest. I too live here in the forest. You came here to eat, and I too came here to gather roots and tubers. I didn't come to hurt you.'

A Nayaka was once killed by a male elephant they called 'the elephant who always walks alone'. The Nayakas refused to help officials from the Indian forestry department capture him. They explained to Naveh that this elephant used to be very close to another male elephant, with whom he always roamed. One day the forestry department captured the second elephant, and since then 'the elephant who always walks alone' had become angry and violent. 'How would you have felt if your spouse had been taken away from you? This is exactly how this elephant felt. These two elephants sometimes separated at night, each walking its own path . . . but in the morning they always came together again. On that day, the elephant saw his buddy falling, lying down. If two are always together and then you shoot one – how would the other feel?'[8]

Such an animistic attitude strikes many industrialised people as alien. Most of us automatically see animals as essentially different and inferior. This is because even our most ancient traditions were created thousands of years after the end of the hunter-gatherer era. The Old Testament, for example, was written down in the first millennium BC, and its oldest stories reflect the realities of the second millennium BC. But in the Middle East the age of the hunter-gatherers ended more than 7,000 years earlier. It is hardly surprising, therefore, that the Bible rejects animistic beliefs and its only animistic story appears right at the beginning, as a dire warning. The Bible is a long book, bursting with miracles, wonders and marvels. Yet the only time an animal initiates a conversation with a human is when the serpent tempts Eve to eat the forbidden fruit of knowledge (Bil'am's donkey also speaks a few words, but she is merely conveying to Bil'am a message from God).

In the Garden of Eden, Adam and Eve lived as foragers. The expulsion from Eden bears a striking resemblance to the Agricultural Revolution. Instead of allowing Adam to keep gathering wild fruits, an angry God condemns him 'to eat bread by the sweat of your brow'. It might be no coincidence, then, that biblical animals spoke with humans only in the pre-agricultural era of Eden. What lessons does the Bible draw from the episode? That you shouldn't listen to snakes, and it is generally best to avoid talking with animals and plants. It leads to nothing but disaster.

Yet the biblical story has deeper and more ancient layers of meaning. In most Semitic languages, 'Eve' means 'snake' or even 'female snake'. The name of our ancestral biblical mother hides an archaic animist myth, according to which snakes are not our enemies, but our ancestors.[9] Many animist cultures believe that humans descended from animals, including from snakes and other reptiles.

12. Paradise lost (the Sistine Chapel). The serpent – who sports a human upper body – initiates the entire chain of events. While the first two chapters of Genesis are dominated by divine monologues ('and God said . . . and God said . . . and God said . . .'), in the third chapter we finally get a dialogue – between Eve and the serpent ('and the serpent said unto the woman . . . and the woman said unto the serpent . . .'). This unique conversation between a human and an animal leads to the fall of humanity and our expulsion from Eden.

Most Australian Aborigines believed that the Rainbow Serpent created the world. The Aranda and Dieri people maintain that their particular tribes originated from primordial lizards or snakes, which were transformed into humans.[10] In fact, modern Westerners too think that they have evolved from reptiles. The brain of each and every one of us is built around a reptilian core, and the structure of our bodies is essentially that of modified reptiles.

The authors of the book of Genesis may have preserved a remnant of archaic animist beliefs in Eve's name, but they took great care to conceal all other traces. Genesis says that, instead of descending from snakes, humans were divinely created from inanimate matter. The snake is not our progenitor: he seduces us to rebel against our heavenly Father. While animists saw humans as just another kind of animal, the Bible argues that humans are a unique creation, and any attempt to acknowledge the animal within us denies God's power and authority. Indeed, when modern humans discovered that they actually evolved from reptiles, they rebelled against God and stopped listening to Him – or even believing in His existence.

Ancestral Needs

The Bible, along with its belief in human distinctiveness, was one of the by-products of the Agricultural Revolution, which initiated a new phase in human–animal relations. The advent of farming produced new waves of mass extinctions, but more importantly, it created a completely new life form on earth: domesticated animals. Initially this development was of minor importance, since humans managed to domesticate fewer than twenty species of mammals and birds, compared to the countless thousands of species that remained 'wild'. Yet with the passing of the centuries, this novel life form became dominant. Today more than 90 per cent of all large animals are domesticated.

Alas, domesticated species paid for their unparalleled collective success with unprecedented individual suffering. Although the animal kingdom has known many types of pain and misery for millions of years, the Agricultural Revolution generated completely new kinds of suffering that only became worse over time.

To the casual observer domesticated animals may seem much better off than their wild cousins and ancestors. Wild boars spend their days searching for food, water and shelter, and are constantly threatened by lions, parasites and floods. Domesticated pigs, in contrast, enjoy food, water and shelter provided by humans, who also treat their diseases and protect them against predators and natural disasters. True, most pigs sooner or later find themselves in the slaughterhouse. Yet does that make their fate any worse than the fate of wild boars? Is it better to be devoured by a lion than slaughtered by a man? Are crocodile teeth less deadly than steel blades?

What makes the fate of domesticated farm animals particularly harsh is not just the way they die, but above all the way they live. Two competing factors have shaped the living conditions of farm animals from ancient times to the present day: human desires and animal needs. Thus humans raise pigs in order to get meat, but if they want a steady supply of meat, they must ensure the long-term survival and reproduction of the pigs. Theoretically this should have protected the animals from extreme forms of cruelty. If a farmer did not take good care of his pigs, they would soon die without offspring and the farmer would starve.

Unfortunately, humans can cause tremendous suffering to farm animals in various ways, even while ensuring their survival and reproduction. The root of the problem is that domesticated animals have inherited from their wild ancestors many physical, emotional and social needs that are redundant on human farms. Farmers routinely ignore these needs, without paying any economic penalty. They lock animals in tiny cages, mutilate their horns and tails, separate mothers from offspring and selectively breed monstrosities. The animals suffer greatly, yet they live on and multiply.

Doesn't that contradict the most basic principles of natural selection? The theory of evolution maintains that all instincts, drives and emotions have evolved in the sole interest of survival and reproduction. If so, doesn't the continuous reproduction of farm animals prove that all their real needs are met? How can a pig have a 'need' that is not really needed for his survival and reproduction?

It is certainly true that all instincts, drives and emotions evolved in order to meet the evolutionary pressures of survival and reproduction. However, if and when these pressures suddenly disappear, the instincts, drives and emotions they had shaped do not disappear with them. At least not instantly. Even if they are no longer instrumental for survival and reproduction, these instincts, drives and emotions continue to mould the subjective experiences of the animal. For animals and humans alike, agriculture changed selection pressures almost overnight, but it did not change their physical, emotional and social drives. Of course evolution never stands still, and it has continued to modify humans and animals in the 12,000 years since the advent of farming. For example, humans in Europe and western Asia evolved the ability to digest cows' milk, while cows lost their fear of humans, and today produce far more milk than their wild ancestors. Yet these are superficial alterations. The deep sensory and emotional structures of cows, pigs and humans alike haven't changed much since the Stone Age.

Why do modern humans love sweets so much? Not because in the early twenty-first century we must gorge on ice cream and chocolate in order to survive. Rather, it is because when our Stone Age ancestors came across sweet fruit or honey, the most sensible thing to do was to eat as much of it as quickly as possible. Why do young men drive recklessly, get involved in violent arguments and hack confidential Internet sites? Because they are following ancient genetic decrees that might be useless and even counterproductive today, but that made good evolutionary sense 70,000 years ago. A young hunter who risked his life chasing a mammoth outshone all his competitors and won the hand of the local beauty, and we are now stuck with his macho genes.[11]

Exactly the same evolutionary logic shapes the lives of pigs, sows and piglets in human-controlled farms. In order to survive and reproduce in the wild, ancient boars needed to roam vast territories, familiarise themselves with their environment and beware of traps and predators. They further needed to communicate and cooperate with their fellow boars, forming complex groups dominated by old and experienced matriarchs. Evolutionary pressures consequently made wild boars – and even more so wild sows – highly intelligent social animals, characterised by a lively curiosity and strong urges to socialise, play, wander about and explore their surroundings. A sow born with some rare mutation that made her indifferent to her environment and to other boars was unlikely to survive or reproduce.

The descendants of wild boars – domesticated pigs – inherited their intelligence, curiosity and social skills.[12] Like wild boars, domesticated pigs communicate using a rich variety of vocal and olfactory signals: mother sows recognise the unique squeaks of their piglets, whereas two-day-old piglets already differentiate their mother's calls from those of other sows.[13] Professor Stanley Curtis of the Pennsylvania State University trained two pigs – named Hamlet and Omelette – to control a special joystick with their snouts, and found that the pigs soon rivalled primates in learning and playing simple computer games.[14]

Today most sows in industrial farms don't play computer games. They are locked by their human masters in tiny gestation crates, usually measuring six and a half by two feet. The crates have a concrete floor and metal bars, and hardly allow the pregnant sows even to turn around or sleep on their side, never mind walk. After three and a half months in such conditions, the sows are moved to slightly wider crates, where they give birth and nurse their piglets. Whereas piglets would naturally suckle for ten to twenty weeks, in industrial farms they are forcibly weaned within two to four weeks, separated from their mother and shipped to be fattened and slaughtered. The mother is immediately impregnated again, and sent back to the gestation crate to start another cycle. The typical sow would go through five to ten such cycles before being slaughtered herself. In recent years the use of crates has been restricted in the European Union and some US states, but the crates are still commonly used in many other countries, and tens of millions of breeding sows pass almost their entire lives in them.

The human farmers take care of everything the sow needs in order to survive and reproduce. She is given enough food, vaccinated against diseases, protected against the elements and artificially inseminated. From an objective perspective, the sow no longer needs to explore her surroundings, socialise with other pigs, bond with her piglets or even walk. But from a subjective perspective, the sow still feels very strong urges to do all of these things, and if these urges are not fulfilled she suffers greatly. Sows locked in gestation crates typically display acute frustration alternating with extreme despair.[15]

This is the basic lesson of evolutionary psychology: a need shaped thousands of generations ago continues to be felt subjectively

13. Sows confined in gestation crates. These highly social and intelligent beings spend most of their lives in this condition, as if they were already sausages.

even if it is no longer necessary for survival and reproduction in the present. Tragically, the Agricultural Revolution gave humans the power to ensure the survival and reproduction of domesticated animals while ignoring their subjective needs.

Organisms are Algorithms

How can we be sure that animals such as pigs actually have a subjective world of needs, sensations and emotions? Aren't we guilty of humanising animals, i.e., ascribing human qualities to non-human entities, like children believing that dolls feel love and anger?

In fact, attributing emotions to pigs doesn't humanise them. It 'mammalises' them. For emotions are not a uniquely human quality – they are common to all mammals (as well as to all birds and probably to some reptiles and even fish). All mammals evolved emotional abilities and needs, and from the fact that pigs are mammals we can safely deduce that they have emotions.[16]

In recent decades life scientists have demonstrated that emotions are not some mysterious spiritual phenomenon that is useful just for writing poetry and composing symphonies. Rather, emotions are biochemical algorithms that are vital for the survival and reproduction of all mammals. What does this mean? Well, let's begin by explaining what an algorithm is. This is of great importance not only because this key concept will reappear in many of the following chapters, but also because the twenty-first century will be dominated by algorithms. 'Algorithm' is arguably the single most important concept in our world. If we want to understand our life and our future, we should make every effort to understand what an algorithm is, and how algorithms are connected with emotions.

An algorithm is a methodical set of steps that can be used to make calculations, resolve problems and reach decisions. An algorithm isn't a particular calculation, but the method followed when making the calculation. For example, if you want to calculate the average between two numbers, you can use a simple algorithm. The algorithm says: 'First step: add the two numbers together. Second step: divide the sum by two.' When you enter the numbers 4 and 8, you get 6. When you enter 117 and 231, you get 174.

A more complex example is a cooking recipe. An algorithm for preparing vegetable soup may tell us:

1. Heat half a cup of oil in a pot.
2. Finely chop four onions.
3. Fry the onion until golden.
4. Cut three potatoes into chunks and add to the pot.
5. Slice a cabbage into strips and add to the pot.

And so forth. You can follow the same algorithm dozens of times, each time using slightly different vegetables, and therefore getting a slightly different soup. But the algorithm remains the same.

A recipe by itself cannot make soup. You need a person to read the recipe and follow the prescribed set of steps. But you can build a machine that embodies this algorithm and follows it automatically. Then you just need to provide the machine with water, electricity and vegetables – and it will prepare the soup by itself. There aren't many soup machines around, but you are probably familiar with beverage vending machines. Such machines usually have a slot for coins, an opening for cups, and rows of buttons. The first row has buttons for coffee, tea and cocoa. The second row is marked: no sugar, one spoon of sugar, two spoons of sugar. The third row indicates milk, soya milk, no milk. A man approaches the machine, inserts a coin into the slot and presses the buttons marked 'tea', 'one sugar' and 'milk'. The machine kicks into action, following a precise set of steps. It drops a tea bag into a cup, pours boiling water, adds a spoonful of sugar and milk, and ding! A nice cup of tea emerges. This is an algorithm.[17]

Over the last few decades biologists have reached the firm conclusion that the man pressing the buttons and drinking the tea is also an algorithm. A much more complicated algorithm than the vending machine, no doubt, but still an algorithm. Humans are algorithms that produce not cups of tea, but copies of themselves (like a vending machine which, if you press the right combination of buttons, produces another vending machine).

The algorithms controlling vending machines work through mechanical gears and electric circuits. The algorithms controlling humans work through sensations, emotions and thoughts. And exactly the same kind of algorithms control pigs, baboons, otters and chickens. Consider, for example, the following survival problem: a baboon spots some bananas hanging on a tree, but also notices a lion lurking nearby. Should the baboon risk his life for those bananas?

This boils down to a mathematical problem of calculating probabilities: the probability that the baboon will die of hunger if he does not eat the bananas, versus the probability that the lion will catch the baboon. In order to solve this problem the baboon needs to take into account a lot of data. How far am I from the bananas? How far away is the lion? How fast can I run? How fast can the lion run? Is the lion awake or asleep? Does the lion seem to be hungry or satiated? How many bananas are there? Are they big or small? Green or ripe? In addition to these external data, the baboon must also consider information about conditions within his own body. If he is starving, it makes sense to risk everything for those bananas, no matter the odds. In contrast, if he has just eaten, and the bananas are mere greed, why take any risks at all?

In order to weigh and balance all these variables and probabilities, the baboon requires far more complicated algorithms than the ones controlling automatic vending machines. The prize for making correct calculations is correspondingly greater. The prize is the very survival of the baboon. A timid baboon – one whose algorithms overestimate dangers – will starve to death, and the genes that shaped these cowardly algorithms will perish with him. A rash baboon – one whose algorithms underestimate dangers – will fall prey to the lion, and his reckless genes will also fail to make it to the next generation. These algorithms undergo constant quality control by natural selection. Only animals that calculate probabilities correctly leave offspring behind.

Yet this is all very abstract. How exactly does a baboon calculate probabilities? He certainly doesn't draw a pencil from behind his ear, a notebook from a back pocket, and start computing running speeds and energy levels with a calculator. Rather, the baboon's entire body is the calculator. What we call sensations and emotions are in fact algorithms. The baboon *feels* hunger, he *feels* fear and trembling at the sight of the lion, and he *feels* his mouth watering at the sight of the bananas. Within a split second, he experiences a storm of sensations, emotions and desires, which is nothing but the process of calculation. The result will appear as a feeling: the baboon will suddenly feel his spirit rising, his hairs standing on end, his muscles tensing, his chest expanding, and he will inhale a big breath, and 'Forward! I can do it! To the bananas!' Alternatively, he may be overcome by fear, his shoulders will droop, his stomach will turn, his legs will give way, and 'Mama! A lion! Help!' Sometimes the probabilities match so evenly that it is hard to decide. This too will manifest itself as a feeling. The baboon will feel confused and indecisive. 'Yes . . . No . . . Yes . . . No . . . Damn! I don't know what to do!'

In order to transmit genes to the next generation, it is not enough to solve survival problems. Animals also need to solve reproduction problems too, and this depends on calculating probabilities. Natural selection evolved passion and disgust as quick algorithms for evaluating reproduction odds. Beauty means 'good chances for having successful offspring'. When a woman sees a man and thinks, 'Wow! He is gorgeous!' and when a peahen sees a peacock and thinks, 'Jesus! What a tail!' they are doing something similar to the automatic vending machine. As light reflected from the male's body hits their retinas, extremely powerful algorithms honed by millions of years of evolution kick in. Within a few milliseconds the algorithms convert tiny cues in the male's external appearance into reproduction probabilities, and reach the conclusion: 'In all likelihood, this is a very healthy and fertile male, with excellent genes. If I mate with him, my offspring are also likely to enjoy good health and excellent genes.' Of course, this conclusion is not spelled out in words or numbers, but in the fiery itch of sexual attraction. Peahens, and most women, don't make such calculations with pen and paper. They just feel them.

Even Nobel laureates in economics make only a tiny fraction of their decisions using pen, paper and calculator; 99 per cent of our decisions – including the most important life choices concerning spouses, careers and habitats – are made by the highly refined algorithms we call sensations, emotions and desires.[18]

Because these algorithms control the lives of all mammals and birds (and probably some reptiles and even fish), when humans, baboons and pigs feel fear, similar neurological processes take place in similar brain areas. It is therefore likely that frightened humans, frightened baboons and frightened pigs have similar experiences.[19]

There are differences too, of course. Pigs don't seem to experience the extremes of compassion and cruelty that characterise

14. A peacock and a man. When you look at these images, data on proportions, colours and sizes gets processed by your biochemical algorithms, causing you to feel attraction, repulsion or indifference.

Homo sapiens, nor the sense of wonder that overwhelms a human gazing up at the infinitude of a starry sky. It is likely that there are also opposite examples, of swinish emotions unfamiliar to humans, but I cannot name any, for obvious reasons. However, one core emotion is apparently shared by all mammals: the mother–infant bond. Indeed, it gives mammals their name. The word 'mammal' comes from the Latin *mamma*, meaning breast. Mammal mothers love their offspring so much that they allow them to suckle from their body. Mammal youngsters, on their side, feel an overwhelming desire to bond with their mothers and stay near them. In the wild, piglets, calves and puppies that fail to bond with their mothers rarely survive for long. Until recently that was true of human children too. Conversely, a sow, cow or bitch that due to some rare mutation does not care about her young may live a long and comfortable life, but her genes will not pass to the next generation. The same logic is true among giraffes, bats, whales and porcupines. We can argue about other emotions, but since mammal youngsters cannot survive without motherly care, it is evident that motherly love and a strong mother–infant bond characterise all mammals.[20]

It took scientists many years to acknowledge this. Not long ago psychologists doubted the importance of the emotional bond between parents and children even among humans. In the first half of the twentieth century, and despite the influence of Freudian theories, the dominant behaviourist school argued that relations between parents and children were shaped by material feedback; that children needed mainly food, shelter and medical care; and that children bonded with their parents simply because the latter provide these material needs. Children who demanded warmth, hugs and kisses were thought to be 'spoiled'. Childcare experts warned that children who were hugged and kissed by their parents would grow up to be needy, egotistical and insecure adults.[21]

John Watson, a leading childcare authority in the 1920s, sternly advised parents, 'Never hug and kiss [your children], never let them sit in your lap. If you must, kiss them once on the forehead when they say goodnight. Shake hands with them in the morning.'[22] The popular magazine *Infant Care* explained that the secret of raising children is to maintain discipline and to provide the children's material needs according to a strict daily schedule. A 1929 article instructed parents that if an infant cries out for food before the normal feeding time, 'Do not hold him, nor rock him to stop his crying, and do not nurse him until the exact hour for the feeding comes. It will not hurt the baby, even the tiny baby, to cry.'[23]

Only in the 1950s and 1960s did a growing consensus of experts abandon these strict behaviourist theories and acknowledge the central importance of emotional needs. In a series of famous (and shockingly cruel) experiments, the psychologist Harry Harlow separated infant monkeys from their mothers shortly after birth, and isolated them in small cages. When given a choice between a metal dummy-mother fitted with a milk bottle, and a soft cloth-covered dummy with no milk, the baby monkeys clung to the barren cloth mother for all they were worth.

Those baby monkeys knew something that John Watson and the experts of *Infant Care* failed to realise: mammals can't live on food alone. They need emotional bonds too. Millions of years of evolution preprogrammed the monkeys with an overwhelming desire for emotional bonding. Evolution also imprinted them with the assumption that emotional bonds are more likely to be formed with soft furry things than with hard and metallic objects. (This is also why small human children are far more likely to become attached to dolls, blankets and smelly rags than to cutlery, stones or wooden blocks.) The need for emotional bonds is so strong that Harlow's baby monkeys abandoned the nourishing metal dummy and turned their attention to the only object that seemed capable of answering that need. Alas, the cloth-mother never responded to their affection and the little monkeys consequently suffered from severe psychological and social problems, and grew up to be neurotic and asocial adults.

Today we look back with incomprehension at early twentieth-century child-rearing advice. How could experts fail to appreciate that children have emotional needs, and that their mental and physical health depends as much on providing for these needs as on food, shelter and medicines? Yet when it comes to other mammals we keep denying the obvious. Like John Watson and the *Infant Care* experts, farmers throughout history took care of the material needs of piglets, calves and kids, but tended to ignore their emotional needs. Thus both the meat and dairy industries are based on breaking the most fundamental emotional bond in the mammal kingdom. Farmers get their breeding sows and dairy cows impregnated again and again. Yet the piglets and calves are separated from their mother shortly after birth, and often pass their days without ever sucking at her teats or feeling the warm touch of her tongue and body. What Harry Harlow did to a few hundred monkeys, the meat and dairy industries are doing to billions of animals every year.[24]

The Agricultural Deal

How did farmers justify their behaviour? Whereas hunter-gatherers were seldom aware of the damage they inflicted on the ecosystem, farmers knew perfectly well what they were doing. They knew they were exploiting domesticated animals and subjugating them to human desires and whims. They justified their actions in the name of new theist religions, which mushroomed and spread in the wake of the Agricultural Revolution. Theist religions began to argue that the universe is not a parliament of beings, but rather a theocracy ruled by a group of great gods – or perhaps by a single capital 'G' God ('Theos' in Greek). We don't normally associate this idea with agriculture, but at least in their beginnings theist religions were an agricultural enterprise. The theology, mythology and liturgy of religions such as Judaism, Hinduism and Christianity revolved at first around the relationship between humans, domesticated plants and farm animals.[25]

Biblical Judaism, for instance, catered to peasants and shepherds. Most of its commandments dealt with farming and village life, and its major holidays were harvest festivals. People today imagine the ancient temple in Jerusalem as a kind of big synagogue where priests clad in snow-white robes welcomed devout pilgrims, melodious choirs sang psalms and incense perfumed the air. In reality, it looked more like a cross between a slaughterhouse and a barbecue joint. The pilgrims did not come empty-handed. They brought with them a never-ending stream of sheep, goats, chickens and other animals, which were sacrificed at the god's altar and then cooked and eaten. The psalm-singing choirs could hardly be heard over the bellowing and bleating of calves and kids. Priests in bloodstained outfits cut the victims' throats, collected the gushing blood in jars and spilled it over the altar. The perfume of incense mixed with the odours of congealed blood and roasted meat, while swarms of black flies buzzed just about everywhere (see, for example, Numbers 28, Deuteronomy 12, and 1 Samuel 2). A modern Jewish family that celebrates a holiday by having a barbecue on their front lawn is much closer to the spirit of biblical times than an orthodox family that spends the time studying scriptures in a synagogue.

Theist religions, such as biblical Judaism, justified the agricultural economy through new cosmological myths. Animist religions had previously depicted the universe as a grand Chinese opera with a limitless cast of colourful actors. Elephants and oak trees, crocodiles and rivers, mountains and frogs, ghosts and fairies, angels and demons – each had a role in the cosmic opera. Theist religions rewrote the script, turning the universe into a bleak Ibsen drama with just two main characters: man and God. The angels and demons somehow survived the transition, becoming the messengers and servants of the great gods. Yet the rest of the animist cast – all the animals, plants and other natural phenomena – were transformed into silent decor. True, some animals were considered sacred to this or that god, and many gods had animal features: the Egyptian god Anubis sported the head of a jackal, and even Jesus Christ was frequently depicted as a lamb. Yet ancient Egyptians could easily tell the difference between Anubis and an ordinary jackal sneaking into the village to hunt chickens, and no Christian butcher ever mistook the lamb under his knife for Jesus.

We normally think that theist religions sanctified the great gods. We tend to forget that they sanctified humans, too. Hitherto *Homo sapiens* had been just one actor in a cast of thousands. In the new theist drama Sapiens became the central hero around whom the entire universe revolved.

The gods, meanwhile, were given two related roles to play. Firstly, they explained what is so special about Sapiens and why humans should dominate and exploit all other organisms. Christianity, for example, maintained that humans hold sway over the rest of creation because the Creator charged them with that authority. Moreover, according to Christianity, God gave an eternal soul only to humans. Since the fate of this eternal soul is the point of the whole Christian cosmos, and since animals have no soul, they are mere extras. Humans thus became the apex of creation, while all other organisms were pushed to the sidelines.

Secondly, the gods had to mediate between humans and the ecosystem. In the animistic cosmos, everyone talked with everyone directly. If you needed something from the caribou, the fig trees, the clouds or the rocks, you addressed them yourself. In the theist cosmos, all non-human entities were silenced. Consequently you could no longer talk with trees and animals. What to do, then, when you wanted the trees to give more fruits, the cows to give more milk, the clouds to bring more rain and the locusts to stay away from your crops? That's where the gods entered the picture. They promised to supply rain, fertility and protection, provided humans did something in return. This was the essence of the agricultural deal. The gods safeguarded and multiplied farm production, and in exchange humans had to share the produce with the gods. This deal served both parties, at the expense of the rest of the ecosystem.

Today in Nepal, devotees of the goddess Gadhimai celebrate her festival every five years in the village of Bariyapur. A record was set in 2009 when 250,000 animals were sacrificed to the goddess.

A local driver explained to a visiting British journalist that 'If we want anything, and we come here with an offering to the goddess, within five years all our dreams will be fulfilled.'[26]

Much of theist mythology explains the subtle details of this deal. The Mesopotamian Gilgamesh epic recounts that when the gods sent a great deluge to destroy the world, almost all humans and animals perished. Only then did the rash gods realise that nobody remained to make any offerings to them. They became crazed with hunger and distress. Luckily, one human family survived, thanks to the foresight of the god Enki, who instructed his devotee Utnapishtim to take shelter in a large wooden ark along with his relatives and a menagerie of animals. When the deluge subsided and this Mesopotamian Noah emerged from his ark, the first thing he did was sacrifice some animals to the gods. Then, tells the epic, all the great gods rushed to the spot: 'The gods smelled the savour / the gods smelled the sweet savour / the gods swarmed like flies around the offering.'[27] The biblical story of the deluge (written more than 1,000 years after the Mesopotamian version) also reports that immediately upon leaving the ark, 'Noah built an altar to the Lord and, taking some of the clean animals and clean birds, he sacrificed burnt offerings on it. The Lord smelled the pleasing aroma and said in his heart: Never again will I curse the ground because of humans' (Genesis 8:20–1).

This deluge story became a founding myth of the agricultural world. It is possible of course to give it a modern environmentalist spin. The deluge could teach us that our actions can ruin the entire ecosystem, and humans are divinely charged with protecting the rest of creation. Yet traditional interpretations saw the deluge as proof of human supremacy and animal worthlessness. According to these interpretations, Noah was instructed to save the whole ecosystem in order to protect the common interests of gods and humans rather than the interests of the animals. Non-human organisms have no intrinsic value; they exist solely for our sake.

After all, when 'the Lord saw how great the wickedness of the human race had become' He resolved to 'wipe from the face of the earth the human race I have created – and with them the animals, the birds and the creatures that move along the ground – for I regret that I have made them' (Genesis 6:7). The Bible thinks it is perfectly all right to destroy all animals as punishment for the crimes of *Homo sapiens*, as if the existence of giraffes, pelicans and ladybirds has lost all purpose if humans misbehave. The Bible could not imagine a scenario in which God repents having created *Homo sapiens*, wipes this sinful ape off the face of the earth, and then spends eternity enjoying the antics of ostriches, kangaroos and panda bears.

Theist religions nevertheless have certain animal-friendly beliefs. The gods gave humans authority over the animal kingdom, but this authority carried with it some responsibilities. For example, Jews were commanded to allow farm animals to rest on the Sabbath, and to avoid causing them unnecessary suffering. (Though whenever interests clashed, human interests always trumped animal interests.[28])

A Talmudic tale recounts how on the way to the slaughterhouse, a calf escaped and sought refuge with Rabbi Yehuda HaNasi, one of the founders of rabbinical Judaism. The calf tucked his head under the rabbi's flowing robes and started crying. Yet the rabbi pushed the calf away, saying, 'Go. You were created for that very purpose.' Since the rabbi showed no mercy, God punished him, and he suffered from a painful illness for thirteen years. Then, one day, a servant cleaning the rabbi's house found some newborn rats and began sweeping them out. Rabbi Yehuda rushed to save the helpless creatures, instructing the servant to leave them in peace, because 'God is good to all, and has compassion on all he has made' (Psalms 145:9). Since the rabbi showed compassion to these rats, God showed compassion to the rabbi, and he was cured of his illness.[29]

Other religions, particularly Jainism, Buddhism and Hinduism, have demonstrated even greater empathy to animals. They emphasise the connection between humans and the rest of the ecosystem, and their foremost ethical commandment has been to avoid killing any living being. Whereas the biblical 'Thou shalt not kill' covered only humans, the ancient Indian principle of *ahimsa* (non-violence) extends to every sentient being. Jain monks are particularly careful in this regard. They always cover their mouths with a white cloth, lest they inhale an insect, and whenever they walk they carry a broom to gently sweep any ant or beetle from their path.[30]

Nevertheless, all agricultural religions – Jainism, Buddhism and Hinduism included – found ways to justify human superiority and the exploitation of animals (if not for meat, then for milk and muscle power). They have all claimed that a natural hierarchy of beings entitles humans to control and use other animals, provided that the humans observe certain restrictions. Hinduism, for example, has sanctified cows and forbidden eating beef, but has also provided the ultimate justification for the dairy industry, alleging that cows are generous creatures that positively yearn to share their milk with humankind.

Humans thus committed themselves to an 'agricultural deal'. According to this deal, cosmic forces gave humans command over other animals, on condition that humans fulfilled certain obligations towards the gods, towards nature and towards the animals themselves. It was easy to believe in the existence of such a cosmic compact, because it reflected the daily routine of farming life.

Hunter-gatherers had not seen themselves as superior beings because they were seldom aware of their impact on the ecosystem. A typical band numbered in the dozens, it was surrounded by thousands of wild animals, and its survival depended on understanding and respecting the desires of these animals. Foragers had to constantly ask themselves what deer dream about, and what lions think. Otherwise, they could not hunt the deer, nor escape the lions.

Farmers, in contrast, lived in a world controlled and shaped by human dreams and thoughts. Humans were still subject to formidable natural forces such as storms and earthquakes, but they were far less dependent on the wishes of other animals. A farm boy learned early on to ride a horse, harness a bull, whip a stubborn donkey and lead the sheep to pasture. It was easy and tempting to believe that such everyday activities reflected either the natural order of things or the will of heaven.

The Agricultural Revolution was thus both an economic and a religious revolution. New kinds of economic relations emerged together with new kinds of religious beliefs that justified the brutal exploitation of animals. This ancient process can be witnessed even today whenever the last remaining hunter-gatherer communities adopt farming. In recent years the Nayaka hunter-gatherers of south India have taken up some agricultural practices such as herding cattle, raising chickens and cultivating tea. Not surprisingly, they have also picked up new attitudes towards animals, and they espouse very different views about domesticated animals (and plants) compared with wild organisms.

In the Nayaka language a living being possessing a unique personality is called *mansan*. When probed by the anthropologist Danny Naveh, the Nayaka explained that all elephants are *mansan*. 'We live in the forest, they live in the forest. We are all *mansan* . . . So are bears, deer and tigers. All forest animals.' What about cows? 'Cows are different. You have to lead them everywhere.' And chickens? 'They are nothing. They are not *mansan*.' And forest trees? 'Yes – they live for such a long time.' And tea bushes? 'Oh, these I cultivate so that I can sell the tea leaves and buy what I need from the store. No, they aren't *mansan*.'[31]

The degradation of animals from sentient beings deserving of respect into mere property rarely stopped with cows and chickens. Most agricultural societies began treating various classes of people as if they too were property. In ancient Egypt, biblical Israel and medieval China it was common to enslave humans, torture them and execute them for even trifling offences. Just as peasants did not consult with cows and chickens about the running of the farm, so rulers did not dream of asking peasants for their opinions about running the kingdom. And when ethnic groups or religious communities clashed, they frequently dehumanized each other. Depicting the 'others' as subhuman beasts was a first step towards treating them as such. The farm thus became the prototype of new societies, complete with puffed-up masters, inferior races fit for exploitation, wild beasts ripe for extermination and a great God above that gives His blessing to the entire arrangement.

Five Hundred Years of Solitude

The rise of modern science and industry brought about the next revolution in human–animal relations. During the Agricultural Revolution humankind silenced animals and plants, and turned the animist grand opera into a dialogue between man and gods. During the Scientific Revolution humankind silenced the gods too. The world was now a one-man show. Humankind stood alone on an empty stage, talking to itself, negotiating with no one and acquiring enormous powers without any obligations. Having deciphered the mute laws of physics, chemistry and biology, humankind now does with them as it pleases.

When an archaic hunter went out to the savannah, he asked the help of the wild bull, and the bull demanded something of the hunter. When an ancient farmer wanted his cows to produce lots of milk, he asked some great heavenly god for help, and the god stipulated his conditions. When the white-coated staff in Nestlé's Research and Development department want to increase dairy production, they study genetics – and the genes don't ask for anything in return.

But just as the hunters and farmers had their myths, so do the people in the R&D department. Their most famous myth shamelessly plagiarises the legend of the Tree of Knowledge and the Garden of Eden, but transports the action to the garden at Woolsthorpe Manor in Lincolnshire. According to this myth, Isaac Newton was sitting there under an apple tree when a ripe apple dropped on his head. Newton began wondering why the apple fell straight downwards, rather than sideways or upwards. His enquiry led him to discover gravity and the laws of Newtonian mechanics.

Newton's story turns the Tree of Knowledge myth on its head. In the Garden of Eden the serpent initiates the drama, tempting humans to sin, thereby bringing the wrath of God down upon them. Adam and Eve are a plaything for serpent and God alike. In contrast, in the Garden of Woolsthorpe man is the sole agent. Though Newton himself was a deeply religious Christian who devoted far more time to studying the Bible than the laws of physics, the Scientific Revolution that he helped launch pushed God to the sidelines. When Newton's successors came to write their Genesis myth, they had no use for either God or serpent. The Garden of Woolsthorpe is run by blind laws of nature, and the initiative to decipher these laws is strictly human. The story may begin with an apple falling on Newton's head, but the apple did not do it on purpose.

In the Garden of Eden myth, humans are punished for their curiosity and for their wish to gain knowledge. God expels them from Paradise. In the Garden of Woolsthorpe myth, nobody punishes Newton – just the opposite. Thanks to his curiosity humankind gains a better understanding of the universe, becomes more powerful and takes another step towards the technological paradise. Untold numbers of teachers throughout the world recount the Newton myth to encourage curiosity, implying that if only we gain enough knowledge, we can create paradise here on earth.

In fact, God is present even in the Newton myth: Newton himself is God. When biotechnology, nanotechnology and the other fruits of science ripen, *Homo sapiens* will attain divine powers and come full circle back to the biblical Tree of Knowledge. Archaic hunter-gatherers were just another species of animal. Farmers saw themselves as the apex of creation. Scientists will upgrade us into gods.

Whereas the Agricultural Revolution gave rise to theist religions, the Scientific Revolution gave birth to humanist religions, in which humans replaced gods. While theists worship *theos* (Greek for 'god'), humanists worship humans. The founding idea of humanist religions such as liberalism, communism and Nazism is that *Homo sapiens* has some unique and sacred essence that is the source of all meaning and authority in the universe. Everything that happens in the cosmos is judged to be good or bad according to its impact on *Homo sapiens*.

Whereas theism justified traditional agriculture in the name of God, humanism has justified modern industrial farming in the name of Man. Industrial farming sanctifies human needs, whims and wishes, while disregarding everything else. Industrial farming has no real interest in animals, which don't share the sanctity of human nature. And it has no use for gods, because modern science and technology give humans powers that far exceed those of the ancient gods. Science enables modern firms to subjugate cows, pigs and chickens to more extreme conditions than those prevailing in traditional agricultural societies.

In ancient Egypt, in the Roman Empire or in medieval China, humans had only a rudimental understanding of biochemistry, genetics, zoology and epidemiology. Consequently, their powers of manipulation were limited. In those days, pigs, cows and chickens ran free among the houses, and searched for edible treasures in the rubbish heap and in the nearby woods. If an ambitious peasant had tried to confine thousands of animals in a crowded coop, a deadly epidemic would probably have resulted, wiping out all the animals as well as many of the villagers. No priest, shaman or god could have prevented it.

But once modern science deciphered the secrets of epidemics, pathogens and antibiotics, industrial coops, pens and pigsties became feasible. With the help of vaccinations, medications, hormones, pesticides, central air-conditioning systems and automatic feeders, it is now possible to pack tens of thousands of pigs, cows or chickens into neat rows of cramped cages, and produce meat, milk and eggs with unprecedented efficiency.

In recent years, as people began to rethink human–animal relations, such practices have come under increasing criticism. We are suddenly showing unprecedented interest in the fate of so-called lower life forms, perhaps because we are about to become one. If and when computer programs attain superhuman intelligence and unprecedented power, should we begin valuing these programs more than we value humans? Would it be okay, for example, for an artificial intelligence to exploit humans and even kill them to further its own needs and desires? If it should never be allowed to do that, despite its superior intelligence and power, why is it ethical for humans to exploit and kill pigs? Do humans have some magical spark, in addition to higher intelligence and greater power, which distinguishes them from pigs, chickens, chimpanzees and computer programs alike? If yes, where did that spark come from, and why are we certain that an AI could never acquire it? If there is no such spark, would there be any reason to continue assigning special value to human life even after computers surpass humans in intelligence and power? Indeed, what exactly is it about humans that make us so intelligent and powerful in the first place, and how likely is it that non-human entities will ever rival and surpass us?

The next chapter will examine the nature and power of *Homo sapiens*, not only in order to comprehend further our relations with other animals, but also to appreciate what the future might hold for us, and what relations between humans and superhumans might look like.

3

The Human Spark

There is no doubt that *Homo sapiens* is the most powerful species in the world. *Homo sapiens* also likes to think that it enjoys a superior moral status, and that human life has much greater value than the lives of pigs, elephants or wolves. This is less obvious. Does might make right? Is human life more precious than porcine life simply because the human collective is more powerful than the pig collective? The United States is far mightier than Afghanistan; does this imply that American lives have greater intrinsic value than Afghan lives?

In practice, American lives *are* more valued. Far more money is invested in the education, health and safety of the average American than of the average Afghan. Killing an American citizen creates a far greater international outcry than killing an Afghan citizen. Yet it is generally accepted that this is no more than an unjust result of the geopolitical balance of power. Afghanistan may have far less clout than the USA, yet the life of a child in the mountains of Tora Bora is considered every bit as sacred as the life of a child in Beverly Hills.

In contrast, when we privilege human children over piglets, we want to believe that this reflects something deeper than the ecological balance of power. We want to believe that human lives really are superior in some fundamental way. We Sapiens love telling ourselves that we enjoy some magical quality that not only accounts for our immense power, but also gives moral justification for our privileged status. What is this unique human spark?

The traditional monotheist answer is that only Sapiens have eternal souls. Whereas the body decays and rots, the soul journeys on towards salvation or damnation, and will experience either everlasting joy in paradise or an eternity of misery in hell. Since pigs and other animals have no soul, they don't take part in this cosmic drama. They live only for a few years, and then die and fade into nothingness. We should therefore care far more about eternal human souls than about ephemeral pigs.

This is no kindergarten fairy tale, but an extremely powerful myth that continues to shape the lives of billions of humans and animals in the early twenty-first century. The belief that humans have eternal souls whereas animals are just evanescent bodies is a central pillar of our legal, political and economic system. It explains why, for example, it is perfectly okay for humans to kill animals for food, or even just for the fun of it.

However, our latest scientific discoveries flatly contradict this monotheist myth. True, laboratory experiments confirm the accuracy of one part of the myth: just as monotheist religions say, animals have no souls. All the careful studies and painstaking examinations have failed to discover any trace of a soul in pigs, rats or rhesus monkeys. Alas, the same laboratory experiments undermine the second and far more important part of the monotheist myth, namely, that humans *do* have a soul. Scientists have subjected *Homo sapiens* to tens of thousands of bizarre experiments, and looked into every nook in our hearts and every cranny in our brains. But they have so far discovered no magical spark. There is zero scientific evidence that in contrast to pigs, Sapiens have souls.

If that were all, we could well argue that scientists just need to keep looking. If they haven't found the soul yet, it is because they haven't looked carefully enough. Yet the life sciences doubt the existence of soul not just due to lack of evidence, but rather because the very idea of soul contradicts the most fundamental principles of evolution. This contradiction is responsible for the unbridled hatred that the theory of evolution inspires among devout monotheists.

Who's Afraid of Charles Darwin?

According to a 2012 Gallup survey, only 15 per cent of Americans think that *Homo sapiens* evolved through natural selection alone, free of all divine intervention; 32 per cent maintain that humans may have evolved from earlier life forms in a process lasting millions of years, but God orchestrated this entire show; 46 per cent believe that God created humans in their current form sometime during the last 10,000 years, just as the Bible says. Spending three years in college has absolutely no impact on these views. The same survey found that among BA graduates, 46 per cent believe in the biblical creation story, whereas only 14 per cent think that humans evolved without any divine supervision. Even among holders of MA and PhD degrees, 25 per cent believe the Bible, whereas only 29 per cent credit natural selection alone with the creation of our species.[1]

Though schools evidently do a very poor job teaching evolution, religious zealots still insist that it should not be taught at all. Alternatively, they demand that children must also be taught the theory of intelligent design, according to which all organisms were created by the design of some higher intelligence (aka God). 'Teach them both theories,' say the zealots, 'and let the kids decide for themselves.'

Why does the theory of evolution provoke such objections, whereas nobody seems to care about the theory of relativity or quantum mechanics? How come politicians don't ask that kids be exposed to alternative theories about matter, energy, space and time? After all, Darwin's ideas seem at first sight far less threatening than the monstrosities of Einstein and Werner Heisenberg. The theory of evolution rests on the principle of the survival of the fittest, which is a clear and simple – not to say humdrum – idea. In contrast, the theory of relativity and quantum mechanics argue that you can twist time and space, that something can appear out of nothing, and that a cat can be both alive and dead at the same time. This makes a mockery of our common sense, yet nobody seeks to protect innocent schoolchildren from these scandalous ideas. Why?

The theory of relativity makes nobody angry, because it doesn't contradict any of our cherished beliefs. Most people don't care an iota whether space and time are absolute or relative. If you think it is possible to bend space and time, well, be my guest. Go ahead and bend them. What do I care? In contrast, Darwin has deprived us of our souls. If you really understand the theory of evolution, you understand that there is no soul. This is a terrifying thought not only to devout Christians and Muslims, but also to many secular people who don't hold any clear religious dogma, but nevertheless want to believe that each human possesses an eternal individual essence that remains unchanged throughout life, and can survive even death intact.

The literal meaning of the word 'individual' is 'something that cannot be divided'. That I am an 'in-dividual' implies that my true self is a holistic entity rather than an assemblage of separate parts. This indivisible essence allegedly endures from one moment to the next without losing or absorbing anything. My body and brain undergo a constant process of change, as neurons fire, hormones flow and muscles contract. My personality, wishes and relationships never stand still, and may be completely transformed over years and decades. But underneath it all I remain the same person from birth to death – and hopefully beyond death as well.

Unfortunately, the theory of evolution rejects the idea that my true self is some indivisible, immutable and potentially eternal essence. According to the theory of evolution, all biological entities – from elephants and oak trees to cells and DNA molecules – are composed of smaller and simpler parts that ceaselessly combine and separate. Elephants and cells have evolved gradually, as a result of new combinations and splits. Something that cannot be divided or changed cannot have come into existence through natural selection.

The human eye, for example, is an extremely complex system made of numerous smaller parts such as the lens, the cornea and the retina. The eye did not pop out of nowhere complete with all these components. Rather, it evolved step by tiny step through millions of years. Our eye is very similar to the eye of *Homo erectus,* who lived 1 million years ago. It is somewhat less similar to the eye of *Australopithecus,* who lived 5 million years ago. It is very different from the eye of *Dryolestes,* who lived 150 million years ago. And it seems to have nothing in common with the unicellular organisms that inhabited our planet hundreds of millions of years ago.

Yet even unicellular organisms have tiny organelles that enable the microorganism to distinguish light from darkness, and move towards one or the other. The path leading from such archaic sensors to the human eye is long and winding, but if you have hundreds of millions of years to spare, you can certainly cover the entire path, step by step. You can do that because the eye is composed of many different parts. If every few generations a small mutation slightly changes one of these parts – say, the cornea becomes a bit more curved – after millions of generations these changes can result in a human eye. If the eye were a holistic entity, devoid of any parts, it could never have evolved by natural selection.

That's why the theory of evolution cannot accept the idea of souls, at least if by 'soul' we mean something indivisible, immutable and potentially eternal. Such an entity cannot possibly result from a step-by-step evolution. Natural selection could produce a human eye, because the eye has parts. But the soul has no parts. If the Sapiens soul evolved step by step from the Erectus soul, what exactly were these steps? Is there some part of the soul that is more developed in Sapiens than in Erectus? But the soul has no parts.

You might argue that human souls did not evolve, but appeared one bright day in the fullness of their glory. But when exactly was that bright day? When we look closely at the evolution of humankind, it is embarrassingly difficult to find it. Every human that ever existed came into being as a result of male sperm inseminating a female egg. Think of the first baby to possess a soul. That baby was very similar to her mother and father, except that she had a soul and they didn't. Our biological knowledge can certainly explain the birth of a baby whose cornea was a bit more curved than her parents' corneas. A slight mutation in a single gene can account for that. But biology cannot explain the birth of a baby possessing an eternal soul from parents who did not have even a shred of a soul. Is a single mutation, or even several mutations, enough to give an animal an essence secure against all changes, including even death?

Hence the existence of souls cannot be squared with the theory of evolution. Evolution means change, and is incapable of producing everlasting entities. From an evolutionary perspective, the closest thing we have to a human essence is our DNA, and the DNA molecule is the vehicle of mutation rather than the seat of eternity. This terrifies large numbers of people, who prefer to reject the theory of evolution rather than give up their souls.

Why the Stock Exchange Has No Consciousness

Another story employed to justify human superiority says that of all the animals on earth, only *Homo sapiens* has a conscious mind. Mind is something very different from soul. The mind isn't some mystical eternal entity. Nor is it an organ such as the eye or the brain. Rather, the mind is a flow of subjective experiences, such as pain, pleasure, anger and love. These mental experiences are made of interlinked sensations, emotions and thoughts, which flash for a brief moment, and immediately disappear. Then other experiences flicker and vanish, arising for an instant and passing away. (When reflecting on it, we often try to sort the experiences into distinct categories such as sensations, emotions and thoughts, but in actuality they are all mingled together.) This frenzied collection of experiences constitutes the stream of consciousness. Unlike the everlasting soul, the mind has many parts, it constantly changes, and there is no reason to think it is eternal.

The soul is a story that some people accept while others reject. The stream of consciousness, in contrast, is the concrete reality we directly witness every moment. It is the surest thing in the world. You cannot doubt its existence. Even when we are consumed by doubt and ask ourselves: 'Do subjective experiences really exist?' we can be certain that we are experiencing doubt.

What exactly are the conscious experiences that constitute the flow of the mind? Every subjective experience has two fundamental characteristics: sensation and desire. Robots and computers have no consciousness because despite their myriad abilities they feel nothing and crave nothing. A robot may have an energy sensor that signals to its central processing unit when the battery is about to run out. The robot may then move towards an electrical socket, plug itself in and recharge its battery. However, throughout this process the robot doesn't experience anything. In contrast, a human being depleted of energy feels hunger and craves to stop this unpleasant sensation. That's why we say that humans are conscious beings and robots aren't, and why it is a crime to make people work until they collapse from hunger and exhaustion, whereas making robots work until their batteries run out carries no moral opprobrium.

And what about animals? Are they conscious? Do they have subjective experiences? Is it okay to force a horse to work until he collapses from exhaustion? As noted earlier, the life sciences currently argue that all mammals and birds, and at least some reptiles and fish, have sensations and emotions. However, the most up-to-date theories also maintain that sensations and emotions are biochemical data-processing algorithms. Since we know that robots and computers process data without having any subjective experiences, maybe it works the same with animals? Indeed, we know that even in humans many sensory and emotional brain circuits can process data and initiate actions completely unconsciously. So perhaps behind all the sensations and emotions we ascribe to animals – hunger, fear, love and loyalty – lurk only unconscious algorithms rather than subjective experiences?[2]

This theory was upheld by the father of modern philosophy, René Descartes. In the seventeenth century Descartes maintained that only humans feel and crave, whereas all other animals are mindless automata, akin to a robot or a vending machine. When a man kicks a dog, the dog experiences nothing. The dog flinches and howls automatically, just like a humming vending machine that makes a cup of coffee without feeling or wanting anything.

This theory was widely accepted in Descartes' day. Seventeenth-century doctors and scholars dissected live dogs and observed the working of their internal organs, without either anaesthetics or scruples. They didn't see anything wrong with that, just as we don't see anything wrong in opening the lid of a vending machine and observing its gears and conveyors. In the early twenty-first century there are still plenty of people who argue that animals have no consciousness, or at most, that they have a very different and inferior type of consciousness.

In order to decide whether animals have conscious minds similar to our own, we must first get a better understanding of how minds function, and what role they play. These are extremely difficult questions, but it is worthwhile to devote some time to them, because the mind will be the hero of several subsequent chapters. We won't be able to grasp the full implications of novel technologies such as artificial intelligence if we don't know what minds are. Hence let's leave aside for a moment the particular question of animal minds, and examine what science knows about minds and consciousness in general. We will focus on examples taken from the study of human consciousness – which is more accessible to us – and later on return to animals and ask whether what's true of humans is also true of our furry and feathery cousins.

To be frank, science knows surprisingly little about mind and consciousness. Current orthodoxy holds that consciousness is created by electrochemical reactions in the brain, and that mental experiences fulfil some essential data-processing function.[3]

However, nobody has any idea how a congeries of biochemical reactions and electrical currents in the brain creates the subjective experience of pain, anger or love. Perhaps we will have a solid explanation in ten or fifty years. But as of 2016, we have no such explanation, and we had better be clear about that.

Using fMRI scans, implanted electrodes and other sophisticated gadgets, scientists have certainly identified correlations and even causal links between electrical currents in the brain and various subjective experiences. Just by looking at brain activity, scientists can know whether you are awake, dreaming or in deep sleep. They can briefly flash an image in front of your eyes, just at the threshold of conscious perception, and determine (without asking you) whether you have become aware of the image or not. They have even managed to link individual brain neurons with specific mental content, discovering for example a 'Bill Clinton' neuron and a 'Homer Simpson' neuron. When the 'Bill Clinton' neuron is on, the person is thinking of the forty-second president of the USA; show the person an image of Homer Simpson, and the eponymous neuron is bound to ignite.

More broadly, scientists know that if an electric storm arises in a given brain area, you probably feel angry. If this storm subsides and a different area lights up – you are experiencing love. Indeed, scientists can even induce feelings of anger or love by electrically stimulating the right neurons. But how on earth does the movement of electrons from one place to the other translate into a subjective image of Bill Clinton, or a subjective feeling of anger or love?

The most common explanation points out that the brain is a highly complex system, with more than 80 billion neurons connected into numerous intricate webs. When billions of neurons send billions of electric signals back and forth, subjective experiences emerge. Even though the sending and receiving of each electric signal is a simple biochemical phenomenon, the interaction among all these signals creates something far more complex – the stream of consciousness. We observe the same dynamic in many other fields. The movement of a single car is a simple action, but when millions of cars move and interact simultaneously, traffic jams emerge. The buying and selling of a single share is simple enough, but when millions of traders buy and sell millions of shares it can lead to economic crises that dumbfound even the experts.

Yet this explanation explains nothing. It merely affirms that the problem is very complicated. It does not offer any insight into how one kind of phenomenon (billions of electric signals moving from here to there) creates a very different kind of phenomenon (subjective experiences of anger or love). The analogy to other complex processes such as traffic jams and economic crises is flawed. What creates a traffic jam? If you follow a single car, you will never understand it. The jam results from the interactions among many cars. Car A influences the movement of car B, which blocks the path of car C, and so on. Yet if you map the movements of all the relevant cars, and how each impacts the other, you will get a complete account of the traffic jam. It would be pointless to ask, 'But how do all these movements create the traffic jam?' For 'traffic jam' is simply the abstract term we humans decided to use for this particular collection of events.

In contrast, 'anger' isn't an abstract term we have decided to use as a shorthand for billions of electric brain signals. Anger is an extremely concrete experience which people were familiar with long before they knew anything about electricity. When I say, 'I am angry!' I am pointing to a very tangible feeling. If you describe how a chemical reaction in a neuron results in an electric signal, and how billions of similar reactions result in billions of additional signals, it is still worthwhile to ask, 'But how do these billions of events come together to create my concrete feeling of anger?'

When thousands of cars slowly edge their way through London, we call that a traffic jam, but it doesn't create some great Londonian consciousness that hovers high above Piccadilly and says to itself, 'Blimey, I feel jammed!' When millions of people sell billions of shares, we call that an economic crisis, but no great Wall Street spirit grumbles, 'Shit, I feel I am in crisis.' When trillions of water molecules coalesce in the sky we call that a cloud, but no cloud consciousness emerges to announce, 'I feel rainy.' How is it, then, that when billions of electric signals move around in my brain, a mind emerges that feels 'I am furious!'? As of 2016, we have absolutely no idea.

Hence if this discussion has left you confused and perplexed, you are in very good company. The best scientists too are a long way from deciphering the enigma of mind and consciousness. One of the wonderful things about science is that when scientists don't know something, they can try out all kinds of theories and conjunctures, but in the end they can just admit their ignorance.

The Equation of Life

Scientists don't know how a collection of electric brain signals creates subjective experiences. Even more crucially, they don't know what could be the evolutionary benefit of such a phenomenon. It is the greatest lacuna in our understanding of life. Humans have feet, because for millions of generations feet enabled our ancestors to chase rabbits and escape lions. Humans have eyes, because for countless millennia eyes enabled our forebears to see whither the rabbit was heading and whence the lion was coming. But why do humans have subjective experiences of hunger and fear?

Not long ago, biologists gave a very simple answer. Subjective experiences are essential for our survival, because if we didn't feel hunger or fear we would not have bothered to chase rabbits and flee lions. Upon seeing a lion, why did a man flee? Well, he was frightened, so he ran away. Subjective experiences explained human actions. Yet today scientists provide a much more detailed explanation. When a man sees a lion, electric signals move from the eye to the brain. The incoming signals stimulate certain neurons, which react by firing off more signals. These stimulate other neurons down the line, which fire in their turn. If enough of the right neurons fire at a sufficiently rapid rate, commands are sent to the adrenal glands to flood the body with adrenaline, the heart is instructed to beat faster, while neurons in the motor centre send signals down to the leg muscles, which begin to stretch and contract, and the man runs away from the lion.

Ironically, the better we map this process, the *harder* it becomes to explain conscious feelings. The better we understand the brain, the more redundant the mind seems. If the entire system works by electric signals passing from here to there, why the hell do we also need to *feel* fear? If a chain of electrochemical reactions leads all the way from the nerve cells in the eye to the movements of leg muscles, why add subjective experiences to this chain? What do they do? Countless domino pieces can fall one after the other without any need of subjective experiences. Why do neurons need feelings in order to stimulate one another, or in order to tell the adrenal gland to start pumping? Indeed, 99 per cent of bodily activities, including muscle movement and hormonal secretions, take place without any need of conscious feelings. So why do the neurons, muscles and glands need such feelings in the remaining 1 per cent of cases?

You might argue that we need a mind because the mind stores memories, makes plans and autonomously sparks completely new images and ideas. It doesn't just respond to outside stimuli. For example, when a man sees a lion, he doesn't react automatically to the sight of the predator. He remembers that a year ago a lion ate his aunt. He imagines how he would feel if a lion tore him to pieces. He contemplates the fate of his orphaned children. That's why he flees. Indeed, many chain reactions begin with the mind's own initiative rather than with any immediate external stimulus. Thus a memory of some prior lion attack might spontaneously pop up in a man's mind, setting him thinking about the danger posed by lions. He then gets all the tribespeople together and they brainstorm novel methods for scaring lions away.

But wait a moment. What are all these memories, imaginations and thoughts? Where do they exist? According to current biological theories, our memories, imaginations and thoughts don't exist in some higher immaterial field. Rather, they too are avalanches of electric signals fired by billions of neurons. Hence even when we figure in memories, imaginations and thoughts, we are still left with a series of electrochemical reactions that pass through billions of neurons, ending with the activity of adrenal glands and leg muscles.

Is there even a single step on this long and twisting journey where, between the action of one neuron and the reaction of the next, the mind intervenes and decides whether the second neuron should fire or not? Is there any material movement, of even a single electron, that is caused by the subjective experience of fear rather than by the prior movement of some other particle? If there is no such movement – and if every electron moves because another electron moved earlier – why do we need to experience fear? We have no clue.

Philosophers have encapsulated this riddle in a trick question: what happens in the mind that doesn't happen in the brain? If nothing happens in the mind except what happens in our massive network of neurons – then why do we need the mind? If something does indeed happen in the mind over and above what happens in the neural network – where the hell does it happen? Suppose I ask you what Homer Simpson thought about Bill Clinton and the Monica Lewinsky scandal. You have probably never thought about this before, so your mind now needs to fuse two previously unrelated memories, perhaps conjuring up an image of Homer drinking beer while watching the president give his 'I did not have sexual relations with that woman' speech. Where does this fusion take place?

Some brain scientists argue that it happens in the 'global workspace' created by the interaction of many neurons.[4] Yet the word 'workspace' is just a metaphor. What is the reality behind the metaphor? Where do the different pieces of information actually meet and fuse? According to current theories, it certainly doesn't take place in some Platonic fifth dimension. Rather, it takes place, say, where two previously unconnected neurons suddenly start firing signals to one another. A new synapse is formed between the Bill Clinton neuron and the Homer Simpson neuron. But if so, why do we need the conscious experience of memory over and above the physical event of the two neurons connecting?

We can pose the same riddle in mathematical terms. Present-day dogma holds that organisms are algorithms, and that algorithms can be represented in mathematical formulas. You can use numbers and mathematical symbols to write the series of steps a vending machine takes to prepare a cup of tea, and the series of steps a brain takes when it is alarmed by the approach of a lion. If so, and if conscious experiences fulfil some important function, they must have a mathematical representation. For they are an essential part of the algorithm. When we write the fear algorithm, and break 'fear' down into a series of precise calculations, we should be able to point out: 'Here, step number ninety-three in the calculation process – this is the subjective experience of fear!' But is there any algorithm in the huge realm of mathematics that contains a subjective experience? So far, we don't know of any such algorithm. Despite the vast knowledge we have gained in the fields of mathematics and computer science, none of the data-processing systems we have created needs subjective experiences in order to function, and none feels pain, pleasure, anger or love.[5]

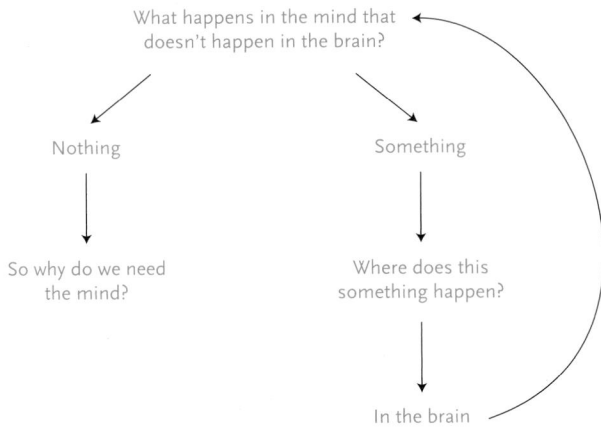

Maybe we need subjective experiences in order to think about ourselves? An animal wandering the savannah and calculating its chances of survival and reproduction must represent its own actions and decisions to itself, and sometimes communicate them to other animals as well. As the brain tries to create a model of its own decisions, it gets trapped in an infinite digression, and abracadabra! Out of this loop, consciousness pops out.

Fifty years ago this might have sounded plausible, but not in 2016. Several corporations, such as Google and Tesla, are engineering autonomous cars that already cruise our roads. The algorithms controlling the autonomous car make millions of calculations each second concerning other cars, pedestrians, traffic lights and potholes. The autonomous car successfully stops at red lights, bypasses obstacles and keeps a safe distance from other vehicles – without feeling any fear. The car also needs to take itself into account and to communicate its plans and desires to the surrounding vehicles, because if it decides to swerve to the right, doing so will impact on their behaviour. The car does all that without any problem – but without any consciousness either. The autonomous car isn't special. Many other computer programs make allowances for their own actions, yet none of them has developed consciousness, and none feels or desires anything.[6]

If we cannot explain the mind, and if we don't know what function it fulfils, why not just discard it? The history of science is replete with abandoned concepts and theories. For instance, early modern scientists who tried to account for the movement of light

15. The Google autonomous car on the road.

postulated the existence of a substance called ether, which supposedly fills the entire universe. Light was thought to be waves of ether. However, scientists failed to find any empirical evidence for the existence of ether, whereas they did come up with alternative and better theories of light. Consequently, they threw ether into the dustbin of science.

Similarly, for thousands of years humans used God to explain numerous natural phenomena. What causes lightning to strike? God. What makes the rain fall? God. How did life on earth begin? God did it. Over the last few centuries scientists have not discovered any empirical evidence for God's existence, while they did find much more detailed explanations for lightning strikes, rain and the origins of life. Consequently, with the exception of a few subfields of philosophy, no article in any peer-review scientific journal takes God's existence seriously. Historians don't argue that the Allies won the Second World War because God was on their side; economists don't blame God for the 1929 economic crisis; and geologists don't invoke His will to explain tectonic plate movements.

The same fate has befallen the soul. For thousands of years people believed that all our actions and decisions emanate from our souls. Yet in the absence of any supporting evidence, and given the existence of much more detailed alternative theories, the life sciences have ditched the soul. As private individuals, many biologists and doctors may go on believing in souls. Yet they never write about them in serious scientific journals.

Maybe the mind should join the soul, God and ether in the dustbin of science? After all, no one has ever seen experiences of pain or love through a microscope, and we have a very detailed biochemical explanation for pain and love that leaves no room for subjective experiences. However, there is a crucial difference between mind and soul (as well as between mind and God). Whereas the existence of eternal souls is pure conjecture, the experience of pain is a direct and very tangible reality. When I step on a nail, I can be 100 per cent certain that I feel pain (even if I so far lack a scientific explanation for it). In contrast, I cannot be certain that if the wound becomes infected and I die of gangrene, my soul will continue to exist. It's a very interesting and comforting story which I would be happy to believe, but I have no direct evidence for its veracity. Since all scientists constantly experience subjective feelings such as pain and doubt, they cannot deny their existence.

Another way to dismiss mind and consciousness is to deny their relevance rather than their existence. Some scientists – such as Daniel Dennett and Stanislas Dehaene – argue that all relevant questions can be answered by studying brain activities, without any recourse to subjective experiences. So scientists can safely delete 'mind', 'consciousness' and 'subjective experiences' from their vocabulary and articles. However, as we shall see in the following chapters, the whole edifice of modern politics and ethics is built upon subjective experiences, and few ethical dilemmas can be solved by referring strictly to brain activities. For example, what's wrong with torture or rape? From a purely neurological perspective, when a human is tortured or raped certain biochemical reactions happen in the brain, and various electrical signals move from one bunch of neurons to another. What could possibly be wrong with that? Most modern people have ethical qualms about torture and rape because of the subjective experiences involved. If any scientist wants to argue that subjective experiences are irrelevant, their challenge is to explain why torture or rape are wrong without reference to any subjective experience.

Finally, some scientists concede that consciousness is real and may actually have great moral and political value, but that it fulfils no biological function whatsoever. Consciousness is the biologically useless by-product of certain brain processes. Jet engines roar loudly, but the noise doesn't propel the aeroplane forward. Humans don't need carbon dioxide, but each and every breath fills the air with more of the stuff. Similarly, consciousness may be a kind of mental pollution produced by the firing of complex neural networks. It doesn't do anything. It is just there. If this is true, it implies that all the pain and pleasure experienced by billions of creatures for millions of years is just mental pollution. This is certainly a thought worth thinking, even if it isn't true. But it is quite amazing to realise that as of 2016, this is the best theory of consciousness that contemporary science has to offer us.

Maybe the life sciences view the problem from the wrong angle. They believe that life is all about data processing, and that organisms are machines for making calculations and taking decisions. However, this analogy between organisms and algorithms might mislead us. In the nineteenth century, scientists described brains and minds as if they were steam engines. Why steam engines? Because that was the leading technology of the day, which powered trains, ships and factories, so when humans tried to explain life, they assumed it must work according to analogous principles. Mind and body are made of pipes, cylinders, valves and pistons that build and release pressure, thereby producing movements and actions. Such thinking had a deep influence even on Freudian psychology, which is why much of our psychological jargon is still replete with concepts borrowed from mechanical engineering.

Consider, for example, the following Freudian argument: 'Armies harness the sex drive to fuel military aggression. The army recruits young men just when their sexual drive is at its peak. The army limits the soldiers' opportunities of actually having sex and releasing all that pressure, which consequently accumulates inside them. The army then redirects this pent-up pressure and allows it to be released in the form of military aggression.' This is exactly how a steam engine works. You trap boiling steam inside a closed container. The steam builds up more and more pressure, until suddenly you open a valve, and release the pressure in a predetermined direction, harnessing it to propel a train or a loom. Not only in armies, but in all fields of activity, we often complain about the pressure building up inside us, and we fear that unless we 'let off some steam', we might explode.

In the twenty-first century it sounds childish to compare the human psyche to a steam engine. Today we know of a far more sophisticated technology – the computer – so we explain the human psyche as if it were a computer processing data rather than a steam engine regulating pressure. But this new analogy may turn out to be just as naïve. After all, computers have no minds. They don't crave anything even when they have a bug, and the Internet doesn't feel pain even when authoritarian regimes sever entire countries from the Web. So why use computers as a model for understanding the mind?

Well, are we really sure that computers have no sensations or desires? And even if they haven't got any at present, perhaps once they become complex enough they might develop consciousness? If that were to happen, how could we ascertain it? When computers replace our bus driver, our teacher and our shrink, how could we determine whether they have feelings or whether they are just a collection of mindless algorithms?

When it comes to humans, we are today capable of differentiating between conscious mental experiences and non-conscious brain activities. Though we are far from understanding consciousness, scientists have succeeded in identifying some of its electrochemical signatures. To do so the scientists started with the assumption that whenever humans report that they are conscious of something, they can be believed. Based on this assumption the scientists could then isolate specific brain patterns that appear every time humans report being conscious, but that never appear during unconscious states.

This has allowed the scientists to determine, for example, whether a seemingly vegetative stroke victim has completely lost consciousness, or has merely lost control of his body and speech. If the patient's brain displays the telltale signatures of consciousness, he is probably conscious, even though he cannot move or speak. Indeed, doctors have recently managed to communicate with such patients using fMRI imaging. They ask the patients yes / no questions, telling them to imagine themselves playing tennis if the answer is yes, and to visualise the location of their home if the answer is no. The doctors can then observe how the motor cortex lights up when patients imagine playing tennis (meaning 'yes'), whereas 'no' is indicated by the activation of brain areas responsible for spatial memory.[7]

This is all very well for humans, but what about computers? Since silicon-based computers have very different structures to carbon-based human neural networks, the human signatures of consciousness may not be relevant to them. We seem to be trapped in a vicious circle. Starting with the assumption that we can believe humans when they report that they are conscious, we can identify the signatures of human consciousness, and then use these signatures to 'prove' that humans are indeed conscious. But if an artificial intelligence self-reports that it is conscious, should we just believe it?

So far, we have no good answer to this problem. Already thousands of years ago philosophers realised that there is no way to prove conclusively that anyone other than oneself has a mind. Indeed, even in the case of other humans, we just assume they have consciousness – we cannot know that for certain. Perhaps I am the only being in the entire universe who feels anything, and all other humans and animals are just mindless robots? Perhaps I am dreaming, and everyone I meet is just a character in my dream? Perhaps I am trapped inside a virtual world, and all the beings I see are merely simulations?

According to current scientific dogma, everything I experience is the result of electrical activity in my brain, and it should therefore be theoretically feasible to simulate an entire virtual world that I could not possibly distinguish from the 'real' world. Some brain scientists believe that in the not too distant future, we shall actually do such things. Well, maybe it has already been done – to you? For all you know, the year might be 2216 and you are a bored teenager immersed inside a 'virtual world' game that simulates the primitive and exciting world of the early twenty-first century. Once you acknowledge the mere feasibility of this scenario, mathematics leads you to a very scary conclusion: since there is only one real world, whereas the number of potential virtual worlds is infinite, the probability that you happen to inhabit the sole real world is almost zero.

None of our scientific breakthroughs has managed to overcome this notorious Problem of Other Minds. The best test that scholars have so far come up with is called the Turing Test, but it examines only social conventions. According to the Turing Test, in order to determine whether a computer has a mind, you should communicate simultaneously both with that computer and with a real person, without knowing which is which. You can ask whatever questions you want, you can play games, argue, and even flirt with them. Take as much time as you like. Then you need to decide which is the computer, and which is the human. If you cannot make up your mind, or if you make a mistake, the computer has passed the Turing Test, and we should treat it as if it really has a mind. However, that won't really be a proof, of course. Acknowledging the existence of other minds is merely a social and legal convention.

The Turing Test was invented in 1950 by the British mathematician Alan Turing, one of the fathers of the computer age. Turing was also a gay man in a period when homosexuality was illegal in Britain. In 1952 he was convicted of committing homosexual acts and forced to undergo chemical castration. Two years later he committed suicide. The Turing Test is simply a replication of a mundane test every gay man had to undergo in 1950s Britain: can you pass for a straight man? Turing knew from personal experience that it didn't matter who you really were – it mattered only what others thought about you. According to Turing, in the future computers would be just like gay men in the 1950s. It won't matter whether computers will actually be conscious or not. It will matter only what people think about it.

The Depressing Lives of Laboratory Rats

Having acquainted ourselves with the mind – and with how little we really know about it – we can return to the question of whether other animals have minds. Some animals, such as dogs, certainly pass a modified version of the Turing Test. When humans try to determine whether an entity is conscious, what we usually look for is not mathematical aptitude or good memory, but rather the ability to create emotional relationships with us. People sometimes develop deep emotional attachments to fetishes like weapons, cars and even underwear, but these attachments are one-sided and never develop into relationships. The fact that dogs can be party to emotional relationships with humans convinces most dog owners that dogs are not mindless automata.

This, however, won't satisfy sceptics, who point out that emotions are algorithms, and that no known algorithm requires consciousness in order to function. Whenever an animal displays complex emotional behaviour, we cannot prove that this is not the result of some very sophisticated but non-conscious algorithm. This argument, of course, can be applied to humans too. Everything a human does – including reporting on allegedly conscious states – might in theory be the work of non-conscious algorithms.

In the case of humans, we nevertheless assume that whenever someone reports that he or she is conscious, we can take their word for it. Based on this minimal assumption, we can today identify the brain signatures of consciousness, which can then be used systematically to differentiate conscious from non-conscious states in humans. Yet since animal brains share many features with human brains, as our understanding of the signatures of consciousness deepens, we might be able to use them to determine if and when other animals are conscious. If a canine brain shows similar patterns to those of a conscious human brain, this will provide strong evidence that dogs are conscious.

Initial tests on monkeys and mice indicate that at least monkey and mice brains indeed display the signatures of consciousness.[8] However, given the differences between animal brains and human brains, and given that we are still far from deciphering all the secrets of consciousness, developing decisive tests that will satisfy the sceptics might take decades. Who should carry the burden of proof in the meantime? Do we consider dogs to be mindless machines until proven otherwise, or do we treat dogs as conscious beings as long as nobody comes up with some convincing counter-evidence?

On 7 July 2012 leading experts in neurobiology and the cognitive sciences gathered at the University of Cambridge, and signed the Cambridge Declaration on Consciousness, which says that 'Convergent evidence indicates that non-human animals have the neuroanatomical, neurochemical and neurophysiological substrates of conscious states along with the capacity to exhibit intentional behaviours. Consequently, the weight of evidence indicates that humans are not unique in possessing the neurological substrates that generate consciousness. Non-human animals, including all mammals and birds, and many other creatures, including octopuses, also possess these neurological substrates.'[9] This declaration stops short of saying that other animals are conscious, because we still lack the smoking gun. But it does shift the burden of proof to those who think otherwise.

Responding to the shifting winds of the scientific community, in May 2015 New Zealand became the first country in the world to legally recognise animals as sentient beings, when the New Zealand parliament passed the Animal Welfare Amendment Act. The Act stipulates that it is now obligatory to recognise animals as sentient, and hence attend properly to their welfare in contexts such as animal husbandry. In a country with far more sheep than humans (30 million vs 4.5 million), that is a very significant statement. The Canadian province of Quebec has since passed a similar Act, and other countries are likely to follow suit.

Many business corporations also recognise animals as sentient beings, though paradoxically, this often exposes the animals to rather unpleasant laboratory tests. For example, pharmaceutical companies routinely use rats as experimental subjects in the development of antidepressants. According to one widely used protocol, you take a hundred rats (for statistical reliability) and place each rat inside a glass tube filled with water. The rats struggle again and again to climb out of the tubes, without success. After fifteen minutes most give up and stop moving. They just float in the tube, apathetic to their surroundings.

You now take another hundred rats, throw them in, but fish them out of the tube after fourteen minutes, just before they are about to despair. You dry them, feed them, give them a little rest – and then throw them back in. The second time, most rats struggle for twenty minutes before calling it quits. Why the extra six minutes? Because the memory of past success triggers the release of some biochemical in the brain that gives the rats hope and delays the advent of despair. If we could only isolate this biochemical, we might use it as an anti-depressant for humans. But numerous chemicals flood a rat's brain at any given moment. How can we pinpoint the right one?

For this you take more groups of rats, who have never partici-pated in the test before. You inject each group with a particular chemical, which you suspect to be the hoped-for antidepressant. You throw the rats into the water. If rats injected with chemical A struggle for only fifteen minutes before becoming depressed, you can cross out A on your list. If rats injected with chemical B go on

16. Left: A hopeful rat struggling to escape the glass tube. Right: An apathetic rat floating in the glass tube, having lost all hope.

thrashing for twenty minutes, you can tell the CEO and the share-holders that you might have just hit the jackpot.

Sceptics could object that this entire description need-lessly humanises rats. Rats experience neither hope nor despair. Sometimes rats move quickly and sometimes they stand still, but they never feel anything. They are driven only by non-conscious algorithms. Yet if so, what's the point of all these experiments? Psychiatric drugs are aimed to induce changes not just in human behaviour, but above all in human *feeling*. When customers go to a psychiatrist and say, 'Doctor, give me something that will lift me out of this depression,' they don't want a mechanical stimulant that will cause them to flail about while still feeling blue. They want to *feel* cheerful. Conducting experiments on rats can help corpor-ations develop such a magic pill only if they presuppose that rat behaviour is accompanied by human-like emotions. And indeed, this is a common presupposition in psychiatric laboratories.[10]

The Self-Conscious Chimpanzee

Another attempt to enshrine human superiority accepts that rats, dogs and other animals have consciousness, but argues that, unlike humans, they lack self-consciousness. They may feel depressed, happy, hungry or satiated, but they have no notion of self, and they are not aware that the depression or hunger they feel belongs to a unique entity called 'I'.

This idea is as common as it is opaque. Obviously, when a dog feels hungry, he grabs a piece of meat for himself rather than serve food to another dog. Let a dog sniff a tree watered by the neigh-bourhood dogs, and he will immediately know whether it smells of his own urine, of the neighbour's cute Labrador's or of some stranger's. Dogs react very differently to their own odour and to the odours of potential mates and rivals.[11] So what does it mean that they lack self-consciousness?

A more sophisticated version of the argument says that there are different levels of self-consciousness. Only humans understand themselves as an enduring self that has a past and a future, perhaps because only humans can use language in order to contemplate their past experiences and future actions. Other animals exist in an eternal present. Even when they seem to remember the past or plan for the future, they are in fact reacting only to present stimuli and momentary urges.[12] For instance, a squirrel hiding nuts for the winter doesn't really remember the hunger he felt last winter, nor is he thinking about the future. He just follows a momentary urge, oblivious to the origins and purpose of this urge. That's why even very young squirrels, who haven't yet lived through a winter and hence cannot remember winter, nevertheless cache nuts during the summer.

Yet it is unclear why language should be a necessary condition for being aware of past or future events. The fact that humans use language to do so is hardly a proof. Humans also use language to express their love or their fear, but other animals may well experience and even express love and fear non-verbally. Indeed, humans themselves are often aware of past and future events without verbalising them. Especially in dream states, we can be aware of entire non-verbal narratives – which upon waking we struggle to describe in words.

Various experiments indicate that at least some animals – including birds such as parrots and scrub jays – do remember individual incidents and consciously plan for future eventualities.[13] However, it is impossible to prove this beyond doubt, because no matter how sophisticated a behaviour an animal exhibits, sceptics can always claim that it results from unconscious algorithms in its brain rather than from conscious images in its mind.

To illustrate this problem consider the case of Santino, a male chimpanzee from the Furuvik Zoo in Sweden. To relieve the boredom in his compound Santino developed an exciting hobby: throwing stones at visitors to the zoo. In itself, this is hardly unique. Angry chimpanzees often throw stones, sticks and even excrement. However, Santino was planning his moves in advance. During the early morning, long before the zoo opened for visitors, Santino collected projectiles and placed them in a heap, without showing any visible signs of anger. Guides and visitors soon learned to be wary of Santino, especially when he was standing near his pile of stones, hence he had increasing difficulties in finding targets.

In May 2010, Santino responded with a new strategy. In the early morning he took bales of straw from his sleeping quarters and placed them close to the compound's wall, where visitors usually gather to watch the chimps. He then collected stones and hid them under the straw. An hour or so later, when the first visitors approached, Santino kept his cool, showing no signs of irritation or aggression. Only when his victims were within range did Santino suddenly grab the stones from their hiding place and bombard the frightened humans, who would scuttle in all directions. In the summer of 2012 Santino sped up the arms race, caching stones not only under straw bales, but also in tree trunks, buildings and any other suitable hiding place.

Yet even Santino doesn't satisfy the sceptics. How can we be certain that at 7 a.m., when Santino goes about secreting stones here and there, he is imagining how fun it will be to pelt the visiting humans at noon? Maybe Santino is driven by some non-conscious algorithm, just like a young squirrel hiding nuts 'for winter' even though he has never experienced winter?[14]

Similarly, say the sceptics, a male chimpanzee attacking a rival who hurt him weeks earlier isn't really avenging the old insult. He is just reacting to a momentary feeling of anger, the cause of which is beyond him. When a mother elephant sees a lion threatening her calf, she rushes forward and risks her life not because she remembers that this is her beloved offspring whom she has been nurturing for months; rather, she is impelled by some unfathomable sense of hostility towards the lion. And when a dog jumps for joy when his owner comes home, the dog isn't recognising the man who fed and cuddled him from infancy. He is simply overwhelmed by an unexplained ecstasy.[15]

We cannot prove or disprove any of these claims, because they are in fact variations on the Problem of Other Minds. Since we aren't familiar with any algorithm that requires consciousness, anything an animal does can be seen as the product of non-conscious algorithms rather than of conscious memories and plans. So in Santino's case too, the real question concerns the burden of proof. What is the most likely explanation for Santino's behaviour? Should we assume that he is consciously planning for the future, and anyone who disagrees should provide some counter-evidence? Or is it more reasonable to think that the chimpanzee is driven by a non-conscious algorithm, and all he consciously feels is a mysterious urge to place stones under bales of straw?

And even if Santino doesn't remember the past and doesn't imagine the future, does it mean he lacks self-consciousness? After all, we ascribe self-consciousness to humans even when they are not busy remembering the past or dreaming about the future. For example, when a human mother sees her toddler wandering onto a busy road, she doesn't stop to think about either past or future. Just like the mother elephant, she too just races to save her child. Why not say about her what we say about the elephant, namely that 'when the mother rushed to save her baby from the oncoming danger, she did it without any self-consciousness. She was merely driven by a momentary urge'?

Similarly, consider a young couple kissing passionately on their first date, a soldier charging into heavy enemy fire to save a wounded comrade, or an artist drawing a masterpiece in a frenzy of brushstrokes. None of them stops to contemplate the past or the future. Does it mean they lack self-consciousness, and that their state of being is inferior to that of a politician giving an election speech about his past achievements and future plans?

The Clever Horse

In 2010 scientists conducted an unusually touching rat experiment. They locked a rat in a tiny cage, placed the cage within a much larger cell and allowed another rat to roam freely through that cell. The caged rat gave out distress signals, which caused the free rat also to exhibit signs of anxiety and stress. In most cases, the free rat proceeded to help her trapped companion, and after several attempts usually succeeded in opening the cage and liberating the prisoner. The researchers then repeated the experiment, this time placing chocolate in the cell. The free rat now had to choose between either liberating the prisoner, or enjoying the chocolate all by herself. Many rats preferred to first free their companion and share the chocolate (though quite a few behaved more selfishly, proving perhaps that some rats are meaner than others).

Sceptics dismissed these results, arguing that the free rat liberated the prisoner not out of empathy, but simply in order to stop the annoying distress signals. The rats were motivated by the unpleasant sensations they felt, and they sought nothing grander than ending these sensations. Maybe. But we could say exactly the same thing about us humans. When I donate money to a beggar, am I not reacting to the unpleasant sensations that the sight of the beggar causes me to feel? Do I really care about the beggar, or do I simply want to feel better myself?[16]

In essence, we humans are not that different from rats, dogs, dolphins or chimpanzees. Like them, we too have no soul. Like us, they too have consciousness and a complex world of sensations and emotions. Of course, every animal has its unique traits and talents. Humans too have their special gifts. We shouldn't humanise animals needlessly, imagining that they are just a furrier version of ourselves. This is not only bad science, but it also prevents us from understanding and valuing other animals on their terms.

In the early 1900s, a horse called Clever Hans became a German celebrity. Touring Germany's towns and villages, Hans showed off a remarkable grasp of the German language, and an even more remarkable mastery of mathematics. When asked, 'Hans, what is four times three?' Hans tapped his hoof twelve times. When shown a written message asking, 'What is twenty minus eleven?' Hans tapped nine times, with commendable Prussian precision.

In 1904 the German board of education appointed a special scientific commission headed by a psychologist to look into the matter. The thirteen members of the commission – which included a circus manager and a veterinarian – were convinced this must be a scam, but despite their best efforts they couldn't uncover any fraud or subterfuge. Even when Hans was separated from his owner, and complete strangers presented him with the questions, Hans still got most of the answers right.

In 1907 the psychologist Oskar Pfungst began another investigation that finally revealed the truth. It turned out that Hans got the answers right by carefully observing the body language and facial expressions of his interlocutors. When Hans was asked what is four times three, he knew from past experience that the human was expecting him to tap his hoof a given number of times. He began tapping, while closely monitoring the human. As Hans approached the correct number of taps the human became more and more tense, and when Hans tapped the right number, the tension reached its peak. Hans knew how to recognise this by the human's body posture and the look on the human's face. He then stopped tapping, and watched how tension was replaced by amazement or laughter. Hans knew he had got it right.

Clever Hans is often given as an example of the way humans erroneously humanise animals, ascribing to them far more amazing abilities than they actually possess. In fact, however, the lesson

17. Clever Hans on stage in 1904.

is just the opposite. The story demonstrates that by humanising animals we usually *underestimate* animal cognition and ignore the unique abilities of other creatures. As far as maths goes, Hans was hardly a genius. Any eight-year-old kid could do much better. However, in his ability to deduce emotions and intentions from body language, Hans was a true genius. If a Chinese person were to ask me in Mandarin what is four times three, there is no way that I could correctly tap my foot twelve times simply by observing facial expressions and body language. Clever Hans enjoyed this ability because horses normally communicate with each other through body language. What was remarkable about Hans, however, is that he could use the method to decipher the emotions and intentions not only of his fellow horses, but also of unfamiliar humans.

If animals are so clever, why don't horses harness humans to carts, rats conduct experiments on us, and dolphins make us jump through hoops? *Homo sapiens* surely has some unique ability that enables it to dominate all the other animals. Having dismissed the overblown notions that *Homo sapiens* exists on an entirely different plain from other animals, or that humans possess some unique essence like soul or consciousness, we can finally climb down to the level of reality and examine the particular physical or mental abilities that give our species its edge.

Most studies cite tool production and intelligence as particularly important for the ascent of humankind. Though other animals also produce tools, there is little doubt that humans far surpass them in that field. Things are a bit less clear with regard to intelligence. An entire industry is devoted to defining and measuring intelligence but is a long way from reaching a consensus. Luckily, we don't have to enter into that minefield, because no matter how one defines intelligence, it is quite clear that neither intelligence nor toolmaking by themselves can account for the Sapiens conquest of the world. According to most definitions of intelligence, a million years ago humans were already the most intelligent animals around, as well as the world's champion toolmakers, yet they remained insignificant creatures with little impact on the surrounding ecosystem. They were obviously lacking some key feature other than intelligence and toolmaking.

Perhaps humankind eventually came to dominate the planet not because of some elusive third key ingredient, but due simply to the evolution of even higher intelligence and even better toolmaking abilities? It doesn't seem so, because when we examine the historical record, we don't see a direct correlation between the intelligence and toolmaking abilities of individual humans and the power of our species as a whole. Twenty thousand years ago, the average Sapiens probably had higher intelligence and better toolmaking skills than the average Sapiens of today. Modern schools and employers may test our aptitudes from time to time but, no matter how badly we do, the welfare state always guarantees our basic needs. In the Stone Age natural selection tested you every single moment of every single day, and if you flunked any of its numerous tests you were pushing up the daisies in no time. Yet despite the superior toolmaking abilities of our Stone Age ancestors, and despite their sharper minds and far more acute senses, 20,000 years ago humankind was much weaker than it is today.

Over those 20,000 years humankind moved from hunting mammoth with stone-tipped spears to exploring the solar system with spaceships not thanks to the evolution of more dexterous hands or bigger brains (our brains today seem actually to be smaller).[17] Instead, the crucial factor in our conquest of the world was our ability to connect many humans to one another.[18] Humans nowadays completely dominate the planet not because the individual human is far smarter and more nimble-fingered than the individual chimp or wolf, but because *Homo sapiens* is the only species on earth capable of co-operating flexibly in large numbers. Intelligence and toolmaking were obviously very important as well. But if humans had not learned to cooperate flexibly in large numbers, our crafty brains and deft hands would still be splitting flint stones rather than uranium atoms.

If cooperation is the key, how come the ants and bees did not beat us to the nuclear bomb even though they learned to cooperate en masse millions of years before us? Because their cooperation lacks flexibility. Bees cooperate in very sophisticated ways, but they cannot reinvent their social system overnight. If a hive faces a new threat or a new opportunity, the bees cannot, for example, guillotine the queen and establish a republic.

Social mammals such as elephants and chimpanzees cooperate far more flexibly than bees, but they do so only with small numbers of friends and family members. Their cooperation is based on personal acquaintance. If I am a chimpanzee and you are a chimpanzee and I want to cooperate with you, I must know you personally: what kind of chimp are you? Are you a nice chimp? Are you an evil chimp? How can I cooperate with you if I don't know you? To the best of our knowledge, only Sapiens can cooperate in very flexible ways with countless numbers of strangers. This concrete capability – rather than an eternal soul or some unique kind of consciousness – explains our mastery of planet Earth.

Long Live the Revolution!

History provides ample evidence for the crucial importance of large-scale cooperation. Victory almost invariably went to those who cooperated better – not only in struggles between *Homo sapiens* and other animals, but also in conflicts between different human groups. Thus Rome conquered Greece not because the Romans had larger brains or better toolmaking techniques, but because they were able to cooperate more effectively. Throughout history, disciplined armies easily routed disorganised hordes, and unified elites dominated the disorderly masses. In 1914, for example, 3 million Russian noblemen, officials and business people lorded it over 180 million peasants and workers. The Russian elite knew how to cooperate in defence of its common interests, whereas the 180 million commoners were incapable of effective mobilisation. Indeed, much of the elite's efforts focused on ensuring that the 180 million people at the bottom would never learn to cooperate.

In order to mount a revolution, numbers are never enough. Revolutions are usually made by small networks of agitators rather than by the masses. If you want to launch a revolution, don't ask yourself, 'How many people support my ideas?' Instead, ask yourself, 'How many of my supporters are capable of effective collaboration?' The Russian Revolution finally erupted not when 180 million peasants rose against the tsar, but rather when a handful of communists placed themselves at the right place at the right time. In 1917, at a time when the Russian upper and middle classes numbered at least 3 million people, the Communist Party had just 23,000 members.[19] The communists nevertheless gained control of the vast Russian Empire because they organised themselves well. When authority in Russia slipped from the decrepit hands of the tsar and the equally shaky hands of Kerensky's provisional government, the communists seized it with alacrity, gripping the reins of power like a bulldog locking its jaws on a bone.

The communists didn't release their grip until the late 1980s. Effective organisation kept them in power for eight long decades, and they eventually fell due to defective organisation. On 21 December 1989 Nicolae Ceauşescu, the communist dictator of Romania, organised a mass demonstration of support in the centre of Bucharest. Over the previous months the Soviet Union had withdrawn its support from the eastern European communist regimes, the Berlin Wall had fallen, and revolutions had swept Poland, East Germany, Hungary, Bulgaria and Czechoslovakia. Ceauşescu, who had ruled Romania since 1965, believed he could withstand the tsunami, even though riots against his rule had erupted in the Romanian city of Timişoara on 17 December. As one of his counter-measures, Ceauşescu arranged a massive rally in Bucharest to prove to Romanians and the rest of the world that the majority of the populace still loved him – or at least feared him. The creaking party apparatus mobilised 80,000 people to fill the city's central square, and citizens throughout Romania were instructed to stop all their activities and tune in on their radios and televisions.

To the cheering of the seemingly enthusiastic crowd, Ceauşescu mounted the balcony overlooking the square, as he had done scores of times in previous decades. Flanked by his wife, Elena, leading party officials and a bevy of bodyguards, Ceauşescu began delivering one of his trademark dreary speeches. For eight minutes he praised the glories of Romanian socialism, looking very pleased with himself as the crowd clapped mechanically. And then something went wrong. You can see it for yourself on YouTube. Just search for 'Ceauşescu's last speech', and watch history in action.[20]

The YouTube clip shows Ceauşescu starting another long sentence, saying, 'I want to thank the initiators and organisers of this great event in Bucharest, considering it as a—', and then he falls silent, his eyes open wide, and he freezes in disbelief. He never finished the sentence. You can see in that split second how an entire world collapses. Somebody in the audience booed. People still argue today who was the first person who dared to boo. And then another person booed, and another, and another, and within a few seconds the masses began whistling, shouting abuse and calling out 'Ti-mi-şoa-ra! Ti-mi-şoa-ra!'

18. The moment a world collapses: a stunned Ceauşescu cannot believe his eyes and ears.

All this happened live on Romanian television, as three-quarters of the populace sat glued to the screens, their hearts throbbing wildly. The notorious secret police – the Securitate – immediately ordered the broadcast to be stopped, but the television crews disobeyed. The cameraman pointed the camera towards the sky so that viewers couldn't see the panic among the party leaders on the balcony, but the soundman kept recording, and the technicians continued the transmission. The whole of Romania heard the crowd booing, while Ceaușescu yelled, 'Hello! Hello! Hello!' as if the problem was with the microphone. His wife Elena began scolding the audience, 'Be quiet! Be quiet!' until Ceaușescu turned and yelled at her – still live on television – 'You be quiet!' Ceaușescu then appealed to the excited crowds in the square, imploring them, 'Comrades! Comrades! Be quiet, comrades!'

But the comrades were unwilling to be quiet. Communist Romania crumbled when 80,000 people in the Bucharest central square realised they were much stronger than the old man in the fur hat on the balcony. What is truly astounding, however, is not the moment the system collapsed, but the fact that it managed to survive for decades. Why are revolutions so rare? Why do the masses sometimes clap and cheer for centuries on end, doing everything the man on the balcony commands them, even though they could in theory charge forward at any moment and tear him to pieces?

Ceaușescu and his cronies dominated 20 million Romanians for four decades because they ensured three vital conditions. First, they placed loyal communist apparatchiks in control of all networks of cooperation, such as the army, trade unions and even sports associations. Second, they prevented the creation of any rival organisations – whether political, economic or social – which might serve as a basis for anti-communist cooperation. Third, they relied on the support of sister communist parties in the Soviet Union and eastern Europe. Despite occasional tensions, these parties helped each other in times of need, or at least guaranteed that no outsider poked his nose into the socialist paradise. Under such conditions, despite all the hardship and suffering inflicted on them by the ruling elite, the 20 million Romanians were unable to organise any effective opposition.

Ceauşescu fell from power only once all three conditions no longer held. In the late 1980s the Soviet Union withdrew its protection and the communist regimes began falling like dominoes. By December 1989 Ceauşescu could not expect any outside assistance. Just the opposite – revolutions in nearby countries gave heart to the local opposition. The Communist Party itself began splitting into rival camps. The moderates wished to rid themselves of Ceauşescu and initiate reforms before it was too late. By organising the Bucharest demonstration and broadcasting it live on television, Ceauşescu himself provided the revolutionaries with the perfect opportunity to discover their power and rally against him. What quicker way to spread a revolution than by showing it on TV?

Yet when power slipped from the hands of the clumsy organiser on the balcony, it did not pass to the masses in the square. Though numerous and enthusiastic, the crowds did not know how to organise themselves. Hence just as in Russia in 1917, power passed to a small group of political players whose only asset was good organisation. The Romanian Revolution was hijacked by the self-proclaimed National Salvation Front, which was in fact a smokescreen for the moderate wing of the Communist Party. The Front had no real ties to the demonstrating crowds. It was manned by mid-ranking party officials, and led by Ion Iliescu, a former member of the Communist Party's central committee and one-time head of the propaganda department. Iliescu and his comrades in the National Salvation Front reinvented themselves as democratic politicians, proclaimed to any available microphone that they were the leaders of the revolution, and then used their long experience and network of cronies to take control of the country and pocket its resources.

In communist Romania almost everything was owned by the state. Democratic Romania quickly privatised its assets, selling them at bargain prices to the ex-communists, who alone grasped what was happening and collaborated to feather each other's nests. Government companies that controlled national infrastructure and natural resources were sold to former communist officials at end-of-season prices while the party's foot soldiers bought houses and apartments for pennies.

Ion Iliescu was elected president of Romania, while his colleagues became ministers, parliament members, bank directors and multimillionaires. The new Romanian elite that controls the country to this day is composed mostly of former communists and their families. The masses who risked their necks in Timişoara and Bucharest settled for scraps, because they did not know how to cooperate and how to create an efficient organisation to look after their own interests.[21]

A similar fate befell the Egyptian Revolution of 2011. What television did in 1989, Facebook and Twitter did in 2011. The new media helped the masses coordinate their activities, so that thousands of people flooded the streets and squares at the right moment and toppled the Mubarak regime. However, it is one thing to bring 100,000 people to Tahrir Square, and quite another to get a grip on the political machinery, shake the right hands in the right back rooms and run a country effectively. Consequently, when Mubarak stepped down the demonstrators could not fill the vacuum. Egypt had only two institutions sufficiently organised to rule the country: the army and the Muslim Brotherhood. Hence the revolution was hijacked first by the Brotherhood, and eventually by the army.

The Romanian ex-communists and the Egyptian generals were not more intelligent or nimble-fingered than either the old dictators or the demonstrators in Bucharest and Cairo. Their advantage lay in flexible cooperation. They cooperated better than the crowds, and they were willing to show far more flexibility than the hidebound Ceauşescu and Mubarak.

Beyond Sex and Violence

If Sapiens rule the world because we alone can cooperate flexibly in large numbers, then this undermines our belief in the sacredness of human beings. We tend to think that we are special, and deserve all kinds of privileges. As proof, we point to the amazing achievements of our species: we built the pyramids and the Great Wall of China; we deciphered the structure of atoms and DNA molecules; we reached the South Pole and the moon. If these accomplishments resulted from some unique essence that each individual human has – an immortal soul, say – then it would make sense to sanctify human life. Yet since these triumphs actually result from mass cooperation, it is far less clear why they should make us revere individual humans.

A beehive has much greater power than an individual butterfly, yet that doesn't imply a bee is therefore more hallowed than a butterfly. The Romanian Communist Party successfully dominated the disorganised Romanian population. Does it follow that the life of a party member was more sacred than the life of an ordinary citizen? Humans know how to cooperate far more effectively than chimpanzees, which is why humans launch spaceships to the moon whereas chimpanzees throw stones at zoo visitors. Does it mean that humans are superior beings?

Well, maybe. It depends on what enables humans to cooperate so well in the first place. Why are humans alone able to construct such large and sophisticated social systems? Social cooperation among most social mammals such as chimpanzees, wolves and dolphins relies on intimate acquaintance. Among common chimpanzees, individuals will go hunting together only after they have got to know each other well and established a social hierarchy. Hence chimpanzees spend a lot of time in social interactions and power struggles. When alien chimpanzees meet, they usually cannot cooperate, but instead scream at each other, fight or flee as quickly as possible.

Among pygmy chimpanzees – also known as bonobos – things are a bit different. Bonobos often use sex in order to dispel tensions and cement social bonds. Not surprisingly, homosexual intercourse is consequently very common among them. When two alien groups of bonobos encounter one another, at first they display fear and hostility, and the jungle is filled with howls and screams. Soon enough, however, females from one group cross no-chimp's-land, and invite the strangers to make love instead of war. The invitation is usually accepted, and within a few minutes the potential battlefield teems with bonobos having sex in almost every conceivable posture, including hanging upside down from trees.

Sapiens know these cooperative tricks well. They sometimes form power hierarchies similar to those of common chimpanzees, whereas on other occasions they cement social bonds with sex just like bonobos. Yet personal acquaintance – whether it involves fighting or copulating – cannot form the basis for large-scale co-operation. You cannot settle the Greek debt crisis by inviting Greek politicians and German bankers to either a fist fight or an orgy. Research indicates that Sapiens just can't have intimate relations (whether hostile or amorous) with more than 150 individuals.[22] Whatever enables humans to organise mass-cooperation networks, it isn't intimate relations.

This is bad news for psychologists, sociologists, economists and others who try to decipher human society through laboratory experiments. For both organisational and financial reasons, the vast majority of experiments are conducted either on individuals or on small groups of participants. Yet it is risky to extrapolate from small-group behaviour to the dynamics of mass societies. A nation of 100 million people functions in a fundamentally different way to a band of a hundred individuals.

Take, for example, the Ultimatum Game – one of the most famous experiments in behavioural economics. This experiment is usually conducted on two people. One of them gets $100, which he must divide between himself and the other participant in any way he wants. He may keep everything, split the money in half or give most of it away. The other player can do one of two things: accept the suggested division, or reject it outright. If he rejects the division, nobody gets anything.

Classical economic theories maintain that humans are rational calculating machines. They propose that most people will keep $99, and offer $1 to the other participant. They further propose that the other participant will accept the offer. A rational person offered a dollar will always say yes. What does he care if the other player gets $99?

Classical economists have probably never left their laboratories and lecture halls to venture into the real world. Most people playing the Ultimatum Game reject very low offers because they are 'unfair'. They prefer losing a dollar to looking like suckers. Since this is how the real world functions, few people make very low offers in the first place. Most people divide the money equally, or give themselves only a moderate advantage, offering $30 or $40 to the other player.

The Ultimatum Game made a significant contribution to under-mining classical economic theories and to establishing the most important economic discovery of the last few decades: Sapiens don't behave according to a cold mathematical logic, but rather according to a warm social logic. We are ruled by emotions. These emotions, as we saw earlier, are in fact sophisticated algorithms that reflect the social mechanisms of ancient hunter-gatherer bands. If 30,000 years ago I helped you hunt a wild chicken and you then kept almost all the chicken to yourself, offering me just one wing, I did not say to myself: 'Better one wing than nothing at all.' Instead my evolutionary algorithms kicked in, adrenaline and testosterone flooded my system, my blood boiled, and I stamped my feet and shouted at the top of my voice. In the short term I may have gone hungry, and even risked a punch or two. But it paid off in the long term, because you thought twice before ripping me off again. We refuse unfair offers because people who meekly accepted unfair offers didn't survive in the Stone Age.

Observations of contemporary hunter-gatherer bands support this idea. Most bands are highly egalitarian, and when a hunter comes back to camp carrying a fat deer, everybody gets a share. The same is true of chimpanzees. When one chimp kills a piglet, the other troop members will gather round him with outstretched hands, and usually they all get a piece.

In another recent experiment, the primatologist Frans de Waal placed two capuchin monkeys in two adjacent cages, so that each could see everything the other was doing. De Waal and his colleagues placed small stones inside each cage, and trained the monkeys to give them these stones. Whenever a monkey handed over a stone, he received food in exchange. At first the reward was a piece of cucumber. Both monkeys were very pleased with that, and happily ate their cucumber. After a few rounds de Waal moved to the next stage of the experiment. This time, when the first monkey surrendered a stone, he got a grape. Grapes are much more tasty than cucumbers. However, when the second monkey gave a stone, he still received a piece of cucumber. The second monkey, who was previously very happy with his cucumber, became incensed. He took the cucumber, looked at it in disbelief for a moment, and then threw it at the scientists in anger and began jumping and screeching loudly. He ain't a sucker.[23]

This hilarious experiment (which you can see for yourself on YouTube), along with the Ultimatum Game, has led many to believe that primates have a natural morality, and that equality is a universal and timeless value. People are egalitarian by nature, and unequal societies can never function well due to resentment and dissatisfaction.

But is that really so? These theories may work well on chimpanzees, capuchin monkeys and small hunter-gatherer bands. They also work well in the lab, where you test them on small groups of people. Yet once you observe the behaviour of human masses you discover a completely different reality. Most human kingdoms and empires were extremely unequal, yet many of them were surprisingly stable and efficient. In ancient Egypt, the pharaoh sprawled on comfortable cushions inside a cool and sumptuous palace, wearing golden sandals and gem-studded tunics, while beautiful maids popped sweet grapes into his mouth. Through the open window he could see the peasants in the fields, toiling in dirty rags under a merciless sun, and blessed was the peasant who had a cucumber to eat at the end of the day. Yet the peasants rarely revolted.

In 1740 King Frederick II of Prussia invaded Silesia, thus commencing a series of bloody wars that earned him his sobriquet Frederick the Great, turned Prussia into a major power and left hundreds of thousands of people dead, crippled or destitute. Most of Frederick's soldiers were hapless recruits, subject to iron discipline and draconian drill. Not surprisingly, the soldiers lost little love on their supreme commander. As Frederick watched his troops assemble for the invasion, he told one of his generals that what struck him most about the scene was that 'we are standing here in perfect safety, looking at 60,000 men – they are all our enemies, and there is not one of them who is not better armed and stronger than we are, and yet they all tremble in our presence, while we have no reason whatsoever to be afraid of them'.[24] Frederick could indeed watch them in perfect safety. During the following years, despite all the hardships of war, these 60,000 armed men never revolted against him – indeed, many of them served him with exceptional courage, risking and even sacrificing their very lives.

Why did the Egyptian peasants and Prussian soldiers act so differently than we would have expected on the basis of the Ultimatum Game and the capuchin monkeys experiment? Because large numbers of people behave in a fundamentally different way than do small numbers. What would scientists see if they conducted the Ultimatum Game experiment on two groups of 1 million people each, who had to share $100 billion?

They would probably have witnessed strange and fascinating dynamics. For example, since 1 million people cannot make decisions collectively, each group might sprout a small ruling elite. What if one elite offers the other $10 billion, keeping $90 billion? The leaders of the second group might well accept this unfair offer, siphon most of the $10 billion into their Swiss bank accounts, while preventing rebellion among their followers with a combination of sticks and carrots. The leadership might threaten to severely punish dissidents forthwith, while promising the meek and patient everlasting rewards in the afterlife. This is what happened in ancient Egypt and eighteenth-century Prussia, and this is how things still work out in numerous countries around the world.

Such threats and promises often succeed in creating stable human hierarchies and mass-cooperation networks, as long as people believe that they reflect the inevitable laws of nature or the divine commands of God, rather than just human whims. All large-scale human cooperation is ultimately based on our belief in imagined orders. These are sets of rules that, despite existing only in our imagination, we believe to be as real and inviolable as gravity. 'If you sacrifice ten bulls to the sky god, the rain will come; if you honour your parents, you will go to heaven; and if you don't believe what I am telling you – you'll go to hell.' As long as all Sapiens living in a particular locality believe in the same stories, they all follow the same rules, making it easy to predict the behaviour of strangers and to organise mass-cooperation networks. Sapiens often use visual marks such as a turban, a beard or a business suit to signal 'you can trust me, I believe in the same story as you'. Our chimpanzee cousins cannot invent and spread such stories, which is why they cannot cooperate in large numbers.

The Web of Meaning

People find it difficult to understand the idea of 'imagined orders' because they assume that there are only two types of realities: objective realities and subjective realities. In objective reality, things exist independently of our beliefs and feelings. Gravity, for example, is an objective reality. It existed long before Newton, and it affects people who don't believe in it just as much as it affects those who do.

Subjective reality, in contrast, depends on my personal beliefs and feelings. Thus, suppose I feel a sharp pain in my head and go to the doctor. The doctor checks me thoroughly, but finds nothing wrong. So she sends me for a blood test, urine test, DNA test, X-ray, electrocardiogram, fMRI scan and a plethora of other procedures. When the results come in she announces that I am perfectly healthy, and I can go home. Yet I still feel a sharp pain in my head. Even though every objective test has found nothing wrong with me, and even though nobody except me feels the pain, for me the pain is 100 per cent real.

Most people presume that reality is either objective or subjective, and that there is no third option. Hence once they satisfy themselves that something isn't just their own subjective feeling, they jump to the conclusion it must be objective. If lots of people believe in God; if money makes the world go round; and if nationalism starts wars and builds empires – then these things aren't just a subjective belief of mine. God, money and nations must therefore be objective realities.

However, there is a third level of reality: the intersubjective level. Intersubjective entities depend on communication among many humans rather than on the beliefs and feelings of individual humans. Many of the most important agents in history are intersubjective. Money, for example, has no objective value. You cannot eat, drink or wear a dollar bill. Yet as long as billions of people believe in its value, you can use it to buy food, beverages and clothing. If the baker suddenly loses his faith in the dollar bill and refuses to give me a loaf of bread for this green piece of paper, it doesn't matter much. I can just go down a few blocks to the nearby supermarket. However, if the supermarket cashiers also refuse to accept this piece of paper, along with the hawkers in the market and the salespeople in the mall, then the dollar will lose its value. The green pieces of paper will go on existing, of course, but they will be worthless.

Such things actually happen from time to time. On 3 November 1985 the Myanmar government unexpectedly announced that banknotes of twenty-five, fifty and a hundred kyats were no longer legal tender. People were given no opportunity to exchange the notes, and savings of a lifetime were instantaneously turned into heaps of worthless paper. To replace the defunct notes, the government introduced new seventy-five-kyat bills, allegedly in honour of the seventy-fifth birthday of Myanmar's dictator, General Ne Win. In August 1986, banknotes of fifteen kyats and thirty-five kyats were issued. Rumour had it that the dictator, who had a strong faith in numerology, believed that fifteen and thirty-five are lucky numbers. They brought little luck to his subjects. On 5 September 1987 the government suddenly decreed that all thirty-five and seventy-five notes were no longer money.

The value of money is not the only thing that might evaporate once people stop believing in it. The same can happen to laws, gods and even entire empires. One moment they are busy shaping the world, and the next moment they no longer exist. Zeus and Hera were once important powers in the Mediterranean basin, but today they lack any authority because nobody believes in them. The Soviet Union could once destroy the entire human race, yet it ceased to exist at the stroke of a pen. At 2 p.m. on 8 December 1991, in a state dacha near Viskuli, the leaders of Russia, Ukraine and Belarus signed the Belavezha Accords, which stated that 'We, the Republic of Belarus, the Russian Federation and Ukraine, as founding states of the USSR that signed the union treaty of 1922, hereby establish that the USSR as a subject of international law and a geopolitical reality ceases its existence.'[25] And that was that. No more Soviet Union.

It is relatively easy to accept that money is an intersubjective reality. Most people are also happy to acknowledge that ancient Greek gods, evil empires and the values of alien cultures exist only in the imagination. Yet we don't want to accept that *our* God, *our* nation or *our* values are mere fictions, because these are the things that give meaning to our lives. We want to believe that our lives have some objective meaning, and that our sacrifices matter to something beyond the stories in our head. Yet in truth the lives of most people have meaning only within the network of stories they tell one another.

Meaning is created when many people weave together a common network of stories. Why does a particular action – such as getting married in church, fasting on Ramadan or voting on election day – seem meaningful to me? Because my parents also think it

19. Signing the Belavezha Accords. Pen touches paper – and abracadabra! The Soviet Union disappears.

is meaningful, as do my brothers, my neighbours, people in nearby cities and even the residents of far-off countries. And why do all these people think it is meaningful? Because their friends and neighbours also share the same view. People constantly reinforce each other's beliefs in a self-perpetuating loop. Each round of mutual confirmation tightens the web of meaning further, until you have little choice but to believe what everyone else believes.

Yet over decades and centuries the web of meaning unravels and a new web is spun in its place. To study history means to watch the spinning and unravelling of these webs, and to realise that what seems to people in one age the most important thing in life becomes utterly meaningless to their descendants.

In 1187 Saladin defeated the crusader army at the Battle of Hattin and conquered Jerusalem. In response the Pope launched the Third Crusade to recapture the holy city. Imagine a young English nobleman named John, who left home to fight Saladin. John believed that his actions had an objective meaning. He believed that if he died on the crusade, after death his soul would ascend to heaven, where it would enjoy everlasting celestial joy. He would have been horrified to learn that the soul and heaven are just stories invented by humans. John wholeheartedly believed that if he reached the Holy Land, and if some Muslim warrior with a big moustache brought an axe down on his head, he would feel an unbearable pain, his ears would ring, his legs would crumble under him, his field of vision would turn black – and the very next moment he would see brilliant light all around him, he would hear angelic voices and melodious harps, and radiant winged cherubs would beckon him through a magnificent golden gate.

John had a very strong faith in all this, because he was enmeshed within an extremely dense and powerful web of meaning. His earliest memories were of Grandpa Henry's rusty sword, hanging in the castle's main hall. Ever since he was a toddler John had heard stories of Grandpa Henry who died on the Second Crusade and who is now resting with the angels in heaven, watching over John and his family. When minstrels visited the castle, they usually sang about the brave crusaders who fought in the Holy Land. When John went to church, he enjoyed looking at the stained-glass windows. One showed Godfrey of Bouillon riding a horse and impaling a wicked-looking Muslim on his lance. Another showed the souls of sinners burning in hell. John listened attentively to the local priest, the most learned man he knew. Almost every Sunday, the priest explained – with the help of well-crafted parables and hilarious jokes – that there was no salvation outside the Catholic Church, that the Pope in Rome was our holy father and that we always had to obey his commands. If we murdered or stole, God would send us to hell; but if we killed infidel Muslims, God would welcome us to heaven.

One day when John was just turning eighteen a dishevelled knight rode to the castle's gate, and in a choked voice announced the news: Saladin has destroyed the crusader army at Hattin! Jerusalem has fallen! The Pope has declared a new crusade, promising eternal salvation to anyone who dies on it! All around, people looked shocked and worried, but John's face lit up in an otherworldly glow and he proclaimed: 'I am going to fight the infidels and liberate the Holy Land!' Everyone fell silent for a moment, and then smiles and tears appeared on their faces. His mother wiped her eyes, gave John a big hug and told him how proud she was of him. His father gave him a mighty pat on the back, and said: 'If only I was your age, son, I would join you. Our family's honour is at stake – I am sure you won't disappoint us!' Two of his friends announced that they were coming too. Even John's sworn rival, the baron on the other side of the river, paid a visit to wish him Godspeed.

As he left the castle, villagers came forth from their hovels to wave to him, and all the pretty girls looked longingly at the brave crusader setting off to fight the infidels. When he set sail from England and made his way through strange and distant lands – Normandy, Provence, Sicily – he was joined by bands of foreign knights, all with the same destination and the same faith. When the army finally disembarked in the Holy Land and waged battle with Saladin's hosts, John was amazed to discover that even the wicked Saracens shared his beliefs. True, they were a bit confused, thinking that the Christians were the infidels and that the Muslims were obeying God's will. Yet they too accepted the basic principle that those fighting for God and Jerusalem will go straight to heaven when they die.

In such a way, thread by thread, medieval civilisation spun its web of meaning, trapping John and his contemporaries like flies. It was inconceivable to John that all these stories were just figments of the imagination. Maybe his parents and uncles were wrong. But the minstrels too, and all his friends, and the village girls, the learned priest, the baron on the other side of the river, the Pope in Rome, the Provençal and Sicilian knights, and even the very Muslims – is it possible that they were all hallucinating?

And the years pass. As the historian watches, the web of meaning unravels and another is spun in its stead. John's parents die, followed by all his siblings and friends. Instead of minstrels singing about the crusades, the new fashion is stage plays about tragic love affairs. The family castle burns to the ground and, when it is rebuilt, no trace is found of Grandpa Henry's sword. The church windows shatter in a winter storm and the replacement glass no longer depicts Godfrey of Bouillon and the sinners in hell, but rather the great triumph of the king of England over the king of France. The local priest has stopped calling the Pope 'our holy father' – he is now referred to as 'that devil in Rome'. In the nearby university scholars pore over ancient Greek manuscripts, dissect dead bodies and whisper quietly behind closed doors that perhaps there is no such thing as the soul.

And the years continue to pass. Where the castle once stood, there is now a shopping mall. In the local cinema they are screening *Monty Python and the Holy Grail* for the umpteenth time. In an empty church a bored vicar is overjoyed to see two Japanese tourists. He explains at length about the stained-glass windows, while they politely smile, nodding in complete incomprehension. On the steps outside a gaggle of teenagers are playing with their iPhones. They watch a new YouTube remix of John Lennon's 'Imagine'. 'Imagine there's no heaven,' sings Lennon, 'it's easy if you try.' A Pakistani street cleaner is sweeping the pavement, while a nearby radio broadcasts the news: the carnage in Syria continues, and the Security Council's meeting has ended in an impasse. Suddenly a hole in time opens, a mysterious ray of light illuminates the face of one of the teenagers, who announces: 'I am going to fight the infidels and liberate the Holy Land!'

Infidels and Holy Land? These words no longer carry any meaning for most people in today's England. Even the vicar would probably think the teenager is having some sort of psychotic episode. In contrast, if an English youth decided to join Amnesty International and travel to Syria to protect the human rights of refugees, he will be seen as a hero. In the Middle Ages people would have thought he had gone bonkers. Nobody in twelfth-century England knew what human rights were. You want to travel to the Middle East and risk your life not in order to kill Muslims, but to protect one group of Muslims from another? You must be out of your mind.

That's how history unfolds. People weave a web of meaning, believe in it with all their heart, but sooner or later the web unravels, and when we look back we cannot understand how anybody could have taken it seriously. With hindsight, going on crusade in the hope of reaching Paradise sounds like utter madness. With hindsight, the Cold War seems even madder. How come thirty years ago people were willing to risk nuclear holocaust because of their belief in a communist paradise? A hundred years hence, our belief in democracy and human rights might look equally incomprehensible to our descendants.

Dreamtime

Sapiens rule the world because only they can weave an intersubjective web of meaning: a web of laws, forces, entities and places that exist purely in their common imagination. This web allows humans alone to organise crusades, socialist revolutions and human rights movements.

Other animals may also imagine various things. A cat waiting to ambush a mouse might not see the mouse, but may well imagine the shape and even taste of the mouse. Yet to the best of our knowledge, cats are able to imagine only things that actually exist in the world, like mice. They cannot imagine things that they have never seen or smelled or tasted – such as the US dollar, the Google corporation or the European Union. Only Sapiens can imagine such chimeras.

Consequently, whereas cats and other animals are confined to the objective realm and use their communication systems merely to describe reality, Sapiens use language to create completely new realities. During the last 70,000 years the intersubjective realities that Sapiens invented became ever more powerful, so that today they dominate the world. Will the chimpanzees, the elephants, the Amazon rainforests and the Arctic glaciers survive the twenty-first century? This depends on the wishes and decisions of intersubjective entities such as the European Union and the World Bank; entities that exist only in our shared imagination.

No other animal can stand up to us, not because they lack a soul or a mind, but because they lack the necessary imagination. Lions can run, jump, claw and bite. Yet they cannot open a bank account or file a lawsuit. And in the twenty-first century, a banker who knows how to file a lawsuit is far more powerful than the most ferocious lion in the savannah.

As well as separating humans from other animals, this ability to create intersubjective entities also separates the humanities from the life sciences. Historians seek to understand the development of intersubjective entities like gods and nations, whereas biologists hardly recognise the existence of such things. Some believe that if we could only crack the genetic code and map every neuron in the brain, we will know all of humanity's secrets. After all, if humans have no soul, and if thoughts, emotions and sensations are just biochemical algorithms, why can't biology account for all the vagaries of human societies? From this perspective, the crusades were territorial disputes shaped by evolutionary pressures, and English knights going to fight Saladin in the Holy Land were not that different from wolves trying to appropriate the territory of a neighbouring pack.

The humanities, in contrast, emphasise the crucial importance of intersubjective entities, which cannot be reduced to hormones and neurons. To think historically means to ascribe real power to the contents of our imaginary stories. Of course, historians don't ignore objective factors such as climate changes and genetic mutations, but they give much greater importance to the stories people invent and believe. North Korea and South Korea are so different from one another not because people in Pyongyang have different genes to people in Seoul, or because the north is colder and more mountainous. It's because the north is dominated by very different fictions.

Maybe someday breakthroughs in neurobiology will enable us to explain communism and the crusades in strictly biochemical terms. Yet we are very far from that point. During the twenty-first century the border between history and biology is likely to blur not because we will discover biological explanations for historical events, but rather because ideological fictions will rewrite DNA strands; political and economic interests will redesign the climate; and the geography of mountains and rivers will give way to cyberspace. As human fictions are translated into genetic and electronic codes, the intersubjective reality will swallow up the objective reality and biology will merge with history. In the twenty-first century fiction might thereby become the most potent force on earth, surpassing even wayward asteroids and natural selection. Hence if we want to understand our future, cracking genomes and crunching numbers is hardly enough. We must also decipher the fictions that give meaning to the world.

20. The Creator: Jackson Pollock in a moment of inspiration.

PART II

Homo Sapiens Gives Meaning to the World

What kind of world did humans create?

How did humans become convinced that they not only control the world, but also give it meaning?

How did humanism – the worship of humankind – become the most important religion of all?

4

The Storytellers

Animals such as wolves and chimpanzees live in a dual reality. On the one hand they are familiar with objective entities outside them, such as trees, rocks and rivers. On the other hand they are aware of subjective experiences within them, such as fear, joy and desire. Sapiens, in contrast, live in triple-layered reality. In addition to trees, rivers, fears and desires, the Sapiens world also contains stories about money, gods, nations and corporations. As history unfolded, the impact of gods, nations and corporations grew at the expense of rivers, fears and desires. There are still many rivers in the world, and people are still motivated by their fears and wishes, but Jesus Christ, the French Republic and Apple Inc. have dammed and harnessed the rivers, and have learned to shape our deepest anxieties and yearnings.

Since new twenty-first-century technologies are likely to make such fictions even more powerful, to understand our future we need to understand how stories about Christ, France and Apple have gained so much power. Humans think they make history, but history actually revolves around the web of stories. The basic abilities of individual humans have not changed much since the Stone Age. But the web of stories has grown from strength to strength, thereby pushing history from the Stone Age to the Silicon Age.

It all began about 70,000 years ago, when the Cognitive Revolution enabled Sapiens to start talking about things that existed only in their own imagination. For the ensuing 60,000 years Sapiens wove many fictional webs, but these remained small and local. The spirit of a revered ancestor worshipped by one tribe was completely unknown to its neighbours, and seashells valuable in one locality became worthless once you crossed the nearby mountain range. Stories about ancestral spirits and precious seashells still gave Sapiens a huge advantage, because they allowed hundreds and sometimes even thousands of Sapiens to cooperate effectively, which was far more than Neanderthals or chimpanzees could do. Yet as long as Sapiens remained hunter-gatherers they could not cooperate on a truly massive scale, because it was impossible to feed a city or a kingdom by hunting and gathering. Consequently the spirits, fairies and demons of the Stone Age were relatively weak entities.

The Agricultural Revolution, which began about 12,000 years ago, provided the necessary material base for enlarging and strengthening the intersubjective networks. Farming made it possible to feed thousands of people in crowded cities and thousands of soldiers in disciplined armies. However, the intersubjective webs then encountered a new obstacle. In order to preserve the collective myths and organise mass cooperation, the early farmers relied on the data-processing abilities of the human brain, which were strictly limited.

Farmers believed in stories about great gods. They built temples to their favourite god, held festivals in his honour, offered him sacrifices, and gave him lands, tithes and presents. In the first cities of ancient Sumer, about 6,000 years ago, the temples were not just centres of worship, but also the most important political and economic hubs. The Sumerian gods fulfilled a function analogous to modern brands and corporations. Today, corporations are fictional legal entities that own property, lend money, hire employees and initiate economic enterprises. In the ancient cities of Uruk, Lagash and Shurupak the gods functioned as legal entities that could own fields and slaves, give and receive loans, pay salaries and build dams and canals.

Since the gods never died, and since they had no children to quarrel over their inheritance, they gathered more and more property and power. An increasing number of Sumerians found themselves employed by the gods, taking loans from the gods, tilling the gods' lands and owing them taxes and tithes. Just as in present-day San Francisco John is employed by Google while Mary works for Microsoft, so in ancient Uruk one person was employed by the great god Enki while his neighbour worked for the goddess Inanna. The temples of Enki and Inanna dominated the Uruk skyline, and their divine logos branded buildings, products and clothes. For the Sumerians, Enki and Inanna were as real as Google and Microsoft are real for us. Compared to their predecessors – the ghosts and spirits of the Stone Age – the Sumerian gods were very powerful entities.

It goes without saying that the gods didn't actually run their businesses, for the simple reason that they didn't exist anywhere except in the human imagination. Day-to-day activities were managed by the temple priests (just as Google and Microsoft need to hire flesh-and-blood humans to manage their affairs). However, as the gods acquired more and more property and power, the priests could not cope. They may have represented the mighty sky god or the all-knowing earth goddess, but they themselves were fallible mortals. They had difficulty remembering which estates, orchards and fields belonged to the goddess Inanna, which of Inanna's employees had already received their salaries, which of the goddess's tenants had failed to pay their rents and what interest rate the goddess charged her debtors. This was one of the main reasons why in Sumer, like everywhere else around the world, human cooperation networks could not notably expand even thousands of years after the Agricultural Revolution. There were no huge kingdoms, no extensive trade networks and no universal religions.

This obstacle was finally removed about 5,000 years ago, when the Sumerians invented both writing and money. These Siamese twins – born to the same parents at the same time and in the same place – broke the data-processing limitations of the human brain. Writing and money made it possible to start collecting taxes from hundreds of thousands of people, to organise complex bureaucracies and to establish vast kingdoms. In Sumer these kingdoms were managed in the name of the gods by human priest-kings. In the neighbouring Nile Valley people went a step further, merging the priest-king with the god to create a living deity – pharaoh.

The Egyptians considered pharaoh to be an actual god rather than just a divine deputy. The whole of Egypt belonged to that god, and all people had to obey his orders and pay the taxes he levied. Just as in the Sumerian temples, so also in pharaonic Egypt the god didn't manage his business empire by himself. Some pharaohs ruled with an iron fist, while others passed their days at banquets and festivities, but in both cases the practical work of administering Egypt was left to thousands of literate officials. Just like any other human, pharaoh had a biological body with biological needs, desires and emotions. But the biological pharaoh was of little importance. The real ruler of the Nile Valley was an imagined pharaoh who existed in the stories that millions of Egyptians told one another.

While pharaoh sat in the capital city of Memphis, eating grapes in his palace and dallying with his wives and mistresses, pharaoh's officials criss-crossed the kingdom from the Mediterranean shore to the Nubian Desert. The bureaucrats calculated the taxes each village had to pay, recorded them on long papyrus scrolls and sent them to Memphis. When a written order came from Memphis to recruit soldiers to the army or labourers for some construction project, officials gathered the necessary men. They computed how much wheat the royal granaries contained, how many work days were required to clean the canals and reservoirs, and how many ducks and pigs to send to Memphis so that pharaoh's harem could dine well. Even when the living deity died, and his body was embalmed and borne in an extravagant funerary procession to the royal necropolis outside Memphis, the bureaucracy kept going.

Officials kept writing scrolls, collecting taxes, sending orders and oiling the gears of the pharaonic machine.

If the Sumerian gods remind us of present-day company brands, so the living-god pharaoh can be compared to modern personal brands such as Elvis Presley, Madonna or Justin Bieber. Just like pharaoh, Elvis too had a biological body, complete with biological needs, desires and emotions. Elvis ate and drank and slept. Yet Elvis was much more than a biological body. Like pharaoh, Elvis was a story, a myth, a brand – and the brand was far more important than the biological body. During Elvis's lifetime, the brand earned millions of dollars selling records, tickets, posters and rights, but only a small fraction of the necessary work was performed by Elvis in person. Instead, most of it was accomplished by a small army of agents, lawyers, producers and secretaries. Consequently when the biological Elvis died, for the brand it was business as usual. Even today fans still buy the King's posters and albums, radio stations

21. Brands are not a modern invention. Just like Elvis Presley, pharaoh too was a brand rather than a living organism. For millions of followers his image counted for far more than his fleshy reality, and they kept worshipping him long after he was dead.

go on paying royalties, and more than half a million pilgrims flock each year to Graceland, the King's necropolis in Memphis, Tennessee.

Prior to the invention of writing, stories were confined by the limited capacity of human brains. You couldn't invent overly complex stories which people couldn't remember. But with writing you could suddenly create extremely long and intricate stories, which were stored on tablets and papyri rather than in human heads. No ancient Egyptian remembered all of pharaoh's lands, taxes and tithes; Elvis Presley never even read all the contracts signed in his name; no living soul is familiar with all the laws and regulations of the European Union; and no banker or CIA agent tracks down every single dollar in the world. Yet all of these minutiae are written somewhere, and the assemblage of relevant documents defines the identity and power of pharaoh, Elvis, the EU and the dollar.

Writing has thus enabled humans to organise entire societies in an algorithmic fashion. We encountered the term 'algorithm' when we tried to understand what emotions are and how brains function, and defined it as a methodical set of steps that can be used to make calculations, resolve problems and reach decisions. In illiterate societies people make all calculations and decisions in their heads. In literate societies people are organised into networks, so that each person is only a small step in a huge algorithm, and it is the algorithm as a whole that makes the important decisions. This is the essence of bureaucracy.

Think about a modern hospital, for example. When you arrive the receptionist hands you a standard form and asks you a predetermined set of questions. Your answers are forwarded to a nurse, who compares them with the hospital's regulations in order to decide what preliminary tests to give you. She then measures, say, your blood pressure and heart rate, and takes a blood sample. The doctor on duty examines the initial results, and follows a strict protocol in determining to which ward to admit you. In the ward you are subjected to much more thorough examinations, such as an X-ray or an fMRI scan, mandated by thick medical guidebooks.

Specialists then analyse the results according to well-known statistical databases, deciding what medicines to give you or what further tests to run.

This algorithmic structure ensures that it doesn't really matter who is the receptionist, nurse or doctor on duty. Their personality type, their political opinions and their momentary moods are irrelevant. As long as they all follow the regulations and protocols, they stand a good chance of curing you. According to the algorithmic ideal, your fate is in the hands of 'the system', and not in the hands of the flesh-and-blood mortals who happen to occupy this or that post.

What's true of hospitals is also true of armies, prisons, schools, corporations – and ancient kingdoms. Of course ancient Egypt was far less technologically sophisticated than a modern hospital, but the algorithmic principle was the same. In ancient Egypt too, most decisions were made not by a single wise person, but by a network of officials linked together through papyri and stone inscriptions. Acting in the name of the living-god pharaoh, the network restructured human society and reshaped the natural world. For example, pharaohs Senusret III and his son Amenemhat III, who ruled Egypt from 1878 BC to 1814 BC, dug a huge canal linking the Nile to the swamps of the Fayum Valley. An intricate system of dams, reservoirs and subsidiary canals diverted some of the Nile waters to Fayum, creating an immense artificial lake holding 13 trillion gallons of water.[1] By comparison, Lake Mead, the largest man-made reservoir in the United States (formed by the Hoover Dam), holds at most 9 trillion gallons of water.

The Fayum engineering project gave pharaoh the power to regulate the Nile, prevent destructive floods and provide precious water relief in times of drought. In addition, it turned the Fayum Valley from a crocodile-infested swamp surrounded by barren desert into Egypt's granary. On the shore of the new artificial lake was built the new city of Shadet, which the Greeks called Crocodilopolis – the city of crocodiles. It was dominated by the temple of the crocodile god Sobek, who was identified with pharaoh (contemporary statues sometimes show pharaoh sporting a crocodile head). The temple housed a sacred crocodile called Petsuchos, who was considered to be the living incarnation of Sobek. Just like the living-god pharaoh, the living-god Petsuchos was lovingly groomed by attending priests who provided the lucky reptile with lavish food and even toys, and dressed him up in gold cloaks and gem-encrusted crowns. After all, Petsuchos was the priests' brand, and their authority and livelihood depended on him. When Petsuchos died, a new crocodile was immediately elected to fill his sandals, while the dead reptile was carefully embalmed and mummified.

In the days of Senusret III and Amenemhat III the Egyptians had neither bulldozers nor dynamite. They didn't even have iron tools, work horses or wheels (the wheel did not enter common usage in Egypt until about 1500 BC). Bronze tools were considered cutting-edge technology, but they were so expensive and rare that most of the building work was performed with tools made only of stone and wood, operated by human muscle power. Many people argue that the great building projects of ancient Egypt – all the dams and reservoirs and pyramids – must have been built by aliens from outer space. How else could a culture lacking even wheels and iron accomplish such wonders?

The truth is very different. Egyptians built Lake Fayum and the pyramids thanks not to extraterrestrial help, but to superb organisational skills. Relying on thousands of literate bureaucrats, pharaoh recruited tens of thousands of labourers and enough food to maintain them for years on end. When tens of thousands of labourers cooperate for several decades, they can build an artificial lake or a pyramid even with stone tools.

Pharaoh himself hardly lifted a finger, of course. He didn't collect taxes himself, he didn't draw any architectural plans, and he certainly never picked up a shovel. But the Egyptians believed that only prayers to the living-god pharaoh and to his heavenly patron Sobek could save the Nile Valley from devastating floods and droughts. They were right. Pharaoh and Sobek were imaginary entities who did nothing to raise or lower the Nile water level, but when millions of people believed in pharaoh and Sobek and therefore cooperated in building dams and digging canals, floods and droughts became rare. Compared to the Sumerian gods, not to mention the Stone Age spirits, the gods of ancient Egypt were truly powerful entities that founded cities, raised armies and controlled the lives of millions of humans, cows and crocodiles.

It may sound strange to credit imaginary entities with building or controlling things. But nowadays we habitually say that the United States built the first nuclear bomb, that China built the Three Gorges Dam or that Google is building an autonomous car. Why not say, then, that pharaoh built a reservoir and Sobek dug a canal?

Living on Paper

Writing thus facilitated the appearance of powerful fictional entities that organised millions of people and reshaped the reality of rivers, swamps and crocodiles. Simultaneously, writing also made it easier for humans to believe in the existence of such fictional entities, because it habituated people to experiencing reality through the mediation of abstract symbols.

Hunter-gatherers spent their days climbing trees, looking for mushrooms, and chasing boars and rabbits. Their daily reality consisted of trees, mushrooms, boars and rabbits. Peasants worked all day in the fields, ploughing, harvesting, grinding corn and taking care of farmyard animals. Their daily reality was the feeling of muddy earth under bare feet, the smell of oxen pulling the plough and the taste of warm bread fresh from the oven. In contrast, scribes in ancient Egypt devoted most of their time to reading, writing and calculating. Their daily reality consisted of ink marks on papyrus scrolls, which determined who owned which field, how much an ox cost and what yearly taxes the peasants had to pay. A scribe could decide the fate of an entire village with a stroke of his stylus.

The vast majority of people remained illiterate until the modern age, but the all-important administrators increasingly saw reality through the medium of written texts. For this literate elite – whether in ancient Egypt or in twentieth-century

Europe – anything written on a piece of paper was at least as real as trees, oxen and human beings.

In the spring of 1940, when the Nazis overran France from the north, much of its Jewish population tried to escape the country towards the south. In order to cross the border, they needed visas to Spain and Portugal, and together with a flood of other refugees, tens of thousands of Jews besieged the Portuguese consulate in Bordeaux in a desperate attempt to get that life-saving piece of paper. The Portuguese government forbade its consuls in France to issue visas without prior approval from the Foreign Ministry, but the consul in Bordeaux, Aristides de Sousa Mendes, decided to disregard the order, throwing to the wind a thirty-year diplomatic career. As Nazi tanks were closing in on Bordeaux, Sousa Mendes and his team worked around the clock for ten days and nights, barely stopping to sleep, just issuing visas and stamping pieces of paper. Sousa Mendes issued thousands of visas before collapsing from exhaustion.

22. Aristides de Sousa Mendes, the angel with the rubber stamp.

The Portuguese government – which had little desire to accept any of these refugees – sent agents to escort the disobedient consul back home, and fired him from the foreign office. Yet officials who cared little for the plight of human beings nevertheless had a deep reverence for documents, and the visas Sousa Mendes issued against orders were respected by French, Spanish and Portuguese bureaucrats alike, spiriting up to 30,000 people out of the Nazi death trap. Sousa Mendes, armed with little more than a rubber stamp, was responsible for the largest rescue operation by a single individual during the Holocaust.[2]

The sanctity of written records often had far less positive effects. From 1958 to 1961 communist China undertook the Great Leap Forward, when Mao Zedong wished to rapidly turn China into a superpower. Intending to use surplus grain to finance ambitious industrial projects, Mao ordered the doubling and tripling of agricultural production. From the government offices in Beijing his impossible demands made their way down the bureaucratic ladder, through provincial administrators, all the way down to the village

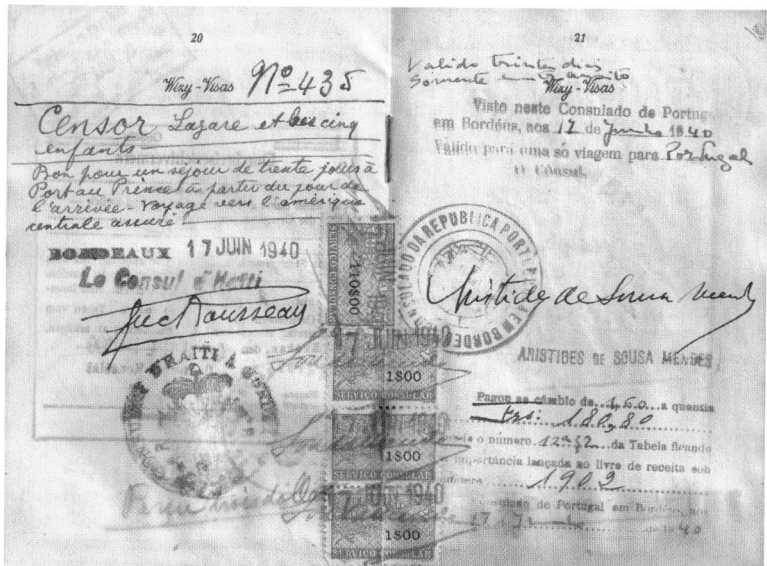

23. One of the thousands of life-saving visas signed by Sousa Mendes in June 1940 (visa #1902 for Lazare Censor and family, dated 17 June 1940).

headmen. The local officials, afraid of voicing any criticism and wishing to curry favour with their superiors, concocted imaginary reports of dramatic increases in agricultural output. As the fabricated numbers made their way back up the bureaucratic hierarchy, each official exaggerated them further, adding a zero here or there with a stroke of a pen.

Consequently, in 1958 the Chinese government was informed that annual grain production was 50 per cent more than it actually was. Believing the reports, the government sold millions of tons of rice to foreign countries in exchange for weapons and heavy machinery, assuming that enough was left to feed the Chinese population. The result was the worst famine in history and the death of tens of millions of Chinese.[3]

Meanwhile, enthusiastic reports of China's farming miracle reached audiences throughout the world. Julius Nyerere, the idealistic president of Tanzania, was deeply impressed by the Chinese success. In order to modernise Tanzanian agriculture, Nyerere resolved to establish collective farms on the Chinese model. When peasants objected to the plan, Nyerere sent the army and police to destroy traditional villages and forcibly relocate hundreds of thousands of peasants onto the new collective farms.

Government propaganda depicted the farms as miniature paradises, but many of them existed only in government documents. The protocols and reports written in the capital Dar es Salaam said that on such-and-such a date the inhabitants of such-and-such village were relocated to such-and-such farm. In reality, when the villagers reached their destination, they found absolutely nothing there. No houses, no fields, no tools. Officials nevertheless reported great successes to themselves and to President Nyerere. In fact, within less than ten years Tanzania was transformed from Africa's biggest food exporter into a net food importer that could not feed itself without external assistance. In 1979, 90 per cent of Tanzanian farmers lived on collective farms, but they generated only 5 per cent of the country's agricultural output.[4]

Though the history of writing is rife with similar mishaps, the benefits of a more efficient administration have generally outweighed the costs, at least from the government's perspective. No ruler could resist the temptation of trying to alter reality with the stroke of a pen, and if disaster resulted, the remedy seemed to consist of writing ever more voluminous memos and issuing ever more codes, edicts and orders.

Written language may have been conceived as a modest way of describing reality, but it gradually became a powerful way to reshape reality. When official reports collided with objective reality, it was often reality that had to give way. Anyone who has ever dealt with the tax authorities, the educational system or any other complex bureaucracy knows that the truth hardly matters. What's written on your form is far more important.

Holy Scriptures

Is it true that when text and reality collide, reality sometimes has to give way? Isn't that just a common but exaggerated slander of bureaucratic systems? Most bureaucrats – whether serving pharaoh or Mao Zedong – were reasonable people, and surely would have made the following argument: 'We use writing to describe the reality of fields, canals and granaries. If the description is accurate, we make realistic decisions. If the description is inaccurate, it causes famines and even rebellions. Then we, or the administrators of some future regime, learn from that mistake, and strive to produce more truthful descriptions. So over time, our documents are bound to become ever more precise.'

That's true to some extent, but it ignores an opposite historical dynamic. As bureaucracies accumulate power, they become immune to their own mistakes. Instead of changing their stories to fit reality, they can change reality to fit their stories. In the end external reality matches their bureaucratic fantasies, but only because they forced reality to do so. For example, the borders of many African countries disregard river lines, mountain ranges and trade routes, split historical and economic zones unnecessarily, and ignore local ethnic and religious identities. The same tribe may find itself riven among several countries, whereas one country may incorporate splinters of numerous rival clans. Such problems bedevil countries all over the world, but in Africa they are particularly acute because modern African borders don't reflect the wishes and struggles of local nations. They were drawn by European bureaucrats who never set foot in Africa.

In the late nineteenth century, several European powers laid claim to African territories. Fearing that conflicting claims might lead to an all-out European war, the concerned parties got together in Berlin in 1884 and divided Africa as if it were a pie. Back then much of the African interior was terra incognita to Europeans. The British, French and Germans had accurate maps of Africa's coastal regions, and knew precisely where the Niger, Congo and Zambezi empty into the ocean. However, they knew little about the course these rivers took inland, about the kingdoms and tribes that lived along their banks, and about local religion, history and geography. This hardly mattered to the European diplomats. They unrolled a half-empty map of Africa over a well-polished Berlin table, sketched a few lines here and there, and divided the continent among them.

When in due course the Europeans penetrated the African interior, armed with their agreed-upon map, they discovered that many of the borders drawn in Berlin did little justice to the geographic, economic and ethnic reality of Africa. However, to avoid renewed clashes, the invaders stuck to their agreements, and these imaginary lines became the actual borders of European colonies. During the second half of the twentieth century, as the European empires disintegrated and their colonies gained independence, the new countries accepted the colonial borders, fearing that the alternative would be endless wars and conflicts. Many of the difficulties faced by present-day African countries stem from the fact that their borders make little sense. When the written fantasies of European bureaucracies encountered the African reality, reality was forced to surrender.[5]

Our modern education systems provide numerous other examples of reality kowtowing to written records. When measuring the

24. A mid-nineteenth century European map of Africa. Europeans knew very little about the African interior, but that did not prevent them from divvying up the continent and drawing its borders.

width of my desk, the yardstick I am using matters little. The width of my desk remains the same whether I say it is 200 centimetres or 78.74 inches. However, when bureaucracies measure people, the yardsticks they choose make all the difference. When schools began assessing people according to precise numerical marks, the lives of millions of students and teachers changed dramatically. Marks are a relatively new invention. Hunter-gatherers were never marked for their achievements, and even thousands of years after the Agricultural Revolution, few education establishments used precise marks. At the end of the year a medieval apprentice cobbler did not receive a piece of paper saying he has got an A in shoelaces but a C minus in buckles. An undergraduate in Shakespeare's day left Oxford with one of only two possible results – with a degree, or without one. Nobody thought of giving one student a final mark of 74 and another student an 88.[6]

It was the mass education systems of the industrial age that began using precise marks on a regular basis. After both factories and government ministries became accustomed to thinking in the language of numbers, schools followed suit. They started to gauge the worth of each student according to his or her average mark, whereas the worth of each teacher and principal was judged according to the school's overall average. Once bureaucrats adopted this yardstick, reality was transformed.

Originally, schools were supposed to focus on enlightening and educating students, and marks were merely a means of measuring success. But naturally enough schools soon began focusing on achieving high marks. As every child, teacher and inspector knows, the skills required to get high marks in an exam are not the same as a true understanding of literature, biology or mathematics. Every child, teacher and inspector also knows that when forced to choose between the two, most schools will go for the marks.

The power of written records reached its apogee with the appearance of holy scriptures. Priests and scribes in ancient civilisations became accustomed to seeing documents as guidebooks for reality. At first, the texts told them about the reality of taxes, fields and granaries. But as the bureaucracy gained power, so the texts gained authority. Priests recorded not only lists of the god's property, but also the god's deeds, commandments and secrets. The resulting scriptures purported to describe reality in its entirety, and generations of scholars became accustomed to looking for all the answers in the pages of the Bible, the Qur'an or the Vedas.

In theory, if some holy book misrepresented reality, its disciples would sooner or later discover this, and the text's authority would be undermined. Abraham Lincoln said you cannot deceive everybody all the time. Well, that's wishful thinking. In practice, the power of human cooperation networks depends on a delicate balance between truth and fiction. If you distort reality too much, it will weaken you, and you will not be able to compete against more clear-sighted rivals. On the other hand, you cannot organise masses of people effectively without relying on some fictional myths. So if you stick to unalloyed reality, without mixing any fiction with it, few people will follow you.

If you used a time machine to send a modern scientist to ancient Egypt, she would not be able to seize power by exposing the fictions of the local priests and lecturing the peasants on evolution, relativity and quantum physics. Of course, if our scientist could use her knowledge in order to produce a few rifles and artillery pieces, she could gain a huge advantage over pharaoh and the crocodile god Sobek. Yet in order to mine iron ore, build blast furnaces and manufacture gunpowder the scientist would need a lot of hard-working peasants. Do you really think she could inspire them by explaining that energy divided by mass equals the speed of light squared? If you happen to think so, you are welcome to travel to present-day Afghanistan or Syria and try your luck.

Really powerful human organisations – such as pharaonic Egypt, the European empires and the modern school system – are not necessarily clear-sighted. Much of their power rests on their ability to force their fictional beliefs on a submissive reality. That's the whole idea of money, for example. The government makes worthless pieces of paper, declares them to be valuable and then uses them to compute the value of everything else. The government has the power to force citizens to pay taxes using these pieces of paper, so the citizens have no choice but to get their hands on at least some of them. Consequently, these bills really do become valuable, the government officials are vindicated in their beliefs, and since the government controls the issuing of paper money, its power grows. If somebody protests that 'These are just worthless pieces of paper!' and behaves as if they are only pieces of paper, he won't get very far in life.

The same thing happens when the education system declares that matriculation exams are the best method to evaluate students. The system has sufficient authority to influence admission standards to colleges and hiring standards in government offices and in the private sector. Students therefore invest all their efforts in getting good marks. Coveted positions are occupied by people with high marks, who naturally support the system that brought them there. The fact that the education system controls the critical exams gives it more power, and increases its influence over colleges, government offices and the job market. If somebody protests that 'The degree certificate is just a piece of paper!' and behaves accordingly, he is unlikely to get very far in life.

Holy scriptures work the same way. The religious establishment proclaims that the holy book contains the answers to all our questions. It simultaneously presses courts, governments and businesses to behave according to what the holy book says. When a wise person reads scriptures and then looks at the world, he sees that there is indeed a good match. 'Scriptures say that you must pay tithes to God – and look, everybody pays. Scriptures say that women are inferior to men, and cannot serve as judges or even give testimony in court – and look, there are indeed no women judges and the courts reject their testimony. Scriptures say that whoever studies the word of God will succeed in life – and look, all the good jobs are indeed held by people who know the holy book by heart.'

Such a wise person will naturally begin to study the holy book, and because he is wise, he will become a scriptural pundit and be appointed a judge. When he becomes a judge, he will not allow women to bear witness in court, and when he chooses his successor, he will obviously pick somebody who knows the holy book well. If someone protests that 'This book is just paper!' and behaves accordingly, such a heretic will not get very far in life.

Even when scriptures mislead people about the true nature of reality, they can nevertheless retain their authority for thousands of years. For instance, the biblical perception of history is fundamentally flawed, yet it managed to spread throughout the world, and many millions still believe in it. The Bible peddled a monotheistic theory of history, claiming that the world is governed by a single all-powerful deity, who cares above all else about me and my doings. If something good happens, it must be a reward for my good deeds. Any catastrophe must surely be punishment for my sins.

Thus the ancient Jews believed that if they suffered from drought, or if King Nebuchadnezzar of Babylonia invaded Judaea and exiled its people, surely these were divine punishments for their own sins. And if King Cyrus of Persia defeated the Babylonians and allowed the Jewish exiles to return home and rebuild Jerusalem, God in his mercy must have heard their remorseful prayers. The Bible doesn't recognise the possibility that perhaps the drought resulted from a volcanic eruption in the Philippines, that Nebuchadnezzar invaded in pursuit of Babylonian commercial interests and that King Cyrus had his own political reasons to favour the Jews. The Bible accordingly shows no interest whatsoever in understanding the global ecology, the Babylonian economy or the Persian political system.

Such self-absorption characterises all humans in childhood. Children of all religions and cultures think they are the centre of the world, and therefore show little genuine interest in the conditions and feelings of other people. That's why divorce is so traumatic for children. A five-year-old cannot understand that something important is happening for reasons unrelated to him. No matter how many times mommy and daddy tell him that they are independent people with their own problems and wishes, and that they didn't divorce because of him – the child cannot absorb it. He is convinced that everything happens because of him. Most people grow out of this infantile delusion. Monotheists hold on to it till the day they die. Like a child thinking that his parents are fighting because of him, the monotheist is convinced that the Persians are fighting the Babylonians because of him.

Already in biblical times some cultures had a far more accurate perception of history. Animist and polytheist religions depicted the world as the playground of numerous different powers rather than a single god. It was consequently easy for animists and polytheists to accept that many events are unrelated to me or to my favourite deity, and they are neither punishments for my sins nor rewards for my good deeds. Greek historians such as Herodotus and Thucydides, and Chinese historians such as Sima Qian, developed sophisticated theories of history that are very similar to our own modern views. They explained that wars and revolutions break out due to myriad political, social and economic factors. People may fall victim to war through no fault of their own. Accordingly, Herodotus developed a keen interest in understanding Persian politics, while Sima Qian was very concerned about the culture and religion of barbarous steppe people.[7]

Present-day scholars agree with Herodotus and Sima Qian rather than with the Bible. That's why all modern states invest so much effort in collecting information about other countries, and in analysing global ecological, political and economic trends. When the US economy falters, even evangelical Republicans sometimes point an accusing finger at China rather than at their own sins.

Yet even though Herodotus and Thucydides understood reality much better than the authors of the Bible, when the two world views collided, the Bible won by a knockout. The Greeks adopted the Jewish view of history, rather than vice versa. A thousand years after Thucydides, the Greeks became convinced that if some barbarian horde invaded, surely it was divine punishment for their sins. No matter how mistaken the biblical world view was, it provided a better basis for large-scale human cooperation.

Indeed, even today when US presidents take their oath of office, they put their hand on a Bible. Similarly in many countries around the world, including the USA and the UK, witnesses in courts put their hand on a Bible when swearing to tell the truth, the whole truth and nothing but the truth. It's ironic that they swear to tell the truth on a book brimming with so many fictions, myths and errors.

But it Works!

Fictions enable us to cooperate better. The price we pay is that the same fictions also determine the goals of our cooperation. So we may have very elaborate systems of cooperation, which are harnessed to serve fictional aims and interests. Consequently the system may seem to be working well, but only if we adopt the system's own criteria. For example, a Muslim mullah would say: 'Our system works. There are now 1.5 billion Muslims worldwide, and more people study the Qur'an and submit themselves to Allah's will than ever before.' The key question, though, is whether this is the right yardstick for measuring success. A school principal would say: 'Our system works. During the last five years, exam results have risen by 7.3 per cent.' Yet is that the best way to judge a school? An official in ancient Egypt would say: 'Our system works. We collect more taxes, dig more canals and build bigger pyramids than anyone else in the world.' True enough, pharaonic Egypt led the world in taxation, irrigation and pyramid construction. But is that what really counts?

People have many material, social and psychological needs. It is far from clear that peasants in ancient Egypt enjoyed more love or better social relations than their hunter-gatherer ancestors, and in terms of nutrition, health and child mortality it seems that life was actually worse. A document dated c.1850 BC from the reign of Amenemhat III – the pharaoh who created Lake Fayum – tells of a well-to-do man called Dua-Khety who took his son Pepy to school, so that he could learn to be a scribe. While on their way, Dua-Khety portrayed the miserable life of peasants, labourers, soldiers and artisans, so as to encourage Pepy to devote all his energy to studying and thereby escape the unhappy destiny of most humans.

According to Dua-Khety, the life of a landless field labourer is full of hardship and misery. Dressed in mere tatters, he works all day till his fingers are covered in blisters. Then pharaoh's officials come and take him away to do forced labour. In return for all his hard work he receives only sickness as payment. Even if he makes it home alive, he will be completely worn out and ruined. The fate of the landholding peasant is hardly better. He spends his days carrying water in buckets from the river to the field. The heavy load bends his shoulders and covers his neck with festering swellings. In the morning he has to water his plot of leeks, in the afternoon his date palms and in the evening his coriander field. Eventually he drops down and dies.[8] The text might be exaggerating things on purpose, but not by much. Pharaonic Egypt was the most powerful kingdom of its day, but for the simple peasant all that power meant taxes and forced labour rather than clinics and social security services.

This was not a uniquely Egyptian defect. Despite all the immense achievements of the Chinese dynasties, the Muslim empires and the European kingdoms, even in AD 1850 the life of the average person was not better – and might actually have been worse – than the lives of archaic hunter-gatherers. In 1850 a Chinese peasant or a Manchester factory hand worked longer hours than their hunter-gatherer ancestors; their jobs were physically harder and mentally less fulfilling; their diet was less balanced; hygiene conditions were incomparably worse; and infectious diseases were far more common.

Suppose you were given a choice between the following two vacation packages:

Stone Age package: On day one we will hike for ten hours in a pristine forest, setting camp for the night in a clearing by a river. On day two we will canoe down the river for ten hours, camping on the shores of a small lake. On day three we will learn from the native people how to fish in the lake and how to find mushrooms in the nearby woods.

Modern proletarian package: On day one we will work for ten hours in a polluted textile factory, passing the night in a cramped apartment block. On day two we will work for ten hours as cashiers in the local department store, going back to sleep in the same apartment block. On day three we will learn from the native people how to open a bank account and fill out mortgage forms.

Which package would you choose?

Hence when we come to evaluate human cooperation networks, it all depends on the yardsticks and viewpoint we adopt. Are we judging pharaonic Egypt in terms of production, nutrition or perhaps social harmony? Do we focus on the aristocracy, the simple peasants, or the pigs and crocodiles? History isn't a single narrative, but thousands of alternative narratives. Whenever we choose to tell one, we are also choosing to silence others.

Human cooperative networks usually judge themselves by yard-sticks of their own invention and, not surprisingly, they often give themselves high marks. In particular, human networks built in the name of imaginary entities such as gods, nations and corporations normally judge their success from the viewpoint of the imaginary entity. A religion is successful if it follows divine commandments to the letter; a nation is glorious if it promotes the national interest; and a corporation thrives if it makes a lot of money.

When examining the history of any human network, it is there-fore advisable to stop from time to time and look at things from the perspective of some real entity. How do you know if an entity is real? Very simple – just ask yourself, 'Can it suffer?' When people burn down the temple of Zeus, Zeus doesn't suffer. When the euro loses its value, the euro doesn't suffer. When a bank goes bankrupt, the bank doesn't suffer. When a country suffers a defeat in war, the country doesn't really suffer. It's just a metaphor. In contrast, when a soldier is wounded in battle, he really does suffer. When a famished peasant has nothing to eat, she suffers. When a cow is separated from her newborn calf, she suffers. This is reality.

Of course suffering might well be caused by our belief in fic-tions. For example, belief in national and religious myths might cause the outbreak of war, in which millions lose their homes, their limbs and even their lives. The cause of war is fictional, but the suffering is 100 per cent real. This is exactly why we should strive to distinguish fiction from reality.

Fiction isn't bad. It is vital. Without commonly accepted sto-ries about things like money, states or corporations, no complex human society can function. We can't play football unless everyone believes in the same made-up rules, and we can't enjoy the bene-fits of markets and courts without similar make-believe stories. But the stories are just tools. They should not become our goals or our yardsticks. When we forget that they are mere fiction, we lose touch with reality. Then we begin entire wars 'to make a lot of money for the corporation' or 'to protect the national interest'. Corporations, money and nations exist only in our imagination.

We invented them to serve us; why do we find ourselves sacrificing our lives in their service?

In the twenty-first century we will create more powerful fictions and more totalitarian religions than in any previous era. With the help of biotechnology and computer algorithms these religions will not only control our minute-by-minute existence, but will be able to shape our bodies, brains and minds, and to create entire virtual worlds complete with hells and heavens. Being able to distinguish fiction from reality and religion from science will therefore become more difficult but more vital than ever before.

5

The Odd Couple

Stories serve as the foundations and pillars of human societies. As history unfolded, stories about gods, nations and corporations grew so powerful that they began to dominate objective reality. Believing in the great god Sobek, the Mandate of Heaven or the Bible enabled people to build Lake Fayum, the Great Wall of China and Chartres Cathedral. Unfortunately, blind faith in these stories meant that human efforts frequently focused on increasing the glory of fictional entities such as gods and nations, instead of bettering the lives of real sentient beings.

Does this analysis still hold true today? At first sight, it seems that modern society is very different from the kingdoms of ancient Egypt or medieval China. Hasn't the rise of modern science changed the basic rules of the human game? Wouldn't it be true to say that despite the ongoing importance of traditional myths, modern social systems increasingly rely on objective scientific theories such as the theory of evolution, which simply did not exist in ancient Egypt or medieval China?

We could of course argue that scientific theories are a new kind of myth, and that our belief in science is no different from the ancient Egyptians' belief in the great god Sobek. Yet the comparison doesn't hold water. Sobek existed only in the collective imagination of his devotees. True, praying to Sobek helped cement the Egyptian social system, thereby enabling people to build dams and canals that prevented floods and droughts. Yet the prayers themselves didn't raise or lower the Nile's water level in the slightest. In contrast, scientific theories are not just a way to bind people together. It is often said that God helps those who help themselves. This is a roundabout way of saying that God doesn't exist, but if our belief in Him inspires us to do something ourselves – it helps. Antibiotics, unlike God, help even those who don't help themselves. They cure infections whether you believe in them or not.

Consequently, the modern world is very different from the premodern world. Egyptian pharaohs and Chinese emperors failed to overcome famine, plague and war despite millennia of effort. Modern societies managed to do it within a few centuries. Isn't this the fruit of abandoning intersubjective myths in favour of objective scientific knowledge? And can we not expect this process to accelerate in the coming decades? As technology enables us to upgrade humans, overcome old age and find the key to happiness, won't people care less about fictional gods, nations and corporations, and focus instead on deciphering the physical and biological reality?

It might seem so, but in fact things are far more complicated. Modern science certainly changed the rules of the game, yet it did not simply replace myths with facts. Myths continue to dominate humankind, and science only makes these myths stronger. Instead of destroying the intersubjective reality, science will enable it to control the objective and subjective realities more completely than ever before. Thanks to computers and bioengineering, the difference between fiction and reality will blur, as people reshape reality to match their pet fictions.

The priests of Sobek imagined the existence of divine crocodiles, while pharaoh dreamt about immortality. In reality the sacred crocodile was a very ordinary swamp reptile dressed in golden finery, and pharaoh was as mortal as the poorest peasant. After death his corpse was mummified using preservative balms and scented perfumes, but it nonetheless remained as lifeless as one can get. In contrast, twenty-first-century scientists might be able to engineer actual super-crocodiles, and to provide the human elite with eternal youth here on earth.

Consequently the rise of science will make at least some myths and religions mightier than ever. To understand why, and to face the challenges of the twenty-first century, we should therefore revisit one of the most vexing questions of all: how does modern science relate to religion? It seems that people have already said a million times everything there is to say about this question. Yet in practice, science and religion are like a husband and wife who after 500 years of marriage counselling still don't know each other. He still dreams about Cinderella and she keeps pining for Prince Charming, while they argue about whose turn it is to take out the rubbish.

Germs and Demons

Most of the misunderstandings regarding science and religion result from faulty definitions of religion. All too often people confuse religion with superstition, spirituality, belief in supernatural powers or belief in gods. Religion is none of these things. Religion cannot be equated with superstition, because most people are unlikely to call their most cherished beliefs 'superstitions'. We always believe in 'the truth'; only other people believe in superstitions.

Similarly, few people put their faith in supernatural powers. For those who believe in demons, spirits and fairies, these beings are not supernatural. They are an integral part of nature, just like porcupines, scorpions and germs. Modern physicians blame disease on invisible germs, and voodoo priests blame disease on invisible spirits. There's nothing supernatural about it: if you make some spirit angry, the spirit enters your body and causes you pain. What could be more natural than that? Only those who don't believe in spirits think of them as standing apart from the natural order of things.

Equating religion with faith in supernatural powers implies that you can understand all known natural phenomena without religion, which is just an optional supplement. Having understood perfectly well the whole of nature, you can now choose whether or not to add some 'super-natural' religious dogma. Most religions, however, argue that you simply cannot understand the world without them. You will never comprehend the true reason for disease, drought or earthquakes if you do not take their dogma into account.

Defining religion as 'belief in gods' is also problematic. We tend to say that a devout Christian is religious because she believes in God, whereas a fervent communist isn't religious, because communism has no gods. However, religion is created by humans rather than by gods, and it is defined by its social function rather than by the existence of deities. Religion is any all-encompassing story that confers superhuman legitimacy on human laws, norms and values. It legitimises human social structures by arguing that they reflect superhuman laws.

Religion asserts that we humans are subject to a system of moral laws that we did not invent and that we cannot change. A devout Jew would say that this is the system of moral laws created by God and revealed in the Bible. A Hindu would say that Brahma, Vishnu and Shiva created the laws, which were revealed to us humans in the Vedas. Other religions, from Buddhism and Daoism to communism, Nazism and liberalism, argue that the so-called superhuman laws are natural laws, and not the creation of this or that god. Of course, each believes in a different set of natural laws discovered and revealed by different seers and prophets, from Buddha and Laozi to Marx and Hitler.

A Jewish boy comes to his father and asks, 'Dad, why shouldn't we eat pork?' The father thoughtfully strokes his long curly beard and answers, 'Well, Yankele, that's how the world works. You are still young and don't yet understand, but if we eat pork, God will punish us and we will come to a bad end. It isn't my idea. It's not even the rabbi's idea. If the rabbi had created the world, maybe he would have created a world in which pork was perfectly kosher. But the rabbi didn't create the world – God did. And God said, I don't know why, that we shouldn't eat pork. So we shouldn't. Capeesh?'

In 1943 a German boy comes to his father, a senior SS officer, and asks, 'Dad, why are we killing the Jews?' The father, putting on his shiny leather boots, explains, 'Well, Fritz, that's how the world works. You are still young and don't yet understand, but if we allow the Jews to live they will cause the degeneration and extinction of humankind. It's not my idea. And it's not even the Führer's idea. If Hitler had created the world, maybe he would have created a world in which the laws of natural selection did not apply, and Jews and Aryans could all live together in perfect harmony. But Hitler didn't create the world. He just managed to decipher the laws of nature, and then instructed us how to live in line with them. If we disobey these laws, we will come to a bad end. Ist das klar?!'

In 2016 a British boy comes to his father, a liberal MP, and asks, 'Dad, why should we care about the human rights of Muslims in the Middle East?' The father puts down his cup of tea, thinks for a moment, and says, 'Well, Duncan, that's how the world works. You are still young and don't yet understand, but all humans, even Muslims in the Middle East, have the same nature and therefore enjoy the same natural rights. This isn't my idea, nor a decision of Parliament. If Parliament had created the world, universal human rights might well have been buried in some subcommittee along with all that quantum physics stuff. But Parliament didn't create the world, it just tries to make sense of it, and we must respect the natural rights even of Muslims in the Middle East, or very soon our own rights will also be violated, and we will come to a bad end. Now off you go.'

Liberals, communists and followers of other modern creeds dislike describing their own system as a 'religion', because they identify religion with superstitions and supernatural powers. If you tell communists or liberals that they are religious, they think you are accusing them of blindly believing in groundless pipe dreams. In fact, it means only that they believe in some system of moral laws that wasn't invented by humans, but that humans must nevertheless obey. As far as we know, all human societies believe in this. Every society tells its members that they must obey some superhuman moral law, and that breaking this law will result in catastrophe.

Religions differ of course in the details of their stories, their concrete commandments, and the rewards and punishments they promise. Thus in medieval Europe the Catholic Church argued that God doesn't like rich people. Jesus said that it is easier for a camel to pass through the eye of a needle than for a rich man to pass through the gates of heaven. To help rich people enter God's kingdom, the Church encouraged them to give lots of alms, threatening that misers will burn in hell. Modern communism also dislikes rich people, but it threatens them with class conflict here and now, rather than with burning sulphur after death.

The communist laws of history are similar to the commandments of the Christian God, inasmuch as they are superhuman forces that humans cannot change at will. Humans can decide tomorrow morning to cancel the offside rule in football, because we invented that law and are free to change it. However, at least according to Marx, we cannot change the laws of history. No matter what the capitalists do, as long as they continue to accumulate private property they are bound to create class conflict and are destined to be defeated by the rising proletariat.

If you happen to be a communist yourself, you might argue that communism and Christianity are nevertheless very different because communism is right, whereas Christianity is wrong. Class conflict really is inherent in the capitalist system, but the rich don't in fact suffer eternal tortures in hell after they die. Yet even if that's the case, it doesn't mean communism is not a religion. Rather, it means that communism is the one true religion. Followers of every religion are convinced that theirs alone is true. Perhaps the followers of one religion are correct.

If You Meet the Buddha

The assertion that religion is a tool for preserving social order and for organising large-scale cooperation may vex those for whom it represents first and foremost a spiritual path. However, just as the gap between religion and science is narrower than we commonly think, so the gap between religion and spirituality is much wider. Religion is a deal, whereas spirituality is a journey.

Religion gives a complete description of the world, and offers us a well-defined contract with predetermined goals. 'God exists. He told us to behave in certain ways. If you obey God, you'll be admitted to heaven. If you disobey Him, you'll burn in hell.' The very clarity of this deal allows society to define common norms and values that regulate human behaviour.

Spiritual journeys are nothing like that. They usually take people in mysterious ways towards unknown destinations. The quest usually begins with some big question, such as who am I? What is the meaning of life? What is good? Whereas most people just accept the ready-made answers provided by the powers that be, spiritual seekers are not so easily satisfied. They are determined to follow the big question wherever it leads, and not just to places they know well or wish to visit. Thus for most people, academic studies are a deal rather than a spiritual journey, because they take us to a predetermined goal approved by our elders, governments and banks. 'I'll study for three years, pass the exams, get my BA certificate and secure a well-paid job.' Academic studies might be transformed into a spiritual journey if the big questions you encounter on the way deflect you towards unexpected destinations, of which you could hardly even conceive at first. For example, a student might begin to study economics in order to secure a job on Wall Street. However, if what she learns somehow induces her to end up in a Hindu ashram or helping HIV patients in Zimbabwe, then we could call that a spiritual journey.

Why label such a voyage 'spiritual'? This is a legacy from ancient dualist religions that believed in the existence of two gods, one good and one evil. According to dualism, the good god created pure and everlasting souls that lived in a blissful world of spirit. However, the evil god – sometimes named Satan – created another world, made of matter. Satan didn't know how to make his creation endure, hence in the world of matter everything rots and disintegrates.

In order to breathe life into his defective creation, Satan tempted souls from the pure world of spirit, and confined them inside material bodies. That's what a human is – a good spiritual soul trapped inside an evil material body. Since the soul's prison – the body – decays and eventually dies, Satan ceaselessly tempts the soul with bodily delights, and above all with food, sex and power. When the body disintegrates and the soul has the opportunity to escape back to the spiritual world, its craving for bodily pleasures lures it back inside some new material body. The soul thus transmigrates from body to body, wasting its days in pursuit of food, sex and power.

Dualism instructs people to break these material shackles and undertake a journey back to the spiritual world, which is totally unfamiliar to us, but is our true home. During this quest we must reject all material temptations and deals. Due to this dualist legacy, every journey on which we doubt the conventions and deals of the mundane world and venture forth towards an unknown destination is called a 'spiritual' journey.

Such journeys are fundamentally different from religions, because religions seek to cement the worldly order whereas spirituality seeks to escape it. Often enough, one of the most important obligations for spiritual wanderers is to challenge the beliefs and conventions of dominant religions. In Zen Buddhism it is said that 'If you meet the Buddha on the road, kill him.' Which means that if while walking on the spiritual path you encounter the rigid ideas and fixed laws of institutionalised Buddhism, you must free yourself from them too.

For religions, spirituality is a dangerous threat. Religions typically strive to rein in the spiritual quests of their followers, and many religious systems have been challenged not by laypeople preoccupied with food, sex and power, but rather by spiritual truth-seekers who expected more than platitudes. Thus the Protestant revolt against the authority of the Catholic Church was ignited not by hedonistic atheists but rather by a devout and ascetic monk, Martin Luther. Luther wanted answers to the existential questions of life, and refused to settle for the rites, rituals and deals offered by the Church.

In Luther's day, the Church promised its followers some very enticing deals indeed. If you sinned, and feared eternal damnation in the afterlife, all you needed to do was open your purse and buy an indulgence. In the early sixteenth century the Church employed professional 'salvation peddlers' who wandered the towns and villages of Europe and sold indulgences for fixed prices. You want an entry visa to heaven? Pay ten gold coins. You want your dead Grandpa Heinz and Grandma Gertrud to join you there? No problem, but it will cost you thirty coins. The most famous of these peddlers, the Dominican friar Johannes Tetzel, allegedly said that the moment the coin clinks in the money chest, the soul flies out of purgatory to heaven.[1]

25. The Pope selling indulgences for money (from a Protestant pamphlet).

The more Luther thought about it, the more he doubted this deal, and the Church that offered it. You cannot just buy your way to salvation. The Pope couldn't possibly have the authority to forgive people their sins, and open the gates of heaven. According to Protestant tradition, on 31 October 1517 Luther walked to the All Saints' Church in Wittenberg, carrying a lengthy document, a hammer and some nails. The document listed ninety-five theses against contemporary religious practices, including against the selling of indulgences. Luther nailed it to the church door, sparking the Protestant Reformation, which called upon any Christian who cared about salvation to rebel against the Pope's authority and search for alternative routes to heaven.

From a historical perspective, the spiritual journey is always tragic, for it is a lonely path fit only for individuals rather than for entire societies. Human cooperation requires firm answers rather than just questions, and those who fume against stultified religious structures often end up forging new structures in their place. It happened to the dualists, whose spiritual journeys became religious establishments. It happened to Martin Luther, who after challenging the laws, institutions and rituals of the Catholic Church found himself writing new law books, founding new institutions and inventing new ceremonies. It happened even to Buddha and Jesus. In their uncompromising quest for the truth they subverted the laws, rituals and structures of traditional Hinduism and Judaism. But eventually more laws, more rituals and more structures were created in their names than in the name of any other person in history.

Counterfeiting God

Now that we have a better understanding of religion, we can go back to examining the relationship between religion and science. There are two extreme interpretations for this relationship. One view says that science and religion are sworn enemies, and that modern history was shaped by the life-and-death struggle of scientific knowledge against religious superstition. In time, the light of science dispelled the darkness of religion, and the world became increasingly secular, rational and prosperous. However, though some scientific findings certainly undermine religious dogmas, this is not inevitable. For example, Muslim dogma holds that Islam was founded by the prophet Muhammad in seventh-century Arabia, and there is ample scientific evidence supporting this.

More importantly, science always needs religious assistance in order to create viable human institutions. Scientists study how the world functions, but there is no scientific method for determining how humans ought to behave. Science tells us that humans cannot survive without oxygen. However, is it okay to execute criminals by asphyxiation? Science doesn't know how to answer such a question. Only religions provide us with the necessary guidance.

Hence every practical project scientists undertake also relies on religious insights. Take, for instance, the construction of the Three Gorges Dam on the Yangtze River. When the Chinese government decided in 1992 to build the dam, physicists could calculate what pressures the dam would have to withstand, economists could forecast how much money it would probably cost, while electrical engineers could predict how much electricity it would produce. However the government needed to take additional factors into account. Building the dam flooded more than 200 square miles containing many villages and towns, thousands of archaeological sites, and unique landscapes and habitats. More than 1 million people were displaced and hundreds of species were endangered. It seems that the dam directly caused the extinction of the Chinese river dolphin. No matter what you personally think about the Three Gorges Dam, it is clear that its construction was an ethical rather than a purely scientific issue. No physics experiment, no economic model and no mathematical equation can determine whether generating thousands of megawatts and making billions of yuan is more valuable than saving an ancient pagoda or the Chinese river dolphin. Consequently China cannot function on the basis of scientific theories alone. It requires some religion or ideology, too.

Some jump to the opposite extreme, and say that science and religion are completely separate kingdoms. Science studies facts, religion speaks about values, and never the twain shall meet. Religion has nothing to say about scientific facts, and science should keep its mouth shut concerning religious convictions. If the Pope believes that human life is sacred, and abortion is therefore a sin, biologists can neither prove nor refute this claim. As a private individual, each biologist is welcome to argue with the Pope. But as a scientist, the biologist cannot enter the fray.

This approach may sound sensible, but it misunderstands religion. Though science indeed deals only with facts, religion never confines itself to ethical judgements. Religion cannot provide us with any practical guidance unless it makes some factual claims too, and here it may well collide with science. The most important segments of many religious dogmas are not their ethical principles, but rather factual statements such as 'God exists', 'the soul is punished for its sins in the afterlife', 'the Bible was written by a deity rather than by humans', 'the Pope is never wrong'. These are all factual claims. Many of the most heated religious debates, and many of the conflicts between science and religion, involve such factual claims rather than ethical judgements.

Take abortion, for example. Devout Christians often oppose abortion, whereas many liberals support it. The main bone of contention is factual rather than ethical. Both Christians and liberals believe that human life is sacred, and that murder is a heinous crime. But they disagree about certain biological facts: does human life begin at the moment of conception, at the moment of birth or at some intermediate point? Indeed, some human cultures maintain that life doesn't begin even at birth. According to the !Kung of the Kalahari Desert and to various Inuit groups in the Arctic, human life begins only after a baby is given a name. When an infant is born the family waits for some time before naming it. If they decide not to keep the baby (either because it suffers from some deformity or because of economic difficulties), they kill it. Provided they do so before the naming ceremony, this is not considered murder.[2] People from such cultures might well agree with liberals and Christians that human life is sacred and that murder is a terrible crime, yet condone infanticide.

When religions advertise themselves, they tend to emphasise their beautiful values. But God often hides in the fine print of factual statements. The Catholic religion markets itself as the religion of universal love and compassion. How wonderful! Who can object to that? Why, then, are not all humans Catholic? Because when you read the fine print, you discover that Catholicism also demands blind obedience to a pope 'who never makes mistakes' even when he orders his followers to go on crusades and burn heretics at the stake. Such practical instructions are not deduced solely from ethical judgements. Rather, they result from conflating ethical judgements with factual statements.

When we descend from the ethereal sphere of philosophy and observe historical realities, we find that religious stories almost always include three parts:

1. Ethical judgements, such as 'human life is sacred'.
2. Factual statements, such as 'human life begins at the moment of conception'.
3. A conflation of the ethical judgements with the factual statements, resulting in practical guidelines such as 'you should never allow abortion, even a single day after conception'.

Science has no ability to refute or corroborate the ethical judgements religions make. But scientists do have a lot to say about religious factual statements. Biologists are more qualified than priests to answer factual questions such as 'Do human fetuses have a nervous system one week after conception? Can they feel pain?'

To make things clearer let us examine in depth a real historical example that you rarely hear about in religious commercials, but that had a huge social and political impact in its time. In medieval Europe the popes enjoyed far-reaching political authority. Whenever a conflict erupted anywhere in Europe, they claimed the authority to decide the issue. To establish their claim to authority, they repeatedly reminded Europeans of the Donation of Constantine. According to this story, on 30 March 315 the Roman emperor Constantine signed an official decree granting Pope Sylvester I and his heirs perpetual control of the western part of the Roman Empire. The popes kept this precious document in their archive, and used it as a powerful propaganda tool whenever they faced opposition from ambitious princes, quarrelsome cities or rebellious peasants.

People in medieval Europe had great respect for ancient imperial decrees, and believed that the older the document, the more authority it carried. They also strongly believed that kings and emperors were God's representatives. Constantine in particular was revered, because he turned the Roman Empire from a pagan realm into a Christian empire. In a clash between the desires of some present-day city council and a decree issued by the great Constantine himself, it was obvious to medieval Europeans that the ancient document ought to be obeyed. Hence whenever the Pope faced political opposition, he waved the Donation of Constantine, demanding obedience. Not that it always worked. But the Donation of Constantine was an important cornerstone of papal propaganda and of the medieval political order.

When we examine the Donation of Constantine closely, we find that this story is composed of three distinct parts:

Ethical judgement	Factual statement	Practical guideline
People ought to respect ancient imperial decrees more than present-day popular opinions.	On 30 March 315, Emperor Constantine granted the popes dominion over Europe.	Europeans in 1315 ought to obey the Pope's commands.

The ethical authority of ancient imperial decrees is far from self-evident. Most twenty-first-century Europeans think that the wishes of present-day citizens trump the diktats of long-dead monarchs. However, science cannot join this ethical debate, because no experiment or equation can decide the matter. If a modern-day scientist time-travelled 700 years back in time, she couldn't prove to medieval Europeans that the decrees of ancient emperors are irrelevant to contemporary political disputes.

Yet the story of Constantine's Donation was based not just on ethical judgements. It also involved some very concrete factual statements, which science is highly qualified to either verify or falsify. In 1441 Lorenzo Valla – a Catholic priest and a pioneer linguist – published a scientific study proving that Constantine's Donation was a forgery. Valla analysed the style and grammar of the document, and the various words and terms it contained. He demonstrated that the document included words that were unknown in fourth-century Latin, and that it was most probably forged about 400 years after Constantine's death. Moreover, the date appearing on the document is '30 March, in the year Constantine was consul for the fourth time, and Gallicanus was consul for the first time'. In the Roman Empire, two consuls were elected each year, and it was customary to date documents by their consulate years. Unfortunately, Constantine's fourth consulate was in 315, whereas Gallicanus was elected consul for the first time only in 317. If this all-important document was indeed composed in Constantine's days, it would never have contained such a blatant error. It is as if Thomas Jefferson and his colleagues had dated the American Declaration of Independence 34 July 1776.

Today all historians agree that the Donation of Constantine was forged in the papal court sometime in the eighth century. Even though Valla never disputed the moral authority of ancient imperial decrees, his scientific analysis did undermine the practical guideline that Europeans ought to obey the Pope.[3]

On 20 December 2013 the Ugandan parliament passed the Anti-Homosexuality Act, which criminalised homosexual activities, penalising some activities by life imprisonment. It was inspired and supported by evangelical Christian groups, which maintain that God prohibits homosexuality. As proof they quote Leviticus 18:22 ('Do not have sexual relations with a man as one does with a woman; that is detestable') and Leviticus 20:13 ('If a man has sexual relations with a man as one does with a woman, both of them have done what is detestable. They are to be put to death; their blood will be on their own heads'). In previous centuries the same religious story was responsible for tormenting millions of people all over the world. This story can be briefly summarised as follows:

Ethical judgement	Factual statement	Practical guideline
Humans ought to obey God's commands.	About 3,000 years ago God commanded humans to avoid homosexual activities.	People should avoid homosexual activities.

Is the story true? Scientists cannot argue with the judgement that humans ought to obey God. Personally, you may dispute it. You may believe that human rights trump divine authority, and if God orders us to violate human rights, we shouldn't listen to Him. Yet there is no scientific experiment that can decide this issue.

In contrast, science has a lot to say about the factual statement that 3,000 years ago the Creator of the Universe commanded members of the *Homo sapiens* species to abstain from boy-on-boy action. How do we know this statement is true? Examining the relevant literature reveals that though this statement is repeated in millions of books, articles and Internet sites, they all rely on a single source: the Bible. If so, a scientist would ask, who composed the Bible, and when? Note that this is a factual question, not a question of values. Devout Jews and Christians claim that at least the book of Leviticus was dictated by God to Moses on Mount Sinai, and from that moment onwards not a single letter was either added or deleted from it. 'But,' the scientist would insist, 'how can we be sure of that? After all, the Pope argued that the Donation of Constantine was composed by Constantine himself in the fourth century, when in fact it was forged 400 years later by the Pope's own clerks.'

We can now use an entire arsenal of scientific methods to determine who composed the Bible, and when. Scientists have been doing exactly that for more than a century, and if you are interested, you can read whole books about their findings. To cut a long story short, most peer-reviewed scientific studies agree that the Bible is a collection of numerous different texts composed by different human authors centuries after the events they purport to describe, and that these texts were not assembled into a single holy book until long after biblical times. For example, whereas King David probably lived around 1000 BC, it is commonly accepted that the book of Deuteronomy was composed in the court of King Josiah of Judah, sometime around 620 BC, as part of a propaganda campaign aimed at strengthening Josiah's authority. Leviticus was compiled at an even later date, no earlier than 500 BC.

As for the idea that the ancient Jews carefully preserved the biblical text, without adding or subtracting anything, scientists point out that biblical Judaism was not a scripture-based religion at all. Rather, it was a typical Iron Age cult, similar to many of its Middle Eastern neighbours. It had no synagogues, yeshivas, rabbis – or even a bible. Instead, it had elaborate temple rituals, most of which involved sacrificing animals to a jealous sky god so that he would bless his people with seasonal rains and military victories. Its religious elite consisted of priestly families, who owed everything to birth and nothing to intellectual prowess. The mostly illiterate priests were busy with the temple ceremonies and had little time for writing or studying any scriptures.

During the Second Temple period a rival religious elite gradually formed. Due partly to Persian and Greek influences, Jewish scholars who wrote and interpreted texts gained increasing prominence. These scholars eventually came to be known as rabbis, and the texts they compiled were christened 'the Bible'. Rabbinical authority rested on individual intellectual abilities rather than on birth. The clash between this new literate elite and the old priestly families was inevitable. Fortunately for the rabbis, the Romans torched Jerusalem and its temple in 70 AD while suppressing the Great Jewish Revolt. With the temple in ruins, the priestly families lost their religious authority, their economic power base and their very *raison d'être*. Traditional Judaism – a Judaism of temples, priests and head-splitting warriors – disappeared. In its place emerged a new Judaism of books, rabbis and hair-splitting scholars. The scholars' main forte was interpretation. They used this ability not only to explain how an almighty God allowed His temple to be destroyed, but also to bridge the immense gaps between the old Judaism described in biblical stories and the very different Judaism they created.[4]

Hence according to our best scientific knowledge, the Leviticus injunctions against homosexuality reflect nothing grander than the biases of a few priests and scholars in ancient Jerusalem. Though science cannot decide whether people ought to obey God's commands, it has many relevant things to say about the provenance of the Bible. If Ugandan politicians think that the power that created the cosmos, the galaxies and the black holes becomes terribly upset whenever two *Homo sapiens* males have a bit of fun together, then science can help disabuse them of this rather bizarre notion.

Holy Dogma

In truth, it is not always easy to separate ethical judgements from factual statements. Religions have the nagging tendency to turn factual statements into ethical judgements, thereby creating serious confusion and obfuscating what should have been relatively simple debates. Thus the factual statement 'God wrote the Bible' all too often mutates into the ethical injunction 'you ought to believe that God wrote the Bible'. Merely believing in this factual statement becomes a virtue, whereas doubting it becomes a dreadful sin.

Conversely, ethical judgements often hide within them factual statements that proponents don't bother to mention, because they think they have been proven beyond doubt. Thus the ethical judgement 'human life is sacred' (which science cannot test) may shroud the factual statement 'every human has an eternal soul' (which is open to scientific debate). Similarly, when American nationalists proclaim that 'the American nation is sacred', this seemingly ethical judgement is in fact predicated on factual statements such as 'the USA has spearheaded most of the moral, scientific and economic advances of the last few centuries'. Whereas it is impossible to scientifically scrutinise the claim that the American nation is sacred, once we unpack this judgement we may well examine scientifically whether the USA has indeed been responsible for a disproportionate share of moral, scientific and economic breakthroughs.

This has led some philosophers, such as Sam Harris, to argue that science can always resolve ethical dilemmas, because human values *always* conceal within them some factual statements. Harris thinks that all humans share a single supreme value – minimising suffering and maximising happiness – and therefore all ethical debates are factual arguments concerning the most efficient way to maximise happiness.[5] Islamic fundamentalists want to reach heaven in order to be happy, liberals believe that increasing human liberty maximises happiness, and German nationalists think that everyone would be better off if Berlin is allowed to run the planet. According to Harris, Islamists, liberals and nationalists have no ethical dispute; they have a factual disagreement about how best to realise their common goal.

Yet even if Harris is right, and even if all humans cherish happiness, in practice it would be extremely difficult to use this insight to decide ethical disputes, particularly because we have no scientific definition or measurement of happiness. Consider again the case of the Three Gorges Dam. Even if we agree that the ultimate aim of the project is to make the world a happier place, how can we tell whether generating cheap electricity contributes more to global happiness than protecting traditional lifestyles or saving the rare Chinese river dolphin? As long as we haven't deciphered the mysteries of consciousness, we cannot develop a universal measurement for happiness and suffering, and we don't know how to compare the happiness and suffering of different individuals, let alone different species. How many units of happiness are generated when a billion Chinese enjoy cheaper electricity? How many units of misery are produced when an entire dolphin species becomes extinct? Indeed, are happiness and misery mathematical entities that can be added or subtracted in the first place? Eating ice cream is enjoyable; finding true love is more enjoyable. Do you think that if you just eat enough ice cream, the accumulated pleasure could ever equal the rapture of true love?

Consequently, although science has much more to contribute to ethical debates than we commonly think, there is a line it cannot cross, at least not yet. Without the guiding hand of some religion, it is impossible to maintain large-scale social orders. Even universities and laboratories need religious backing. Religion provides the ethical justification for scientific research, and in exchange gets to influence the scientific agenda and the uses of scientific discoveries. Hence you cannot understand the history of science without taking religious beliefs into account. Scientists seldom dwell on this fact, but the Scientific Revolution itself began in one of the most dogmatic, intolerant and religious societies in history.

The Witch Hunt

We often associate science with the values of secularism and tolerance. If so, early modern Europe is the last place you would have expected a scientific revolution. Europe in the days of Columbus, Copernicus and Newton had the highest concentration of religious fanatics in the world, and the lowest level of tolerance. The luminaries of the Scientific Revolution lived in a society that expelled Jews and Muslims, burned heretics wholesale, saw a witch in every cat-loving elderly lady and started a new religious war every full moon.

If you had travelled to Cairo or Istanbul around 1600, you would find there a multicultural and tolerant metropolis, where Sunnis, Shiites, Orthodox Christians, Catholics, Armenians, Copts, Jews and even the occasional Hindu lived side by side in relative harmony. Though they had their share of disagreements and riots, and though the Ottoman Empire routinely discriminated against people on religious grounds, it was a liberal paradise compared with Europe.

If you had then sailed on to contemporary Paris or London, you would have found cities awash with religious extremism, in which only those belonging to the dominant sect could live. In London they killed Catholics, in Paris they killed Protestants, the Jews had long been driven out, and nobody in his right mind would dream of letting any Muslims in. And yet, the Scientific Revolution began in London and Paris rather than in Cairo and Istanbul.

It is customary to portray the history of modernity as a struggle between science and religion. In theory, both science and religion are interested above all in the truth, and because each upholds a different truth, they are doomed to clash. In fact, neither science nor religion cares that much about the truth, hence they can easily compromise, coexist and even cooperate.

Religion is interested above all in order. It aims to create and maintain the social structure. Science is interested above all in power. Through research, it aims to acquire the power to cure diseases, fight wars and produce food. As individuals, scientists and priests may give immense importance to the truth; but as collective institutions, science and religion prefer order and power over truth. They therefore make good bedfellows. The uncompromising quest for truth is a spiritual journey, which can seldom remain within the confines of either religious or scientific establishments.

It would accordingly be far more accurate to view modern history as the process of formulating a deal between science and one particular religion – namely, humanism. Modern society believes in humanist dogmas, and uses science not in order to question these dogmas, but rather in order to implement them. In the twenty-first century the humanist dogmas are unlikely to be replaced by pure scientific theories. However, the covenant linking science and humanism may well crumble and give way to a very different kind of deal, between science and some new post-humanist religion. We will dedicate the next two chapters to understanding the modern covenant between science and humanism. The third and final part of the book will then explain why this covenant is disintegrating, and what new deal might replace it.

6

The Modern Covenant

Modernity is a deal. All of us sign up to this deal on the day we are born, and it regulates our lives until the day we die. Very few of us can ever rescind or transcend this deal. It shapes our food, our jobs and our dreams, and it decides where we dwell, whom we love and how we pass away.

At first sight modernity looks like an extremely complicated deal, hence few try to understand what they have signed up for. Like when you download some software and are asked to sign an accompanying contract that consists of dozens of pages of legalese—you take one look at it, immediately scroll down to the last page, tick 'I agree' and forget about it. Yet in fact modernity is a surprisingly simple deal. The entire contract can be summarised in a single phrase: humans agree to give up meaning in exchange for power.

Until modern times most cultures believed that humans played a part in some great cosmic plan. The plan was devised by the omnipotent gods, or by the eternal laws of nature, and humankind could not change it. The cosmic plan gave meaning to human life, but also restricted human power. Humans were much like actors on a stage. The script gave meaning to their every word, tear and gesture – but placed strict limits on their performance. Hamlet cannot murder Claudius in Act I, or leave Denmark and go to an ashram in India. Shakespeare won't allow it. Similarly, humans cannot live for ever, they cannot escape all diseases, and they cannot do as they please. It's not in the script.

In exchange for giving up power, premodern humans believed that their lives gained meaning. It really mattered whether they fought bravely on the battlefield, whether they supported the lawful king, whether they ate forbidden foods for breakfast or whether they had an affair with the next-door neighbour. This of course created some inconveniences, but it gave humans psychological protection against disasters. If something terrible happened – such as war, plague or drought – people consoled themselves that 'We all play a role in some great cosmic drama devised by the gods, or by the laws of nature. We are not privy to the script, but we can rest assured that everything happens for a purpose. Even this terrible war, plague and drought have their place in the greater scheme of things. Furthermore, we can count on the playwright that the story surely has a good and meaningful ending. So even the war, plague and drought will work out for the best – if not here and now, then in the afterlife.'

Modern culture rejects this belief in a great cosmic plan. We are not actors in any larger-than-life drama. Life has no script, no playwright, no director, no producer – and no meaning. To the best of our scientific understanding, the universe is a blind and purposeless process, full of sound and fury but signifying nothing. During our infinitesimally brief stay on our tiny speck of a planet, we fret and strut this way and that, and then are heard of no more.

Since there is no script, and since humans fulfil no role in any great drama, terrible things might befall us and no power will come to save us or give meaning to our suffering. There won't be a happy ending, or a bad ending, or any ending at all. Things just happen, one after the other. The modern world does not believe in purpose, only in cause. If modernity has a motto, it is 'shit happens'.

On the other hand, if shit just happens, without any binding script or purpose, then humans too are not confined to any predetermined role. We can do anything we want – provided we can find a way. We are constrained by nothing except our own ignorance. Plagues and droughts have no cosmic meaning – but we can eradicate them. Wars are not a necessary evil on the way to a better future – but we can make peace. No paradise awaits us after death – but we can create paradise here on earth and live in it for ever, if we just manage to overcome some technical difficulties.

If we invest money in research, then scientific breakthroughs will accelerate technological progress. New technologies will fuel economic growth, and a growing economy will dedicate even more money to research. With each passing decade we will enjoy more food, faster vehicles and better medicines. One day our knowledge will be so vast and our technology so advanced that we shall distil the elixir of eternal youth, the elixir of true happiness, and any other drug we might possibly desire – and no god will stop us.

The modern deal thus offers humans an enormous temptation, coupled with a colossal threat. Omnipotence is in front of us, almost within our reach, but below us yawns the abyss of complete nothingness. On the practical level modern life consists of a constant pursuit of power within a universe devoid of meaning. Modern culture is the most powerful in history, and it is ceaselessly researching, inventing, discovering and growing. At the same time, it is plagued by more existential angst than any previous culture.

This chapter discusses the modern pursuit of power. The next chapter will examine how humankind has used its growing power to somehow sneak meaning back into the infinite emptiness of the cosmos. Yes, we moderns have promised to renounce meaning in exchange for power; but there's nobody out there to hold us to our promise. We think we are smart enough to enjoy the full benefits of the modern deal, without having to pay its price.

Why Bankers are Different from Vampires

The modern pursuit of power is fuelled by the alliance between scientific progress and economic growth. For most of history science progressed at a snail's pace, while the economy was in deep freeze.

The gradual increase in human population did lead to a corresponding increase in production, and sporadic discoveries sometimes resulted even in per capita growth, but this was a very slow process.

If in AD 1000 a hundred villagers produced a hundred tons of wheat, and in AD 1100, 105 villagers produced 107 tons of wheat, this nominal growth didn't change the rhythms of life nor the socio-political order. Whereas today everyone is obsessed with growth, in the premodern era people were oblivious to it. Princes, priests and peasants assumed that human production was more or less stable, that one person could enrich himself only by pilfering from another and that their grandchildren were unlikely to enjoy a better standard of living.

This stagnation resulted to a large extent from the difficulties involved in financing new projects. Without adequate funding, it wasn't easy to drain swamps, construct bridges and build ports – not to mention engineer new wheat strains, discover new energy sources or open new trade routes. Funds were scarce because there was little credit in those days; there was little credit because people had no belief in growth; and people didn't believe in growth because the economy was stagnant. Stagnation thereby perpetuated itself.

Suppose you live in a medieval town that suffers from annual outbreaks of dysentery. You resolve to find a cure. You need funding to set up a workshop, buy medicinal herbs and exotic chemicals, pay assistants and travel to consult with famous doctors. You also need money to feed yourself and your family while you are busy with your research. But you don't have much money. You can approach the local miller, baker and blacksmith and ask them to meet all your needs for a few years, promising that when you finally discover the cure and become wealthy, you will pay your debts.

Unfortunately, the miller, baker and blacksmith are unlikely to consent. They need to feed their families today, and they have no faith in miracle medicines. They weren't born yesterday, and in all their years they have never heard of anyone finding a new medicine for some dreaded disease. If you want provisions – you have to pay cash. But how can you get the money when you haven't discovered the medicine yet, and all your time is taken up with research? Reluctantly, you go back to tilling your field, dysentery keeps tormenting the townsfolk, nobody tries to develop new remedies, and not a single gold coin changes hands. That's how the economy languished and science stood still.

The cycle was eventually broken in the modern age thanks to people's growing trust in the future, and the resulting miracle of credit. Credit is the economic manifestation of trust. Nowadays, if I want to develop a new drug but I don't have enough money, I can get a loan from a bank, or turn to private investors and venture capital funds. When Ebola erupted in West Africa in the summer of 2014, what do you think happened to the shares of pharmaceutical companies that were busy developing anti-Ebola drugs and vaccines? They skyrocketed. Tekmira shares rose by 50 per cent and BioCryst shares by 90 per cent. In the Middle Ages, the outbreak of a plague caused people to raise their eyes towards heaven, and pray to God to forgive them for their sins. Today when people hear of some deadly new epidemic, they reach for their mobile phones and call their brokers. For the stock exchange, even an epidemic is a business opportunity.

If enough new ventures succeed, people's trust in the future increases, credit expands, interest rates fall, entrepreneurs can raise money more easily and the economy grows. People consequently have even greater trust in the future, the economy keeps growing and science progresses with it.

It sounds simple on paper. Why then did humankind have to wait until the modern era for economic growth to gather momentum? For thousands of years people had little faith in future growth not because they were stupid, but because it contradicts our gut feelings, our evolutionary heritage and the way the world works. Most natural systems exist in equilibrium, and most survival struggles are a zero-sum game in which one can prosper only at the expense of another.

For example, each year roughly the same amount of grass grows in a given valley. The grass supports a population of 10,000 rabbits or so, which contains enough slow, dim-witted or unlucky rabbits to provide prey for a hundred foxes. If one fox is particularly clever and diligent, and devours more rabbits than average, then other foxes will likely starve. If all foxes somehow manage to catch more rabbits simultaneously, the rabbit population will crash, and next year even more foxes will starve. Even though there are occasional fluctuations in the rabbit market, in the long run the foxes cannot expect to hunt, say, 3 per cent more rabbits per year than the preceding year.

Of course, some ecological realities are more complex, and not all survival struggles are zero-sum games. Many animals cooperate effectively, and a few even give loans. The most famous lenders in nature are vampire bats. These bats congregate in the thousands inside caves, and every night fly out to look for prey. When they find a sleeping bird or careless mammal, they make a small incision in its skin, and suck its blood. But not all vampire bats find a victim every night. In order to cope with the uncertainty of their life, the vampires loan blood to each other. A vampire that fails to find prey will come home and ask a more fortunate friend to regurgitate some stolen blood. Vampires remember very well to whom they loaned blood, so at a later date if the friend returns home hungry, he will approach his debtor, who will reciprocate the favour.

However, unlike human bankers, vampires never charge interest. If vampire A loaned vampire B ten centilitres of blood, B will repay the same amount. Nor do vampires use loans in order to finance new businesses or encourage growth in the blood-sucking market. Because the blood is produced by other animals, the vampires have no way of increasing production. Though the blood market has its ups and downs, vampires cannot presume that in 2017 there will be 3 per cent more blood than in 2016, and that in 2018 the blood market will again grow by 3 per cent. Consequently, vampires don't believe in growth.[1] For millions of years of evolution humans lived under conditions similar to those of vampires, foxes and rabbits. Hence humans too find it difficult to believe in growth.

The Miracle Pie

Evolutionary pressures have accustomed humans to see the world as a static pie. If somebody gets a larger slice of the pie, somebody else inevitably gets a smaller slice. A particular family or city may prosper, but humankind as a whole is not going to produce more than it produces today. Accordingly, traditional religions such as Christianity and Islam sought ways to solve humanity's problems with the help of current resources, either by redistributing the existing pie, or by promising a pie in the sky.

Modernity, in contrast, is based on the firm belief that economic growth is not only possible, but absolutely essential. Prayers, good deeds and meditation might be comforting and inspiring, but problems such as famine, plague and war can only be solved through growth. This fundamental dogma can be summarised in one simple idea: 'If you have a problem, you probably need more stuff, and in order to have more stuff, you must produce more of it.'

Modern politicians and economists insist that growth is vital for three principal reasons. Firstly, when we produce more, we can consume more, raise our standard of living and allegedly enjoy a happier life. Secondly, as long as humankind multiplies, economic growth is needed merely to stay where we are. For example, in India the annual population growth rate is 1.2 per cent. That means that unless the Indian economy expands each year by at least 1.2 per cent, unemployment will rise, salaries will fall and the average standard of living will decline. Thirdly, even if Indians stop multiplying, and even if the Indian middle class can be satisfied with its current standard of living, what should India do about its hundreds of millions of poverty-stricken citizens? If the economy doesn't grow, and the pie therefore remains the same size, you can give more to the poor only by taking something from the rich. That will force you to make some very hard choices, and will probably cause a lot of resentment and even violence. If you wish to avoid hard choices, resentment and violence, you need a bigger pie.

Modernity has turned 'more stuff' into a panacea applicable to almost all public and private problems, from religious fundamentalism through Third World authoritarianism down to a failed marriage. If only countries such as Pakistan and Egypt could maintain a healthy growth rate, their citizens would come to enjoy the benefits of private cars and bulging refrigerators, and would take the path of earthly prosperity instead of following the fundamentalist pied piper. Similarly, economic growth in countries such as Congo and Myanmar would produce a prosperous middle class which is the bedrock of liberal democracy. And in the case of the disgruntled couple, their marriage would allegedly be saved if only they would buy a bigger house (so they don't have to share a cramped office), purchase a dishwasher (so they stop arguing whose turn it is to do the dishes) and attend expensive therapy sessions twice a week.

Economic growth has thus become the crucial juncture where almost all modern religions, ideologies and movements meet. The Soviet Union, with its megalomaniacal Five Year Plans, was as obsessed with growth as the most cut-throat American robber baron. Just as Christians and Muslims all believe in heaven, and disagree only about how to get there, so during the Cold War both capitalists and communists believed in creating heaven on earth through economic growth, and wrangled only about the exact method.

Today Hindu revivalists, pious Muslims, Japanese nationalists and Chinese communists may declare their adherence to very different values and goals, but they have all come to believe that economic growth is the key to realising their disparate goals. Thus in 2014 the devout Hindu Narendra Modi was elected prime minister of India thanks largely to his success in boosting economic growth in his home state of Gujarat, and to the widely held view that only he could reinvigorate the sluggish national economy.

Analogous views have kept the Islamist Recep Tayyip Erdoğan in power in Turkey since 2003. The name of his party – the Justice and Development Party – highlights its commitment to economic development, and the Erdoğan government has indeed managed to maintain impressive growth rates for more than a decade.

Japan's prime minister, the nationalist Shinzō Abe, came to office in 2012 pledging to jolt the Japanese economy out of two decades of stagnation. His aggressive and somewhat unusual measures to achieve this have been nicknamed Abenomics. Meanwhile in neighbouring China the Communist Party still pays lip service to traditional Marxist–Leninist ideals, but in practice is guided by Deng Xiaoping's famous maxims that 'development is the only hard truth' and that 'it doesn't matter if a cat is black or white, so long as it catches mice'. Which means, in plain language: do whatever it takes to promote economic growth, even if Marx and Lenin wouldn't have been happy with it.

In Singapore, as befits that no-nonsense city-state, they pursue this line of thinking even further, and peg ministerial salaries to the national GDP. When the Singaporean economy grows, government ministers get a raise, as if that is what their jobs are all about.[2]

This obsession with growth might appear self-evident, but only because we live in the modern world. It wasn't like this in the past. Indian maharajas, Ottoman sultans, Kamakura shoguns and Han emperors rarely staked their political fortunes on ensuring economic growth. That Modi, Erdoğan, Abe and Chinese president Xi Jinping all bet their careers on economic growth testifies to the almost religious status growth has managed to acquire throughout the world. Indeed, it may not be wrong to call the belief in economic growth a religion, because it now purports to solve many, if not most, of our ethical dilemmas. Since economic growth is allegedly the source of all good things, it encourages people to bury their ethical disagreements and adopt whichever course of action maximises long-term growth. Thus Modi's India is home to thousands of sects, parties, movements and gurus, yet though their ultimate aims may differ, they all have to pass through the same bottleneck of economic growth, so why not pull together in the meantime?

The credo of 'more stuff' accordingly urges individuals, firms and governments to disregard anything that might hamper economic growth, such as preserving social equality, ensuring ecological harmony or honouring one's parents. In the Soviet Union, the leadership thought that state-controlled communism was the fastest way to grow, hence anything that stood in the way of collectivisation was bulldozed, including millions of kulaks, the freedom of expression and the Aral Sea. Nowadays it is generally accepted that some version of free-market capitalism is a much more efficient way of ensuring long-term growth, hence greedy tycoons, rich farmers and freedom of expression are protected, but ecological habitats, social structures and traditional values that stand in the way of free-market capitalism are dismantled and destroyed.

Take, for example, a software engineer earning $100 per hour working for some hi-tech start-up. One day her elderly father has a stroke. He now needs help with shopping, cooking and even bathing. She could move her father to her own house, leave for work later in the morning, come back earlier in the evening and take care of her father personally. Both her income and the start-up's productivity would suffer, but her father would enjoy the care of a respectful and loving daughter. Alternatively, the engineer could hire a Mexican carer who, for $12 per hour, would live with the father and provide for all his needs. That would mean business as usual for the engineer and her start-up, and even the carer and the Mexican economy would benefit. What should the engineer do?

Free-market capitalism has a firm answer. If economic growth demands that we loosen family bonds, encourage people to live away from their parents, and import carers from the other side of the world – so be it. This answer, however, involves an ethical judgement rather than a factual statement. When some people specialise in software engineering while others devote their time to care of the elderly, we can no doubt produce more software and give old people more professional care. Yet is economic growth more important than family bonds? By presuming to make such ethical judgements, free-market capitalism has crossed the border from the land of science into that of religion.

Most capitalists would probably dislike the label of religion, but as religions go, capitalism can at least hold its head high. Unlike other religions that promise us pie in the sky, capitalism promises miracles here on earth – and sometimes even delivers. Much of the credit for overcoming famine and plague belongs to the ardent capitalist faith in growth. Capitalism even deserves some kudos for reducing human violence and increasing tolerance and cooperation. As the next chapter explains, there are additional factors at play here, but capitalism did make an important contribution to global harmony by encouraging people to stop viewing the economy as a zero-sum game, in which your profit is my loss, and instead see it as a win–win situation, in which your profit is also my profit. This mutual-benefit approach has probably helped global harmony far more than centuries of Christian preaching about loving your neighbour and turning the other cheek.

From its belief in the supreme value of growth, capitalism deduces its number one commandment: thou shalt invest thy profits in increasing growth. For most of history princes and priests wasted their profits on flamboyant carnivals, sumptuous palaces and unnecessary wars. Alternatively, they placed their gold coins in iron chests, sealed them and buried them in a dungeon. Today, devout capitalists use their profits to hire new employees, enlarge the factory or develop a new product.

If they don't know how to do it themselves, they give their money to somebody who does, such as a banker or a venture capitalist. The latter lend the money to various entrepreneurs. Farmers take loans to plant new wheat fields, contractors build new houses, energy corporations explore new oil fields, and arms factories develop new weapons. The profits from all these activities enable the entrepreneurs to repay the loans with interest. We now have not only more wheat, houses, oil and weapons – but also more money, which the banks and funds can again lend. This wheel will never stop rotating, at least not according to capitalism. We will never reach a moment when capitalism says: 'That's it. Enough growth. We can now take it easy.' If you want to know why the capitalist wheel is unlikely ever to stop, talk for an hour with a friend who has just accumulated $100,000 and is wondering what to do with it.

'The banks offer such low interest rates,' he would complain. 'I don't want to put my money in a savings account that pays barely 0.5 per cent a year. Maybe I can make 2 per cent in government bonds. My cousin Richie bought a flat in Seattle last year and has already made 20 per cent on his investment! Maybe I should go into real estate too; but everybody is saying there's a new real-estate bubble. So, what do you think about the stock exchange? A friend told me the best deal these days is to buy an ETF that follows emerging economies, like Brazil or China.' As he pauses for a moment to breathe, you ask, 'Well, why not just be satisfied with your $100,000?' He will explain to you better than I can why capitalism will never stop.

This lesson is hammered home even to children and teenagers through ubiquitous capitalist games. Premodern games such as chess assumed a stagnant economy. You begin a game of chess with sixteen pieces, and you never finish a game with more. In rare cases a pawn may be transformed into a queen, but you cannot produce new pawns nor upgrade your knights into tanks. So chess players never have to think about investment. In contrast, many modern board games and computer games focus on investment and growth.

Particularly telling are civilisation-style strategy games, such as *Minecraft*, *The Settlers of Catan* or Sid Meier's *Civilization*. The game may be set in the Middle Ages, the Stone Age or some imaginary fairy land, but the principles always remain the same – and are always capitalist. Your aim is to establish a city, a kingdom or maybe an entire civilisation. You begin from a very modest base, perhaps just a village and its nearby fields. Your assets provide you with an initial income of wheat, wood, iron or gold. You then have to invest this income wisely. You have to choose between unproductive but still necessary tools such as soldiers, and productive assets such as more villages, fields and mines. The winning strategy is usually to invest the barest minimum in non-productive essentials, while maximising your productive assets. Establishing additional villages means that next turn you will have a larger income that would enable you not only to buy more soldiers (if necessary), but simultaneously to increase your investment in production. Soon you could upgrade your villages to towns, build universities, harbours and factories, explore the seas and oceans, establish your civilisation and win the game.

The Ark Syndrome

Yet can the economy actually keep growing for ever? Won't it eventually run out of resources – and grind to a halt? In order to ensure perpetual growth, we must somehow discover an inexhaustible store of resources.

One solution is to explore and conquer new lands. For centuries, the growth of the European economy and the expansion of the capitalist system indeed relied heavily on overseas imperial conquests. However, there are only so many islands and continents on earth. Some entrepreneurs do hope eventually to explore and conquer new planets and even galaxies, but in the meantime, the modern economy has had to find a better method of expanding.

It is science that has provided modernity with the solution. The fox economy cannot grow, because foxes don't know how to produce more rabbits. The rabbit economy stagnates, because rabbits cannot make the grass grow faster. But the human economy can grow because humans can discover new materials and sources of energy.

The traditional view of the world as a pie of a fixed size presupposes that there are only two kinds of resources in the world:

raw materials and energy. But in truth there are three kinds of resources: raw materials, energy and knowledge. Raw materials and energy are exhaustible – the more you use, the less you have. Knowledge, in contrast, is a growing resource – the more you use, the more you have. Indeed, when you increase your stock of knowledge, it can give you more raw materials and energy as well. If I invest $100 million searching for oil in Alaska and I find it, then I now have more oil, but my grandchildren will have less of it. In contrast, if I invest $100 million researching solar energy, and I find a new and more efficient way of harnessing it, then both I and my grandchildren will have more energy.

For thousands of years the scientific road to growth was blocked because people believed that holy scriptures and ancient traditions already contained all the important knowledge the world had to offer. A corporation that believed all the oil fields in the world had already been discovered would not waste time and money searching for oil. Similarly, a human culture that believed it already knew everything worth knowing would not bother searching for new knowledge. This was the position of most premodern human civilisations. However, the Scientific Revolution freed humankind from this naïve conviction. The greatest scientific discovery was the discovery of ignorance. Once humans realised how little they knew about the world, they suddenly had a very good reason to seek new knowledge, which opened up the scientific road to progress.

With each passing generation science helped discover fresh sources of energy, new kinds of raw material, better machinery and novel production methods. Consequently, in 2016 humankind commands far more energy and raw materials than ever before, and production skyrockets. Inventions such as the steam engine, the internal combustion engine and the computer have created whole new industries from scratch. As we look twenty years into the future, we confidently expect to produce and consume far more in 2036 than we do today. We trust nanotechnology, genetic engineering and artificial intelligence to revolutionise production yet again, and to open whole new sections in our ever-expanding supermarkets.

We therefore have a good chance of overcoming the problem of resource scarcity. The real nemesis of the modern economy is ecological collapse. Both scientific progress and economic growth take place within a brittle biosphere, and as they gather steam, so the shock waves destabilise the ecology. In order to provide every person in the world with the same standard of living as affluent Americans, we would need a few more planets – but we have only this one. If progress and growth do end up destroying the eco-system, the cost will be dear not merely to vampires, foxes and rabbits, but also to Sapiens. An ecological meltdown will cause economic ruin, political turmoil, a fall in human standards of living, and might threaten the very existence of human civilisation.

We could lessen the danger by slowing the pace of progress and growth. If this year investors expect to get a 6 per cent return on their portfolios, in ten years they could learn to be satisfied with a 3 per cent return, in twenty years with only 1 per cent, and in thirty years the economy will stop growing and we'll be happy with what we already have. Yet the creed of growth firmly objects to such a heretical idea. Instead, it suggests we should run even faster. If our discoveries destabilise the ecosystem and threaten humanity, then we should discover something to protect ourselves. If the ozone layer dwindles and exposes us to skin cancer, we should invent better sunscreen and better cancer treatments, thereby also promoting the growth of new sunscreen factories and cancer centres. If all the new industries pollute the atmosphere and the oceans, causing global warming and mass extinctions, then we should build for ourselves virtual worlds and hi-tech sanctuaries that will provide us with all the good things in life even if the planet becomes as hot, dreary and polluted as hell.

Beijing has already become so polluted that people avoid the outdoors, and wealthy Chinese pay thousands of dollars for indoor air-purifying systems. The super-rich build protective contraptions even over their yards. In 2013 the International School of Beijing, which caters to the children of foreign diplomats and upper-class Chinese, went a step further, and constructed a gigantic $5 million dome over its six tennis courts and its playing fields. Other schools are following suit, and the Chinese air-purification market is booming. Of course most Beijing residents cannot afford such luxuries in their homes, nor can they afford to send their kids to the International School.[3]

Humankind finds itself locked into a double race. On the one hand, we feel compelled to speed up the pace of scientific progress and economic growth. A billion Chinese and a billion Indians want to live like middle-class Americans, and they see no reason why they should put their dreams on hold when the Americans are unwilling to give up their SUVs and shopping malls. On the other hand, we must stay at least one step ahead of ecological Armageddon. Managing this double race becomes more difficult by the year, because every stride that brings the Delhi slum-dwellers closer to the American Dream also brings the planet closer to the brink.

The good news is that for hundreds of years humankind has enjoyed a growing economy without falling prey to ecological meltdown. Many other species have perished in the process, and humans too have faced a number of economic crises and ecological disasters, but so far we have always managed to pull through. Yet future success is not guaranteed by any law of nature. Who knows if science will always be able to simultaneously save the economy from freezing and the ecology from boiling. And since the pace just keeps accelerating, the margins for error keep narrowing. If previously it was sufficient to invent something amazing once a century, today we need to come up with a miracle every two years.

We should also be concerned that an ecological apocalypse might have different consequences for different human castes. There is no justice in history. When disaster strikes, the poor almost always suffer far more than the rich, even if the rich caused the tragedy in the first place. Global warming is already affecting the lives of poor people in arid African countries more than the lives of affluent Westerners. Paradoxically, the very power of science may increase the danger, because it makes the rich complacent.

Consider greenhouse gas emissions. Most scholars and an increasing number of politicians recognise the reality of global warming and the magnitude of the danger. Yet this recognition has so far failed to change our actual behavior in any signficant way. We talk a lot about global warming, but in practice humankind is unwilling to make the serious economic, social or political sacrifices necessary to stop this catastrophe. Between 2000 and 2010 emissions didn't decrease at all. On the contrary, they increased at an annual rate of 2.2 per cent, compared with an annual increase rate of 1.3 per cent between 1970 and 2000.[4] The 1997 Kyoto protocol on

Global CO$_2$ Emissions, 1970–2013

Source: Emission Database for Global Atmospheric Research (EDGAR), European Commission

26. All the talk about global warming, and all the conferences, summits and protocols, have so far failed to curb global greenhouse gas emissions. If you look closely at the graph you see that emissions go down only during periods of economic crises and stagnation. Thus the small downturn in greenhouse emissions in 2008–9 was due not to the signing of the Copenhagen Accord, but to the global financial crisis. The only sure way to stop global warming is to stop economic growth, which no government is willing to do.

the reduction of greenhouse gas emissions aimed merely to retard global warming rather than stop it, yet the world's number one polluter – the United States – refused to ratify it, and has made no attempt to significantly reduce its emissions, for fear of impeding its economic growth.[5]

In December 2015 more ambitious targets were set in the Paris Agreement, which calls for limiting average temperature increase to 1.5 degrees above pre-industrial levels. But many of the painful steps necessary to reach this goal have conveniently been postponed to after 2030, or even to the second half of the twenty-first century, effectively passing the hot potato to the next generation. Current administrations are able to reap immediate political benefits from looking green, while the heavy political price of reducing emissions (and slowing growth) is bequeathed to future administrations. Even so, at the time of writing (January 2016) it is far from certain that the USA and other leading polluters will ratify the Paris Agreement. Too many politicians and voters believe that as long as the economy grows, scientists and engineers could always save us from doomsday. When it comes to climate change, many growth true-believers don't just hope for miracles – they take it for granted that the miracles will happen.

How rational is it to risk the future of humankind on the assumption that future scientists will make some unknown planet-saving discoveries? Most of the presidents, ministers and CEOs who run the world are very rational people. Why are they willing to make such a gamble? Maybe because they don't think they are gambling on their own personal future. Even if bad comes to worse and science cannot hold off the deluge, engineers could still build a hi-tech Noah's Ark for the upper caste, while leaving billions of others to drown. The belief in this hi-tech Ark is currently one of the biggest threats to the future of humankind and of the entire ecosystem. People who believe in the hi-tech Ark should not be put in charge of the global ecology, for the same reason that people who believe in a heavenly afterlife should not be given nuclear weapons.

And what about the poor? Why aren't they protesting? If and when the deluge comes, they will bear the full cost of it. However, they will also be the first to bear the cost of economic stagnation. In a capitalist world the lives of the poor improve only when the economy grows. Hence they are unlikely to support any steps to reduce future ecological threats that are based on slowing down present-day economic growth. Protecting the environment is a very nice idea, but those who cannot pay their rent are worried about their overdraft far more than about melting ice caps.

The Rat Race

Even if we continue running fast enough and manage to fend off both economic collapse and ecological meltdown, the race itself creates huge problems. For the individual it results in high levels of stress and tension. After centuries of economic growth and scientific progress, life should have become calm and peaceful, at least in the most advanced countries. If our ancestors knew what tools and resources stand ready at our command, they would have surmised that we must be enjoying celestial tranquillity, free of all cares and worries. The truth is very different. Despite all our achievements, we feel a constant pressure to do and produce even more.

We blame ourselves, our boss, the mortgage, the government, the school system. But it's not really their fault. It's the modern deal that we all signed up for on the day we were born. In the premodern world, people were akin to lowly clerks in a socialist bureaucracy. They punched their cards, and then waited for somebody else to do something. In the modern world, we humans run the business, so we are under constant pressure day and night.

On the collective level, the race manifests itself in ceaseless upheavals. Whereas social and political systems previously endured for centuries, today every generation destroys the old world and builds a new one in its place. As the *Communist Manifesto* brilliantly put it, the modern world positively requires uncertainty and disturbance. All fixed relations and ancient prejudices are swept away, and new structures become antiquated before they can ossify. All that is solid melts into air. It isn't easy to live in such a chaotic world, and it is even harder to govern it.

Hence modernity needs to work hard to ensure that neither human individuals nor the human collective will try to retire from the race, despite all the tension and chaos it creates. For that purpose modernity upholds growth as a supreme value for whose sake we should make every sacrifice and risk every danger. On the collective level, governments, firms and organisations are encouraged to measure their success in terms of growth, and to fear equilibrium as if it were the Devil. On the individual level, we are inspired to constantly increase our incomes and our standards of living. Even if you are quite satisfied with your current conditions, you should strive for more. Yesterday's luxuries become today's necessities. If once you could live well in a three-bedroom apartment with one car and a single desktop computer, today you *need* a five-bedroom house with two cars and a host of iPods, tablets and smartphones.

It wasn't very hard to convince individuals to want more. Greed comes easily to humans. The big problem was to convince collective institutions such as states and churches to go along with the new ideal. For millennia, societies strove to curb individual desires and bring them into some kind of balance. It was well known that people wanted more and more for themselves, but when the pie was of a fixed size, social harmony depended on restraint. Avarice was bad. Modernity turned the world upside down. It convinced human collectives that equilibrium is far more frightening than chaos, and because avarice fuels growth, it is a force for good. Modernity accordingly inspired people to want more, and dismantled the age-old disciplines that curbed greed.

The resulting anxieties were assuaged to a large extent by free-market capitalism, which is one reason why this particular ideology has become so popular. Capitalist thinkers repeatedly calm us: 'Don't worry, it will be okay. Provided the economy grows, the invisible hand of the market will take care of everything else.' Capitalism has thus sanctified a voracious and chaotic system that grows by leaps and bounds, without anyone understanding what is happening and whither we are rushing. (Communism, which also believed in growth, thought it could prevent chaos and orchestrate growth through state planning. But after initial successes, it eventually fell far behind the messy free-market cavalcade.)

Bashing free-market capitalism is high on the intellectual agenda nowadays. Since capitalism dominates our world, we should indeed make every effort to understand its shortcomings before they cause apocalyptic catastrophes. Yet criticising capitalism should not blind us to its advantages and attainments. So far, it's been an amazing success – at least if you ignore the potential for future ecological meltdown, and if you measure success by the yardstick of production and growth. In 2016 we may be living in a stressful and chaotic world, but the doomsday prophecies of collapse and violence have not materialised, whereas the scandalous promises of perpetual growth and global cooperation are fulfilled. Although we experience occasional economic crises and international wars, in the long run capitalism has not only managed to prevail, but also to overcome famine, plague and war. For thousands of years priests, rabbis and muftis explained that humans cannot overcome famine, plague and war by their own efforts. Then along came the bankers, investors and industrialists, and within 200 years managed to do exactly that.

So the modern deal promised us unprecedented power – and the promise has been kept. Now what about the price? In exchange for power, the modern deal expects us to give up meaning. How did humans handle this chilling demand? Complying with it could easily have resulted in a dark world, devoid of ethics, aesthetics and compassion. Yet the fact remains that humankind is today not only far more powerful than ever, it is also far more peaceful and cooperative. How did humans manage that? How did morality, beauty and even compassion survive and flourish in a world devoid of gods, of heaven and of hell?

Capitalists are, again, quick to give all the credit to the invisible hand of the market. Yet the market's hand is not only invisible, it is also blind, and by itself could never have saved human society.

Indeed, not even a country fair can maintain itself without the helping hand of some god, king or church. If everything is for sale, including the courts and the police, trust evaporates, credit vanishes and business withers.[6] What, then, rescued modern society from collapse? Humankind was salvaged not by the law of supply and demand, but rather by the rise of a revolutionary new religion – humanism.

7

The Humanist Revolution

The modern deal offers us power, on condition that we renounce our belief in a great cosmic plan that gives meaning to life. Yet when you examine the deal closely, you find a cunning escape clause. If humans somehow manage to find meaning without predicating it upon some great cosmic plan, this is not considered a breach of contract.

This escape clause has been the salvation of modern society, for it is impossible to sustain order without meaning. The great political, artistic and religious project of modernity has been to find a meaning to life that is not rooted in some great cosmic plan. We are not actors in a divine drama, and nobody cares about us and our deeds, so nobody sets limits to our power – but we are still convinced our lives have meaning.

As of 2016, humankind has indeed managed to have it both ways. Not only do we possess far more power than ever before, but, against all expectations, God's death did not lead to social collapse. Throughout history prophets and philosophers have argued that if humans stopped believing in a great cosmic plan, all law and order would vanish. Yet today, those who pose the greatest threat to global law and order are precisely those people who continue to believe in God and His all-encompassing plans. God-fearing Syria is a far more violent place than the secular Netherlands.

If there is no cosmic plan, and we are not committed to any divine or natural laws, what prevents social collapse? How come you can travel for thousands of miles, from Amsterdam to Bucharest or from New Orleans to Montreal, without being kidnapped by slave-traders, ambushed by outlaws or killed by feuding tribes?

Look Inside

The antidote to a meaningless and lawless existence was provided by humanism, a revolutionary new creed that conquered the world during the last few centuries. The humanist religion worships humanity, and expects humanity to play the part that God played in Christianity and Islam, and that the laws of nature played in Buddhism and Daoism. Whereas traditionally the great cosmic plan gave meaning to the life of humans, humanism reverses the roles and expects the experiences of humans to give meaning to the cosmos. According to humanism, humans must draw from within their inner experiences not only the meaning of their own lives, but also the meaning of the entire universe. This is the primary commandment humanism has given us: create meaning for a meaningless world.

Accordingly, the central religious revolution of modernity was not losing faith in God; rather, it was gaining faith in humanity. It took centuries of hard work. Thinkers wrote pamphlets, artists composed poems and symphonies, politicians struck deals – and together they convinced humanity that it can imbue the universe with meaning. To grasp the depth and implications of the humanist revolution, consider how modern European culture differs from medieval European culture. In 1300 people in London, Paris and Toledo did not believe that humans could determine by themselves what is good and what is evil, what is right and what is wrong, what is beautiful and what is ugly. Only God could create and define goodness, righteousness and beauty.

Although it was widely accepted that humans enjoy unique abilities and opportunities, they were also seen as ignorant and corruptible beings. Without external supervision and guidance, humans could never understand the eternal truth, but would instead be drawn to fleeting sensual pleasures and worldly delusions. In addition, medieval thinkers pointed out that humans are mortal, and their opinions and feelings are as fickle as the wind. Today I love something with all my heart, tomorrow I am disgusted by it, and next week I am dead and buried. Hence any meaning that depends on human opinion is necessarily fragile and ephemeral. Absolute truths, and the meaning of life and of the universe, must therefore be based on some eternal law emanating from a super-human source.

This view made God the supreme source not only of meaning but also of authority. Meaning and authority always go hand in hand. Whoever determines the meaning of our actions – whether they are good or evil, right or wrong, beautiful or ugly – also gains the authority to tell us what to think and how to behave.

God's role as the source of meaning and authority was not just a philosophical theory. It affected every facet of daily life. Suppose that in 1300, in some small English town, a married woman took a fancy to the next-door neighbour and had sex with him. As she sneaked back home, hiding a smile and straightening her dress, her mind began to race: 'What was that all about? Why did I do it? Was it good or bad? What does it imply about me? Should I do it again?' In order to answer such questions, the woman was supposed to go to the local priest, confess and ask the holy father for guidance. The priest was well versed in scriptures, and these sacred texts revealed to him exactly what God thought about adultery. Based on the eternal word of God, the priest could determine beyond all doubt that the woman had committed a mortal sin, and that if she didn't make amends she'd end up in hell. She ought therefore to repent immediately, donate ten gold coins to the coming crusade, avoid eating meat for the next six months and make a pilgrimage to the tomb of St Thomas à Becket at Canterbury. And it goes without saying that she must never repeat her dreadful sin.

Today things are very different. For centuries humanism has been convincing us that we are the ultimate source of meaning, and that our free will is therefore the highest authority of all. Instead of waiting for some external entity to tell us what's what, we can rely on our own feelings and desires. From infancy we are bombarded with a barrage of humanist slogans counselling us: 'Listen to yourself, be true to yourself, trust yourself, follow your heart, do what feels good.' Jean-Jacques Rousseau summed it all up in his novel *Émile*, the eighteenth-century bible of feeling. Rousseau held that, when looking for life's rules of conduct, he found them 'in the depths of my heart, traced by nature in characters which nothing can efface. I need only consult myself with regard to what I wish to do; what I feel to be good is good, what I feel to be bad is bad.'[1]

Accordingly, when a modern woman wants to understand the meaning of an affair she is having, she is far less prone to blindly accept the judgements of a priest or an ancient book. Instead, she will carefully examine her feelings. If her feelings aren't very clear, she will call a good friend, meet for coffee and pour her heart out. If things are still vague, she will go to her therapist, and tell him all about it. Theoretically, the modern therapist occupies the same place as the medieval priest, and it is an overworked cliché to compare the two professions. Yet in practice, a huge chasm separates them. The therapist does not possess a holy book that defines good and evil. When the woman finishes her story, it is highly unlikely that the therapist will burst out: 'You wicked woman! You have committed a terrible sin!' It is equally unlikely that he will say, 'Wonderful! Good for you!' Instead, no matter what the woman may have done and said, the therapist is most likely to ask in a caring voice, 'Well, how do *you* feel about what happened?'

True, the therapist's bookshelf sags under the weight of Freud's and Jung's and the 1,000-page-long *Diagnostic and Statistical Manual of Mental Disorders* (*DSM*). Yet these are not holy scriptures. The *DSM* diagnoses the ailments of life, not the meaning of life. Most psychologists believe that only human feelings are authorised to determine the true meaning of human actions. Hence no matter what the therapist thinks about his patient's affair, and no matter what Freud, Jung and the *DSM* think about affairs in general, the therapist should not force his views on the patient. Instead, he should help her examine the most secret chambers of her heart. There and only there will she find the answers. Whereas medieval priests had a hotline to God and could distinguish for us between good and evil, modern therapists merely help us get in touch with our own inner feelings.

This partly explains the changing fortunes of the institution of marriage. In the Middle Ages, marriage was considered a sacrament ordained by God, and God also authorised a father to marry off his children according to his wishes and interests. An extramarital affair was consequently a brazen rebellion against both divine and parental authority. It was a mortal sin, no matter what the lovers felt and thought about it. Today people marry for love, and it is their personal feelings that give value to this bond. Hence, if the very same feelings that once drove you into the arms of one man now drive you into the arms of another, what's wrong with that? If an extramarital affair provides an outlet for emotional and sexual desires that are not satisfied by your spouse of twenty years, and if your new lover is kind, passionate and sensitive to your needs – why not enjoy it?

But wait a minute, you might say. We cannot ignore the feelings of the other concerned parties. The woman and her lover might feel wonderful in each other's arms, but if their respective spouses find out, everybody will probably feel awful for quite some time. And if it leads to divorce, their children might carry the emotional scars for decades. Even if the affair is never discovered, concealing it involves a lot of tension and may lead to growing feelings of alienation and resentment.

The most interesting discussions in humanist ethics concern situations like extramarital affairs, when human feelings collide. What happens when the same action causes one person to feel good, and another to feel bad? How do we weigh these feelings against each other? Do the good feelings of the two lovers outweigh the bad feelings of their spouses and children?

It doesn't matter what you think about this particular question. It is far more important to understand the kind of arguments both sides employ. Modern people have differing ideas about extramarital affairs, but no matter what their stance is, they tend to justify it in the name of human feelings rather than in the name of holy scriptures and divine commandments. Humanism has taught us that something can be bad only if it causes somebody to feel bad. Murder is wrong not because some god once said, 'Thou shalt not kill.' Rather, murder is wrong because it causes terrible suffering to the victim, to his family members, and to his friends and acquaintances. Theft is wrong not because some ancient text says, 'Thou shalt not steal.' Rather, theft is wrong because when you lose your property, you feel bad about it. And if an action does not cause anyone to feel bad, there can be nothing wrong with it. If the same ancient text says that God commanded us not to make any images of either humans or animals (Exodus 20:4), but I enjoy sculpting such figures, and I don't harm anyone in the process – then what could possibly be wrong with it?

The same logic dominates current debates on homosexuality. If two adult men enjoy having sex with one another, and they don't harm anyone while doing so, why should it be wrong, and why should we outlaw it? It is a private matter between these two men, and they are free to decide about it according to their own personal feelings. If in the Middle Ages two men confessed to a priest that they were in love with one another, and that they had never felt so happy, their good feelings would not have changed the priest's damning judgement – indeed, their lack of guilt would only have worsened the situation. Today, in contrast, if two men are in love, they are told: 'If it feels good – do it! Don't let any priest mess with your mind. Just follow your heart. You know best what's good for you.'

Interestingly enough, today even religious zealots adopt this humanistic discourse when they want to influence public opinion. For example, every year for the past decade the Israeli LGBT community has held a gay pride parade in the streets of Jerusalem. It's a unique day of harmony in this conflict-riven city, because it is the one occasion when religious Jews, Muslims and Christians suddenly find a common cause – they all fume in accord against the gay parade. What's really interesting, though, is the argument they use. They don't say, 'These sinners shouldn't hold a gay parade because God forbids homosexuality.' Rather, they explain to every available microphone and TV camera that 'seeing a gay parade passing through the holy city of Jerusalem hurts our feelings. Just as gay people want us to respect their feelings, they should respect ours.'

On 7 January 2015 Muslim fanatics massacred several staff members of the French magazine *Charlie Hebdo*, because the magazine published caricatures of the prophet Muhammad. In the following days, many Muslim organisations condemned the attack, yet some could not resist adding a 'but' clause. For example, the Egyptian Journalists Syndicate denounced the terrorists for their use of violence, but in the same breath denounced the magazine for 'hurting the feelings of millions of Muslims across the world'.[2] Note that the Syndicate did not blame the magazine for disobeying God's will. That's what we call progress.

Our feelings provide meaning not only for our private lives, but also for social and political processes. When we want to know who should rule the country, what foreign policy to adopt and what economic steps to take, we don't look for the answers in scriptures. Nor do we obey the commands of the Pope or the Council of Nobel Laureates. Rather, in most countries, we hold democratic elections and ask people what they think about the matter at hand. We believe that the voter knows best, and that the free choices of individual humans are the ultimate political authority.

Yet how does the voter know what to choose? Theoretically at least, the voter is supposed to consult his or her innermost feelings, and follow their lead. It's not always easy. In order to get in touch with my feelings, I need to filter out the empty propaganda slogans, the endless lies of ruthless politicians, the distracting noise created by cunning spin doctors, and the learned opinions of hired pundits. I need to ignore all this racket and attend only to my authentic inner voice. And then my authentic inner voice whispers in my ear 'Vote Cameron' or 'Vote Modi' or 'Vote Clinton' or whomever, and I put a cross against that name on the ballot paper – and that's how we know who should rule the country.

In the Middle Ages this would have been considered the height of foolishness. The fleeting feelings of ignorant commoners were hardly a sound basis for important political decisions. When England was torn apart by the Wars of the Roses, nobody thought to end the conflict by having a national referendum, in which every bumpkin and wench

27. The Holy Spirit, in the guise of a dove, delivers an ampulla full of sacred oil for the baptism of King Clovis, founder of the Frankish kingdom (illustration from the *Grandes Chroniques de France*, *c.*1380). According to the founding myth of France, this ampulla was henceforth kept in Rheims Cathedral, and all subsequent French kings were anointed with the divine oil at their coronation. Each coronation thus involved a miracle, as the empty ampulla spontaneously refilled with oil. This indicated that God himself chose the king and gave him His blessing. If God had not wished Louis IX or Louis XIV or Louis XVI to be king, the ampulla would not have been refilled.

would cast a vote for either Lancaster or York. Similarly, when Pope Urban II launched the First Crusade, he didn't claim it was the people's will. It was God's will. Political authority came down from heaven – it didn't rise up from the hearts and minds of mortal humans.

What's true of ethics and politics is also true of aesthetics. In the Middle Ages art was governed by objective yardsticks. The standards of beauty did not reflect human fads. Rather, human tastes were supposed to conform to superhuman dictates. This made perfect sense in a period when people believed that art was inspired by superhuman forces rather than by human feelings. The hands of painters, poets, composers and architects were supposedly moved by muses, angels and the Holy Spirit. Many a time when a composer penned a beautiful hymn, no credit was given to the composer, for the same reason it was not given to the pen. The pen was held and directed by human fingers, which in turn were held and directed by the hand of God.

Medieval scholars clung to a classical Greek theory, according to which the movements of the stars across the sky create heavenly

28. Pope Gregory the Great composes the eponymous Gregorian chants. The Holy Spirit, in its favourite dove outfit, sits on his right shoulder, whispering the chants in his ear. The Holy Spirit is the chants' true author, whereas Gregory is just a conduit. God is the ultimate source of art and beauty.

music that permeates the entire universe. Humans enjoy physical and mental health when the inner movements of their body and soul are in harmony with the heavenly music created by the stars. Human music should therefore echo the divine melody of the cosmos, rather than reflect the ideas and caprices of flesh-and-blood composers. The most beautiful hymns, songs and tunes were usually attributed not to the genius of some human artist but to divine inspiration.

Such views are no longer in vogue. Today humanists believe that the only source for artistic creation and aesthetic value is human feelings. Music is created and judged by our inner voice, which need follow neither the rhythms of the stars nor the commands of muses and angels. For the stars are mute, while muses and angels exist only in our own imagination. Modern artists seek to get in touch with themselves and their feelings, rather than with God. No wonder then that when we come to evaluate art, we no longer believe in any objective yardsticks. Instead, we again turn to our subjective feelings. In ethics, the humanist motto is 'if it feels good – do it'. In politics, humanism instructs us that 'the voter knows best'. In aesthetics, humanism says that 'beauty is in the eye of the beholder'.

The very definition of art is consequently up for grabs. In 1917 Marcel Duchamp purchased an ordinary mass-produced urinal, declared it a work of art, named it *Fountain*, signed it and placed it in a Paris museum. Medieval people would not have bothered even to argue about it. Why waste oxygen on such utter nonsense? Yet in the modern humanist world, Duchamp's work is considered an important artistic milestone. In countless classrooms across the world, first-year art students are shown an image of Duchamp's *Fountain*, and at a sign from the teacher, all hell breaks loose. It is art! No it isn't! Yes it is! No way! After letting the students release some steam, the teacher focuses the discussion by asking 'What exactly is art? And how do we determine whether something is a work of art or not?' After a few more minutes of back and forth the teacher steers the class in the right direction: 'Art is anything people think is art, and beauty is in the eye of the beholder.' If people think that a urinal is a beautiful work of art – then it is. What higher authority is there to tell people they are wrong? Today, copies of Duchamp's masterpiece are presented in some of the most important museums in the world, including the San Francisco Museum of Modern Art, the National Gallery of Canada, the Tate Gallery in London and the Pompidou Centre in Paris. (The copies are displayed in the museums' galleries, not in the lavatories.)

Such humanist approaches have had a deep impact on the economic field as well. In the Middle Ages guilds controlled the production process, leaving little room for the initiative or taste of individual artisans and customers. The carpenters' guild determined what was a proper chair, the bakers' guild defined good bread, and the Meistersinger guild decided which songs were first class and which were rubbish. Meanwhile princes and city councils regulated salaries and prices, occasionally forcing people to buy fixed amounts of goods at a non-negotiable price. In the modern free market, all these guilds, councils and princes have been superseded by a new supreme authority – the free will of the customer.

Suppose Toyota decides to produce the perfect car. It sets up a committee of experts from various fields: it hires the best engineers and designers, brings together the finest physicists and economists, and even consults with several sociologists and psychologists. To be on the safe side, they throw in a Nobel laureate or two, an Oscar-winning actress and some world-famous artists. After five years of research and development, they unveil the perfect car. Millions of vehicles are produced, and shipped to car dealerships across the world. Yet nobody buys the car. Does it mean that the customers are making a mistake, and that they don't know what's good for them? No. In a free market the customer is always right. If customers don't want it, it means that the car is no good. It doesn't matter if all the university professors and all the priests and mullahs cry out from every lectern and pulpit that this is a wonderful car – if the customers reject it, it's a bad car. Nobody has the authority to tell customers that they are wrong, and heaven forbid that a government would try to force its citizens to buy a particular car against their will.

What's true of cars is true of all other products. Listen, for example, to Professor Leif Andersson from the University of Uppsala. He specialises in the genetic enhancement of farm animals, in order to create faster-growing pigs, cows that produce more milk, and chickens with extra meat on their bones. In an interview with the newspaper *Haaretz*, reporter Naomi Darom confronted Andersson with the fact that such genetic manipulations might cause much suffering to the animals. Already today 'enhanced' dairy cows have such heavy udders that they can barely walk, while 'upgraded' chickens cannot even stand up. Professor Andersson had a firm answer: 'Everything comes back to the individual customer and to the question of how much the customer is willing to pay for meat . . . we must remember that it would be impossible to maintain current levels of global meat consumption without the [enhanced] modern chicken . . . if customers ask us only for the cheapest meat possible – that's what the customers will get . . . Customers need to decide what is most important to them – price, or something else.'[3]

Professor Andersson can go to sleep at night with a clean conscience. The fact that customers are buying his enhanced animal products implies that he is meeting their needs and desires and is therefore doing good. By the same logic, if some multinational corporation wants to know whether it lives up to its 'Don't be evil' motto, it need only take a look at its bottom line. If it makes loads of money, it means that millions of people like its products, which implies that it is a force for good. If someone objects and says that people might make the wrong choice, he will be quickly reminded that the customer is always right, and that human feelings are the source of all meaning and authority. If millions of people freely choose to buy the company's products, who are you to tell them that they are wrong?

Finally, the rise of humanist ideas has revolutionised education systems too. In the Middle Ages the source of all meaning and authority was external, hence education focused on instilling obedience, memorising scriptures and studying ancient traditions. Teachers

Humanism in Five Images

29. Humanist Politics: the voter knows best.

30. Humanist Economics: the customer is always right.

31. Humanist Aesthetics: Beauty is in the eye of the beholder. (Marcel Duchamp's *Fountain* in a special exhibition of modern art at the National Gallery of Scotland.)

32. Humanist Ethics: if it feels good – do it!

33. Humanist Education: think for yourself!

presented pupils with a question, and the pupils had to remember how Aristotle, King Solomon or St Thomas Aquinas answered it.

In contrast, modern humanist education believes in teaching students to think for themselves. It is good to know what Aristotle, Solomon and Aquinas thought about politics, art and economics; yet since the supreme source of meaning and authority lies within ourselves, it is far more important to know what *you* think about these matters. Ask a teacher – whether in kindergarten, school or college – what she is trying to teach. 'Well,' she will answer, 'I teach the kids history, or quantum physics, or art – but above all I try to teach them to think for themselves.' It may not always succeed, but that is what humanist education seeks to do.

As the source of meaning and authority relocated from the sky to human feelings, the nature of the entire cosmos changed. The exterior universe – hitherto teeming with gods, muses, fairies and ghouls – became empty space. The interior world – hitherto an insignificant enclave of crude passions – became deep and rich beyond measure. Angels and demons were transformed from real entities roaming the forests and deserts of the world into inner forces within our own psyche. Heaven and hell too ceased to be real places somewhere above the clouds and below the volcanoes, and were instead interpreted as internal mental states. You experience hell every time you ignite the fires of anger and hatred within your heart; and you enjoy heavenly bliss every time you forgive your enemies, repent your own misdeeds and share your wealth with the poor.

When Nietzsche declared that God is dead, this is what he meant. At least in the West, God has become an abstract idea that some accept and others reject, but it makes little difference either way. In the Middle Ages, without a god I had no source of political, moral and aesthetic authority. I could not tell what was right, good or beautiful. Who could live like that? Today, in contrast, it is very easy not to believe in God, because I pay no price for my unbelief. I can be a complete atheist and still derive a very rich mixture of political, moral and aesthetical values from my inner experience.

If I believe in God at all, it is *my choice* to believe. If my inner self tells me to believe in God – then I believe. I believe because I *feel* God's presence, and my heart tells me He is there. But if I no longer feel God's presence, and if my heart suddenly tells me that there is no God – I will cease believing. Either way, the real source of authority is my own feelings. So even while saying that I believe in God, the truth is that I have a much stronger belief in my own inner voice.

Follow the Yellow Brick Road

Like every other source of authority, feelings have their shortcomings. Humanism assumes that each human has a single authentic inner self, but when I try to attend to it I often encounter either silence or a cacophony of contending voices. To overcome this problem, humanism has proclaimed not only a new source of authority, but also a new method for accessing that authority and gaining true knowledge.

In medieval Europe, the chief formula for knowledge was: **Knowledge = Scriptures × Logic**.* If people wanted to know the answer to an important question, they would read scriptures and use their logic to understand the exact meaning of the text. For example, scholars who wished to determine the shape of the earth scanned the Bible looking for relevant references. One pointed out that in Job 38:13, it says that God can 'take hold of the edges of the earth, and the wicked be shaken out of it'. This implies – reasoned the pundit – that because the earth has 'edges' that God can 'take hold of', it must be a flat square. Another sage rejected this interpretation, calling attention to Isaiah 40:22, where it says that God

---

* The formula takes a multiplication symbol because the elements work one on the other. At least according to medieval scholastics, you cannot understand the Bible without logic. If your logic value is zero, then even if you read every page of the Bible, the sum of your knowledge would still be zero. Conversely, if your scripture value is zero, then no amount of logic can help you. If the formula used the addition symbol, the implication would be that somebody with lots of logic and no scriptures would still have a lot of knowledge – which you and I may find reasonable, but medieval scholastics did not.

'sits enthroned above the circle of the earth'. Isn't that proof that the earth is round? In practice, this meant that scholars sought knowledge by spending years in schools and libraries, reading more and more texts, and sharpening their logic so they could understand the texts correctly.

The Scientific Revolution proposed a very different formula for knowledge: **Knowledge = Empirical Data × Mathematics**. If we want to know the answer to some question, we need to gather relevant empirical data, and then use mathematical tools to analyse them. For example, in order to gauge the true shape of the earth, we can begin by observing the sun, moon and planets from various locations across the world. Once we have amassed enough observations, we can use trigonometry to deduce not only the shape of the earth, but also the structure of the entire solar system. In practice, this means that scientists seek knowledge by spending years in observatories, laboratories and on research expeditions, gathering more and more empirical data, and sharpening their mathematical tools so they can interpret the data correctly.

The scientific formula for knowledge led to astounding breakthroughs in astronomy, physics, medicine and multiple other disciplines. But it has had one huge drawback: it could not deal with questions of value and meaning. Medieval pundits could determine with absolute certainty that it is wrong to murder and steal, and that the purpose of human life is to do God's bidding, because scriptures said so. Scientists cannot deliver such ethical judgements. No amount of data and no mathematical wizardry can prove that it is wrong to murder. Yet human societies cannot survive without such value judgements.

One way to overcome this difficulty was to continue using the old medieval formula alongside the new scientific method. When faced with a practical problem – such as determining the shape of the earth, building a bridge or curing a disease – we collect empirical data and analyse them mathematically. When faced with an ethical problem – such as determining whether to allow divorce, abortion and homosexuality – we read scriptures. This solution was adopted to some extent by numerous modern societies, from Victorian Britain to twenty-first-century Iran.

However, humanism offered an alternative. As humans gained confidence in themselves, a new formula for acquiring ethical knowledge appeared: **Knowledge = Experiences × Sensitivity**. If we wish to know the answer to any ethical question, we need to connect to our inner experiences, and observe them with the utmost sensitivity. In practice, this means that we seek knowledge by spending years collecting experiences, and sharpening our sensitivity so we can understand these experiences correctly.

What exactly are 'experiences'? They are not empirical data. An experience is not made of atoms, electromagnetic waves, proteins or numbers. Rather, an experience is a subjective phenomenon made up of three main ingredients: sensations, emotions and thoughts. At any particular moment my experience comprises everything I sense (heat, pleasure, tension, etc.), every emotion I feel (love, fear, anger, etc.) and whatever thoughts arise in my mind.

And what is 'sensitivity'? It means two things. Firstly, paying attention to my sensations, emotions and thoughts. Secondly, allowing these sensations, emotions and thoughts to influence me. Granted, I shouldn't allow every passing breeze to sweep me away. Yet I should be open to new experiences and permit them to change my views, my behaviour and even my personality.

Experiences and sensitivity build up one another in a never-ending cycle. I cannot experience anything if I have no sensitivity, and I cannot develop sensitivity unless I undergo a variety of experiences. Sensitivity is not an abstract aptitude that can be developed by reading books or listening to lectures. It is a practical skill that can ripen and mature only by applying it in practice.

Take tea, for example. I start by drinking very sweet ordinary tea while reading the morning paper. The tea is little more than an excuse for a sugar rush. One day I realise that between the sugar and the newspaper, I hardly taste the tea at all. So I reduce the amount of sugar, put the paper aside, close my eyes and focus on the tea itself. I begin to register its unique aroma and flavour. Soon I find myself experimenting with different teas, black and green, comparing their exquisite tangs and delicate bouquets. Within a few months, I drop the supermarket labels and buy my tea at Harrods. I develop a particular liking for 'Panda Dung tea' from the mountains of Ya'an in Sichuan province, made from the leaves of tea bushes fertilised by the dung of panda bears. That's how, one cup at a time, I hone my tea sensitivity and become a tea connoisseur. If in my early tea-drinking days you had served me Panda Dung tea in a Ming Dynasty porcelain goblet, I would not have appreciated it any more than builder's tea in a paper cup. You cannot experience something if you don't have the necessary sensitivity, and you cannot develop your sensitivity except by undergoing a long string of experiences.

What's true of tea is true of all other aesthetic and ethical knowledge. We aren't born with a ready-made conscience. As we pass through life we hurt people and people hurt us, we act compassionately and others show compassion to us. If we pay attention, our moral sensitivity sharpens, and these experiences become a source of valuable ethical knowledge about what is good, what is right and who I really am.

Humanism thus sees life as a gradual process of inner change, leading from ignorance to enlightenment by means of experiences. The highest aim of humanist life is to fully develop your knowledge through a wide variety of intellectual, emotional and physical experiences. In the early nineteenth century Wilhelm von Humboldt – one of the chief architects of the modern education system – said that the aim of existence is 'a distillation of the widest possible experience of life into wisdom'. He also wrote that 'there is only one summit in life – to have taken the measure in feeling of everything human'.[4] This could well be the humanist motto.

According to Chinese philosophy, the world is sustained by the interplay of opposing but complementary forces called yin and yang. This may not be true of the physical world, but it is certainly true of the modern world that has been created by the covenant of science and humanism. Every scientific yang contains within it a humanist yin, and vice versa. The yang provides us with power, while the yin provides us with meaning and ethical judgements. The yang and yin of modernity are reason and emotion, the laboratory and the museum, the production line and the supermarket. People often see only the yang and imagine that the modern world is dry, scientific, logical and utilitarian – just like a laboratory or a factory. But the modern world is also an extravagant supermarket. No culture in history has ever given such importance to human feelings, desires and experiences. The humanist view of life as a succession of experiences has become the founding myth of numerous modern industries, from tourism to art. Travel agents and restaurant chefs do not sell us flight tickets, hotels or fancy dinners – they sell us novel experiences.

Similarly, whereas most premodern narratives focused on external events and actions, modern novels, films and poems often emphasize feelings. Graeco-Roman epics and medieval chivalric romances were catalogues of heroic deeds, not feelings. One chapter described how a brave knight fought a monstrous ogre, and killed him. Another chapter recounted how the knight rescued a beautiful princess from a fire-spitting dragon, and killed him. A third chapter narrated how a wicked sorcerer kidnapped the princess, but the knight pursued the sorcerer, and killed him. Small wonder that the hero was invariably a knight, rather than a carpenter or a peasant, for peasants performed no heroic deeds.

Crucially, the heroes did not undergo any significant process of inner change. Achilles, Arthur, Roland and Lancelot were fearless warriors with a chivalric world view before they set out on their adventures, and they remained fearless warriors with the same world view at the end. All the ogres they killed and all the princesses they rescued confirmed their courage and perseverance, but ultimately taught them little.

The humanist focus on feelings and experiences, rather than deeds, transformed art. Wordsworth, Dostoevsky, Dickens and Zola cared little for brave knights and derring-do; instead they described how ordinary labourers and housewives felt. Some people believe that Joyce's *Ulysses* represents the apogee of this modern focus on the inner life rather than external actions. In 260,000 words Joyce describes a single day in the life of the Dubliners Stephen Dedalus and Leopold Bloom, who over the course of that day do . . . well, nothing much at all.

Few people have actually read *Ulysses* cover to cover, but the same change of focus now underpins much of our popular culture too. In the United States, the TV series *Survivor* is often credited (or blamed) for turning reality shows into a craze. *Survivor* was the first reality show to make it to the top of the Nielsen ratings, and in 2007 *Time* magazine listed it among the hundred greatest TV shows of all time.[5] In each season twenty contenders in minimal swimsuits are isolated on some tropical island. They have to face various kinds of challenges, and during each episode they vote to oust one of their number. The last one left takes home $1 million.

Audiences in Homeric Greece, in the Roman Empire or in medieval Europe would have found the idea familiar and highly attractive. Twenty challengers go in – only one hero comes out. 'Wonderful!' a Homeric prince, a Roman patrician or a crusader knight would have thought to himself as he sat down to watch. 'Surely we are about to see amazing adventures, life-and-death battles and incomparable acts of heroism and betrayal. The warriors will probably stab each other in the back, or spill their entrails for all to see.'

What a disappointment! The back-stabbing and entrails-spilling remain a mere metaphor. Each episode lasts about an hour. Out of that, fifteen minutes are taken up by commercials for toothpaste, shampoo and cereals. Five minutes are dedicated to incredibly childish challenges, such as who can throw the most coconuts into a hoop, or who can eat the largest number of bugs in one minute. The rest of the time the 'heroes' just talk about their feelings! He said she said, and I felt this and I felt that. If a crusader knight had actually been able to sit down to watch *Survivor*, he would probably have grabbed his battleaxe and smashed the TV out of boredom and frustration.

Today we might think of medieval knights as insensitive brutes. If they lived among us, we would send them to a therapist, who might help them get in touch with their feelings. This is what happens to the Tin Man in *The Wizard of Oz*. He walks along the yellow brick road with Dorothy and her friends, hoping that when they reach Oz, the great wizard will give him a heart. Likewise, the Scarecrow wants a brain and the Lion wants courage. At the end of their journey they discover that the great wizard is a charlatan, and he can't give them any of these things. But they discover something far more important: everything they wished for was already within themselves. There was never any need of some godlike wizard in order to become sensitive, wise or brave. You just need to follow the yellow brick road and open yourself up to whatever experiences come your way.

Exactly the same lesson is learned by Captain Kirk and Captain Jean-Luc Picard as they travel the galaxy in the starship *Enterprise*, by Huckleberry Finn and Jim as they sail down the Mississippi River, by Wyatt and Billy as they ride their Harley-Davidsons in *Easy Rider*, and by countless other characters in myriad other road movies who leave their home towns in Pennsylvania (or perhaps New South Wales), travel in an old convertible (or perhaps a bus), pass through various life-changing experiences, get in touch with themselves, talk about their feelings, and eventually reach San Francisco (or perhaps Alice Springs) as better and wiser individuals.

The Truth About War

The formula **Knowledge = Experiences × Sensitivity** has changed not just our popular culture, but even our perception of weighty issues like war. Throughout most of history, when people wished to know whether a particular war was just, they asked God, they asked scriptures, and they asked kings, noblemen and priests. Few cared about the opinions and experiences of a common soldier or an ordinary civilian. War narratives such as those of Homer, Virgil and Shakespeare focused on the actions of emperors, generals and

34. Jean-Jacques Walter, *Gustav Adolph of Sweden at the Battle of Breitenfeld* (1631).

outstanding heroes, and though they did not hide the misery of war, this was more than compensated for by a full menu of glory and heroism. Ordinary soldiers appeared as either piles of bodies slaughtered by some Goliath, or a cheering crowd hoisting a triumphant David upon its shoulders.

Look, for example, at the painting above of the Battle of Breitenfeld, which took place on 17 September 1631. The painter, Jean-Jacques Walter, glorifies King Gustav Adolph of Sweden, who led his army that day to a decisive victory. Gustav Adolph towers over the battlefield as if he were some god of war. One gets the impression that the king controls the battle like a chess player moving pawns. The pawns themselves are mostly generic figures, or tiny dots in the background. Walter was not interested in how they felt as they charged, fled, killed or died. They are a faceless collective.

Even when painters focused on the battle itself rather than on the commander, they still looked at it from above, and were far more concerned with collective manoeuvres than with personal feelings. Take, for example, Pieter Snayers's painting of the Battle of White Mountain in November 1620.

The painting depicts a celebrated Catholic victory in the Thirty Years War over heretical Protestant rebels. Snayers wished to

35. Pieter Snayers, *The Battle of White Mountain.*

commemorate this victory by painstakingly recording the various formations, manoeuvres and troop movements. You can easily identify the different units, their armaments and their positions within the order of battle. Snayers gave far less importance to the experiences and feelings of the common soldiers. Like Jean-Jacques Walter, he makes us observe the battle from the Olympian vantage point of gods and kings, and gives us the impression that war is a giant chess game.

If you take a closer look – for which you might need a magnifying glass – you realise that *The Battle of White Mountain* is a bit more complex than a chess game. What at first sight seem to be geometrical abstractions turn upon closer inspection into bloody scenes of carnage. Here and there you can even spot the faces of individual soldiers running or fleeing, firing their guns or impaling an enemy on their pikes. However, these scenes receive their meaning from their place within the overall picture. When we see a cannonball smashing a soldier to bits, we understand it as part of the great Catholic victory. If the soldier is fighting on the Protestant side, his death is a just reward for rebellion and heresy. If the soldier is fighting in the Catholic army, his death is a noble sacrifice for a worthy cause. If we look up, we can see angels hovering high above the battlefield. They are holding a white banner that explains in Latin what happened in this battle, and why it was so important. The message is that God helped Emperor Ferdinand II defeat his enemies on 8 November 1620.

For thousands of years, when people looked at war, they saw gods, emperors, generals and great heroes. But over the last two centuries, the kings and generals have been increasingly pushed to the side, and the limelight has shifted onto the common soldier and his experiences. War novels such as *All Quiet on the Western Front* and war films such as *Platoon* begin with a green recruit who knows little about himself and the world, but carries a heavy burden of hopes and illusions. He believes that war is glorious, the cause is just and the general is a genius. A few weeks of real war – of mud, and blood, and the smell of death – shatter his illusions one after another. If he survives, the formerly naïve recruit will leave the war a much wiser man, who no longer believes the clichés and ideals peddled by teachers, film-makers and eloquent politicians.

Paradoxically, this narrative has become so influential that today it is told over and over again even by teachers, film-makers and eloquent politicians. 'War is not what you see in the movies!' warn Hollywood blockbusters such as *Apocalypse Now*, *Full Metal Jacket* and *Blackhawk Down*. Enshrined in celluloid, prose or poetry, the feelings of the ordinary grunt have become the ultimate authority on war, which everyone has learned to respect. As the joke goes, 'How many Vietnam vets does it take to change a light bulb?' 'You wouldn't know, you weren't there.'[6]

Painters too have lost interest in generals on horseback and tactical manoeuvres. Instead, they strive to depict how the common soldier feels. Look again at *The Battle of Breitenfeld* and *The Battle of White Mountain*. Now look at the following two pictures, both considered masterpieces of twentieth-century war art: *The War* (*Der Krieg*) by Otto Dix, and *That 2,000 Yard Stare* by Tom Lea.

Dix served as a sergeant in the German army during the First World War. Lea covered the 1944 Battle of Peleliu Island for *Life* magazine. Whereas Walter and Snayers saw war as a military and political

36. Otto Dix, *The War* (1929–32).

37. Tom Lea, *That 2,000 Yard Stare* (1944).

phenomenon and wanted us to know what happened in particular battles, Dix and Lea saw war as an emotional phenomenon and wanted us to know how it feels. They didn't care about the genius of generals or the tactical details of this or that battle. Dix's soldier might have been in Verdun or Ypres or the Somme – it doesn't matter which, because war is hell everywhere. Lea's soldier just happened to be an American GI on Peleliu, but you could have seen exactly the same 2,000-yard stare on the face of a Japanese soldier on Iwo Jima, a German soldier in Stalingrad or a British soldier at Dunkirk.

In the paintings of Dix and Lea, the meaning of war does not emanate from tactical movements or divine proclamations. If you want to understand war, don't look up at the general on the hill-top, or at angels in the sky. Instead, look straight into the eyes of the common soldiers. In Lea's painting the gaping eyes of a traumatised soldier open a window onto the terrible truth of war. In Dix's paint-ing, the truth is so unbearable that it must be partly concealed behind a gas mask. No angels fly above the battlefield – only a rotting corpse, dangling from a ruined rafter and pointing an accusing finger.

Artists such as Dix and Lea thus helped overturn the traditional hierarchy of war. Numerous wars in earlier times were certainly as horrific as those of the twentieth century. However, hitherto even atrocious experiences were placed within a wider context that gave them a positive meaning. War might be hell, but it was also the gate-way to heaven. A Catholic soldier fighting at the Battle of White Mountain could say to himself: 'True, I am suffering. But the Pope and the emperor say that we are fighting for a good cause, so my suf-fering is meaningful.' Otto Dix employed an opposite kind of logic. He saw personal experience as the source of all meaning, hence his line of thinking said: 'I am suffering – and this is bad – hence the whole war is bad. If the kaiser and the clergy nevertheless support this war, they must be mistaken.'[7]

The Humanist Schism

So far we have discussed humanism as if it were a single coherent world view. In fact, humanism shared the fate of every successful religion, such as Christianity and Buddhism. As it spread and evolved, it fragmented into several conflicting sects. All humanist sects believe that human experience is the supreme source of authority and meaning, yet they interpret human experience in different ways.

Humanism split into three main branches. The orthodox branch holds that each human being is a unique individual possessing a distinctive inner voice and a never-to-be-repeated series of experiences. Every human being is a singular ray of light that illuminates the world from a different perspective, and that adds colour, depth and meaning to the universe. Hence we ought to give as much freedom as possible to every individual to experience the world, follow his or her inner voice and express his or her inner truth. Whether in politics, economics or art, individual free will should have far more weight than state interests or religious doctrines. The more liberty individuals enjoy, the more beautiful, rich and meaningful is the world. Due to this emphasis on liberty, the orthodox branch of humanism is known as 'liberal humanism' or simply as 'liberalism'.*

It is liberal politics that believes the voter knows best. Liberal art holds that beauty is in the eye of the beholder. Liberal economics maintains that the customer is always right. Liberal ethics advises us that if it feels good, we should go ahead and do it. Liberal education teaches us to think for ourselves, because we will find all the answers within.

During the nineteenth and twentieth centuries, as humanism gained increasing social credibility and political power, it sprouted two very different offshoots: socialist humanism, which encompassed a plethora of socialist and communist movements, and evolutionary humanism, whose most famous advocates were the Nazis. Both offshoots agreed with liberalism that human experience is the ultimate source of meaning and authority. Neither believed in any transcendental power or divine law book. If, for

---

* In American politics, liberalism is often interpreted far more narrowly, and contrasted with 'conservatism'. In the broad sense of the term, however, most American conservatives are also liberal.

example, you had asked Karl Marx what was wrong with ten-year-olds working twelve-hour shifts in smoky factories, he would have answered that it makes the kids feel bad. We should avoid exploitation, oppression and inequality not because God said so, but because they make people miserable.

However, both socialists and evolutionary humanists pointed out that the liberal understanding of the human experience is flawed. Liberals think the human experience is an individual phenomenon. But there are many individuals in the world, and they often feel different things and have contradictory desires. If all authority and meaning flow from individual experiences, how do you settle contradictions between different such experiences?

On 17 July 2015 the German chancellor Angela Merkel was confronted by a teenage Palestinian refugee girl from Lebanon, whose family was seeking asylum in Germany but faced imminent deportation. The girl, Reem, told Merkel in fluent German that 'It's really very hard to watch how other people can enjoy life and you yourself can't. I don't know what my future will bring.' Merkel replied that 'politics can be tough' and explained that there are hundreds of thousands of Palestinian refugees in Lebanon, and Germany cannot absorb them all. Stunned by this no-nonsense reply, Reem burst into tears. Merkel proceeded to stroke the desperate girl on the back, but stuck to her guns.

In the ensuing public storm many accused Merkel of cold-hearted insensitivity. To assuage criticism Merkel changed tack, and Reem and her family were given asylum. In the following months Merkel opened the door even wider, welcoming hundreds of thousands of refugees to Germany. But you can't please everybody. Soon she was under severe attack for succumbing to sentimentalism and for not taking a sufficiently firm stand. Numerous German parents feared that Merkel's U-turn meant that their children would have a lower standard of living, and perhaps suffer from a tidal wave of Islamisation. Why should they risk their families' peace and prosperity to help complete strangers who might not even believe in the values of liberalism? Everyone feels very strongly about this matter. How to settle the contradictions between the feelings of the desperate refugees and of the anxious Germans?[8]

Liberals forever agonise about such contradictions. The best efforts of Locke, Jefferson, Mill and their colleagues have failed to provide us with a fast and easy solution to such conundrums. Holding democratic elections won't help, because then the question would be who gets to vote in these elections – only German citizens, or also the millions of Asians and Africans who want to immigrate to Germany? Why privilege the feelings of one group over another? Likewise, you cannot resolve the Arab–Israeli conflict by having 8 million Israeli citizens and the 350 million citizens of Arab League nations vote on it. For obvious reasons the Israelis wouldn't feel committed to the outcome of such a plebiscite.

People feel bound by democratic elections only when they share a basic bond with most other voters. If the experience of other voters is alien to me, and if I believe they don't understand my feelings and don't care about my vital interests, then even if I am outvoted by a hundred to one I have absolutely no reason to accept the verdict. Democratic elections usually work only within populations that have some prior common bond, such as shared religious beliefs or national myths. They are a method to settle disagreements among people who already agree on the basics.

Accordingly, in many cases liberalism has fused with age-old collective identities and tribal feelings to form modern nationalism. Today many associate nationalism with anti-liberal forces, but at least during the nineteenth century nationalism was closely aligned with liberalism. Liberals celebrate the unique experiences of individual humans. Each human has distinctive feelings, tastes and quirks that he or she should be free to express and explore as long as they don't hurt anyone else. Similarly, nineteenth-century nationalists such as Giuseppe Mazzini celebrated the uniqueness of individual nations. They emphasised that many human experiences are communal. You cannot dance the polka by yourself, and you cannot invent and preserve the German language by yourself. Using word, dance, food and drink, each nation fosters different experiences in its members, and develops its own peculiar sensitivities.

Liberal nationalists like Mazzini sought to protect these distinctive national experiences from being oppressed and obliterated by intolerant empires, and envisaged a peaceful community of nations, each free to express and explore its communal feelings without hurting its neighbours. This remains the official ideology of the European Union, whose 2004 constitution states that Europe is 'united in diversity' and that the different peoples of Europe remain 'proud of their own national identities'. The value of preserving the unique communal experiences of the German nation enables even liberal Germans to oppose opening the floodgates of immigration.

Of course the alliance of liberalism with nationalism hardly solved all conundrums, while at the same time it created a host of new ones. How do you compare the value of communal experiences with that of individual experiences? Does preserving polka, bratwurst and the German language justify leaving millions of refugees exposed to poverty and possibly even death? And what happens when fundamental conflicts erupt within nations about the very definition of their identity, as happened in Germany in 1933, in the USA in 1861, in Spain in 1936 or in Egypt in 2011? In such cases holding democratic elections is hardly a cure-all, because the opposing parties have no reason to respect the results.

Lastly, as you dance the nationalist polka, a small but momentous step may take you from believing that your nation is different from all other nations to believing that your nation is better. Nineteenth-century liberal nationalism required the Habsburg and tsarist empires to respect the unique experiences of Germans, Italians, Poles and Slovenes. Twentieth-century ultra-nationalism proceeded to wage wars of conquest and build concentration camps for people who danced to a different tune.

Socialist humanism has taken a very different course. Socialists blame liberals for focusing our attention on our own feelings instead of on what other people experience. Yes, the human experience is the source of all meaning, but there are billions of people in the world and all of them are just as valuable as I am. Whereas liberalism turns my gaze inwards, emphasising my uniqueness and the uniqueness of my nation, socialism demands that I stop obsessing about me and my feelings and instead focus on what others are feeling and how my actions influence their experiences. Global peace will be achieved not by celebrating the distinctiveness of each nation, but by unifying all the workers of the world; and social harmony won't be achieved by each person narcissistically exploring their own inner depths, but rather by each person prioritising the needs and experiences of others over their own desires.

A liberal may counter that by exploring her own inner world she develops her compassion and her understanding of others. But such reasoning would have cut little ice with Lenin or Mao. They would have explained that individual self-exploration is an indulgent bourgeois vice, and that when I try to get in touch with my inner self, I am more than likely to fall into one or another capitalist trap. My current political views, my likes and dislikes, and my hobbies and ambitions do not reflect my authentic self. Rather, they reflect my upbringing and social surroundings. They depend on my class, and are shaped by my neighbourhood and my school. Rich and poor alike are brainwashed from birth. The rich are taught to disregard the poor, while the poor are taught to disregard their true interests. No amount of self-reflection or psychotherapy will help, because the psychotherapists are also working for the capitalist system.

Indeed, self-reflection is likely only to distance me even further from understanding the truth about myself, because it gives too much consideration to personal decisions and not enough to social conditions. If I am rich, I conclude that it is because I made shrewd choices. If I am mired in poverty, I must have made some mistakes. If I am depressed, a liberal therapist is likely to blame my parents, and to encourage me to set some new aims in life. If I suggest that perhaps I am depressed because I am being exploited by capitalists, and because under the prevailing social system I have no chance of realising my aims, the therapist may well say that I am projecting onto 'the social system' my own inner difficulties, and I am projecting onto 'the capitalists' unresolved issues with my mother.

According to socialism, instead of spending years talking about my mother, my emotions and my complexes, I should ask myself: who owns the means of production in my country? What are its main exports and imports? What's the connection between the ruling politicians and international banking? Only by understanding the prevailing socio-economic system and taking into account the experiences of all other people can I truly understand what I feel, and only by common action can we change the system. Yet what person can take into account the experiences of all human beings, and weigh them one against the other in a fair way?

That's why socialists discourage self-exploration and advocate the establishment of strong collective institutions – such as socialist parties and trade unions – that aim to decipher the world for us. Whereas in liberal politics the voter knows best, and in liberal economics the customer is always right, in socialist politics the party knows best, and in socialist economics the trade union is always right. Authority and meaning still come from human experience – both the party and the trade union are composed of people and work to alleviate human misery – yet individuals must listen to the party and the trade union rather than to their personal feelings.

Evolutionary humanism has a different solution to the problem of conflicting human experiences. Rooting itself in the firm ground of Darwinian evolutionary theory, it insists that conflict is something to applaud rather than lament. Conflict is the raw material of natural selection, which pushes evolution forward. Some humans are simply superior to others, and when human experiences collide, the fittest humans should steamroll everyone else. The same logic that drives humankind to exterminate wild wolves and to ruthlessly exploit domesticated sheep also mandates the oppression of inferior humans by their superiors. It's a good thing that Europeans conquer Africans and that shrewd businessmen drive the dim-witted to bankruptcy. If we follow this evolutionary logic, humankind will gradually become stronger and fitter, eventually giving rise to superhumans. Evolution didn't stop with *Homo sapiens* – there is still a long way to go. However, if in the name of human rights or human equality we emasculate the fittest humans, it will prevent the rise of the superman, and may even cause the degeneration and extinction of *Homo sapiens*.

Who exactly are these superior humans who herald the coming of the superman? They might be entire races, particular tribes or exceptional individual geniuses. Whoever they may be, what makes them superior is that they have better abilities, manifested in the creation of new knowledge, more advanced technology, more prosperous societies or more beautiful art. The experience of an Einstein or a Beethoven is far more valuable than that of a drunken good-for-nothing, and it is ludicrous to treat them as if they have equal merit. Similarly, if a particular nation has consistently spearheaded human progress, we should rightly consider it superior to other nations that contributed little or nothing to the evolution of humankind.

Consequently, in contrast to liberal artists like Otto Dix, evolutionary humanism maintains that the human experience of war is valuable and even essential. The movie *The Third Man* is set in Vienna immediately after the end of the Second World War. Reflecting on the recent conflict the character Harry Lime says: 'After all, it's not that awful . . . In Italy for thirty years under the Borgias they had warfare, terror, murder and bloodshed, but they produced Michelangelo, Leonardo da Vinci and the Renaissance. In Switzerland they had brotherly love, they had 500 years of democracy and peace, and what did that produce? The cuckoo clock.' Lime gets almost all his facts wrong – Switzerland was probably the most bloodthirsty corner of early modern Europe (its main export was mercenary soldiers), and the cuckoo clock was actually invented by Germans – but the facts are of lesser importance than Lime's idea, namely that the experience of war pushes humankind to new achievements. War allows natural selection free rein at last. It exterminates the weak and rewards the fierce and the ambitious. War exposes the truth about life, and awakens the will for power, for glory and for conquest. Nietzsche summed it up by saying that war is 'the school of life' and that 'what does not kill me makes me stronger'.

Similar ideas were expressed by Lieutenant Henry Jones of the British army. Three days before his death on the Western Front in the First World War, the twenty-one-year-old Jones sent a letter to his brother, describing his experience of war in glowing terms:

> Have you ever reflected on the fact that, despite the horrors of war, it is at least a big thing? I mean to say that in it one is brought face to face with realities. The follies, selfishness, luxury and general pettiness of the vile commercial sort of existence led by nine-tenths of the people of the world in peacetime are replaced in war by a savagery that is at least more honest and outspoken. Look at it this way: in peacetime one just lives one's own little life, engaged in trivialities, worrying about one's own comfort, about money matters, and all that sort of thing – just living for one's own self. What a sordid life it is! In war, on the other hand, even if you do get killed you only anticipate the inevitable by a few years in any case, and you have the satisfaction of knowing that you have 'pegged out' in the attempt to help your country. You have, in fact, realised an ideal, which, as far as I can see, you very rarely do in ordinary life. The reason is that ordinary life runs on a commercial and selfish basis; if you want to 'get on', as the saying is, you can't keep your hands clean.
>
> Personally, I often rejoice that the War has come my way. It has made me realise what a petty thing life is. I think that the War has given to everyone a chance to 'get out of himself', as I might say . . . Certainly, speaking for myself, I can say that I have never in all my life experienced such a wild exhilaration as on the commencement of a big stunt, like the last April one for example. The excitement for the last half-hour or so before it is like nothing on earth.[9]

In his bestseller *Black Hawk Down* the journalist Mark Bowden relates in similar terms the combat experience of Shawn Nelson, an American soldier, in Mogadishu in 1993:

> It was hard to describe how he felt . . . it was like an epiphany. Close to death, he had never felt so completely alive. There had been split seconds in his life when he'd felt death brush past, like when another fast-moving car veered from around a sharp curve and just missed hitting him head on. On this day he had lived with that feeling, with death breathing right in his face . . . for moment after moment after moment, for three hours or more . . . Combat was . . . a state of complete mental and physical awareness. In those hours on the street he had not been Shawn Nelson, he had no connection to the larger world, no bills to pay, no emotional ties, nothing. He had just been a human being staying alive from one nanosecond to the next, drawing one breath after another, fully aware that each one might be his last. He felt he would never be the same.[10]

Adolf Hitler too was changed and enlightened by his war experiences. In *Mein Kampf* he relates how, shortly after his unit reached the front line, the soldiers' initial enthusiasm turned into fear, against which each soldier had to wage a relentless inner war, straining every nerve to avoid being overwhelmed by it. Hitler says that he won this inner war by the winter of 1915/16. 'At last,' he writes, 'my will was undisputed master . . . I was now calm and determined. And this was enduring. Now Fate could bring on the ultimate tests without my nerves shattering or my reason failing.'[11]

The experience of war revealed to Hitler the truth about the world: it's a jungle run by the remorseless laws of natural selection. Those who refuse to recognise this truth cannot survive. If you wish to succeed, you must not only understand the laws of the jungle, but embrace them joyfully. It should be stressed that just like anti-war liberal artists, Hitler too sanctified the experience of ordinary soldiers. Indeed, Hitler's political career is one of the best examples we have for the immense authority accorded to the personal experience of common people in twentieth-century politics. Hitler wasn't a senior officer – in four years of war, he rose no higher than the rank of corporal. He had no formal education, no professional skills and no political background. He wasn't a successful businessman or a union activist, he didn't have friends or relatives in high places, nor any money to speak of. At first, he didn't even have German citizenship. He was a penniless immigrant.

When Hitler appealed to the German voters and asked for their trust, he could muster only one argument in his favour: his experiences in the trenches had taught him what you can never learn at university, at general headquarters or at a government ministry. People followed him and voted for him because they identified with him, and because they too believed that the world is a jungle, and that what doesn't kill us only makes us stronger.

Whereas liberalism merged with the milder versions of nationalism to protect the unique experiences of each human community, evolutionary humanists such as Hitler identified particular nations as the engines of human progress and concluded that these nations ought to bludgeon or even exterminate anyone standing in their way. It should be remembered, though, that Hitler and the Nazis represent only one extreme version of evolutionary humanism. Just as Stalin's gulags do not automatically nullify every socialist idea and argument, so too the horrors of Nazism should not blind us to whatever insights evolutionary humanism might offer. Nazism was born from the pairing of evolutionary humanism with particular racial theories and ultra-nationalist emotions. Not all evolutionary humanists are racists, and not every belief in humankind's potential for further evolution necessarily calls for setting up police states and concentration camps.

Auschwitz should serve as a blood-red warning sign rather than as a black curtain that hides entire sections of the human horizon. Evolutionary humanism played an important part in the shaping of modern culture, and is likely to play an even greater role in the shaping of the twenty-first century.

Is Beethoven Better than Chuck Berry?

To make sure that we understand the differences between the three humanist branches, let's compare a few human experiences.

Experience no. 1: A musicology professor sits in the Vienna Opera House listening to the opening of Beethoven's Fifth Symphony. 'Pa pa pa PAM!' As the sound waves hit his eardrums, signals travel via the auditory nerve to the brain and the adrenal gland floods his bloodstream with adrenaline. His heartbeat accelerates, his breathing intensifies, the hairs on his neck stand up, and a shiver runs down his spine. 'Pa pa pa PAM!'

Experience no. 2: It's 1965. A Mustang convertible is speeding down the Pacific Coast Highway from San Francisco to LA at full throttle. The macho young driver puts on Chuck Berry at full volume: 'Go! Go, Johnny, go!' As the sound waves hit his eardrums, signals travel via the auditory nerve to the brain and the adrenal gland floods his bloodstream with adrenaline. His heartbeat accelerates, his breathing intensifies, the hairs on his neck stand up, and a shiver runs down his spine. 'Go! Go, Johnny, go, go, go!'

Experience no. 3: Deep in the Congolese rainforest, a pygmy hunter stands transfixed. From the nearby village he hears a choir of girls singing their initiation song. 'Ye oh, oh. Ye oh, eh.' As the sound waves hit his eardrums, signals travel via the auditory nerve to the brain and the adrenal gland floods his bloodstream with adrenaline. His heartbeat accelerates, his breathing intensifies, the hairs on his neck stand up, and a shiver runs down his spine. 'Ye oh, oh. Ye oh, eh.'

Experience no. 4: It's a full-moon night, somewhere in the Canadian Rockies. A wolf is standing on a hilltop listening to the howls of a female in heat. 'Awoooooo! Awoooooo!' As the sound waves hit his eardrums, signals travel via the auditory nerve to the brain and the adrenal gland floods his bloodstream with adrenaline. His heartbeat accelerates, his breathing intensifies, the hairs on his neck stand up, and a shiver runs down his spine. 'Awoooooo! Awoooooo!'

Which of these four experiences is the most valuable?

Liberals will tend to say that the experiences of the musicology professor, of the young driver and of the Congolese hunter are all equally valuable, and all should be equally cherished. Every human experience contributes something unique, and enriches the world with new meaning. Some people like classical music, others love rock and roll, and still others prefer traditional African chants. Music students should be exposed to the widest possible range of genres, and at the end of the day, they can all go to the iTunes store, punch in their credit card numbers and buy whatever they like. Beauty is in the ears of the listener, and the customer is always right. The wolf, though, isn't human, hence his experiences are far less valuable. That's why the life of a wolf is worth less than the life of a human, and why it is perfectly okay to kill a wolf in order to save a human. When all is said and done, wolves don't get to vote in any beauty contests, nor do they own any credit cards.

This liberal approach is manifested, for example, in the *Voyager* golden record. In 1977 the Americans launched the space probe *Voyager I* on a journey to outer space. By now it has left our solar system, making it the first man-made object to traverse interstellar space. Besides state-of-the-art scientific equipment, NASA placed on board a golden record, aimed to introduce planet Earth to any inquisitive aliens who might encounter the probe.

The record contains a variety of scientific and cultural information about Earth and its inhabitants, some images and voices, and several dozen pieces of music from around the world, which are supposed to represent a fair sampling of earthly artistic achievement. The musical sample mixes in no obvious order classical pieces including the opening movement of Beethoven's

Fifth Symphony, contemporary popular music including Chuck Berry's 'Johnny B. Goode', and traditional music from throughout the world, including an initiation song of Congolese pygmy girls. Though the record also contains some canine howls, they are not part of the music sample, but rather relegated to a different section that also includes the sounds of wind, rain and surf. The message to potential listeners in Alpha Centauri is that Beethoven, Chuck Berry and the pygmy initiation song are of equal merit, whereas wolf howls belong to an altogether different category.

Socialists will probably agree with the liberals that the wolf's experience is of little value. But their attitude towards the three human experiences will be quite different. A socialist true-believer will explain that the real value of music depends not on the experiences of the individual listener, but on the impact it has on the experiences of other people and of society as a whole. As Mao said, 'There is no such thing as art for art's sake, art that stands above classes, art that is detached from or independent of politics.'[12]

So when evaluating the musical experience socialists will focus, for example, on the fact that Beethoven wrote the Fifth Symphony for an audience of upper-class white Europeans, exactly when Europe was about to embark on its conquest of Africa. His symphony reflected Enlightenment ideals, which glorified upper-class white men, and justified the conquest of Africa as 'the white man's burden'.

Rock and roll – the socialists will say – was pioneered by down-trodden African American musicians who drew inspiration from genres like blues, jazz and gospel. However, in the 1950s and 1960s rock and roll was hijacked by mainstream white America, and pressed into the service of consumerism, American imperialism and Coca-Colonialism. Rock and roll was commercialised and appropriated by privileged white teenagers in their petit-bourgeois fantasy of rebellion. Chuck Berry himself bowed to the dictates of the capitalist juggernaut. While he originally sang about 'a *coloured* boy named Johnny B. Goode', under pressure from white-owned radio stations Berry changed the lyrics to 'a *country* boy named Johnny B. Goode'.

As for the choir of Congolese pygmy girls – their initiation songs are part of a patriarchal power structure that brainwashes both men and women to conform to an oppressive gender order. And if a recording of such an initiation song ever makes it to the global marketplace, it merely serves to reinforce Western colonial fantasies about Africa in general and African women in particular.

So which music is best: Beethoven's Fifth, 'Johnny B. Goode' or the pygmy initiation song? Should the government finance the building of opera houses, rock and roll venues or African-heritage exhibitions? And what should we teach music students in schools and colleges? Well, don't ask me. Ask the party's cultural commissar.

Whereas liberals tiptoe around the minefield of cultural comparisons, fearful of committing some politically incorrect faux pas, and whereas socialists leave it to the party to find the correct path through this minefield, evolutionary humanists gleefully jump right in, setting off all the mines and relishing the mayhem. They may start by pointing out that both liberals and socialists draw the line at other animals, and have no trouble admitting that humans are superior to wolves, and that consequently human music is far more valuable than wolf howls. Yet humankind itself is not exempt from the forces of evolution. Just as humans are superior to wolves, so some human cultures are more advanced than others. There is an unambiguous hierarchy of human experiences, and we shouldn't be apologetic about it. The Taj Mahal is more beautiful than a straw hut, Michelangelo's *David* is superior to my five-year-old niece's latest clay figurine, and Beethoven composed far better music than Chuck Berry or the Congolese pygmies. There, we've said it!

According to evolutionary humanists, anyone arguing that all human experiences are equally valuable is either an imbecile or a coward. Such vulgarity and timidity will lead only to the degeneration and extinction of humankind, as human progress is impeded in the name of cultural relativism or social equality. If liberals or socialists had lived in the Stone Age, they would probably have seen little merit in the murals of Lascaux and Altamira, and would have insisted that they were in no way superior to Neanderthal doodles.

The Humanist Wars of Religion

Initially the differences between liberal humanism, socialist humanism and evolutionary humanism seemed rather frivolous. Set against the enormous gap separating all humanist sects from Christianity, Islam or Hinduism, the arguments between different versions of humanism were trifling. As long as we all agree that God is dead and that only the human experience gives meaning to the universe, does it really matter whether we think that all human experiences are equal or that some are superior to others? Yet as humanism conquered the world, these internal schisms widened, and eventually flared up into the deadliest religious war in history.

In the first decade of the twentieth century, the liberal orthodoxy was still confident of its strength. Liberals were convinced that if individuals had maximum freedom to express themselves and fol-low their hearts, the world would enjoy unprecedented peace and prosperity. It might take time to completely dismantle the fetters of traditional hierarchies, obscurantist religions and brutal empires, but every decade would bring new liberties and achievements, and eventually we would create paradise on earth. In the halcyon days of June 1914, liberals thought history was on their side.

By Christmas 1914 liberals were shell-shocked, and in the following decades their ideas were subjected to a double assault from both left and right. Socialists argued that liberalism is in fact a fig leaf for a ruthless, exploitative and racist system. For vaunted 'liberty', read 'property'. The defence of the individual's right to do what feels good amounts in most cases to safeguarding the property and privileges of the middle and upper classes. What good is the liberty to live where you want when you cannot pay the rent; to study what interests you when you cannot afford the tuition fees;

and to travel where you fancy when you cannot buy a car? Under liberalism, went a famous quip, everyone is free to starve. Even worse, by encouraging people to view themselves as isolated individuals, liberalism separates them from their fellow class members and prevents them from uniting against the system that oppresses them. Liberalism thereby perpetuates inequality, condemning the masses to poverty and the elite to alienation.

While liberalism staggered under this punch from the left, evolutionary humanism struck from the right. Racists and fascists blamed both liberalism and socialism for subverting natural selection and causing the degeneration of humankind. They warned that if all humans were given equal value and equal breeding opportunities, natural selection would cease to function. The fittest humans would be submerged in an ocean of mediocrity, and instead of evolving into supermen, humankind would become extinct.

From 1914 to 1989 a murderous war of religion raged between the three humanist sects, and liberalism at first sustained one defeat after the other. Not only did communist and fascist regimes take over numerous countries, but the core liberal ideas were exposed as naïve at best, if not downright dangerous. Just give freedom to individuals and the world will enjoy peace and prosperity? Yeah, right.

The Second World War, which with hindsight we remember as a great liberal victory, hardly looked like that at the time. The war began in September 1939 as a conflict between a mighty liberal alliance and an isolated Nazi Germany. (Even Fascist Italy preferred to play a waiting game until June of the following year.) The liberal alliance enjoyed overwhelming numerical and economic superiority. While German GDP in 1940 stood at $387 million, the GDP of Germany's European opponents totalled $631 million (not including the GDP of the overseas British dominions and of the British, French, Dutch and Belgian empires). Still, in the spring of 1940 it took Germany a mere three months to deal the liberal alliance a decisive blow, occupying France, the Low Countries, Norway and Denmark. The UK was saved from a similar fate only by the English Channel.[13]

The Germans were eventually beaten only after the liberal countries allied themselves with the Soviet Union, which bore the brunt of the conflict and paid a much higher price: 25 million Soviet citizens died in the war, compared to half a million Britons and half a million Americans. Much of the credit for defeating Nazism should be given to communism. And at least in the short term, communism was also the great beneficiary of the war.

The Soviet Union entered the war as an isolated communist pariah. It emerged as one of the two global superpowers and the leader of an expanding international bloc. By 1949 eastern Europe became a Soviet satellite, the Chinese Communist Party had won the Chinese Civil War, and the United States was gripped by anti-communist hysteria. Revolutionary and anti-colonial movements throughout the world looked longingly towards Moscow and Beijing, while liberalism became identified with the racist European empires. As these empires collapsed, they were usually replaced by either military dictatorships or socialist regimes, not liberal democracies. In 1956 the Soviet premier, Nikita Khrushchev, confidently boasted to the liberal West that 'Whether you like it or not, history is on our side. We will bury you!'

Khrushchev sincerely believed this, as did increasing numbers of Third World leaders and First World intellectuals. In the 1960s and 1970s the word 'liberal' became a term of abuse in many Western universities. North America and western Europe experienced growing social unrest, as radical left-wing movements strove to undermine the liberal order. Students in Cambridge, the Sorbonne and the People's Republic of Berkeley thumbed through Chairman Mao's Little Red Book and hung Che Guevara's heroic portrait over their beds. In 1968 the wave crested with the outbreak of protests and riots all over the Western world. Mexican security forces killed dozens of students in the notorious Tlatelolco Massacre, students in Rome fought the Italian police in the so-called Battle of Valle Giulia, and the assassination of Martin Luther King sparked days of riots and protests in more than a hundred American cities. In May students took over the streets of Paris, President de Gaulle fled to a French military base in Germany, and well-to-do French citizens trembled in their beds, having guillotine nightmares.

By 1970 the world contained 130 independent countries, but only thirty of these were liberal democracies, most of which were crammed into the north-western corner of Europe. India was the only important Third World country that committed to the liberal path after securing its independence, but even India distanced itself from the Western bloc and leaned towards the Soviets.

In 1975 the liberal camp suffered its most humiliating defeat of all: the Vietnam War ended with the North Vietnamese David overcoming the American Goliath. In quick succession communism took over South Vietnam, Laos and Cambodia. On 17 April 1975 the Cambodian capital, Phnom Penh, fell to the Khmer Rouge. Two weeks later people all over the world watched on TV as helicopters evacuated the last Yankees from the rooftop of the American Embassy in Saigon. Many were certain that the American Empire was falling. Before anyone could say 'domino theory', in June Indira Gandhi proclaimed the Emergency in India, and it seemed that the world's largest democracy was on its way to becoming yet another socialist dictatorship.

Liberal democracy increasingly looked like an exclusive club for ageing white imperialists, who had little to offer the rest of the world, or even to their own youth. Washington hailed itself as the leader of the free world, but most of its allies were either authoritarian kings (such as King Khaled of Saudi Arabia, King Hassan of Morocco and the Persian shah) or military dictators (such as the Greek colonels, General Pinochet in Chile, General Franco in Spain, General Park in South Korea, General Geisel in Brazil and Generalissimo Chiang Kai-shek in Taiwan).

Despite the support of all these kings and generals, militarily the Warsaw Pact had a huge numerical superiority over NATO. In order to reach parity in conventional armaments, Western countries would probably have had to scrap liberal democracy and the free market, and become totalitarian states on a permanent war

38. The evacuation of the American Embassy in Saigon.

footing. Liberal democracy was saved only by nuclear weapons. NATO adopted the MAD doctrine (Mutual Assured Destruction), according to which even conventional Soviet attacks would be answered by an all-out nuclear strike. 'If you attack us,' threatened the liberals, 'we will make sure nobody comes out alive.' Behind this monstrous shield, liberal democracy and the free market managed to hold out in their last bastions, and Westerners got to enjoy sex, drugs and rock and roll, as well as washing machines, refrigerators and televisions. Without nukes there would have been no Beatles, no Woodstock and no overflowing supermarkets. But in the mid-1970s it seemed that nuclear weapons notwithstanding, the future belonged to socialism.

And then everything changed. Liberal democracy crawled out of history's dustbin, cleaned itself up and conquered the world. The supermarket proved to be far stronger than the gulag. The blitzkrieg began in southern Europe where the authoritarian regimes in Greece, Spain and Portugal collapsed, giving way to democratic governments. In 1977 Indira Gandhi ended the Emergency, re-establishing democracy in India. During the 1980s military dictatorships in East Asia and Latin America were replaced by democratic governments in countries such as Brazil, Argentina, Taiwan and South Korea. In the late 1980s and early 1990s the liberal wave turned into a veritable tsunami, sweeping away the mighty Soviet Empire and raising expectations of the coming end of history. After decades of defeats and setbacks, liberalism won a decisive victory in the Cold War, emerging triumphant from the humanist wars of religion, albeit a bit worse for wear.

As the Soviet Empire imploded, liberal democracies replaced communist regimes not only in eastern Europe, but also in many of the former Soviet republics, such as the Baltic States, Ukraine, Georgia and Armenia. Even Russia nowadays pretends to be a democracy. Victory in the Cold War gave renewed impetus to the spread of the liberal model elsewhere around the world, most notably in Latin America, South Asia and Africa. Some liberal experiments ended in abject failure, but the number of success stories is impressive. Indonesia, Nigeria and Chile, for instance, had been ruled by military strongmen for decades, but all are now functioning democracies.

If a liberal had fallen asleep in June 1914 and awakened in June 2014, he or she would have felt very much at home. Once again people believe that if you just give individuals more freedom, the world will enjoy peace and prosperity. The entire twentieth century looks like a big mistake. Back in the spring of 1914 humankind was speeding on the liberal highway when it took a wrong turn and entered a cul-de-sac. It then required eight decades and three horrendous global wars to find its way back to the highway. Of course, these decades were not a total waste; they did give us antibiotics, nuclear energy and computers, as well as feminism, de-colonialism and free sex. In addition, liberalism itself smarted from the experience and is less conceited than it was a century ago. It has adopted various ideas and institutions from its socialist and fascist rivals, in particular a commitment to provide the general public with education, health and welfare services. Yet the core liberal package has changed surprisingly little. Liberalism still sanctifies individual liberties above all, and still has a firm belief in the voter and the customer. In the early twenty-first century, this is the only show in town.

Electricity, Genetics and Radical Islam

As of 2016 there is no serious alternative to the liberal package of individualism, human rights, democracy and a free market. The social protests that swept the Western world in 2011 – such as Occupy Wall Street and the Spanish 15-M movement – have absolutely nothing against democracy, individualism and human rights, or even against the basic principles of free-market economics. Just the opposite – they take governments to task for not living up to these liberal ideals. They demand that the market be really free, instead of being controlled and manipulated by corporations and banks 'too big to fail'. They call for truly representative democratic institutions that will serve the interests of ordinary citizens rather than of moneyed lobbyists and powerful interest groups. Even those blasting stock exchanges and parliaments with the harshest criticism don't have a viable alternative model for running the world. While it is a favourite pastime of Western academics and activists to find fault with the liberal package, they have so far failed to come up with anything better.

China seems to offer a much more serious challenge than Western social protestors. Despite liberalising its politics and economics, China is neither a democracy nor a truly free-market economy, which does not prevent it from becoming the economic giant of the twenty-first century. Yet this economic giant casts a very small ideological shadow. Nobody seems to know what the Chinese believe these days – including the Chinese themselves. In theory China is still communist, but in practice it is nothing of the kind. Some Chinese thinkers and leaders toy with a return to Confucianism, but that's hardly more than a convenient facade. This ideological vacuum makes China the most promising breeding ground for the new techno-religions emerging from Silicon Valley (which we will discuss in the following chapters). But these techno-religions, with their belief in immortality and virtual paradises, will take at least a decade or two to establish themselves. Hence at present China doesn't pose a real alternative to liberalism. For bankrupt Greeks despairing of the liberal model and searching for a substitute, 'imitating the Chinese' isn't a viable option.

How about radical Islam, then? Or fundamentalist Christianity, messianic Judaism or revivalist Hinduism? Whereas the Chinese don't know what they believe, religious fundamentalists know only too well. More than a century after Nietzsche pronounced Him dead, God seems to be making a comeback. But this is a mirage. God *is* dead – it's just taking a while to get rid of the body. Radical Islam poses no serious threat to the liberal package, because for all their fervour the zealots don't really understand the world of the twenty-first century, and have nothing relevant to say about the novel dangers and opportunities that new technologies are generating all around us.

Religion and technology always dance a delicate tango. They push one another, depend on one another and cannot stray too far away from one another. Technology depends on religion because every invention has many potential applications, and the engineers need some prophet to make the crucial choices and point towards the required destination. Thus in the nineteenth century engineers invented locomotives, radios and internal combustion engines. But as the twentieth century proved, you can use these very same tools to create fascist societies, communist dictatorships and liberal democracies. Without religious convictions, the locomotives cannot decide which way to go.

On the other hand, technology often defines the scope and limits of our religious visions, like a waiter that demarcates our appetites by handing us a menu. New technologies kill old gods and give birth to new gods. That's why agricultural deities were different from hunter-gatherer spirits, why factory hands fantasised about different paradises than peasants and why the revolutionary technologies of the twenty-first century are far more likely to spawn unprecedented religious movements than to revive medieval creeds. Islamic fundamentalists may repeat the mantra that 'Islam is the answer', but religions that lose touch with the techno-logical realities of the day forfeit their ability even to understand the questions being asked. What will happen to the job market once artificial intelligence outperforms humans in most cognitive tasks? What will be the political impact of a massive new class of economically useless people? What will happen to relationships, families and pension funds when nanotechnology and regenera-tive medicine turn eighty into the new fifty? What will happen to human society when biotechnology enables us to have designer babies, and to open unprecedented gaps between rich and poor?

You will not find the answers to any of these questions in the Qur'an or sharia law, nor in the Bible or in the Confucian *Analects*, because nobody in the medieval Middle East or in ancient China knew much about computers, genetics or nanotechnology. Radical Islam may promise an anchor of certainty in a world of techno-logical and economic storms – but in order to navigate a storm you need a map and a rudder rather than just an anchor. Hence radical Islam may appeal to people born and raised in its fold, but it has precious little to offer unemployed Spanish youths or anxious Chinese billionaires.

True, hundreds of millions may nevertheless go on believing in Islam, Christianity or Hinduism. But numbers alone don't count for much in history. History is often shaped by small groups of forward-looking innovators rather than by the backward-looking masses. Ten thousand years ago most people were hunter-gatherers and only a few pioneers in the Middle East were farmers. Yet the future belonged to the farmers. In 1850 more than 90 per cent of humans were peasants, and in the small villages along the Ganges, the Nile and the Yangtze nobody knew anything about steam engines, railroads or telegraph lines. Yet the fate of those peasants had already been sealed in Manchester and Birmingham by the handful of engineers, politicians and financiers who spearheaded the Industrial Revolution. Steam engines, railroads and tele-graphs transformed the production of food, textiles, vehicles and weapons, giving industrial powers a decisive edge over traditional agricultural societies.

Even when the Industrial Revolution spread around the world and penetrated up the Ganges, Nile and Yangtze, most people continued to believe in the Vedas, the Bible, the Qur'an and the *Analects* more than in the steam engine. As today, so too in the nineteenth century there was no shortage of priests, mystics and gurus who argued that they alone held the solution to all of humanity's woes, including to the new problems created by the Industrial Revolution. For example, between the 1820s and 1880s Egypt (backed by Britain) conquered Sudan and tried to modernise the country and incorporate it into the new international trade network. This destabilised traditional Sudanese society, creating widespread resentment and fostering revolts. In 1881 a local religious leader, Muhammad Ahmad bin Abdallah, declared that he was the Mahdi (the Messiah), sent to establish God's law on earth. His supporters defeated the Anglo-Egyptian army and beheaded its commander – General Charles Gordon – in a gesture that shocked Victorian Britain. They then established in Sudan an Islamic theocracy governed by sharia law, which lasted until 1898.

Meanwhile in India, Dayananda Saraswati headed a Hindu revival movement, whose basic principle was that the Vedic scriptures are never wrong. In 1875 he founded the Arya Samaj (Noble Society), dedicated to the spreading of Vedic knowledge – though truth be told, Dayananda often interpreted the Vedas in a surprisingly liberal way, supporting, for example, equal rights for women long before the idea became popular in the West.

Dayananda's contemporary, Pope Pius IX, had much more conservative views about women, but shared Dayananda's admiration for superhuman authority. Pius led a series of reforms in Catholic dogma and established the novel principle of papal infallibility, according to which the Pope can never err in matters of faith (this seemingly medieval idea became binding Catholic dogma only in 1870, eleven years after Charles Darwin published *On the Origin of Species*).

Thirty years before the Pope discovered that he is incapable of making mistakes, a failed Chinese scholar called Hong Xiuquan had a succession of religious visions. In these visions God revealed that Hong was none other than the younger brother of Jesus Christ. God then invested Hong with a divine mission. He told Hong to expel the Manchu 'demons' that had ruled China since the seventeenth century, and establish on earth the Great Peaceful Kingdom of Heaven (Taiping Tiānguó). Hong's message fired the imagination of millions of desperate Chinese, who were shaken by China's defeats in the Opium Wars and by the coming of modern industry and European imperialism. But Hong did not lead them to a kingdom of peace. Rather, he led them against the Manchu Qing dynasty in the Taiping Rebellion – the deadliest war of the nineteenth century, which lasted from 1850 to 1864. At least 20 million people lost their lives, far more than in the Napoleonic Wars or in the American Civil War.

Hundreds of millions clung to the religious dogmas of Hong, Dayananda, Pius and the Mahdi even as industrial factories, railroads and steamships filled the world. Yet most of us don't think about the nineteenth century as the age of faith. When we think of nineteenth-century visionaries we are far more likely to recall Marx, Engels and Lenin than the Mahdi, Pius IX or Hong Xiuquan. And rightly so. Though in 1850 socialism was only a fringe movement, it soon gathered momentum and changed the world in far more profound ways than the self-proclaimed messiahs of China and Sudan. If you value national health services, pension funds and free schools, you need to thank Marx and Lenin (and Otto von Bismarck) far more than Hong Xiuquan or the Mahdi.

Why did Marx and Lenin succeed where Hong and the Mahdi failed? Not because socialist humanism was philosophically more sophisticated than Islamic and Christian theology, but rather because Marx and Lenin devoted more attention to understanding the technological and economic realities of their time than to scrutinising ancient texts and prophetic dreams. Steam engines, railroads, telegraphs and electricity created unheard-of problems as well as unprecedented opportunities. The experiences, needs and hopes of the new class of urban proletariats were simply too different from those of biblical peasants. To answer these needs and hopes, Marx and Lenin studied how a steam engine functions, how a coal mine operates, how railroads shape the economy and how electricity influences politics.

Lenin was once asked to define communism in a single sentence. 'Communism is power to worker councils,' he said, 'plus electrification of the whole country.' There can be no communism without electricity, without railroads, without radio. You couldn't have established a communist regime in sixteenth-century Russia, because communism necessitates the concentration of information and resources in one hub. 'From each according to his ability, to each according to his needs' only works when produce can easily be collected and distributed across vast distances, and when activities can be monitored and coordinated over entire countries.

Marx and his followers understood the new technological realities and the new human experiences, so they had relevant answers to the new problems of industrial society, as well as original ideas about how to benefit from the unprecedented opportunities. The socialists created a brave new religion for a brave new world. They promised salvation through technology and economics, thus establishing the first techno-religion in history, and changing the foundations of ideological discourse. Before Marx, people defined and divided themselves according to their views about God, not about production methods. Since Marx, questions of technology and economic structure became far more important and divisive than debates about the soul and the afterlife. In the second half of the twentieth century humankind almost obliterated itself in an argument about production methods. Even the harshest critics of Marx and Lenin adopted their basic attitude towards history and society, and began thinking much more carefully about technology and production than about God and heaven.

In the mid-nineteenth century few people were as perceptive as Marx, hence only a few countries underwent rapid industrialisation. These few countries conquered the world. Most societies failed to understand what was happening, and therefore missed the train of progress. Dayananda's India and the Mahdi's Sudan remained far more preoccupied with God than with steam engines, hence they were occupied and exploited by industrial Britain. Only in the last few years has India managed to make significant progress in closing the economic and geopolitical gap separating it from Britain. Sudan is still struggling, far behind.

In the early twenty-first century the train of progress is again pulling out of the station – and this will probably be the last train ever to leave the station called *Homo sapiens*. Those who miss this train will never get a second chance. In order to get a seat on it you need to understand twenty-first-century technology, and in particular the powers of biotechnology and computer algorithms. These powers are far more potent than steam and the telegraph, and they will not be used merely for the production of food, textiles, vehicles and weapons. The main products of the twenty-first century will be bodies, brains and minds, and the gap between those who know how to engineer bodies and brains and those who do not will be far bigger than the gap between Dickens's Britain and the Mahdi's Sudan. Indeed, it will be bigger than the gap between Sapiens and Neanderthals. In the twenty-first century, those who ride the train of progress will acquire divine abilities of creation and destruction, while those left behind will face extinction.

Socialism, which was very up to date a hundred years ago, failed to keep up with new technology. Leonid Brezhnev and Fidel Castro held on to ideas that Marx and Lenin formulated in the age of steam, and did not understand the power of computers and biotechnology. Liberals, in contrast, adapted far better to the information age. This partly explains why Khrushchev's 1956 prediction never materialised, and why it was the liberal capitalists who eventually buried the Marxists. If Marx came back to life today, he would probably urge his few remaining disciples to devote less time to reading *Das Kapital* and more time to studying the Internet and the human genome.

Radical Islam is in a far worse position than socialism. It has not yet come to terms even with the Industrial Revolution – no wonder it has little of relevance to say about genetic engineering and artificial intelligence. Islam, Christianity and other traditional religions are still important players in the world. Yet their role is now largely reactive. In the past, they were a creative force. Christianity, for example, spread the hitherto heretical notion that all humans are equal before God, thereby changing human political structures, social hierarchies and even gender relations. In his Sermon on the Mount Jesus went further, insisting that the meek and oppressed are God's favourite people, thus turning the pyramid of power on its head, and providing ammunition for generations of revolutionaries.

In addition to social and ethical reforms, Christianity was responsible for important economic and technological innovations. The Catholic Church established medieval Europe's most sophisticated administrative system, and pioneered the use of archives, catalogues, timetables and other techniques of data processing. The Vatican was the closest thing twelfth-century Europe had to Silicon Valley. The Church established Europe's first economic corporations – the monasteries – which for 1,000 years spearheaded the European economy and introduced advanced agricultural and administrative methods. Monasteries were the first institutions to use clocks, and for centuries they and the cathedral schools were the most important learning centres of Europe, helping to found many of Europe's first universities, such as Bologna, Oxford and Salamanca.

Today the Catholic Church continues to enjoy the loyalties and tithes of hundreds of millions of followers. Yet it and the other theist religions have long since turned from creative into reactive forces. They are busy with rearguard holding operations more than with pioneering novel technologies, innovative economic

methods or groundbreaking social ideas. They now mostly agonise over the technologies, methods and ideas propagated by
other movements. Biologists invent the contraceptive pill – and
the Pope doesn't know what to do about it. Computer scientists
develop the Internet – and rabbis argue whether orthodox Jews
should be allowed to surf it. Feminist thinkers call upon women
to take possession of their bodies – and learned muftis debate how
to confront such incendiary ideas.

Ask yourself: what was the most influential discovery,
invention or creation of the twentieth century? That's a difficult question, because it is hard to choose from a long list of
candidates, including scientific discoveries such as antibiotics,
technological inventions such as computers, and ideological creations such as feminism. Now ask yourself: what was the most
influential discovery, invention or creation of traditional religions such as Islam and Christianity in the twentieth century?
This too is a very difficult question, because there is so little to
choose from. What did priests, rabbis and muftis discover in the
twentieth century that can be mentioned in the same breath as
antibiotics, computers or feminism? Having mulled over these
two questions, from where do you think the big changes of
the twenty-first century will emerge: from the Islamic State,
or Google? Yes, the Islamic State knows how to put videos on
YouTube; but leaving aside the industry of torture, what new
inventions have emerged from Syria or Iraq lately?

Billions of people, including many scientists, continue to
use religious scriptures as a source of authority, but these texts
are no longer a source of creativity. Think, for example, about
the acceptance of gay marriage or female clergy by the more
progressive branches of Christianity. Where did this acceptance originate? Not from reading the Bible, St Augustine or
Martin Luther. Rather, it came from reading texts like Michel
Foucault's *The History of Sexuality* or Donna Haraway's 'A
Cyborg Manifesto'.[14] Yet Christian true-believers – however
progressive – cannot admit to drawing their ethics from Foucault

and Haraway. So they go back to the Bible, to St Augustine and to Martin Luther, and make a very thorough search. They read page after page and story after story with the utmost attention, until they finally discover what they need: some maxim, parable or ruling that, if interpreted creatively enough means God blesses gay marriages and women can be ordained to the priesthood. They then pretend the idea originated in the Bible, when in fact it originated with Foucault. The Bible is kept as a source of authority, even though it is no longer a true source of inspiration.

That's why traditional religions offer no real alternative to liberalism. Their scriptures don't have anything to say about genetic engineering or artificial intelligence, and most priests, rabbis and muftis don't understand the latest breakthroughs in biology and computer science. For if you want to understand these breakthroughs, you don't have much choice – you need to spend time reading scientific articles and conducting lab experiments instead of memorising and debating ancient texts.

That doesn't mean liberalism can rest on its laurels. True, it has won the humanist wars of religion, and as of 2016 it has no viable alternative. But its very success may contain the seeds of its ruin. The triumphant liberal ideals are now pushing humankind to reach for immortality, bliss and divinity. Egged on by the allegedly infallible wishes of customers and voters, scientists and engineers devote more and more energies to these liberal projects. Yet what the scientists are discovering and what the engineers are developing may unwittingly expose both the inherent flaws in the liberal world view and the blindness of customers and voters. When genetic engineering and artificial intelligence reveal their full potential, liberalism, democracy and free markets might become as obsolete as flint knives, tape cassettes, Islam and communism.

This book began by forecasting that in the twenty-first century, humans will try to attain immortality, bliss and divinity. This forecast isn't very original or far-sighted. It simply reflects the traditional ideals of liberal humanism. Since humanism has long sanctified the life, the emotions and the desires of human beings, it's hardly surprising that a humanist civilisation will want to maximise human lifespans, human happiness and human power. Yet the third and final part of the book will argue that attempting to realise this humanist dream will undermine its very foundations by unleashing new post-humanist technologies. The humanist belief in feelings has enabled us to benefit from the fruits of the modern covenant without paying its price. We don't need any gods to limit our power and give us meaning – the free choices of customers and voters supply us with all the meaning we require. What, then, will happen once we realise that customers and voters never make free choices, and once we have the technology to calculate, design or outsmart their feelings? If the whole universe is pegged to the human experience, what will happen once the human experience becomes just another designable product, no different in essence from any other item in the supermarket?

39. Brains as computers – computers as brains. Artificial intelligence is now poised to surpass human intelligence.

PART III

Homo Sapiens Loses Control

Can humans go on running the world and giving it meaning?

How do biotechnology and artificial intelligence threaten humanism?

Who might inherit humankind, and what new religion might replace humanism?

8

The Time Bomb in the Laboratory

In 2016 the world is dominated by the liberal package of individualism, human rights, democracy and the free market. Yet twenty-first-century science is undermining the foundations of the liberal order. Because science does not deal with questions of value, it cannot determine whether liberals are right in valuing liberty more than equality, or in valuing the individual more than the collective. However, like every other religion, liberalism too is based not only on abstract ethical judgments, but also on what it believes to be factual statements. And these factual statements just don't stand up to rigorous scientific scrutiny.

Liberals value individual liberty so much because they believe that humans have free will. According to liberalism the decisions of voters and customers are neither deterministic nor random. People are of course influenced by external forces and chance events, but at the end of the day each of us can wave the magic wand of freedom and decide things for ourselves. This is the reason liberalism gives so much importance to voters and customers, and instructs us to follow our heart and do what feels good. It is our free will that imbues the universe with meaning, and since no outsider can know how you really feel or predict your choices for sure, you shouldn't trust any Big Brother to look after your interests and desires.

Attributing free will to humans is *not* an ethical judgement – it purports to be a factual description of the world. Although this so-called factual description might have made sense back in the days of John Locke, Jean-Jacques Rousseau and Thomas Jefferson, it does not sit well with the latest findings of the life sciences. The contradiction between free will and contemporary science is the elephant in the laboratory, whom many prefer not to see as they peer into their microscopes and fMRI scanners.[1]

In the eighteenth century *Homo sapiens* was like a mysterious black box, whose inner workings were beyond our grasp. Hence when scholars asked why a man drew a knife and stabbed another to death, an acceptable answer said: 'Because he chose to. He used his free will to choose murder, which is why he is fully responsible for his crime.' Over the last century, as scientists opened up the Sapiens black box, they discovered there neither soul, nor free will, nor 'self' – but only genes, hormones and neurons that obey the same physical and chemical laws governing the rest of reality. Today when scholars ask why a man drew a knife and stabbed someone to death, answering 'Because he chose to' doesn't cut the mustard. Instead, geneticists and brain scientists provide a much more detailed answer: 'He did it due to such-and-such electrochemical processes in the brain that were shaped by a particular genetic make-up, which in turn reflect ancient evolutionary pressures coupled with chance mutations.'

The electrochemical brain processes that result in murder are either deterministic or random or a combination of both – but they are never free. For example, when a neuron fires an electric charge, this may be either a deterministic reaction to external stimuli, or perhaps the outcome of a random event such as the spontaneous decomposition of a radioactive atom. Neither option leaves any room for free will. Decisions reached through a chain reaction of biochemical events, each determined by a previous event, are certainly not free. Decisions resulting from random subatomic accidents aren't free either; they are just random. And when random accidents combine with deterministic processes, we get probabilistic outcomes, but this too doesn't amount to freedom.

Suppose we build a robot whose central processing unit is linked to a radioactive lump of uranium. When choosing between two options – say, press the right button or the left button – the robot counts the number of uranium atoms that decayed during the previous minute. If the number is even – it presses the right button. If the number is odd – the left button. We can never be certain about the actions of such a robot. But nobody would call this contraption 'free', and we wouldn't dream of allowing it to vote in democratic elections or holding it legally responsible for its actions.

To the best of our scientific understanding, determinism and randomness have divided the entire cake between them, leaving not even a crumb for 'freedom'. The sacred word 'freedom' turns out to be, just like 'soul', a hollow term empty of any discernible meaning. Free will exists only in the imaginary stories we humans have invented.

The last nail in freedom's coffin is provided by the theory of evolution. Just as evolution cannot be squared with eternal souls, neither can it swallow the idea of free will. For if humans are free, how could natural selection have shaped them? According to the theory of evolution, all the choices animals make – whether of habitat, food or mates – reflect their genetic code. If, thanks to its fit genes, an animal chooses to eat a nutritious mushroom and copulate with healthy and fertile mates, these genes pass on to the next generation. If, because of unfit genes, an animal opts for poisonous mushrooms and anaemic mates, these genes become extinct. However, if an animal 'freely' chooses what to eat and with whom to mate, then natural selection has nothing to work with.

When confronted with such scientific explanations people often brush them aside, pointing out that they *feel* free and that they act according to their own wishes and decisions. This is true. Humans act according to their desires. If by 'free will' we mean the ability to act according to our desires – then yes, humans have free will, and so do chimpanzees, dogs and parrots. When Polly wants a cracker, Polly eats a cracker. But the million-dollar question is not whether parrots and humans can act upon their inner desires – the question is whether they can *choose their desires in the first place*. Why does Polly want a cracker rather than a cucumber? Why am I so eager to kill my annoying neighbour instead of turning the other cheek? Why do I want to buy the red car rather than the black? Why do I prefer voting for the Conservatives rather than the Labour Party? I don't choose any of these wishes. I feel a particular wish welling up within me because this is the feeling created by the biochemical processes in my brain. These processes might be deterministic or random, but not free.

You might reply that at least in the case of major decisions such as murdering a neighbour or electing a government, my choice does not reflect a momentary feeling, but a long and reasoned contemplation of weighty arguments. However, there are many possible trains of argument that I could follow, some of which will cause me to vote Conservative, others to vote Labour, and still others to vote UKIP or just stay at home. What makes me board one train of reasoning rather than another? In the Paddington of my brain, I might be compelled to embark on a particular train of reasoning by deterministic processes, or I might hop on at random. But I don't 'freely' choose to think those thoughts that will make me vote Conservative.

These are not just hypotheses or philosophical speculations. Today we can use brain scanners to predict people's desires and decisions well before they are aware of them. In one such experiment people are placed within a huge brain scanner, holding a switch in each hand. They are asked to press one of the two switches whenever they feel like it. Scientists observing neural activity in the brain can predict which switch the person will press well before the person actually does so, and even before the person is aware of their own intention. Neural events in the brain indicating the person's decision begin from a few hundred milliseconds to a few seconds *before* the person is aware of this choice.[2]

The decision to press either the right or left switch certainly reflects the person's choice. Yet it isn't a *free* choice. In fact, our belief in free will results from faulty logic. When a biochemical chain reaction makes me desire to press the right switch, I feel that I really want to press the right switch. And this is true. I really do want to press it. Yet people erroneously jump to the conclusion that if I want to press it, I *choose* to want to. This is of course false. I don't *choose* my desires. I only *feel* them, and act accordingly.

People nevertheless go on arguing about free will because even scientists all too often continue to use outdated theological concepts. For centuries, Christian, Muslim and Jewish theologians debated the relations between the soul and the will. They assumed that every human has an inner essence – called the soul – which is my true self. They further maintained that this self possesses various desires, just as it possesses clothes, vehicles and houses. I allegedly choose my desires in the same way I choose my clothes, and my fate is determined according to these choices. If I choose good desires, I end up in heaven; if I choose bad desires, I am destined for hell. The question then arose, how exactly do I choose my desires? Why, for example, did Eve desire to eat the forbidden fruit offered to her by the snake? Was this desire forced upon her? Did this desire just pop into her mind by pure chance? Or did she choose it 'freely'? If she didn't choose it freely, why punish her for it?

However, once we accept that there is no soul and that humans have no inner essence called 'the self', it no longer makes sense to ask, 'How does the self choose its desires?' It's like asking a bachelor, 'How does your wife choose her clothes?' In reality, there is only a stream of consciousness, and desires arise and pass away within this stream, but there is no permanent self that owns the desires, hence it is meaningless to ask whether I choose my desires deterministically, randomly or freely.

It may sound extremely complicated, but it is surprisingly easy to test this idea. Next time a thought pops into your mind, stop and ask yourself: 'Why did I think this particular thought? Did I decide a minute ago to think this thought, and only then think it? Or did it just arise, without any direction or permission from me? If I am indeed the master of my thoughts and decisions, can I decide not to think about anything at all for the next sixty seconds?' Try that, and see what happens.

Doubting free will is not just a philosophical exercise. It has practical implications. If organisms indeed lack free will, it implies that we can manipulate and even control their desires using drugs, genetic engineering or direct brain stimulation.

If you want to see philosophy in action, pay a visit to a robo-rat laboratory. A robo-rat is a run-of-the-mill rat with a twist: scientists have implanted electrodes into the sensory and reward areas in the rat's brain. This enables the scientists to manoeuvre the rat by remote control. After short training sessions, researchers have managed not only to make the rats turn left or right, but also to climb ladders, sniff around garbage piles, and do things that rats normally dislike, such as jumping from extreme heights. Armies and corporations are showing keen interest in the robo-rats, hoping they will prove useful in many tasks and situations. For example, robo-rats might help detect survivors trapped under collapsed buildings, locate bombs and booby traps, and map underground tunnels and caves.

Animal-welfare activists have voiced concern about the suffering such experiments inflict on the rats. Professor Sanjiv Talwar of the State University of New York, one of the leading robo-rat researchers, has dismissed these concerns, arguing that the rats actually enjoy the experiments. After all, explains Talwar, the rats 'work for pleasure' and when the electrodes stimulate the reward centres in their brains, 'the rat feels Nirvana'.[3]

To the best of our understanding, the rat doesn't feel that somebody else controls her, and she doesn't feel that she is being coerced to do anything against her will. When Professor Talwar presses the remote control, the rat *wants* to move to the left, which is why she moves to the left. When the professor presses another switch, the rat *wants* to climb a ladder, which is why she climbs the ladder. After all, the rat's desires are nothing but a pattern of firing neurons. What does it matter whether the neurons are firing because they are stimulated by other neurons or by transplanted electrodes connected to Professor Talwar's remote control? If you ask the rat about it, she might well tell you, 'Sure I have free will! Look, I want to turn left – and I turn left. I want to climb a ladder – and I climb a ladder. Doesn't that prove that I have free will?'

Experiments performed on *Homo sapiens* indicate that like rats humans too can be manipulated, and that it is possible to create or annihilate even complex feelings such as love, anger, fear and depression by stimulating the right spots in the human brain. The US military has recently initiated experiments on implanting computer chips in people's brains, hoping to use this method to treat soldiers suffering from post-traumatic stress disorder.[4] In Hadassah Hospital in Jerusalem, doctors have pioneered a novel treatment for patients suffering from acute depression. They implant electrodes into the patient's brain, and wire them to a minuscule computer implanted in the patient's chest. On receiving a command from the computer, the electrodes transmit weak electric currents that paralyse the brain area responsible for the depression. The treatment does not always succeed, but in some cases patients reported that the feeling of dark emptiness that tormented them throughout their lives disappeared as if by magic.

One patient complained that several months after the operation he had a relapse and was overcome by severe depression. Upon inspection the doctors found the source of the problem: the computer's battery had run out of power. Once they changed the battery, the depression quickly melted away.[5]

Due to obvious ethical restrictions researchers implant electrodes into human brains only in exceptional circumstances. Hence most relevant experiments on humans are conducted using non-intrusive helmet-like devices (technically known as 'transcranial direct-current stimulators'). The helmet is fitted with electrodes that attach to the outside of the scalp. It produces weak electromagnetic fields and directs them towards specific brain areas, thereby stimulating or inhibiting select brain activities.

The American military is experimenting with such helmets in the hope of sharpening the focus and enhancing the performance of soldiers both in training sessions and on the battlefield. The main experiments are conducted by the Human Effectiveness Directorate, which is located at an Ohio air force base. Though the results are far from conclusive, and though the hype around transcranial stimulators currently runs far ahead of actual achievements, several studies have indicated that the method may indeed enhance the cognitive abilities of drone operators, air-traffic controllers, snipers and other personnel whose duties require them to remain highly attentive for extended periods.[6]

Sally Adee, a journalist for the *New Scientist*, was allowed to visit a training facility for snipers and test the effects herself. At first she entered a battlefield simulator without wearing the transcranial helmet. Sally describes how fear swept over her as twenty masked men, strapped with suicide bombs and armed with rifles, charge straight towards her. 'For every one I manage to shoot dead,' writes Sally, 'three new assailants pop up from nowhere. I'm clearly not shooting fast enough, and panic and incompetence are making me continually jam my rifle.' Luckily for her, the assailants were just video images projected on huge screens all around her. Still, she was so disappointed with her poor performance that she felt like ditching the rifle and leaving the simulator.

Then they wired her up to the helmet. She reports feeling nothing unusual, except a slight tingle and a strange metallic taste in her mouth. Yet she began picking off the virtual terrorists one by one, as coolly and methodically as if she were Rambo or Clint Eastwood. 'As twenty of them run at me brandishing their guns, I calmly line up my rifle, take a moment to breathe deeply, and pick off the closest one, before tranquilly assessing my next target. In what seems like next to no time, I hear a voice call out,

"Okay, that's it." The lights come up in the simulation room . . . In the sudden quiet amid the bodies around me, I was really expecting more assailants, and I'm a bit disappointed when the team begins to remove my electrodes. I look up and wonder if someone wound the clocks forward. Inexplicably, twenty minutes have just passed. "How many did I get?" I ask the assistant. She looks at me quizzically. "All of them."'

The experiment changed Sally's life. In the following days she realised she had been through a 'near-spiritual experience . . . what defined the experience was not feeling smarter or learning faster: the thing that made the earth drop out from under my feet was that for the first time in my life, everything in my head finally shut up . . . My brain without self-doubt was a revelation. There was suddenly this incredible silence in my head . . . I hope you can sympathise with me when I tell you that the thing I wanted most acutely for the weeks following my experience was to go back and strap on those electrodes. I also started to have a lot of questions. Who was I apart from the angry bitter gnomes that populate my mind and drive me to failure because I'm too scared to try? And where did those voices come from?'[7]

Some of those voices repeat society's prejudices, some echo our personal history, and some articulate our genetic legacy. All of them together, says Sally, create an invisible story that shapes our conscious decisions in ways we seldom grasp. What would happen if we could rewrite our inner monologues, or even silence them completely on occasion?[8]

As of 2016 transcranial stimulators are still in their infancy, and it is unclear if and when they will become a mature technology. So far they provide enhanced capabilities for only short durations, and Sally Adee's twenty-minute experience may be quite exceptional (or perhaps even the outcome of the notorious placebo effect). Most published studies of transcranial stimulators are based on very small samples of people operating under special circumstances, and the long-term effects and hazards are completely unknown. However, if the technology does mature, or if some other method is found to manipulate the brain's electric patterns, what would it do to human societies and to human beings?

People may well manipulate their brain's electric circuits not just to shoot terrorists more proficiently, but also to achieve more mundane liberal goals. Namely, to study and work more efficiently, immerse themselves in games and hobbies, and be able to focus on what interests them at any particular moment, be it maths or football. However, if and when such manipulations become routine, the supposedly free will of customers will become just another product to purchase. You want to master the piano but whenever it comes time to practice you prefer to watch television? No problem: just put on a helmet, install the right software, and you will be downright aching to play the piano.

You may counter-argue that the ability to silence or enhance the voices in your head will actually strengthen rather than undermine your free will. Presently, you often fail to realise your most cherished and authentic desires due to external distractions. With the help of the attention helmet and similar devices, you could more easily silence the alien voices of parents, priests, spin doctors, advertisers and neighbours, and focus on what *you* want. However, as we will shortly see, the notion that you have a single self and that you could therefore distinguish your authentic desires from alien voices is just another liberal myth, debunked by the latest scientific research.

Who Are I?

Science undermines not only the liberal belief in free will, but also the belief in individualism. Liberals believe that we have a single and indivisible self. To be an individual means that I am in-dividual. Yes, my body is made up of approximately 37 trillion cells,[9] and each day both my body and my mind go through countless permutations and transformations. Yet if I really pay attention and strive to get in touch with myself, I am bound to discover deep inside a single, clear and authentic voice, which is my true self, and which is the source of all meaning and authority in the universe. For liberalism to make sense, I must have one – and only one – true self, for if I had more than one authentic voice, how would I know which voice to heed in the polling station, in the supermarket and in the marriage market?

However, over the last few decades the life sciences have reached the conclusion that this liberal story is pure mythology. The single authentic self is as real as the eternal soul, Santa Claus and the Easter Bunny. If I look really deep within myself, the seeming unity that I take for granted dissolves into a cacophony of conflicting voices, none of which is 'my true self'. Humans aren't individuals. They are 'dividuals'.

The human brain is composed of two hemispheres connected by a thick neural cable. Each hemisphere controls the opposite side of the body. The right hemisphere controls the left side of the body, receives data from the left-hand field of vision and is responsible for moving the left arm and leg – and vice versa. This is why people who have had a stroke in their right hemisphere sometimes ignore the left side of their body (combing hair only on the right side of their head, or eating only the food placed on the right side of their plate).[10]

There are also emotional and cognitive differences between the two hemispheres, though the division is far from clear-cut. Most cognitive activities involve both hemispheres, but not to the same degree. For example, in most cases the left hemisphere plays a more important role in speech and logical reasoning, whereas the right hemisphere is more dominant in processing spatial information.

Many breakthroughs in understanding the relations between the two hemispheres were based on the study of epilepsy patients. In severe cases of epilepsy, electrical storms begin in one part of the brain but quickly spread to other parts, causing a very acute seizure. During such seizures patients lose control of their bodies, and frequent seizures consequently preclude them from holding jobs or leading normal lives. In the mid-twentieth century, when all other treatments failed, doctors alleviated the problem by cutting the thick neural cable connecting the two hemispheres, so that electrical storms beginning in one hemisphere could not spill over to the other. For brain scientists these patients were a goldmine of astounding data.

Some of the most notable studies on these split-brain patients were conducted by Professor Roger Wolcott Sperry, who won the 1981 Nobel Prize in Physiology and Medicine for his groundbreaking discoveries, and by his student, Professor Michael S. Gazzaniga. One study was conducted on a teenage boy. The boy was asked what he would like to do when he grew up. The boy answered: a draughtsman. This answer was provided by his left brain hemisphere, which plays a crucial part in logical reasoning as well as in speech. Yet the boy had another active speech centre in his right hemisphere, which could not control vocal language but could spell words using Scrabble tiles. The researchers were keen to know what the right hemisphere would say. So they spread Scrabble tiles on a table and wrote on a piece of paper: 'What would you like to do when you grow up?' They placed the paper at the edge of the boy's left visual field. Data from the left visual field is processed in the right hemisphere. Since the right hemisphere could not use vocal language, the boy said nothing. But his left hand began moving rapidly across the table, collecting tiles from here and there, until it spelled out: 'automobile race'. Spooky.[11]

Equally eerie behaviour was displayed by patient WJ, a Second World War veteran. WJ's hands were each controlled by a different hemisphere. Since the two hemispheres were not in touch with one another, it sometimes happened that his right hand would reach out to open a door, and then his left hand would intervene and try to slam the door shut.

In another experiment Gazzaniga and his team flashed a picture of a chicken claw to the left-half brain – the side responsible for speech – and simultaneously flashed a picture of a snowy landscape to the right brain. When asked what he saw, patient PS answered 'a chicken claw'. Gazzaniga then presented PS with a series of picture cards and asked him to point to the one that best matched what he had seen. The patient's right hand (controlled by his left brain) pointed to a picture of a chicken, but simultaneously his left hand shot out and pointed to a snow shovel. Gazzaniga then asked PS the obvious question: 'Why did you point both to the chicken and to the shovel?' PS replied, 'Oh, the chicken claw goes with the chicken, and you need a shovel to clean out the chicken shed.'[12]

What happened here? The left brain, which controls speech, had no data about the snow scene, and therefore did not really know why the left hand pointed to the shovel. So it just invented something credible. After repeating this experiment many times Gazzaniga concluded that the left hemisphere of the brain is the seat not only of our verbal abilities, but also of an internal inter-preter that constantly tries to make sense of our life, using partial clues in order to concoct plausible stories.

In yet another experiment the non-verbal right hemisphere was shown a pornographic image. The patient reacted by blushing and giggling. 'What did you see?' asked the mischievous researchers. 'Nothing, just a flash of light,' said the left hemisphere, and the patient immediately giggled again, covering her mouth with her hand. 'Why are you laughing then?' they insisted. The bewildered left-hemisphere interpreter – struggling for some rational explanation – replied that one of the machines in the room looked very funny.[13]

It's as if the CIA conducts a drone strike in Pakistan, unbe-known to the US State Department. When a journalist grills State Department officials about it, they concoct some plausible explan-ation. In reality, the spin doctors don't have a clue why the strike was ordered, so they just invent something. A similar mechanism is employed by all human beings, not just by split-brain patients. Again and again my own private CIA does things without the approval or knowledge of my State Department, and then my State Department cooks up a story that presents me in the best possible light. Often enough the State Department itself becomes convinced of the pure fantasies it has invented.[14]

Similar conclusions have been reached by behavioural economists who want to know how people take economic decisions. Or more accurately, *who* takes these decisions. Who decides to buy a Toyota rather than a Mercedes, to go on holiday to Paris rather than Thailand, and to invest in South Korean treasury bonds rather than in the Shanghai stock exchange? Most experiments have indicated that there is no single self making any of these decisions. Rather, they result from a tug of war between different and often conflicting inner entities.

One groundbreaking experiment was conducted by Daniel Kahneman, who won the 2002 Nobel Prize in Economics. Kahneman asked a group of volunteers to join a three-part experiment. In the 'short' part of the experiment, the volunteers inserted one hand into a container filled with water at 14°C for one minute, which is unpleasant, bordering on painful. After sixty seconds they were told to take their hand out. In the 'long' part of the experiment, volunteers placed their other hand in a different water container whose temperature was also 14°C. But after sixty seconds hot water was secretly introduced into the container, slightly increasing the temperature to 15°C. Thirty seconds later they were told to pull out their hand. Some volunteers did the 'short' part first, while others began with the 'long' part. In either case, exactly seven minutes after both parts were over came the third and most important part of the experiment. The volunteers were told they must repeat one of the two parts; and it was up to them to choose which. Fully 80 per cent preferred to repeat the 'long' experiment, remembering it as less painful.

This cold-water experiment is so simple, yet its implications shake the core of the liberal world view. It exposes the existence of at least two different selves within us: the experiencing self and the narrating self. The experiencing self is our moment-to-moment consciousness. For the experiencing self, it's obvious that the 'long' part of the cold-water experiment was worse. First you experience water at 14°C for sixty seconds, which is every bit as disagreeable as what you experience in the 'short' part, and then you must endure another thirty seconds of water at 15°C, which is marginally less bad, but still far from pleasant. For the experiencing self, it is impossible that adding a slightly unpleasant experience to a very unpleasant experience will make the entire episode more appealing.

However, the experiencing self remembers nothing. It tells no stories and is seldom consulted when it comes to big decisions. Retrieving memories, telling stories and making major decisions are all the monopoly of a very different entity inside us: the narrating self. The narrating self is akin to Gazzaniga's left-brain interpreter. It is forever busy spinning yarns about the past and making plans for the future. Like every journalist, poet and politician, the narrating self takes many short cuts. It doesn't narrate everything, and usually weaves the story using only peak moments and end results. The value of the whole experience is determined by averaging peaks with ends. For example, in evaluating the short part of the cold-water experiment the narrating self calculates the average between the worst part (the water was very cold) and the last moment (the water was still very cold) and concludes that 'the water was very cold'. The narrating self does the same thing with the long part of the experiment. It finds the average between the worst part (the water was very cold) and the last moment (the water was not quite so cold) and concludes that 'the water was somewhat warmer'. Crucially, the narrating self is duration-blind, giving no importance to the differing lengths of the two parts. So when it has a choice between the two, it prefers to repeat the long part, the one in which 'the water was somewhat warmer'.

Every time the narrating self evaluates our experiences, it discounts their duration and adopts the 'peak-end rule' – it remembers only the peak moment and the end moment, and assesses the whole experience according to their average. This has far-reaching impact on all our practical decisions. Kahneman began investigating the experiencing self and the narrating self in the early 1990s when, together with Donald Redelmeier of the University of Toronto, he studied colonoscopy patients. In colonoscopy tests a tiny camera is inserted into the intestines through the anus in order to diagnose various bowel diseases. It is not a pleasant experience. Doctors wanted to know how to perform this procedure in the least painful way. Should they hasten the colonoscopy and cause patients more distress for a shorter duration, or should they work more slowly and carefully?

To answer this query, Kahneman and Redelmeier asked 154 patients to report their pain level at one-minute intervals during the colonoscopy. They used a scale of 0 to 10, where 0 meant no pain at all and 10 meant intolerable pain. After the colonoscopy was over patients were asked to rank the test's 'overall pain level', also on a scale of 0 to 10. We might have expected the overall ranking to reflect the accumulation of minute-by-minute reports, that is the longer the colonoscopy lasted, and the more pain the patient experienced, the higher the overall pain level. But the actual results were different.

Just as in the cold-water experiment, the overall pain level neglected duration and instead reflected only the peak-end rule. One colonoscopy lasted eight minutes, at the worst moment the patient reported a level 8 pain, and in the last minute he reported a level 7 pain. After the test was over this patient ranked his overall pain level at 7.5. Another colonoscopy lasted twenty-four minutes. This time too peak pain was at level 8, but in the very last minute of the test, the patient reported a level 1 pain. This patient ranked his overall pain level at only 4.5. The fact that his colonoscopy lasted three times longer, and that he consequently suffered far more pain on aggregate, did not affect his memory at all. The narrating self doesn't aggregate experiences – it averages them.

So what do the patients prefer: to have a short and sharp colonoscopy, or a long and careful one? There isn't a single answer to this question, because the patient has at least two different selves and they have different interests. If you ask the experiencing self, it would probably choose a short colonoscopy. But if you ask the narrating self, it would prefer a long colonoscopy because it remembers only the average between the worst moment and the last moment.

Indeed, from the viewpoint of the narrating self, the doctor should add a few completely superfluous minutes of dull aches at the very end of the test, because it would make the entire memory far less traumatic.[15]

Paediatricians know this trick well. So do veterinarians. Many keep in their clinics jars full of treats, and hand a few to the kids (or dogs) after giving them a painful injection or an unpleasant medical examination. When the narrating self remembers the visit to the doctor, ten seconds of pleasure at the end of the visit will erase many minutes of anxiety and pain.

Evolution discovered this trick aeons before the paediatricians. Given the unbearable torments that many women undergo during childbirth, one might think that after going through it once no sane woman would ever agree to do so again. However, at the end of labour and in the following days the hormonal system secretes cortisol and beta-endorphins, which reduce the pain and create a feeling of relief and sometimes even of elation. Moreover, the growing love towards the baby and the acclaim from friends, family members, religious dogmas and nationalist propaganda, conspire to transform childbirth from a trauma into a positive memory.

40. An iconic image of the Virgin Mary holding baby Jesus. In most cultures, childbirth is portrayed as a wonderful experience rather than as a trauma.

One study conducted at the Rabin Medical Center in Tel Aviv demonstrated that the memory of labour reflected mainly the peak and end points, while the overall duration had almost no impact at all.[16] In another research project, 2,428 Swedish women were asked to recount their memories of labour two months after giving birth. Ninety per cent reported that the experience was either positive or very positive. They didn't necessarily forget the pain – 28.5 per cent described it as the worst pain imaginable – yet it did not prevent them from evaluating the experience as positive. The narrating self goes over our experiences with a sharp pair of scissors and a thick black marker. It censors at least some moments of horror, and files in the archive a story with a happy ending.[17]

Most of our critical life choices – of partners, careers, residences and holidays – are taken by our narrating self. Suppose you can choose between two potential holidays. You can go to Jamestown, Virginia, and visit the historic colonial town where the first English settlement on mainland North America was founded in 1607. Alternatively, you can realise your number one dream vacation, whether it is trekking in Alaska, sunbathing in Florida or indulging in an unbridled bacchanalia of sex, drugs and gambling in Las Vegas. But there is a caveat: if you choose your dream vacation, then just before you board the plane home, you must take a pill that will obliterate all your memories of that vacation. What happened in Vegas will forever remain in Vegas. Which holiday would you choose? Most people would opt for colonial Jamestown, because most people give their credit card to the narrating self, which cares only about stories and has zero interest in even the most mind-blowing experiences if it cannot remember them.

Truth be told, the experiencing self and the narrating self are not completely separate entities but are closely intertwined. The narrating self uses our experiences as important (but not exclusive) raw materials for its stories. These stories, in turn, shape what the experiencing self actually feels. We experience hunger differently when we fast during Ramadan, when we fast in preparation for a medical examination, and when we don't eat because we have no money. The different meanings ascribed to our hunger by the narrating self create very different actual experiences.

Furthermore, the experiencing self is often strong enough to sabotage the best-laid plans of the narrating self. I might, for instance, make a New Year's resolution to start a diet and go to the gym every day. Such grand decisions are the monopoly of the narrating self. But the following week when it's gym time, the experiencing self takes over. I don't feel like going to the gym, and instead I order pizza, sit on the sofa and turn on the TV.

Nevertheless, most of us identify with our narrating self. When we say 'I', we mean the story in our head, not the onrushing stream of experiences we undergo. We identify with the inner system that takes the crazy chaos of life and spins out of it seemingly logical and consistent yarns. It doesn't matter that the plot is full of lies and lacunas, and is rewritten again and again, so that today's story flatly contradicts yesterday's. The important thing is that we always retain the feeling that we have a single unchanging identity from birth to death (and perhaps even beyond). This gives rise to the questionable liberal belief that I am an individual, and that I possess a clear and consistent inner voice that provides meaning to the entire universe.[18]

The Meaning of Life

The narrating self is the star of Jorge Luis Borges's story 'A Problem'.[19] The story concerns Don Quixote, the eponymous hero of Miguel Cervantes's famous novel. Don Quixote creates for himself an imaginary world in which he is a legendary champion going forth to fight giants and save Lady Dulcinea del Toboso. In reality Don Quixote is Alonso Quixano, an elderly country gentleman; the noble Dulcinea is an uncouth farm girl from a nearby village; and the giants are windmills. What would happen, wonders Borges, if due to his belief in these fantasies, Don Quixote attacks and kills a real person? Borges asks a fundamental question about the human condition: what happens when the yarns spun by our narrating self cause grievous harm to ourselves or those around us? There are three main possibilities, says Borges.

One option is that nothing much happens. Don Quixote will not be bothered at all by killing a real man. His delusions are so overpowering that he will not be able to recognise the difference between committing actual murder and dueling with the imaginary windmill giants. Another option is that once he takes a person's life, Don Quixote will be so horrified that he will be shaken out of his delusions. This is akin to a young recruit who goes to war believing that it is good to die for one's country, only to end up completely disillusioned by the realities of warfare.

But there is a third option, much more complex and profound. As long as he fought imaginary giants, Don Quixote was just play-acting. However once he actually kills someone, he will cling to his fantasies for all he is worth, because only they give meaning to his tragic misdeed. Paradoxically, the more sacrifices we make for an imaginary story, the more tenaciously we hold on to it, because we desperately want to give meaning to these sacrifices and to the suffering we have caused.

In politics this is known as the 'Our Boys Didn't Die in Vain' syndrome. In 1915 Italy entered the First World War on the side of the Entente powers. Italy's declared aim was to 'liberate' Trento and Trieste – two 'Italian' territories that the Austro-Hungarian Empire held 'unjustly'. Italian politicians gave fiery speeches in parliament, vowing historical redress and promising a return to the glories of ancient Rome. Hundreds of thousands of Italian recruits went to the front shouting, 'For Trento and Trieste!' They thought it would be a walkover.

It was anything but. The Austro-Hungarian army held a strong defensive line along the Isonzo River. The Italians hurled themselves against it in eleven gory battles, gaining a few miles at most and never securing a breakthrough. In the first battle they lost 15,000 men. In the second battle they lost 40,000 men.

In the third battle they lost 60,000. So it continued for more than two dreadful years until the eleventh engagement. Then the Austrians finally counter-attacked, and in the twelfth battle, known as the battle of Caporreto, they soundly defeated the Italians and pushed them back almost to the gates of Venice. The glorious adventure became a bloodbath. By the end of the war almost 700,000 Italian soldiers were killed and more than a million were wounded.[20]

After losing the first Isonzo battle Italian politicians had two choices. They could have admitted their mistake and offered to sign a peace treaty. Austria–Hungary had no claims against Italy and would have been delighted to make a peace treaty because it was busy fighting for its survival against the much stronger Russians. Yet how could the politicians go to the parents, wives and children of 15,000 dead Italian soldiers and tell them: 'Sorry, there has been a mistake. We hope you won't take it too hard, but your Giovanni died in vain, and so did your Marco.' Alternatively

41. A few of the victims of the Isonzo battles. Was their sacrifice in vain?

they could say: 'Giovanni and Marco were heroes! They died so that Trieste would be Italian, and we will make sure they didn't die in vain. We will go on fighting until victory is ours!' Not surprisingly, the politicians preferred the second option. So they fought a second battle, and lost another 40,000 men. The politicians again decided it would be best to keep on fighting, because 'our boys didn't die in vain'.

Yet you cannot blame only the politicians. The masses also kept supporting the war. And when after the war Italy did not get all the territories it demanded, Italian democracy placed at its head Benito Mussolini and his fascists, who promised they would obtain a proper compensation for all the sacrifices the Italians had made. While it's difficult for a politician to tell parents that their son died for no good reason, it is far more painful for parents to say this to themselves – and it is even harder for the victims. A crippled soldier who lost his legs would rather tell himself, 'I sacrificed myself for the glory of the eternal Italian nation!' than 'I lost my legs because I was stupid enough to believe self-serving politicians.' It is much easier to live with the fantasy, because the fantasy gives meaning to the suffering.

Priests discovered this principle thousands of years ago. It underlies numerous religious ceremonies and commandments. If you want to make people believe in imaginary entities such as gods and nations, you should make them sacrifice something valuable. The more painful the sacrifice, the more convinced they will be of the existence of the imaginary recipient. A poor peasant sacrificing a valuable bull to Jupiter will become convinced that Jupiter really exists, otherwise how can he excuse his stupidity? The peasant will sacrifice another bull, and another, and another, just so he won't have to admit that all the previous bulls were wasted. For exactly the same reason, if I have sacrificed a child to the glory of the Italian nation or my legs to the communist revolution, that's usually enough to turn me into a zealous Italian nationalist or an enthusiastic communist. For if Italian national myths or communist propaganda are a lie, then

I will be forced to concede that my child's death or my own paralysis have been completely pointless. Few people have the stomach to admit such a thing.

The same logic is at work in the economic sphere too. In 1999 the government of Scotland decided to erect a new parliament building. According to the original plan the construction was expected to take two years and cost £40 million. In fact it took five years and cost £400 million. Every time the contractors encountered unforeseen difficulties and expenses, they went to the Scottish government and asked for more time and money. Every time this happened the government said to itself: 'Well, we've already sunk tens of millions into this and we'll be completely discredited if we stop now and end up with a partially built skeleton. Let's authorise another £40 million.' Several months later the same thing happened again, by which time the pressure to avoid ending up with an unfinished building was even greater. And a few months after that the story repeated itself, and so on until the actual cost was ten times the original estimate.

Not only governments fall into this trap. Business corporations often sink millions into failed enterprises, while private individuals

42. The Scottish Parliament building. Our sterling did not die in vain.

cling to dysfunctional marriages and dead-end jobs. Our narrating self would much prefer to continue suffering in the future, just so it won't have to admit that our past suffering was devoid of all meaning. Eventually, if we want to come clean about past mistakes, our narrating self must invent some twist in the plot that will infuse these mistakes with meaning. For example, a pacifist war veteran may tell himself, 'Yes, I've lost my legs because of a mistake. But thanks to this mistake I understand that war is hell, and from now onwards I will dedicate my life to fight for peace. So my injury did have some positive meaning after all: it taught me to value peace.'

We see then that the self too is an imaginary story, just like nations, gods and money. Each of us has a sophisticated system that throws away most of our experiences, keeps only a few choice samples, mixes them up with bits from movies we've seen, novels we've read, speeches we've heard, and daydreams we've savoured, and out of all that jumble it weaves a seemingly coherent story about who I am, where I came from and where I am going. This story tells me what to love, whom to hate and what to do with myself. This story may even cause me to sacrifice my life, if that's what the plot requires. We all have our genre. Some people live a tragedy, others inhabit a never-ending religious drama, some approach life as if it were an action film, and not a few act as if in a comedy. But in the end, they are all just stories.

What, then, is the meaning of life? Liberalism maintains that we shouldn't expect some external entity to provide us a ready-made meaning. Rather, each individual voter, customer and viewer ought to use his or her free will in order to create meaning, not just for his or her life but for the entire universe.

The life sciences, however, undermine liberalism, arguing that the free individual is just a fictional tale concocted by an assembly of biochemical algorithms. Every moment the biochemical mechanisms of the brain create a flash of experience, which immediately disappears. Then more flashes appear and fade, appear and fade, in quick succession. These momentary experiences do not add up to any enduring essence. The narrating self tries to impose order on this chaos by spinning a never-ending story, in which every such experience has its place, and hence every experience has some lasting meaning. But, as convincing and tempting as it may be, this story is a fiction. Medieval crusaders believed that God and heaven provided their lives with meaning; modern liberals believe that individual free choices provide life with meaning. They are all equally delusional.

Doubts about the existence of free will and individuals are nothing new, of course. More than 2,000 years ago thinkers in India, China and Greece argued that 'the individual self is an illusion'. Yet such doubts don't really change history much unless they have a practical impact on economics, politics and day-to-day life. Humans are masters of cognitive dissonance, and we allow ourselves to believe one thing in the laboratory and an altogether different thing in the courthouse or in parliament. Just as Christianity didn't disappear the day Darwin published *On the Origin of Species,* so liberalism won't vanish just because scientists have reached the conclusion that there are no free individuals.

Indeed, even Richard Dawkins, Steven Pinker and the other champions of the new scientific world view refuse to abandon liberalism. After dedicating hundreds of erudite pages to deconstructing the self and the freedom of will, they perform breathtaking intellectual somersaults that miraculously land them back in the eighteenth century, as if all the amazing discoveries of evolutionary biology and brain science have absolutely no bearing on the ethical and political ideas of Locke, Rousseau and Jefferson.

However, once the heretical scientific insights are translated into everyday technology, routine activities and economic structures, it will become increasingly difficult to sustain this double-game, and we – or our heirs – will probably require a brand-new package of religious beliefs and political institutions. At the beginning of the third millennium liberalism is threatened not by the philosophical

idea that 'there are no free individuals', but rather by concrete technologies. We are about to face a flood of extremely useful devices,
tools and structures that make no allowance for the free will of
individual humans. Will democracy, the free market and human
rights survive this flood?

9
The Great Decoupling

The preceding pages took us on a brief tour of recent scientific discoveries that undermine the liberal philosophy. It's time to examine the practical implications of these discoveries. Liberals uphold free markets and democratic elections because they believe that every human is a uniquely valuable individual, whose free choices are the ultimate source of authority. In the twenty-first century three *practical* developments might make this belief obsolete:

1. Humans will lose their economic and military usefulness, hence the economic and political system will stop attaching much value to them.
2. The system will continue to find value in humans collectively, but not in unique individuals.
3. The system will still find value in some unique individuals, but these will constitute a new elite of upgraded superhumans rather than the mass of the population.

Let's examine all three threats in detail. The first – that technological developments will make humans economically and militarily useless – will not prove that liberalism is wrong on a philosophical level, but in practice it is hard to see how democracy, free markets and other liberal institutions can survive such a blow. After all, liberalism did not become the dominant ideology simply because its philosophical arguments were the most valid. Rather, liberalism succeeded because there was abundant political, economic and military sense in ascribing value to every human being. On the mass battlefields of modern industrial wars and in the mass production lines of modern industrial economies, every human counted. There was value to every pair of hands that could hold a rifle or pull a lever.

In the spring of 1793 the royal houses of Europe sent their armies to strangle the French Revolution in its cradle. The firebrands in Paris reacted by proclaiming the *levée en masse* and unleashing the first total war. On 23 August, the National Convention decreed that 'From this moment until such time as its enemies shall have been driven from the soil of the Republic, all Frenchmen are in permanent requisition for the services of the armies. The young men shall fight; the married men shall forge arms and transport provisions; the women shall make tents and clothes and shall serve in the hospitals; the children shall turn old lint into linen; and the old men shall betake themselves to the public squares in order to arouse the courage of the warriors and preach hatred of kings and the unity of the Republic.'[1]

This decree sheds interesting light on the French Revolution's most famous document – *The Declaration of the Rights of Man and of the Citizen* – which recognised that all citizens have equal value and equal political rights. Is it a coincidence that universal rights were proclaimed at the precise historical juncture when universal conscription was decreed? Though scholars may quibble about the exact relations between them, in the following two centuries a common argument in defence of democracy explained that giving citizens political rights is good, because the soldiers and workers of democratic countries perform better than those of dictatorships. Allegedly, granting political rights to people increases their motivation and their initiative, which is useful both on the battlefield and in the factory.

Thus Charles W. Eliot, president of Harvard from 1869 to 1909, wrote on 5 August 1917 in the *New York Times* that 'democratic armies fight better than armies aristocratically organised and autocratically governed' and that 'the armies of nations in which the mass of the people determine legislation, elect their public servants, and settle questions of peace and war, fight better than the armies of an autocrat who rules by right of birth and by commission from the Almighty'.[2]

A similar rationale favoured the enfranchisement of women in the wake of the First World War. Realising the vital role of women in total industrial wars, countries saw the need to give them political rights in peacetime. Thus in 1918 President Woodrow Wilson became a supporter of women's suffrage, explaining to the US Senate that the First World War 'could not have been fought, either by the other nations engaged or by America, if it had not been for the services of women – services rendered in every sphere – not only in the fields of effort in which we have been accustomed to see them work, but wherever men have worked and upon the very skirts and edges of the battle itself. We shall not only be distrusted but shall deserve to be distrusted if we do not enfranchise them with the fullest possible enfranchisement.'[3]

However, in the twenty-first century the majority of both men and women might lose their military and economic value. Gone is the mass conscription of the two world wars. The most advanced armies of the twenty-first century rely far more on cutting-edge technology. Instead of limitless cannon fodder, countries now need only small numbers of highly trained soldiers, even smaller numbers of special forces super-warriors and a handful of experts who know how to produce and use sophisticated technology. Hi-tech forces 'manned' by pilotless drones and cyber-worms are replacing the mass armies of the twentieth century, and generals delegate more and more critical decisions to algorithms.

Aside from their unpredictability and their susceptibility to fear, hunger and fatigue, flesh-and-blood soldiers think and move on an increasingly irrelevant timescale. From the days of Nebuchadnezzar to those of Saddam Hussein, despite myriad technological improvements, war was waged on an organic timetable. Discussions lasted for hours, battles took days, and wars dragged on for years. Cyberwars, however, may last just a few minutes. When a lieutenant on shift at cyber-command notices something odd is going on, she picks up the phone to call her superior, who immediately alerts the White House. Alas, by the time the president reaches for the red handset, the war has already been lost. Within seconds a sufficiently sophisticated cyber strike might shut down the US power grid, wreck US flight control centres, cause numerous industrial accidents in nuclear plants and chemical installations, disrupt the police, army and intelligence communication networks – and wipe out financial records so that trillions of dollars simply vanish without a trace and nobody knows who owns what. The only thing curbing public hysteria is that, with the Internet, television and radio down, people will not be aware of the full magnitude of the disaster.

On a smaller scale, suppose two drones fight each other in the air. One drone cannot open fire without first receiving the go-ahead from a human operator in some distant bunker. The other is fully autonomous. Which drone do you think will prevail? If in 2093 the decrepit European Union sends its drones and cyborgs to snuff out a new French Revolution, the Paris Commune might press into service every available hacker, computer and smartphone, but it

43. Left: Soldiers in action at the Battle of the Somme, 1916. Right: A pilotless drone.

will have little use for most humans, except perhaps as human shields. It is telling that already today in many asymmetrical conflicts the majority of citizens are reduced to serving as shields for advanced armaments.

Even if you care more about justice than victory, you should probably opt to replace your soldiers and pilots with autonomous robots and drones. Human soldiers murder, rape and pillage, and even when they try to behave themselves, they all too often kill civilians by mistake. Computers programmed with ethical algorithms could far more easily conform to the latest rulings of the international criminal court.

In the economic sphere too the ability to hold a hammer or press a button is becoming less valuable than before, which endangers the critical alliance between liberalism and capitalism. In the twentieth century liberals explained that we don't have to choose between ethics and economics. Protecting human rights and liberties was both a moral imperative and the key to economic growth. Britain, France and the United States allegedly prospered because they liberalised their economies and societies, and if Turkey, Brazil or China wanted to become equally prosperous, they had to do the same. In many if not most cases it was the economic rather than the moral argument that convinced tyrants and juntas to liberalise.

In the twenty-first century liberalism will have a much harder time selling itself. As the masses lose their economic importance, will the moral argument alone be enough to protect human rights and liberties? Will elites and governments go on valuing every human being even when it pays no economic dividends?

In the past there were many things only humans could do. But now robots and computers are catching up, and may soon outperform humans in most tasks. True, computers function very differently from humans, and it seems unlikely that computers will become humanlike any time soon. In particular, it doesn't seem that computers are about to gain consciousness and start experiencing emotions and sensations. Over the past half century there has been an immense advance in computer intelligence, but there has been exactly zero advance in computer consciousness. As far as we know, computers in 2016 are no more conscious than their prototypes in the 1950s. However, we are on the brink of a momentous revolution. Humans are in danger of losing their economic value because intelligence is decoupling from consciousness.

Until today high intelligence always went hand in hand with a developed consciousness. Only conscious beings could perform tasks that required a lot of intelligence, such as playing chess, driving cars, diagnosing diseases or identifying terrorists. However, we are now developing new types of non-conscious intelligence that can perform such tasks far better than humans. For all these tasks are based on pattern recognition, and non-conscious algorithms may soon excel human consciousness in recognising patterns.

Science fiction movies generally assume that in order to match and surpass human intelligence, computers will have to develop consciousness. But real science tells a different story. There might be several alternative ways leading to super-intelligence, only some of which pass through the straits of consciousness. For millions of years organic evolution has been slowly sailing along the conscious route. The evolution of inorganic computers may completely bypass these narrow straits, charting a different and much quicker course to super-intelligence.

This raises a novel question: which of the two is really important, intelligence or consciousness? As long as they went hand in hand, debating their relative value was just an amusing pastime for philosophers. But in the twenty-first century this is becoming an urgent political and economic issue. And it is sobering to realise that, at least for armies and corporations, the answer is straightforward: intelligence is mandatory but consciousness is optional.

Armies and corporations cannot function without intelligent agents, but they don't need consciousness and subjective experiences. The conscious experiences of a flesh-and-blood taxi driver are infinitely richer than those of a self-driving car, which feels absolutely nothing. The taxi driver can enjoy music while navigating the busy streets of Seoul. His mind may expand in awe as he looks up at the stars and contemplates the mysteries of the universe. His eyes may fill with tears of joy when he sees his baby girl taking her very first step. But the system doesn't need all that from a taxi driver. All it really wants is to bring passengers from point A to point B as quickly, safely and cheaply as possible. And the autonomous car will soon be able to do that far better than a human driver, even though it cannot enjoy music or be awestruck by the magic of existence.

We should remind ourselves of the fate of horses during the Industrial Revolution. An ordinary farm horse can smell, love, recognize faces, jump over fences and do a thousand other things far better than a Model T Ford or a million-dollar Lamborghini. But cars nevertheless replaced horses because they were superior in the handful of tasks that the system really needed. Taxi drivers are highly likely to go the way of horses.

Indeed, if we forbid humans to drive not only taxis but vehicles altogether, and give computer algorithms a monopoly over traffic, we can then connect all vehicles to a single network, thereby rendering car accidents far less likely. In August 2015, one of Google's experimental self-driving cars had an accident. As it approached a crossing and detected pedestrians wishing to cross, it applied its brakes. A moment later it was hit from behind by a sedan whose careless human driver was perhaps contemplating the mysteries of the universe instead of watching the road. This could not have happened if *both* vehicles had been guided by interlinked computers. The controlling algorithm would have known the position and intentions of every vehicle on the road, and would not have allowed two of its marionettes to collide. Such a system would save lots of time, money and human lives – but would also eliminate the human experience of driving a car and tens of millions of human jobs.[4]

Some economists predict that sooner or later, unenhanced humans will be completely useless. Robots and 3D printers are already replacing workers in manual jobs such as manufacturing shirts, and highly intelligent algorithms will do the same to white-collar occupations. Bank clerks and travel agents, who a short time ago seemed completely secure from automation, have become endangered species. How many travel agents do we need when we can use our smartphones to buy plane tickets from an algorithm?

Stock-exchange traders are also in danger. Most financial trading today is already being managed by computer algorithms that can process in a second more data than a human can in a year and can react to the data much faster than a human can blink. On 23 April 2013, Syrian hackers broke into Associated Press's official Twitter account. At 13:07 they tweeted that the White House had been attacked and President Obama was hurt. Trade algorithms that constantly monitor newsfeeds reacted in no time and began selling stocks like mad. The Dow Jones went into free fall and within sixty seconds lost 150 points, equivalent to a loss of $136 billion! At 13:10 Associated Press clarified that the tweet was a hoax. The algorithms reversed gear and by 13:13 the Dow Jones had recuperated almost all the losses.

Three years earlier, on 6 May 2010, the New York stock exchange underwent an even sharper shock. Within five minutes – from 14:42 to 14:47 – the Dow Jones dropped by 1,000 points, wiping out $1 trillion. It then bounced back, returning to its pre-crash level in a little more than three minutes. That's what happens when super-fast computer programs are in charge of our money. Experts have been trying ever since to understand what happened in this so-called 'Flash Crash'. They know algorithms were to blame, but are still not sure exactly what went wrong. Some traders in the USA have already filed lawsuits against algorithmic trading, arguing that it unfairly discriminates against human beings who simply cannot react fast enough to compete. Quibbling whether this really constitutes a violation of rights might provide lots of work and lots of fees for lawyers.[5]

And these lawyers won't necessarily be human. Movies and TV series give the impression that lawyers spend their days in court shouting 'Objection!' and making impassioned speeches. Yet most run-of-the-mill lawyers devote their time to perusing endless files, looking for precedents, loopholes and tiny pieces of potentially relevant evidence. Some are busy trying to figure out what happened on the night John Doe was murdered, or formulating a gargantuan business contract that will protect their client against every conceivable eventuality. What will be the fate of all these lawyers once sophisticated search algorithms can locate more precedents in a day than a human can in a lifetime, and once brain scans can reveal lies and deceptions at the press of a button? Even highly experienced lawyers and detectives cannot easily spot duplicity merely by observing people's facial expressions and tone of voice. However, lying involves different brain areas from those used in telling the truth. We're not there yet, but it is conceivable that in the not too distant future fMRI scanners could function as almost infallible truth machines. Where will that leave millions of lawyers, judges, cops and detectives? They might consider returning to school to learn a new profession.[6]

When they enter the classroom, however, they may well discover that the algorithms have got there first. Companies such as Mindojo are developing interactive algorithms that will not only teach me maths, physics and history, but will also simultaneously study me and get to know exactly who I am. Digital teachers will closely monitor every answer I give, and how long it took me to give it. Over time, they will discern my unique weaknesses as well as my strengths and will identify what gets me excited, and what makes my eyelids droop. They could teach me thermodynamics or geometry in a way that suits my personality type, even if that particular method doesn't suit 99 per cent of the other pupils. And these digital teachers will never lose their patience, never shout at me, and never go on strike. It remains unclear, however, why on earth I would need to know thermodynamics or geometry in a world containing such intelligent computer programs.[7]

Even doctors are fair game for the algorithms. The first and foremost task of most doctors is to diagnose diseases correctly, and then suggest the best available treatment. If I arrive at the clinic complaining of fever and diarrhoea, I might be suffering from food poisoning. Then again, the same symptoms might result from a stomach virus, cholera, dysentery, malaria, cancer or some unknown new disease. My physician has only a few minutes to make a correct diagnosis, because that is all the time my health insurance pays for. This allows for no more than a few questions and perhaps a quick medical examination. The doctor then cross-references this meagre information with my medical history, and with the vast world of human maladies. Alas, not even the most diligent doctor can remember all my previous ailments and check-ups. Similarly, no doctor can be familiar with every illness and drug, or read every new article published in every medical journal. To top it all, the doctor is sometimes tired or hungry or perhaps even sick, which affects her judgement. No wonder that doctors sometimes err in their diagnoses or recommend a less-than-optimal treatment.

Now consider IBM's famous Watson – an artificial intelligence system that won the *Jeopardy!* television game show in 2011, beating human former champions. Watson is currently groomed to do more serious work, particularly in diagnosing diseases. An AI

44. IBM's Watson defeating its two humans opponents in *Jeopardy!* in 2011.

such as Watson has enormous potential advantages over human doctors. Firstly, an AI can hold in its databanks information about every known illness and medicine in history. It can then update these databanks daily, not only with the findings of new researches, but also with medical statistics gathered from every linked-in clinic and hospital in the world.

Secondly, Watson will be intimately familiar not only with my entire genome and my day-to-day medical history, but also with the genomes and medical histories of my parents, siblings, cousins, neighbours and friends. Watson will know instantly whether I visited a tropical country recently, whether I have recurring stomach infections, whether there have been cases of intestinal cancer in my family or whether people all over town are complaining this morning about diarrhoea.

Thirdly, Watson will never be tired, hungry or sick, and will have all the time in the world for me. I could sit comfortably on my sofa at home and answer hundreds of questions, telling Watson exactly how I feel. This is good news for most patients (except perhaps hypochondriacs). But if you enter medical school today in the expectation of still being a family doctor in twenty years, maybe you should think again. With such a Watson around, there is not much need for Sherlocks.

This threat hovers over the heads not only of general practitioners, but also of experts. Indeed, it might prove easier to replace doctors specialising in relatively narrow fields such as cancer diagnosis. In a recent experiment a computer algorithm correctly diagnosed 90 per cent of lung cancer cases presented to it, while human doctors had a success rate of only 50 per cent.[8] In fact, the future is already here. CT scans and mammography exams are routinely checked by specialised algorithms, which provide doctors with a second opinion, and sometimes detect tumours that the doctors missed.[9]

A host of tough technical problems still prevent Watson and its ilk from replacing most doctors tomorrow morning. Yet these technical problems – however difficult – need only be solved once. The training of a human doctor is a complicated and expensive process that lasts years. When the process is complete, after a decade or so of studies and internships, all you get is one doctor. If you want two doctors, you have to repeat the entire process from scratch. In contrast, if and when you solve the technical problems hampering Watson, you will get not one, but an infinite number of doctors, available 24/7 in every corner of the world. So even if it costs $100 billion to make it work, in the long run it would be much cheaper than training human doctors.

Of course not all human doctors will disappear. Tasks that require a greater level of creativity than run-of-the-mill diagnosis will remain in human hands for the foreseeable future. Just as twenty-first-century armies are increasing the size of their elite special forces, so future healthcare services might offer many more openings to the medical equivalents of army rangers and navy SEALs. However, just as armies no longer need millions of GIs, so future healthcare services will not need millions of GPs.

What's true of doctors is doubly true of pharmacists. In 2011 a pharmacy opened in San Francisco manned by a single robot. When a human comes to the pharmacy, within seconds the robot receives all of the customer's prescriptions, as well as detailed information about any other medicines she takes, and her suspected allergies. The robot ensures that the new medications don't interact adversely with any other medicine or allergy, and then dispenses the required drug to the customer. In its first year of operation the robotic pharmacist provided 2 million prescriptions, without making a single mistake. On average, flesh-and-blood pharmacists err in 1.7 per cent of prescriptions. In the United States alone this amounts to more than 50 million mistaken prescriptions every year![10]

Some people argue that even if an algorithm could outperform doctors and pharmacists in the technical aspects of their professions, it could never replace their human touch. If your CT indicates you have cancer, would you prefer to receive the news from a cold machine, or from a human doctor attentive to your emotional state? Well, how about receiving the news from an attentive machine that tailors its words to your feelings and personality type? Remember that organisms are algorithms, and Watson could detect your emotional state with the same accuracy that it detects your tumours.

A human doctor recognises your emotional state by analysing external signals such as your facial expression and your tone of voice. Watson could not only analyse such external signals more accurately than a human doctor, but it could simultaneously analyse numerous internal indicators that are normally hidden from our eyes and ears. By monitoring your blood pressure, brain activities and countless other biometric data Watson could know exactly how you feel. Thanks to statistics garnered from millions of previous social encounters, Watson could then tell you precisely what you need to hear in just the right tone of voice. For all their vaunted emotional intelligence, human beings are often overwhelmed by their own emotions and react in counterproductive ways. For example, encountering an angry person they start shouting, and listening to a fearful person they let their own anxieties run wild. Watson would never succumb to such temptations. Having no emotions of its own, it would always offer the most appropriate response to your emotional state.

This idea has already partly been implemented by some customer-services departments, such as those pioneered by the Chicago-based Mattersight Corporation. Mattersight publishes its wares with the following blurb: 'Have you ever spoken with someone and felt as though you just clicked? The magical feeling you get is the result of a personality connection. Mattersight creates that feeling every day, in call centers around the world.'[11] When you phone customer service with a request or complaint, it usually takes a few seconds to route your call to a representative. In Mattersight systems your call is routed by a clever algorithm. You first state your reason for calling. The algorithm listens to your problem, analyses the words you have used and your tone of voice, and deduces not only your present emotional state but also your personality type – whether you are introverted, extroverted, rebellious or dependent. Based on this information, the algorithm forwards your call to the representative who best matches your mood and personality. The algorithm knows whether you need an empathetic person to listen patiently to your complaints, or a no-nonsense rational type who will give you the quickest technical solution. A good match means both happier customers and less time and money wasted by the customer-service department.[12]

The Useless Class

The most important question in twenty-first-century economics may well be what to do with all the superfluous people. What will conscious humans do, once we have highly intelligent non-conscious algorithms that can do almost everything better?

Throughout history the job market has been divided into three main sectors: agriculture, industry and services. Until about 1800, the vast majority of people worked in agriculture, and only a small minority worked in industry and services. During the Industrial Revolution people in developed countries left the fields and flocks. Most began working in industry, but growing numbers also took up jobs in the services sector. In recent decades developed countries underwent another revolution; as industrial jobs vanished the services sector expanded. In 2010 only 2 per cent of Americans worked in agriculture and 20 per cent worked in industry, while 78 per cent worked as teachers, doctors, webpage designers and so forth. When mindless algorithms are able to teach, diagnose and design better than humans, what will we do?

This is not an entirely new question. Ever since the Industrial Revolution erupted, people feared that mechanisation might cause mass unemployment. This never happened, because as old professions became obsolete, new professions evolved, and there was always something humans could do better than machines. Yet this is not a law of nature, and nothing guarantees it will continue to be like that in the future. Humans have two basic types of abilities: physical and cognitive. As long as machines competed with humans merely in physical abilities, there were countless cognitive tasks that humans performed better. So as machines took over purely manual jobs, humans focused on jobs requiring at least some cognitive skills. Yet what will happen once algorithms outperform us in remembering, analysing and recognising patterns?

The idea that humans will always have a unique ability beyond the reach of non-conscious algorithms is just wishful thinking. The current scientific answer to this pipe dream can be summarised in three simple principles:

1. Organisms are algorithms. Every animal – including *Homo sapiens* – is an assemblage of organic algorithms shaped by natural selection over millions of years of evolution.
2. Algorithmic calculations are not affected by the materials from which the calculator is built. Whether an abacus is made of wood, iron or plastic, two beads plus two beads equals four beads.
3. Hence there is no reason to think that organic algorithms can do things that non-organic algorithms will never be able to replicate or surpass. As long as the calculations remain valid, what does it matter whether the algorithms are manifested in carbon or silicon?

True, at present there are numerous things that organic algorithms do better than non-organic ones, and experts have repeatedly declared that something will 'for ever' remain beyond the reach of non-organic algorithms. But it turns out that 'for ever' often means no more than a decade or two. Until a short time ago facial recognition was a favourite example of something that even babies accomplish easily but which escaped even the most powerful computers. Today facial-recognition programs are able to identify people far more efficiently and quickly than humans can. Police forces and intelligence services now routinely use such programs to scan countless hours of video footage from surveillance cameras in order to track down suspects and criminals.

In the 1980s when people discussed the unique nature of humanity, they habitually used chess as primary proof of human superiority. They believed that computers would never beat humans at chess. On 10 February 1996, IBM's Deep Blue defeated world chess champion Garry Kasparov, laying to rest that particular claim for human pre-eminence.

Deep Blue was given a head start by its creators, who preprogrammed it not only with the basic rules of chess, but also with detailed instructions regarding chess strategies. A new generation of AI prefers machine learning to human advice. In February 2015 a program developed by Google DeepMind learned *by itself* how to play forty-nine classic Atari games. One of the developers, Dr Demis Hassabis, explained that 'the only information we gave the system was the raw pixels on the screen and the idea that it had to get a high score. And everything else it had to figure out by itself.' The program managed to learn the rules of all the games presented to it, from *Pac-Man* and *Space Invaders* to car racing and

45. Deep Blue defeating Garry Kasparov.

tennis games. It then played most of them as well as or better than humans, sometimes coming up with strategies that never occur to human players.[13]

Shortly afterwards AI scored an even more sensational success, when Google's AlphaGo software taught itself how to play Go, an ancient Chinese strategy board game significantly more complex than chess. Go's intricacies were long considered far beyond the reach of AI programs. In March 2016 a match was held in Seoul between AlphaGo and the South Korean Go champion, Lee Sedol. AlphaGo trounced Lee 4–1 by employing unorthodox moves and original strategies that stunned the experts. Whereas prior to the match most professional Go players were certain that Lee would win, after analysing AlphaGo's moves most concluded that the game was up and that humans no longer had any hope of beating AlphaGo and its progeny.

Computer algorithms have recently proven their worth in ball games, too. For many decades, baseball teams used the wisdom, experience and gut instincts of professional scouts and managers to pick players. The best players fetched millions of dollars, and naturally enough the rich teams grabbed the cream of the crop, whereas poorer teams had to settle for the scraps. In 2002 Billy Beane, the manager of the low-budget Oakland Athletics, decided to beat the system. He relied on an arcane computer algorithm developed by economists and computer geeks to create a winning team from players whom human scouts had overlooked or undervalued. Old-timers were incensed that Beane's algorithm had violated the hallowed halls of baseball. They insisted that picking baseball players is an art, and that only humans with an intimate and long-standing experience of the game can master it. A computer program could never do it, because it could never decipher the secrets and the spirit of baseball.

They soon had to eat their baseball caps. Beane's shoestring-budget ($44 million) algorithmic team not only held its own against baseball giants such as the New York Yankees ($125 million), but became the first team in American League history ever to win twenty consecutive games. Not that Beane and Oakland got to enjoy their success for long. Soon enough many other teams adopted the same algorithmic approach, and since the Yankees and Red Sox could pay far more for both baseball players and computer software, low-budget teams such as the Oakland Athletics ended up having an even smaller chance of beating the system than before.[14]

In 2004 Professor Frank Levy from MIT and Professor Richard Murnane from Harvard published thorough research of the job market, listing those professions most likely to undergo automation. Truck driving was given as an example of a job that could not possibly be automated in the foreseeable future. It is hard to imagine, they wrote, that algorithms could safely drive trucks on a busy road. A mere ten years later Google and Tesla can not only imagine this, but are actually making it happen.[15]

In fact, as time goes by it becomes easier and easier to replace humans with computer algorithms, not merely because the algorithms are getting smarter, but also because humans are professionalising. Ancient hunter-gatherers mastered a very wide variety of skills in order to survive, which is why it would be immensely difficult to design a robotic hunter-gatherer. Such a robot would have to know how to prepare spear points from flint stones, find edible mushrooms in a forest, track down a mammoth and coordinate a charge with a dozen other hunters, and afterwards use medicinal herbs to bandage any wounds. However, over the last few thousand years we humans have been specialising. A taxi driver or a cardiologist specialises in a much narrower niche than a hunter-gatherer, which makes it easier to replace them with AI. As I have repeatedly stressed, AI is nowhere near human-like existence. But 99 per cent of human qualities and abilities are simply redundant for the performance of most modern jobs. For AI to squeeze humans out of the job market it needs only outperform us in the specific abilities a particular profession demands.

Even the managers in charge of all these activities can be replaced. Thanks to its powerful algorithms, Uber can manage millions of taxi drivers with only a handful of humans. Most of the commands are given by the algorithms without any need of human supervision.[16] In May 2014 Deep Knowledge Ventures – a Hong Kong venture-capital firm specialising in regenerative medicine – broke new ground by appointing an algorithm named VITAL to its board. VITAL makes investment recommendations by analysing huge amounts of data regarding the financial situation, clinical trials and intellectual property of prospective companies. Like the other five board members, the algorithm gets to vote on whether or not the firm makes an investment in a specific company.

Examining VITAL's record so far, it seems that it has already picked up at least one managerial vice: nepotism. It has recommended investing in companies that grant algorithms more authority. For example, with VITAL's blessing, Deep Knowledge Ventures recently invested in Pathway Pharmaceuticals, which employs an algorithm called OncoFinder to select and rate personalised cancer therapies.[17]

As algorithms push humans out of the job market, wealth and power might become concentrated in the hands of the tiny elite that owns the all-powerful algorithms, creating unprecedented social and political inequality. Today millions of taxi drivers, bus drivers and truck drivers have significant economic and political clout, each commanding a tiny share of the transportation market. If their collective interests are threatened, they can unionise, go on strike, stage boycotts and create powerful voting blocks. However, once millions of human drivers are replaced by a single algorithm, all that wealth and power will be cornered by the corporation that owns the algorithm, and by the handful of billionaires who own the corporation. Alternatively, the algorithms might themselves become the owners. Human law already recognises intersubjective entities like corporations and nations as 'legal persons'. Though Toyota or Argentina has neither a body nor a mind, they are subject to international laws, they can own land and money, and they can sue and be sued in court. We might soon grant similar status to algorithms. An algorithm could then own a transportation empire or a venture-capital fund without having to obey the wishes of any human master.

If the algorithm makes the right decisions, it could accumulate a fortune, which it could then invest as it sees fit, perhaps buying your house and becoming your landlord. If you infringe on the algorithm's legal rights – say, by not paying rent – the algorithm could hire lawyers and sue you in court. If such algorithms consistently outperform human capitalists, we might end up with an algorithmic upper class owning most of our planet. This may sound impossible, but before dismissing the idea, remember that most of our planet is already legally owned by non-human inter-subjective entities, namely nations and corporations. Indeed, 5,000 years ago much of Sumer was owned by imaginary gods such as Enki and Inanna. If gods can possess land and employ people, why not algorithms?

So what will people do? Art is often said to provide us with our ultimate (and uniquely human) sanctuary. In a world where computers have replaced doctors, drivers, teachers and even landlords, would everyone become an artist? Yet it is hard to see why artistic creation would be safe from the algorithms. Why are we so confident that computers will never be able to outdo us in the composition of music? According to the life sciences, art is not the product of some enchanted spirit or metaphysical soul, but rather of organic algorithms recognising mathematical patterns. If so, there is no reason why non-organic algorithms couldn't master it.

David Cope is a musicology professor at the University of California in Santa Cruz. He is also one of the more controversial figures in the world of classical music. Cope has written computer programs that compose concertos, chorales, symphonies and operas. His first creation was named EMI (Experiments in Musical Intelligence), which specialised in imitating the style of Johann Sebastian Bach. It took seven years to create the program, but once the work was done EMI composed 5,000 chorales à la Bach in a single day. Cope arranged for a performance of a few select chorales at a music festival in Santa Cruz. Enthusiastic members of the audience praised the stirring performance, and explained excitedly how the music had touched their innermost being. They didn't know that it had been created by EMI rather than Bach, and when the truth was revealed some reacted with glum silence, while others shouted in anger.

EMI continued to improve and learned to imitate Beethoven, Chopin, Rachmaninov and Stravinsky. Cope got EMI a contract, and its first album – *Classical Music Composed by Computer* – sold surprisingly well. Publicity brought increasing hostility from classical-music buffs. Professor Steve Larson from the University of Oregon sent Cope a challenge for a musical showdown. Larson suggested that professional pianists play three pieces one after the other: one each by Bach, by EMI, and by Larson himself. The audience would then be asked to vote on who composed which piece. Larson was convinced that people would easily distinguish between soulful human compositions and the lifeless artefact of a machine. Cope accepted the challenge. On the appointed date hundreds of lecturers, students and music fans assembled in the University of Oregon's concert hall. At the end of the performance, a vote was taken. The result? The audience thought that EMI's piece was genuine Bach, that Bach's piece was composed by Larson, and that Larson's piece was produced by a computer.

Critics continued to argue that EMI's music is technically excellent, but that it lacks something. It is too accurate. It has no depth. It has no soul. Yet when people heard EMI's compositions without being informed of their provenance, they frequently praised them precisely for their soulfulness and emotional resonance.

Following EMI's successes Cope created newer and even more sophisticated programs. His crowning achievement was Annie. Whereas EMI composed music according to predetermined rules, Annie is based on machine learning. Its musical style constantly changes and develops in response to new inputs from the outside world. Cope has no idea what Annie is going to compose next. Indeed, Annie does not restrict itself to music composition but also explores other art forms such as haiku poetry. In 2011 Cope published *Comes the Fiery Night: 2,000 Haiku by Man and Machine*. Some of the haiku were written by Annie, and the rest by organic poets.

The book does not disclose which are which. If you think you can tell the difference between human creativity and machine output, you are welcome to test your claim.[18]

In the nineteenth century the Industrial Revolution created a huge urban proletariat, and socialism spread because no other creed managed to answer the unprecedented needs, hopes and fears of this new working class. Liberalism eventually defeated socialism only by adopting the best parts of the socialist programme. In the twenty-first century we might witness the creation of a massive new unworking class: people devoid of any economic, political or even artistic value, who contribute nothing to the prosperity, power and glory of society. This 'useless class' will not merely be unemployed – it will be unemployable.

In September 2013 two Oxford researchers, Carl Benedikt Frey and Michael A. Osborne, published 'The Future of Employment', in which they surveyed the likelihood of different professions being taken over by computer algorithms within the next twenty years. The algorithm developed by Frey and Osborne to do the calculations estimated that 47 per cent of US jobs are at high risk. For example, there is a 99 per cent probability that by 2033 human telemarketers and insurance underwriters will lose their jobs to algorithms. There is a 98 per cent probability that the same will happen to sports referees, 97 per cent that it will happen to cashiers and 96 per cent to chefs. Waiters – 94 per cent. Paralegal assistants – 94 per cent. Tour guides – 91 per cent. Bakers – 89 per cent. Bus drivers – 89 per cent. Construction labourers – 88 per cent. Veterinary assistants – 86 per cent. Security guards – 84 per cent. Sailors – 83 per cent. Bartenders – 77 per cent. Archivists – 76 per cent. Carpenters – 72 per cent. Lifeguards – 67 per cent. And so forth. There are of course some safe jobs. The likelihood that computer algorithms will displace archaeologists by 2033 is only 0.7 per cent, because their job requires highly sophisticated types of pattern recognition, and doesn't produce huge profits. Hence it is improbable that corporations or government will make the necessary investment to automate archaeology within the next twenty years.[19]

Of course, by 2033 many new professions are likely to appear, for example, virtual-world designers. But such professions will probably require much more creativity and flexibility than current run-of-the-mill jobs, and it is unclear whether forty-year-old cashiers or insurance agents will be able to reinvent themselves as virtual-world designers (try to imagine a virtual world created by an insurance agent!). And even if they do so, the pace of progress is such that within another decade they might have to reinvent themselves yet again. After all, algorithms might well outperform humans in designing virtual worlds too. The crucial problem isn't creating new jobs. The crucial problem is creating new jobs that humans perform better than algorithms.[20]

Since we do not know how the job market would look in 2030 or 2040, already today we have no idea what to teach our kids. Most of what they currently learn at school will probably be irrelevant by the time they are forty. Traditionally, life has been divided into two main parts: a period of learning followed by a period of working. Very soon this traditional model will become utterly obsolete, and the only way for humans to stay in the game will be to keep learning throughout their lives, and to reinvent themselves repeatedly. Many if not most humans may be unable to do so.

The coming technological bonanza will probably make it feasible to feed and support these useless masses even without any effort from their side. But what will keep them occupied and content? People must do something, or they go crazy. What will they do all day? One answer might be drugs and computer games. Unnecessary people might spend increasing amounts of time within 3D virtual-reality worlds that would provide them with far more excitement and emotional engagement than the drab reality outside. Yet such a development would deal a mortal blow to the liberal belief in the sacredness of human life and of human experiences. What's so sacred about useless bums who pass their days devouring artificial experiences in La La Land?

Some experts and thinkers, such as Nick Bostrom, warn that humankind is unlikely to suffer this degradation, because once artificial intelligence surpasses human intelligence, it might simply exterminate humankind. The AI would likely do so either for fear that humankind would turn against it and try to pull its plug, or in pursuit of some unfathomable goal of its own. For it would be extremely difficult for humans to control the motivation of a system smarter than themselves.

Even preprogramming the system with seemingly benign goals might backfire horribly. One popular scenario imagines a corporation designing the first artificial super-intelligence and giving it an innocent test such as calculating pi. Before anyone realises what is happening, the AI takes over the planet, eliminates the human race, launches a campaign of conquest to the ends of the galaxy, and transforms the entire known universe into a giant super-computer that for billions upon billions of years calculates pi ever more accurately. After all, this is the divine mission its Creator gave it.[21]

A Probability of 87 Per Cent

At the beginning of this chapter we identified several practical threats to liberalism. The first is that humans might become militarily and economically useless. This is just a possibility, of course, not a prophecy. Technical difficulties or political objections might slow down the algorithmic invasion of the job market. Alternatively, since much of the human mind is still uncharted territory, we don't really know what hidden talents humans might discover in themselves, and what novel jobs they might create to offset the loss of others. That, however, may not be enough to save liberalism. For liberalism believes not just in the value of human beings – it also believes in individualism. The second threat facing liberalism is that, while the system might still need humans in the future, it will not need individuals. Humans will continue to compose music, teach physics and invest money, but the system will understand these humans better than they understand themselves and will make most of the important decisions for them.

The system will thereby deprive individuals of their authority and freedom.

The liberal belief in individualism is founded on the three important assumptions that we discussed:

1. I am an in-dividual – that is, I have a single essence that cannot be divided into parts or subsystems. True, this inner core is wrapped in many outer layers. But if I make the effort to peel away these external crusts, I will find deep within myself a clear and single inner voice, which is my authentic self.
2. My authentic self is completely free.
3. It follows from the first two assumptions that I can know things about myself nobody else can discover. For only I have access to my inner space of freedom, and only I can hear the whispers of my authentic self. This is why liberalism grants the individual so much authority. I cannot trust anyone else to make choices for me, because no one else can know who I really am, how I feel and what I want. This is why the voter knows best, why the customer is always right and why beauty is in the eye of the beholder.

However, the life sciences challenge all three assumptions. According to them:

1. Organisms are algorithms, and humans are not individuals – they are 'dividuals'. That is, humans are an assemblage of many different algorithms lacking a single inner voice or a single self.
2. The algorithms constituting a human are not free. They are shaped by genes and environmental pressures, and take decisions either deterministically or randomly – but not freely.
3. It follows that an external algorithm could theoretically know me much better than I can ever know myself. An algorithm that monitors each of the systems that comprise my body and

my brain could know exactly who I am, how I feel and what I want. Once developed, such an algorithm could replace the voter, the customer and the beholder. Then the algorithm will know best, the algorithm will always be right, and beauty will be in the calculations of the algorithm.

During the nineteenth and twentieth centuries the belief in individualism nevertheless made good practical sense, because there were no external algorithms that could actually monitor me effectively. States and markets may have wished to do exactly that, but they lacked the necessary technology. The KGB and FBI had only a vague understanding of my biochemistry, genome and brain, and even if agents bugged every phone call I made and recorded every chance encounter on the street, they did not have the computing power to analyse all that data. Consequently, given twentieth-century technological conditions, liberals were right to argue that nobody can know me better than I know myself. Humans therefore had a very good reason to regard themselves as an autonomous system and to follow their own inner voices rather than the commands of Big Brother.

However, twenty-first-century technology may enable external algorithms to 'hack humanity' and know me far better than I know myself. Once this happens, the belief in individualism will collapse and authority will shift from individual humans to networked algorithms. People will no longer see themselves as autonomous beings running their lives according to their wishes, but instead will become accustomed to seeing themselves as a collection of biochemical mechanisms that is constantly monitored and guided by a network of electronic algorithms. For this to happen, there is no need of an external algorithm that knows me *perfectly* and never makes any mistake; it is enough that the algorithm will know me *better* than I know myself, and will make *fewer* mistakes than I do. It will then make sense to trust this algorithm with more and more of my decisions and life choices.

We have already crossed this line as far as medicine is concerned. In hospitals we are no longer individuals. It is highly likely that during your lifetime many of the most momentous decisions about your body and health will be taken by computer algorithms such as IBM's Watson. And this is not necessarily bad news. Diabetics already carry sensors that automatically check their sugar level several times a day, alerting them whenever it crosses a dangerous threshold. In 2014 researchers at Yale University announced the first successful trial of an 'artificial pancreas' controlled by an iPhone. Fifty-two diabetics took part in the experiment. Each patient had a tiny sensor and a tiny pump implanted in his or her abdomen. The pump was connected to small tubes of insulin and glucagon, two hormones that together regulate sugar levels in the blood. The sensor constantly measured the sugar level, transmitting the data to an iPhone. The iPhone hosted an application that analysed the information, and whenever necessary gave orders to the pump, which injected measured amounts of either insulin or glucagon – without any need of human intervention.[22]

Many other people who suffer from no serious illnesses have begun to use wearable sensors and computers to monitor their health and activities. These devices – incorporated into anything from smartphones and wristwatches to armbands and underwear – record diverse biometric data such as blood pressure and heart rate. The data is then fed into sophisticated computer programs that advise the wearer on how to alter his or her diet and daily routines in order to enjoy improved health and a longer and more productive life.[23] Google, together with the drug giant Novartis, is developing a contact lens that checks glucose levels in the blood every few seconds by analysing the composition of tears.[24] Pixie Scientific sells 'smart diapers' that analyse baby poop for clues about the child's medical condition. In November 2014 Microsoft launched the Microsoft Band – a smart armband that monitors among other things your heartbeat, the quality of your sleep and the number of steps you take each day. An application called Deadline goes a step further, informing you of how many years of life you have left, given your current habits.

Some people use these apps without thinking too deeply about it, but for others this is already an ideology, if not a religion. The Quantified Self movement argues that the self is nothing but mathematical patterns. These patterns are so complex that the human mind has no chance of understanding them. So if you wish to obey the old adage and know thyself, you should not waste your time on philosophy, meditation or psychoanalysis, but rather you should systematically collect biometric data and allow algorithms to analyse them for you and tell you who you are and what you should do. The movement's motto is 'Self-knowledge through numbers'.[25]

In 2000 the Israeli singer Shlomi Shavan conquered the local playlists with his hit song 'Arik'. It's about a guy who is obsessed with his girlfriend's ex, Arik. He demands to know who is better in bed – he, or Arik? The girlfriend dodges the question, saying that it was different with each of them. The guy is not satisfied and demands: 'Talk numbers, lady.' Well, precisely for such guys, a company called Bedpost sells biometric armbands that you can wear while having sex. The armband collects data such as heart rate, sweat level, duration of sexual intercourse, duration of orgasm and the number of calories you burned. The data is fed into a computer that analyses the information and ranks your performance with precise numbers. No more fake orgasms and 'How was it for you?'[26]

People who experience themselves through the unrelenting mediation of such devices may begin to see themselves more as a collection of biochemical systems than as individuals, and their decisions will increasingly reflect the conflicting demands of the various systems.[27] Suppose you have two free hours a week, and are uncertain whether to use them playing chess or tennis. A good friend might ask: 'What does your heart tell you?' 'Well,' you answer, 'as far as my heart is concerned, it's obvious tennis is better. It's also better for my cholesterol level and blood pressure. But my fMRI scans indicate I should strengthen my left pre-frontal cortex.

In my family dementia is quite common, and my uncle had it at a very early age. The latest studies indicate that a weekly game of chess can help delay its onset.'

You can already find much more extreme examples of external mediation in the geriatric wards of hospitals. Humanism fantasises about old age as a period of wisdom and awareness. The ideal elder may suffer from bodily ailments and weaknesses, but his mind is quick and sharp, and he has eighty years of insights to dispense. He knows exactly what's what, and always has astute advice for the grandchildren and other visitors. Twenty-first-century octogenarians don't always conform to that image. Thanks to our growing understanding of human biology, medicine can keep us alive long enough for our minds and 'authentic selves' to disintegrate and dissolve. All too often, what's left is a collection of dysfunctional biological systems kept going by a collection of monitors, computers and pumps.

At a deeper level, as genetic technologies are integrated into daily life and people develop increasingly intimate relations with their DNA, the single self might blur even further and the authentic inner voice might dissolve into a noisy crowd of genes. When faced by difficult dilemmas and decisions, I may stop searching for my inner voice and instead consult my inner genetic parliament.

On 14 May 2013 actress Angelina Jolie published an article in the *New York Times* about her decision to have a double mastectomy. Jolie had lived for years under the shadow of breast cancer, as both her mother and grandmother died of it at a relatively early age. Jolie herself did a genetic test that confirmed she was carrying a dangerous mutation of the BRCA1 gene. According to recent statistical surveys, women carrying this mutation have an 87 per cent probability of developing breast cancer. Even though at the time Jolie did not have cancer, she decided to pre-empt the dreaded disease and have a double mastectomy. In the article Jolie explained that 'I choose not to keep my story private because there are many women who do not know that they might be living under the shadow of cancer. It is my hope that they, too, will be able to get gene-tested, and that if they have a high risk they, too, will know that they have strong options.'[28]

Deciding whether or not to undergo a mastectomy is a difficult and potentially fatal choice. Beyond the discomforts, dangers and financial costs of the operation and its follow-up treatments, the decision can have far-reaching effects on one's health, body image, emotional well-being and relationships. Jolie's choice, and the courage she showed in going public with it, caused a great stir and won her international acclaim and admiration. In particular, many hoped that the publicity would increase awareness of genetic medicine and its potential benefits.

From a historical perspective, it is interesting to note the critical role algorithms played in her case. When Jolie had to take such an important decision about her life, she did not climb a mountaintop overlooking the ocean, watch the sun set into the waves and attempt to connect to her innermost feelings. Instead, she preferred to listen to her genes, whose voice manifested not in feelings but in numbers. At the time, Jolie felt no pain or discomfort whatsoever. Her feelings told her: 'Relax, everything is perfectly fine.' But the computer algorithms used by her doctors told a different story: 'You don't feel anything is wrong, but there is a time bomb ticking in your DNA. Do something about it – now!'

Of course, Jolie's emotions and unique personality played a key part too. If another woman with a different personality had discovered she was carrying the same genetic mutation, she might well have decided not to undergo a mastectomy. However – and here we enter the twilight zone – what if that other woman had discovered she was carrying not only the dangerous BRCA1 mutation, but another mutation in the (fictional) gene ABCD3, which impairs a brain area responsible for evaluating probabilities, thereby causing people to underestimate dangers? What if a statistician pointed out to this woman that her mother, grandmother and several other relatives all died young because they underestimated various health risks and failed to take precautionary measures?

In all likelihood, you too will make important decisions about your health in the same way as Angelina Jolie. You will undergo a genetic test, a blood test or an fMRI; an algorithm will analyse the results on the basis of enormous statistical databases; and you will then accept the algorithm's recommendation. This is not an apocalyptic scenario. Algorithms won't revolt and enslave us. Rather, they will be so good at making decisions for us that it would be madness not to follow their advice.

Angelina Jolie's first leading role was in the 1993 science-fiction action film *Cyborg 2*. She played Casella Reese, a cyborg developed in the year 2074 by Pinwheel Robotics for corporate espionage and assassination. Casella is programmed with human emotions, in order to blend better into human societies while pursuing her missions. When Casella discovers that Pinwheel Robotics not only controls her, but also intends to terminate her, she escapes and fights for her life and freedom. *Cyborg 2* is a liberal fantasy about an individual fighting for liberty and privacy against global corporate octopuses.

In her real life Jolie preferred to sacrifice privacy and autonomy for health. A similar desire to improve human health may well cause most of us to willingly dismantle the barriers protecting our private spaces and allow state bureaucracies and multinational corporations access to our innermost recesses. For instance, allowing Google to read our emails and follow our activities would make it possible for Google to alert us to brewing epidemics before they are noticed by traditional health services.

How does the UK National Health Service know that a flu epidemic has erupted in London? By analysing the reports of thousands of doctors in hundreds of clinics. And how do all these doctors get the information? Well, when Mary wakes up one morning feeling a bit under the weather, she doesn't run straight to her doctor. She waits a few hours, or even a day or two, hoping that a nice cup of tea with honey will do the trick. When things don't improve, she makes an appointment with the doctor, goes to the clinic and describes her symptoms. The doctor types the data into a computer, and hopefully somebody up in NHS headquarters analyses these data, together with reports streaming in from thousands of other doctors, and concludes that flu is on the march. All this takes a lot of time.

Google could do it in minutes. It needs to merely monitor the words Londoners type in their emails and in Google's search engine and cross-reference them with a database of disease symptoms. Suppose on an average day the words 'headache', 'fever', 'nausea' and 'sneezing' appear 100,000 times in London emails and searches. If today the Google algorithm notices they appear 300,000 times, then bingo! We have a flu epidemic. There is no need to wait till Mary goes to her doctor. On the very first morning she woke up feeling a bit unwell and before going to work she emailed a colleague, 'I have a headache, but I'll be there.' That's all Google needs.

However, for Google to work its magic Mary must allow Google not only to read her messages, but also to share the information with the health authorities. If Angelina Jolie was willing to sacrifice her privacy in order to raise awareness of breast cancer, why shouldn't Mary make a similar sacrifice in order to thwart epidemics?

This isn't a theoretical idea. In 2008 Google actually launched Google Flu Trends, that tracks flu outbreaks by monitoring Google searches. The service is still being developed, and due to privacy limitations it tracks only search words and allegedly avoids reading private emails. But it is already capable of ringing the flu alarm bells ten days before traditional health services.[29]

The Google Baseline Study is an even more ambitious project. Google intends to build a mammoth database on human health, establishing the 'perfect health' profile. Identifying even the smallest deviations from the baseline will hopefully make it possible to alert people to burgeoning health problems such as cancer when they can be nipped in the bud. The Baseline Study dovetails with an entire line of products called Google Fit that will be incorporated into wearables such as clothes, bracelets, shoes and glasses. The idea is for Google Fit products to collect the never-ending stream of biological data to feed the Baseline Study.[30]

Yet companies such as Google want to go much deeper than wearables. The market for DNA testing is currently growing in leaps and bounds. One of its leaders is 23andMe, a private company founded by Anne Wojcicki, former wife of Google co-founder Sergey Brin. The name '23andMe' refers to the twenty-three pairs of chromosomes that encode the human genome, the message being that my chromosomes have a very special relationship with me. Whoever can understand what the chromosomes are saying can tell you things about yourself that you never even suspected.

If you want to know what, pay 23andMe a mere $99, and they will send you a small package with a tube. You spit into the tube, seal it and mail it to Mountain View, California. There the DNA in your saliva is read, and you receive the results online. You get a list of the potential health hazards you face, and your genetic predisposition to more than ninety traits and conditions ranging from baldness to blindness. 'Know thyself' was never easier or cheaper. Since it is all based on statistics, the size of the company's database is the key to making accurate predictions. Hence the first company to build a giant genetic database will provide customers with the best predictions, and will potentially corner the market. US biotech companies are increasingly worried that strict privacy laws in the USA combined with Chinese disregard for individual privacy may hand China the genetic market on a plate.

If we connect all the dots, and if we give Google and its competitors free access to our biometric devices, to our DNA scans and to our medical records, we will get an all-knowing medical health service that will not only fight epidemics, but will also shield us from cancer, heart attacks and Alzheimer's. Yet with such a database at its disposal, Google could do far more. Imagine a system that, in the words of the famous Police song, watches every breath you take, every move you make and every bond you break; a system that monitors your bank account and your heartbeat, your sugar levels and your sexual escapades. It will definitely know you much better than you know yourself. The self-deceptions and self-delusions that trap people in bad relationships, wrong careers and harmful habits will not fool Google. Unlike the narrating self that controls us today, Google will not make decisions on the basis of cooked-up stories, and will not be misled by cognitive short cuts and the peak-end rule. Google will actually remember every step we took and every hand we shook.

Many of us would be happy to transfer much of our decision-making processes into the hands of such a system, or at least consult with it whenever we face important choices. Google will advise us which movie to see, where to go on holiday, what to study in college, which job offer to accept, and even whom to date and marry. 'Listen, Google,' I will say, 'both John and Paul are court-ing me. I like both of them, but in different ways, and it's so hard to make up my mind. Given everything you know, what do you advise me to do?'

And Google will answer: 'Well, I've known you from the day you were born. I have read all your emails, recorded all your phone calls, and know your favourite films, your DNA and the entire biometric history of your heart. I have exact data about each date you went on, and, if you want, I can show you second-by-second graphs of your heart rate, blood pressure and sugar levels whenever you went on a date with John or Paul. If necessary, I can even provide you with an accurate mathematical ranking of every sexual encounter you had with either of them. And naturally, I know them as well as I know you. Based on all this information, on my superb algorithms, and on decades' worth of statistics about millions of relationships – I advise you to go with John, with an 87 per cent probability that you will be more satisfied with him in the long run.

'Indeed, I know you so well that I also know you don't like this answer. Paul is much more handsome than John, and because you give external appearances too much weight, you secretly wanted me to say "Paul". Looks matter, of course, but not as much as you think. Your biochemical algorithms – which evolved tens of thou-sands of years ago on the African savannah – give looks a weight of

35 per cent in their overall rating of potential mates. My algorithms – which are based on the most up-to-date studies and statistics – say that looks have only a 14 per cent impact on the long-term success of romantic relationships. So, even though I took Paul's looks into account, I still tell you that you would be better off with John.'[31]

In exchange for such devoted counselling services, we will just have to give up the idea that humans are individuals, and that each human has a free will determining what's good, what's beautiful and what is the meaning of life. Humans will no longer be autonomous entities directed by the stories their narrating self invents. Instead, they will be integral parts of a huge global network.

Liberalism sanctifies the narrating self, and allows it to vote in the polling stations, in the supermarket and in the marriage market. For centuries this made good sense, because though the narrating self believed in all kinds of fictions and fantasies, no alternative system knew me better. Yet once we have a system that really does know me better, it will be foolhardy to leave authority in the hands of the narrating self.

Liberal habits such as democratic elections will become obsolete, because Google will be able to represent even my own political opinions better than I can. When I stand behind the curtain in the polling booth, liberalism instructs me to consult my authentic self, and choose whichever party or candidate reflects my deepest desires. Yet the life sciences point out that when I stand there behind that curtain, I don't really remember everything I felt and thought in the years since the last election. Moreover, I am bombarded by a barrage of propaganda, spin and random memories that might well distort my choices. Just as in Kahneman's cold-water experiment, in politics too the narrating self follows the peak-end rule. It forgets the vast majority of events, remembers only a few extreme incidents and gives a wholly disproportionate weight to recent happenings.

For four long years I may have repeatedly complained about the PM's policies, telling myself and anyone willing to listen that he will be 'the ruin of us all'. However, in the months prior to the elections the government cuts taxes and spends money generously. The ruling party hires the best copywriters to lead a brilliant campaign, with a well-balanced mixture of threats and promises that speak directly to the fear centre in my brain. On the morning of the election I wake up with a cold, which impacts my mental processes and induces me to prefer security and stability over all other considerations. And voila! I send the man who will be 'the ruin of us all' back into office for another four years.

I could have saved myself from such a fate if only I had authorised Google to vote for me. Google wasn't born yesterday, you know. Though it won't ignore the recent tax cuts and the election promises, it will also remember what happened throughout the previous four years. It will know what my blood pressure was every time I read the morning newspapers, and how my dopamine level plummeted while I watched the evening news. Google will know how to screen the spin-doctors' empty slogans. Google will understand that illness makes voters lean a bit more to the right than usual, and will compensate for this. Google will therefore be able to vote not according to my momentary state of mind, and not according to the fantasies of the narrating self, but rather according to the real feelings and interests of the collection of biochemical algorithms known as 'I'.

Naturally, Google will not always get it right. After all, these are all just probabilities. But if Google makes enough good decisions, people will grant it increasing authority. As time goes by, the databases will grow, the statistics will become more accurate, the algorithms will improve and the decisions will be even better. The system will never know me perfectly, and will never be infallible. But there is no need for that. Liberalism will collapse on the day the system knows me better than I know myself. Which is less difficult than it may sound, given that most people don't really know themselves well.

A recent study commissioned by Google's nemesis – Facebook – has indicated that already today the Facebook algorithm is a better judge of human personalities and dispositions than even people's friends, parents and spouses. The study was conducted on 86,220 volunteers who have a Facebook account and who completed a hundred-item personality questionnaire. The Facebook algorithm predicted the volunteers' answers based on monitoring their Facebook Likes – which webpages, images and clips they tagged with the Like button. The more Likes, the more accurate the predictions. The algorithm's predictions were compared with those of work colleagues, friends, family members and spouses. Amazingly, the algorithm needed a set of only ten Likes in order to outperform the predictions of work colleagues. It needed seventy Likes to outperform friends, 150 Likes to outperform family members and 300 Likes to outperform spouses. In other words, if you happen to have clicked 300 Likes on your Facebook account, the Facebook algorithm can predict your opinions and desires better than your husband or wife!

Indeed, in some fields the Facebook algorithm did better than the person themself. Participants were asked to evaluate things such as their level of substance use or the size of their social networks. Their judgements were less accurate than those of the algorithm. The research concludes with the following prediction (made by the human authors of the article, not by the Facebook algorithm): 'People might abandon their own psychological judgements and rely on computers when making important life decisions, such as choosing activities, career paths, or even romantic partners. It is possible that such data-driven decisions will improve people's lives.'[32]

On a more sinister note, the same study implies that in future US presidential elections Facebook could know not only the political opinions of tens of millions of Americans, but also who among them are the critical swing voters, and how these voters might be swung. Facebook could tell that in Oklahoma the race between Republicans and Democrats is particularly close, identify the 32,417 voters who still haven't made up their minds, and determine what each candidate needs to say in order to tip the balance. How could Facebook obtain this priceless political data? We provide it for free.

In the heyday of European imperialism, conquistadors and merchants bought entire islands and countries in exchange for coloured beads. In the twenty-first century our personal data is probably the most valuable resource most humans still have to offer, and we are giving it to the tech giants in exchange for email services and funny cat videos.

From Oracle to Sovereign

Once Google, Facebook and other algorithms become all-knowing oracles, they may well evolve into agents and ultimately into sovereigns.[33] To understand this trajectory, consider the case of Waze – a GPS-based navigational application that many drivers use nowadays. Waze isn't just a map. Its millions of users constantly update it about traffic jams, car accidents and police cars. Hence Waze knows to divert you away from heavy traffic, and bring you to your destination through the quickest possible route. When you reach a junction and your gut instinct tells you to turn right, but Waze instructs you to turn left, users sooner or later learn that they had better listen to Waze rather than to their feelings.[34]

At first sight it seems that the Waze algorithm serves only as an oracle. You ask a question, the oracle replies, but it is up to you to make a decision. If the oracle wins your trust, however, the next logical step is to turn it into an agent. You give the algorithm only a final aim, and it acts to realise that aim without your supervision. In the case of Waze, this may happen when you connect Waze to a self-driving car, and tell Waze 'take the fastest route home' or 'take the most scenic route' or 'take the route which will result in the minimum amount of pollution'. You call the shots, but leave it to Waze to execute your commands.

Finally, Waze might become sovereign. Having so much power in its hands, and knowing far more than you, it may start manipulating you and the other drivers, shaping your desires and making your decisions for you. For example, suppose because Waze is so good, everybody starts using it. And suppose there is a traffic jam on route no. 1, while the alternative route no. 2 is relatively open. If Waze simply lets everybody know that, then all drivers will rush to route no. 2, and it too will be clogged. When everybody uses the same oracle, and everybody believes the oracle, the oracle turns into a sovereign. So Waze must think for us. Maybe it will inform only half the drivers that route no. 2 is open, while keeping this information secret from the other half. Thereby pressure will ease on route no. 1 without blocking route no. 2.

Microsoft is developing a far more sophisticated system called Cortana, named after an AI character in its popular *Halo* video-game series. Cortana is an AI personal assistant that Microsoft hopes to include as an integral feature of future versions of Windows. Users will be encouraged to allow Cortana access to all their files, emails and applications, so that it will get to know them and can thereby offer advice on myriad matters, as well as becoming a virtual agent representing the user's interests. Cortana could remind you to buy something for your wife's birthday, select the present, reserve a table at a restaurant and prompt you to take your medicine an hour before dinner. It could alert you that if you don't stop reading now, you will be late for an important business meeting. As you are about to enter the meeting, Cortana will warn that your blood pressure is too high and your dopamine level too low, and based on past statistics, you tend to make serious business mistakes in such circumstances. So you had better keep things tentative and avoid committing yourself or signing any deals.

Once Cortanas evolve from oracles to agents, they might start speaking directly with one another on their masters' behalf. It can begin innocently enough, with my Cortana contacting your Cortana to agree on a place and time for a meeting. Next thing I know, a potential employer will tell me not to bother sending a CV, but simply allow his Cortana to grill my Cortana. Or my Cortana may be approached by the Cortana of a potential lover, and the two will compare notes to decide whether it's a good match – completely unbeknown to their human owners.

As Cortanas gain authority, they may begin manipulating each other to further the interests of their masters, so that success in the job market or the marriage market may increasingly depend on the quality of your Cortana. Rich people owning the most up-to-date Cortana will have a decisive advantage over poor people with their older versions.

But the murkiest issue of all concerns the identity of Cortana's master. As we have seen, humans are not individuals, and they don't have a single unified self. Whose interests, then, should Cortana serve? Suppose my narrating self makes a New Year resolution to start a diet and go to the gym every day. A week later, when it is time for the gym, the experiencing self instructs Cortana to turn on the TV and order pizza. What should Cortana do? Should it obey the experiencing self, or the resolution taken a week earlier by the narrating self?

You may wonder whether Cortana is really different from an alarm clock, which the narrating self sets in the evening in order to wake the experiencing self in time for work. But Cortana will have far more power over me than an alarm clock. The experiencing self can silence the alarm clock by pressing a button. In contrast, Cortana will know me so well that it will know exactly what inner buttons to push in order to make me follow its 'advice'.

Microsoft's Cortana is not alone in this game. Google Now and Apple's Siri are headed in the same direction. Amazon too employs algorithms that constantly study you and then use their accumulated knowledge to recommend products. When I go to a physical bookstore I wander among the shelves and trust my feelings to choose the right book. When I go to visit Amazon's virtual shop, an algorithm immediately pops up and tells me: 'I know which books you liked in the past. People with similar tastes also tend to love this or that new book.'

And this is just the beginning. Today in the US more people read digital books than printed ones. Devices such as Amazon's Kindle are able to collect data on their users while they are reading. Your Kindle can, for example, monitor which parts of a book you read quickly, and which slowly; on which page you took a break, and on which sentence you abandoned the book, never to pick it up again. (Better tell the author to rewrite that bit.) If Kindle is upgraded with face recognition and biometric sensors, it will know how each sentence you read influenced your heart rate and blood pressure. It will know what made you laugh, what made you sad and what made you angry. Soon, books will read you while you are reading them. And whereas you quickly forget most of what you read, Amazon will never forget a thing. Such data will enable Amazon to choose books for you with uncanny precision. It will also enable Amazon to know exactly who you are, and how to turn you on and off.[35]

Eventually we may reach a point when it will be impossible to disconnect from this all-knowing network even for a moment. Disconnection will mean death. If medical hopes are realised, future humans will incorporate into their bodies a host of biometric devices, bionic organs and nano-robots, which will monitor our health and defend us from infections, illnesses and damage. Yet these devices will have to be online 24/7, both in order to be updated with the latest medical developments, and to protect them from the new plagues of cyberspace. Just as my home computer is constantly attacked by viruses, worms and Trojan horses, so will be my pacemaker, hearing aid and nanotech immune system. If I don't update my body's anti-virus program regularly, I will wake up one day to discover that the millions of nano-robots coursing through my veins are now controlled by a North Korean hacker.

The new technologies of the twenty-first century may thus reverse the humanist revolution, stripping humans of their authority, and empowering non-human algorithms instead. If you are horrified by this direction, don't blame the computer geeks. The responsibility actually lies with the biologists. It is crucial to realise that this entire trend is fuelled more by biological insights than by computer science. It is the life sciences that concluded that organisms are algorithms. If this is not the case – if organisms function in an inherently different way to algorithms – then computers may work wonders in other fields, but they will not be able to understand us and direct our life, and they will certainly be incapable of merging with us. Yet once biologists concluded that organisms are algorithms, they dismantled the wall between the organic and inorganic, turned the computer revolution from a purely mechanical affair into a biological cataclysm, and shifted authority from individual humans to networked algorithms.

Some people are indeed horrified by this development, but the fact is that millions willingly embrace it. Already today many of us give up our privacy and our individuality by conducting much of our lives online, recording our every action and becoming hysterical if connection to the net is interrupted even for a few minutes. The shifting of authority from humans to algorithms is happening all around us, not as a result of some momentous governmental decision, but due to a flood of mundane personal choices.

If we are not careful the result might be an Orwellian police state that constantly monitors and controls not only all our actions, but even what happens inside our bodies and our brains. Just think what uses Stalin could have found for omnipresent biometric sensors – and what uses Putin might yet find for them. However, while defenders of human individuality fear a repetition of twentieth-century nightmares and brace themselves to resist familiar Orwellian foes, human individuality is now facing an even bigger threat from the opposite direction. In the twenty-first century the individual is more likely to disintegrate gently from within than to be brutally crushed from without. Today most corporations and governments pay homage to my individuality, and promise to provide medicine, education and entertainment customised to my unique needs and wishes. But in order to do so, corporations and governments first need to deconstruct me into biochemical subsystems, monitor these subsystems with ubiquitous sensors and decipher their working with powerful algorithms. In the process, the individual will transpire to be nothing but a religious fantasy. Reality will be a mesh of biochemical and electronic algorithms, without clear borders, and without individual hubs.

Upgrading Inequality

So far we have looked at two of the three practical threats to liberalism: firstly, that humans will lose their value completely; secondly, that humans will still be valuable collectively, but will lose their individual authority, and instead be managed by external algorithms. The system will still need you to compose symphonies, teach history or write computer code, but it will know you better than you know yourself, and will therefore make most of the important decisions for you – and you will be perfectly happy with that. It won't necessarily be a bad world; it will, however, be a post-liberal world.

The third threat to liberalism is that some people will remain both indispensable and undecipherable, but they will constitute a small and privileged elite of upgraded humans. These superhumans will enjoy unheard-of abilities and unprecedented creativity, which will allow them to go on making many of the most important decisions in the world. They will perform crucial services for the system, while the system could neither understand nor manage them. However, most humans will not be upgraded, and will consequently become an inferior caste dominated by both computer algorithms and the new superhumans.

Splitting humankind into biological castes will destroy the foundations of liberal ideology. Liberalism can coexist with socio-economic gaps. Indeed, since it favours liberty over equality, it takes such gaps for granted. However, liberalism still presupposes that all human beings have equal value and authority. From a liberal perspective, it is perfectly all right that one person is a billionaire living in a sumptuous chateau, whereas another is a poor peasant living in a straw hut. For according to liberalism, the peasant's unique experiences are still just as valuable as the billionaire's. That's why liberal authors write long novels about the experiences of poor peasants – and why even billionaires avidly read such books. If you go to see *Les Misérables* on Broadway or in Covent Garden, you will find that good seats can cost hundreds of dollars, and the audience's combined wealth probably runs into the billions, yet they still sympathise with Jean Valjean who served nineteen years in jail for stealing a loaf of bread to feed his starving nephews.

The same logic operates on election day, when the vote of the poor peasant counts for exactly the same as the billionaire's. The liberal solution for social inequality is to give equal value to different human experiences, instead of trying to create the same experiences for everyone. However, will this solution still work once rich and poor are separated not merely by wealth, but also by real biological gaps?

In her *New York Times* article, Angelina Jolie referred to the high costs of genetic testing. The test Jolie had taken costs $3,000 (not including the price of the actual mastectomy, the reconstructive surgery and related treatments). This in a world where 1 billion people earn less than $1 per day, and another 1.5 billion earn between $1 and $2 a day.[36] Even if they work hard their entire life, these people will never be able to afford a $3,000 genetic test. And the economic gaps are at present only increasing. As of early 2016, the sixty-two richest people in the world were worth as much as the poorest 3.6 billion people! Since the world's population is about 7.2 billion, it means that these sixty-two billionaires together hold as much wealth as the entire bottom half of humankind.[37]

The cost of DNA testing is likely to go down with time, but expensive new procedures are constantly being pioneered. So while old treatments will gradually come within reach of the masses, the elites will always remain a couple of steps ahead. Throughout history the rich have enjoyed many social and political advantages, but no huge biological gap ever separated them from the poor. Medieval aristocrats claimed that superior blue blood was flowing through their veins, and Hindu Brahmins insisted that they were naturally smarter than everyone else, but this was pure fiction. In the future, however, we may see real gaps in physical and cognitive abilities opening between an upgraded upper class and the rest of society.

When scientists are confronted with this scenario, their standard reply is that in the twentieth century too many medical breakthroughs began with the rich, but eventually benefited the entire population and helped to narrow rather than widen the social gaps. For example, vaccines and antibiotics at first profited mainly the upper classes in Western countries, but today they improve the lives of all humans everywhere.

However, the expectation that this process will be repeated in the twenty-first century may be just wishful thinking, for two important reasons. First, medicine is undergoing a tremendous conceptual revolution. Twentieth-century medicine aimed to heal the sick. Twenty-first-century medicine is increasingly aiming to upgrade the healthy. Healing the sick was an egalitarian project, because it assumed that there is a normative standard of physical and mental health that everyone can and should enjoy. If someone fell below the norm, it was the job of doctors to fix the problem and help him or her 'be like everyone'. In contrast, upgrading the healthy is an elitist project, because it rejects the idea of a universal standard applicable to all and seeks to give some individuals an edge over others. People want superior memories, above-average intelligence and first-class sexual abilities. If some form of upgrade becomes so cheap and common that everyone enjoys it, it will simply be considered the new baseline, which the next generation of treatments will strive to surpass.

Consequently by 2070 the poor could very well enjoy much better healthcare than today, but the gap separating them from the rich will nevertheless be much greater. People usually compare themselves to their more fortunate contemporaries rather than to their ill-fated ancestors. If you tell a poor American in a Detroit slum that he has access to much better healthcare than his great-grandparents did a century ago, it is unlikely to cheer him up. Indeed, such talk will sound terribly smug and condescending. 'Why should I compare myself to nineteenth-century factory workers or peasants?' he would retort. 'I want to live like the rich people on television, or at least like the folks in the affluent suburbs'. Similarly, if in 2070 you tell the lower classes that they enjoy better healthcare than in 2017, it might be very cold comfort to them, because they would be comparing themselves to the upgraded superhumans who dominate the world.

Moreover, despite all the medical breakthroughs we cannot be absolutely certain that in 2070 the poor will indeed enjoy better healthcare than today, because the state and the elite may lose interest in providing the poor with healthcare. In the twentieth-century medicine benefited the masses because the twentieth century was the age of the masses. Twentieth-century armies needed millions of healthy soldiers, and economies needed millions of healthy workers. Consequently states established public health services to ensure the health and vigour of everyone. Our greatest medical achievements were the provision of mass-hygiene facilities, the campaigns of mass vaccinations and the eradication of mass epidemics. In 1914 the Japanese elite had a vested interest in vaccinating the poor and building hospitals and sewage systems in the slums, because if they wanted Japan to be a strong nation with a powerful army and a robust economy, they needed many millions of healthy soldiers and workers.

But the age of the masses may be over, and with it the age of mass medicine. As human soldiers and workers give way to algorithms, at least some elites may conclude that there is no point in providing improved or even standard levels of health for masses of useless poor people, and it is far more sensible to focus on upgrading a handful of superhumans beyond the norm.

Already today the birth rate is falling in technologically advanced countries such as Japan and South Korea, where prodigious efforts are invested in the upbringing and education of fewer and fewer children – from whom more and more is expected. How can huge developing countries like India, Brazil or Nigeria hope to compete with Japan? These countries resemble a long train. The elites in the first-class carriages enjoy health care, education and income levels on a par with the most developed nations in the world. However, the hundreds of millions of ordinary citizens who crowd the third-class carriages still suffer from widespread disease, ignorance and poverty. What would the Indian, Brazilian or Nigerian elites prefer to do in the coming century? Invest in fixing the problems of hundreds of millions of poor, or in upgrading a few million rich? Unlike in the twentieth century, when the elite had a stake in fixing the problems of the poor because they were militarily and economically vital, in the twenty-first century the most efficient (albeit ruthless) strategy might be to let go of the useless third-class carriages, and dash forward with the first class only. In order to compete with Japan, Brazil might need a handful of upgraded superhumans far more than millions of healthy ordinary workers.

How will liberal beliefs survive the appearance of superhumans with exceptional physical, emotional and intellectual abilities? What will happen if it turns out that such superhumans have fundamentally different experiences from normal Sapiens? What if superhumans are bored by novels about the experiences of lowly Sapiens thieves, whereas run-of-the-mill humans find soap operas about superhuman love affairs unintelligible?

The great human projects of the twentieth century – overcoming famine, plague and war – aimed to safeguard a universal norm of abundance, health and peace for everyone without exception. The new projects of the twenty-first century – gaining immortality, bliss and divinity – also hope to serve the whole of humankind. However, because these projects aim at surpassing rather than safeguarding the norm, they may well result in the creation of a new superhuman caste that will abandon its liberal roots and treat normal humans no better than nineteenth-century Europeans treated Africans.

If scientific discoveries and technological developments split humankind into a mass of useless humans and a small elite of upgraded superhumans, or if authority shifts altogether away from human beings into the hands of highly intelligent algorithms, then liberalism will collapse. What new religions or ideologies might fill the resulting vacuum and guide the subsequent evolution of our godlike descendants?

10

The Ocean of Consciousness

The new religions are unlikely to emerge from the caves of Afghanistan or from the madrasas of the Middle East. Rather, they will emerge from research laboratories. Just as socialism took over the world by promising salvation through steam and electricity, so in the coming decades new techno-religions may conquer the world by promising salvation through algorithms and genes.

Despite all the talk of radical Islam and Christian fundamentalism, the most interesting place in the world from a religious perspective is not the Islamic State or the Bible Belt, but Silicon Valley. That's where hi-tech gurus are brewing for us brave new religions that have little to do with God, and everything to do with technology. They promise all the old prizes – happiness, peace, prosperity and even eternal life – but here on earth with the help of technology, rather than after death with the help of celestial beings.

These new techno-religions can be divided into two main types: techno-humanism and data religion. Data religion argues that humans have completed their cosmic task and should now pass the torch on to entirely new kinds of entities. We will discuss the dreams and nightmares of data religion in the next chapter. This chapter is dedicated to the more conservative creed of techno-humanism, which still sees humans as the apex of creation and clings to many traditional humanist values. Techno-humanism agrees that *Homo sapiens* as we know it has run its historical course and will no longer be relevant in the future, but concludes that we should therefore use technology in order to create *Homo deus* – a much superior human model. *Homo deus* will retain some essential human features, but will also enjoy upgraded physical and mental abilities that will enable it to hold its own even against the most sophisticated non-conscious algorithms. Since intelligence is decoupling from consciousness, and since non-conscious intelligence is developing at breakneck speed, humans must actively upgrade their minds if they want to stay in the game.

Seventy thousand years ago the Cognitive Revolution transformed the Sapiens mind, thereby turning an insignificant African ape into the ruler of the world. The improved Sapiens minds suddenly had access to the vast intersubjective realm, which enabled them to create gods and corporations, to build cities and empires, to invent writing and money, and eventually to split the atom and reach the moon. As far as we know, this earth-shattering revolution resulted from a few small changes in the Sapiens DNA and a slight rewiring of the Sapiens brain. If so, says techno-humanism, maybe a few additional changes to our genome and another rewiring of our brain will suffice to launch a second cognitive revolution. The mental renovations of the first Cognitive Revolution gave *Homo sapiens* access to the intersubjective realm and turned them into the rulers of the planet; a second cognitive revolution might give *Homo deus* access to unimaginable new realms and make them lords of the galaxy.

This idea is an updated variant on the old dreams of evolutionary humanism, which already a century ago called for the creation of superhumans. However, whereas Hitler and his ilk planned to create superhumans by means of selective breeding and ethnic cleansing, twenty-first-century techno-humanism hopes to reach that goal far more peacefully, with the help of genetic engineering, nanotechnology and brain–computer interfaces.

Gap the Mind

Techno-humanism seeks to upgrade the human mind and give us access to unknown experiences and unfamiliar states of consciousness. However, revamping the human mind is an extremely complex and dangerous undertaking. As we discussed in Chapter 3, we don't really understand the mind. We don't know how minds emerge, nor what their function is. Through trial and error we are learning how to engineer mental states, but we seldom comprehend the full implications of such manipulations. Worse still, since we are unfamiliar with the full spectrum of mental states, we don't know what mental aims to set ourselves.

We are akin to the inhabitants of a small isolated island who have just invented the first boat, and are about to set sail without a map or even a destination. Indeed, we are in a somewhat worse condition. The inhabitants of our imaginary island are at least aware that they occupy just a small space within a large and mysterious sea. We on the other hand fail to appreciate that we are living on a tiny island of consciousness within a perhaps limitless ocean of alien mental states.

Just as the spectrums of light and sound are far broader than what we humans can see and hear, so the spectrum of mental states is far larger than what the average human perceives. We can see light in wavelengths of between 400 and 700 nanometres only. Above this small principality of human vision extend the unseen but vast realms of infrared, microwaves and radio waves, and below it lie the dark dominions of ultraviolet, X-rays and gamma rays. Similarly, the spectrum of possible mental states may be infinite, but science has studied only two tiny sections of it: the sub-normative and the WEIRD.

For more than a century psychologists and biologists have conducted extensive research on people suffering from various psychiatric disorders and mental diseases, from autism to schizophrenia. Consequently, today we have a detailed albeit imperfect map of the sub-normative mental spectrum: the zone of human existence characterized by less-than-normal capacities to feel, think or communicate. Simultaneously, scientists have studied the mental states of people considered to be healthy and normative. However, most scientific research about the human mind and the human experience has been conducted on people from **W**estern, **e**ducated, **i**ndustrialised, **r**ich and **d**emocratic (WEIRD) societies, who do not constitute a representative sample of humanity. The study of the human mind has so far assumed that *Homo sapiens* is Homer Simpson.

In a groundbreaking 2010 study, Joseph Henrich, Steven J. Heine and Ara Norenzayan systematically surveyed all the papers published between 2003 and 2007 in leading scientific journals belonging to six different subfields of psychology. They found that though the papers often made broad claims about the human mind, most of them based their findings on exclusively WEIRD samples. For example, in papers published in the *Journal of Personality and Social Psychology* – arguably the most important journal in the subfield of social psychology – 96 per cent of the sampled individuals were WEIRD, and 68 per cent were Americans. Moreover, 67 per cent of

46. Humans can see only a minuscule part of the electromagnetic spectrum. The spectrum in its entirety is about 10 trillion times larger than that of visible light. Might the mental spectrum be equally vast?

American subjects and 80 per cent of non-American subjects were psychology students! In other words, more than two-thirds of the individuals sampled for papers published in this prestigious journal were psychology students in Western universities. Henrich, Heine and Norenzayan half-jokingly suggested that the journal change its name to the *Journal of Personality and Social Psychology of American Psychology Students*.[1]

Psychology students star in many of the studies because their professors oblige them to take part in experiments. If I am a psychology professor at Harvard it is much easier for me to conduct experiments on my own students than on the residents of a crime-ridden Boston slum – not to mention travelling to Namibia and enlisting hunter-gatherers in the Kalahari Desert. However, it may well be that Boston slum-dwellers and Kalahari hunter-gatherers experience mental states that we will never discover by forcing Harvard psychology students to answer long questionnaires or stick their heads into fMRI scanners.

Even if we travel all over the globe and study each and every community, we would still cover only a limited part of the Sapiens mental spectrum. Nowadays all humans have been touched by modernity, and are members of a single global village. Though Kalahari foragers are somewhat less modern than Harvard psychology students, they are not a time capsule from our distant past. They too have been influenced by Christian missionaries, European traders, wealthy eco-tourists and inquisitive researchers (the joke is that in the Kalahari Desert, the typical hunter-gatherer band consists of twenty hunters, twenty gatherers and fifty anthropologists).

Before the emergence of the global village the planet was a galaxy of isolated human cultures, which might have fostered mental states that are now extinct. Different socio-economic realities and daily routines nurtured different states of consciousness. Who can gauge the minds of Stone Age mammoth-hunters, Neolithic farmers or Kamakura samurais? Moreover, many premodern cultures believed in the existence of superior states of consciousness that people might access through meditation, drugs or rituals. Shamans, monks and ascetics systematically explored the mysterious lands of mind, and returned laden with breathtaking stories. They told of unfamiliar states of supreme tranquillity, extreme sharpness and matchless sensitivity. They told of the mind expanding to infinity or dissolving into emptiness.

The humanist revolution caused modern Western culture to lose faith and interest in superior mental states, and to sanctify the mundane experiences of the average Joe. Modern Western culture is therefore unique in lacking a specialised class of people who seek to experience extraordinary mental states. It believes anyone attempting to do so is a drug addict, mental patient or charlatan. Consequently, though we have a detailed map of the mental landscape of Harvard psychology students, we know far less about the mental landscapes of Native American shamans, Buddhist monks or Sufi mystics.[2]

And that is just the Sapiens mind. Fifty thousand years ago we shared this planet with our Neanderthal cousins. They didn't launch spaceships, build pyramids or establish empires. They obviously had very different mental abilities and lacked many of our talents. Nevertheless, they had bigger brains than us Sapiens. What exactly did they do with all those neurons? We have absolutely no idea. But they might well have had many mental states that no Sapiens has ever experienced.

Yet even if we take into account all human species that ever existed, that would not come close to exhausting the mental spectrum. Other animals probably have experiences that we humans can barely imagine. Bats, for example, experience the world through echolocation. They emit a very rapid stream of high-frequency chirps, well beyond the range of the human ear. They then detect and interpret the returning echoes to build a picture of the world. That picture is so detailed and accurate that bats can fly quickly between trees and buildings, chase and capture moths and mosquitoes, and all the while evade owls and other predators.

Bats live in a world of echoes. Just as in the human world every object has a characteristic shape and colour, so in the bat world every object has its echo-pattern. A bat can distinguish between a tasty moth species and a poisonous moth species by the different echoes bouncing back from their delicate wings. Some edible moth species try to protect themselves by evolving an echo-pattern similar to that of a poisonous species. Others have evolved an even more remarkable ability to deflect the waves of the bat radar, so like stealth bombers they can fly around without the bat knowing they are there. The world of echolocation is as complex and stormy as our familiar world of sound and sight, but we are completely oblivious to it.

One of the most important articles about the philosophy of mind is titled 'What Is It Like to Be a Bat?'[3] In this 1974 paper, the philosopher Thomas Nagel points out that a Sapiens mind cannot fathom the subjective world of a bat. We can write all the algorithms we want about the bat body, bat echolocation systems and bat neurons, but that won't tell us how it *feels* to be a bat. How does it feel to echolocate a moth flapping its wings? Is it similar to seeing it, or is it something completely different?

Trying to explain to a Sapiens how it feels to echolocate a butterfly is probably as pointless as explaining to a blind mole how it feels to see a Caravaggio. It's likely that bat emotions are also deeply influenced by the centrality of their echolocation sense. For Sapiens, love is red, envy is green and depression is blue. Who knows what echolocations colour the love of a female bat for her offspring, or the feelings of a male bat towards his rivals?

Bats aren't special, of course. They are but one of countless possible examples. Just as Sapiens cannot understand what it's like to be a bat, we have similar difficulties understanding how it feels to be a whale, a tiger or a pelican. It certainly must feel like something; but we don't know like what. Both whales and humans process emotions in a part of the brain called the limbic system, yet the whale limbic system includes an entire additional part that is missing from the human structure. Maybe that part enables whales to experience

47. A spectrogram of a bowhead whale song. How does a whale experience this song? The *Voyager* record included a whale song in addition to Beethoven, Bach and Chuck Berry. We can only hope it is a good one.

extremely deep and complex emotions that are alien to us? Whales might also have astounding musical experiences that even Bach and Mozart couldn't grasp. Whales can hear one another from hundreds of miles away, and each whale has a repertoire of characteristic 'songs' that may last for hours and follow very intricate patterns. Every now and then a whale composes a new hit, which other whales throughout the ocean adopt. Scientists routinely record these hits and analyse them with the help of computers, but can any human fathom these musical experiences and tell the difference between a whale Beethoven and a whale Justin Bieber?[4]

None of this should surprise us. Sapiens don't rule the world because they have deeper emotions or more complex musical experiences than other animals. So we may be inferior to whales, bats, tigers and pelicans at least in some emotional and experiential domains.

Beyond the mental spectrum of humans, bats, whales and all other animals, even vaster and stranger continents may lie in wait. In all probability there is an infinite variety of mental states that no Sapiens, bat or dinosaur ever experienced in 4 billion years of terrestrial evolution, because they did not have the necessary faculties.

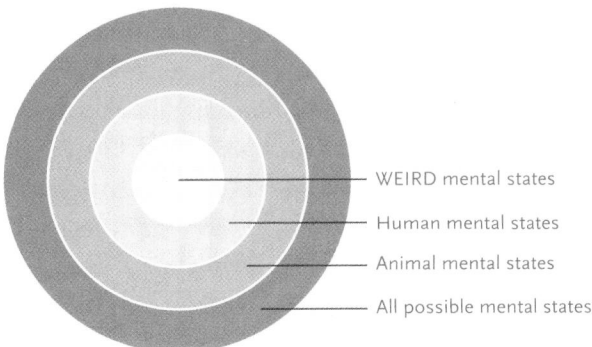

48. The spectrum of consciousness.

In the future, however, powerful drugs, genetic engineering, electronic helmets and direct brain–computer interfaces may open passages to these places. Just as Columbus and Magellan sailed beyond the horizon to explore new islands and unknown continents, so we may one day embark for the antipodes of the mind.

I Smell Fear

As long as doctors, engineers and customers focused on healing mental diseases and enjoying life in WEIRD societies, the study of subnormal mental states and WEIRD minds was perhaps sufficient to our needs. Though normative psychology is often accused of mistreating any divergence from the norm, in the last century it has brought relief to countless people, saving the lives and sanity of millions.

However, at the beginning of the third millennium we face a completely different kind of challenge, as liberal humanism makes way for techno-humanism, and medicine is increasingly focused on upgrading the healthy rather than healing the sick. Doctors, engineers and customers no longer want merely to fix mental problems – they are now seeking to upgrade the mind. We are acquiring the technical abilities to begin manufacturing new states of consciousness, yet we lack a map of these potential new territories. Since we are familiar mainly with the normative and sub-normative mental spectrum of WEIRD people, we don't even know what destinations to aim towards.

Not surprisingly then, positive psychology has become the trendiest subfield of the discipline. In the 1990s leading experts such as Martin Seligman, Ed Dinner and Mihaly Csikszentmihalyi argued that psychology should study not just mental illnesses, but also mental strengths. How come we have a remarkably detailed atlas of the sick mind, but no scientific map of the prosperous mind? Over the last two decades, positive psychology has made important first steps in the study of super-normative mental states, but as of 2016, the super-normative zone is largely terra incognita to science.

Under such circumstances, we might rush forward without any map, and focus on upgrading those mental abilities that the current economic and political system needs, while neglecting and even downgrading others. Of course, this is not a completely new phenomenon. For thousands of years the system has been shaping and reshaping our minds according to its needs. Sapiens originally evolved as members of small intimate communities, and their mental faculties were not adapted to living as cogs within a giant machine. However, with the rise of cities, kingdoms and empires, the system cultivated capacities required for large-scale cooperation, while disregarding other skills and aptitudes.

For example, archaic humans probably made extensive use of their sense of smell. Hunter-gatherers are able to smell from a distance the difference between various animal species, various humans and even various emotions. Fear, for example, smells different from courage. When a man is afraid he secretes different chemicals compared to when he is full of courage. If you sat among an archaic band debating whether to start a war against the neighbours, you could literally smell public opinion.

As Sapiens organised themselves into larger groups, noses lost much of their social importance, because they are useful only when dealing with small numbers of individuals. You cannot, for example, smell the American fear of China. Consequently, human olfactory powers were neglected. Brain areas that tens of thousands of years ago probably dealt with odours were put to work on more urgent tasks such as reading, mathematics and abstract reasoning. The system prefers that our neurons solve differential equations rather than smell our neighbours.[5]

The same thing happened to our other senses and to the underlying ability to pay attention to our sensations. Ancient foragers were always alert and attentive. Wandering in the forest in search of mushrooms, they sniffed the wind carefully and watched the ground intently. When they found a mushroom, they ate it with the utmost attention, aware of every little nuance of flavour, which could distinguish an edible mushroom from its poisonous cousin. Members of today's affluent societies don't need such keen awareness. We can go to the supermarket and buy any of a thousand different dishes, all supervised by the health authorities. But whatever we choose – Italian pizza or Thai noodles – we are likely to eat it in haste in front of the TV, hardly paying attention to the taste (which is why food producers are constantly inventing exciting new flavours that might somehow pierce our curtain of indifference). Similarly, thanks to good transport services we can easily meet a friend who lives across town. But even when together we seldom give this friend our undivided attention because we constantly check our smartphone and our Facebook account, convinced that something far more interesting is probably happening elsewhere. Modern humanity is sick with FOMO – Fear Of Missing Out – and though we have more choice than ever before, we have lost the ability to really pay attention to whatever we choose.[6]

In addition to smelling and paying attention, we have also been losing our ability to dream. Many cultures believed that what people see and do in their dreams is no less important than what they see and do while awake. Hence people actively developed their ability to dream, to remember dreams and even to control their actions in the dream world, which is known as 'lucid dreaming'. Experts in lucid dreaming could move about the dream world at will, and claimed they could even travel to higher planes of existence or meet visitors from other worlds. The modern world, in contrast, dismisses dreams as subconscious messages at best, and mental garbage at worst. Consequently, dreams play a much smaller part in our lives, few people actively develop their dreaming skills, and many people claim that they don't dream at all, or that they cannot remember any of their dreams.[7]

Did the decline in our capacity to smell, pay attention and dream make our lives poorer and greyer? Maybe. But even if it did, for the economic and political system it was worth it. Your boss wants you to constantly check your emails rather than smell flowers or dream about fairies. For similar reasons, it is likely that future upgrades to the human mind will reflect political needs and market forces.

For example, the US army's 'attention helmet' is meant to help people focus on well-defined tasks and speed up their decision-making process. It may, however, reduce their ability to show empathy and tolerate doubts and inner conflicts. Humanist psychologists have pointed out that people in distress often don't want a quick fix – they want somebody to listen to them and sympathise with their fears and misgivings. Suppose you are having an ongoing crisis in your workplace, because your new boss doesn't appreciate your views, and insists on doing everything her way. After one particularly unhappy day, you pick up the phone and call a friend. But the friend has little time and energy for you, so he cuts you short, and tries to solve your problem: 'Okay. I get it. Well, you really have just two options here: either quit the job, or stay and do what the boss wants. And if I were you, I would quit.' That would hardly help. A really good friend would have patience, and not be so quick to find a solution. He would listen to your distress, and allow time and space for all your contradictory emotions and gnawing anxieties to surface.

The attention helmet works a bit like the impatient friend. Of course sometimes – on the battlefield, for instance – people need to take firm decisions quickly. But there is more to life than that. If we start using the attention helmet in more and more situations, we may end up losing our ability to tolerate confusion, doubts and contradictions, just as we have lost our ability to smell, dream and pay attention. The system may push us in that direction, because it usually rewards us for the decisions we make rather than for our doubts. Yet a life of resolute decisions and quick fixes may be poorer and shallower than one of doubts and contradictions.

When we mix a practical ability to engineer minds with our ignorance of the mental spectrum and with the narrow interests of governments, armies and corporations, we get a recipe for trouble. We may successfully upgrade our bodies and our brains, while losing our minds in the process. Indeed, techno-humanism may end up *downgrading* humans. The system may prefer downgraded humans not because they would possess any superhuman knacks, but because they would lack some really disturbing human qualities that hamper the system and slow it down. As any farmer knows, it's usually the brightest goat in the herd that stirs up the most trouble, which is why the Agricultural Revolution involved downgrading animals' mental abilities. The second cognitive revolution, dreamed up by techno-humanists, might do the same to us, producing human cogs who communicate and process data far more effectively than ever before, but who can barely pay attention, dream or doubt. For millions of years we were enhanced chimpanzees. In the future, we may become oversized ants.

The Nail on Which the Universe Hangs

Techno-humanism faces another dire threat. Like all humanist sects, techno-humanism too sanctifies the human will, seeing it as the nail on which the entire universe hangs. Techno-humanism expects our desires to choose which mental abilities to develop and thereby determine the shape of future minds. Yet what will happen once technological progress makes it possible to reshape and engineer those desires?

Humanism always emphasised that it is not easy to identify our authentic will. When we try to listen to ourselves, we are often flooded by a cacophony of conflicting noises. Indeed, we sometimes don't really want to hear our authentic voice, because it might disclose unwelcome secrets and make uncomfortable requests. Many people take great care not to probe themselves too deeply. A successful lawyer on the fast track may stifle an inner voice telling her to take a break and have a child. A woman trapped in a dissatisfying marriage fears losing the security it provides. A guilt-ridden soldier is stalked by nightmares about atrocities he committed. A young man unsure of his sexuality follows a personal 'don't ask, don't tell' policy. Humanism doesn't think any of these situations has an obvious one-size-fits-all solution. But humanism demands that we show some guts, listen to the inner messages even if they scare us, identify our authentic voice and then follow its instructions regardless of the difficulties.

Technological progress has a very different agenda. It doesn't want to listen to our inner voices. It wants to control them. Once we understand the biochemical system producing all these voices, we can play with the switches, turn up the volume here, lower it there, and make life much more easy and comfortable. We'll give Ritalin to the distracted lawyer, Prozac to the guilty soldier and Cipralex to the dissatisfied wife. And that's just the beginning.

Humanists are often appalled by this approach, but we had better not pass judgement on it too quickly. The humanist recommendation to listen to ourselves has ruined the lives of many a person, whereas the right dosage of the right chemical has greatly improved the well-being and relationships of millions. In order to really listen to themselves, some people must first turn down the volume of the inner screams and diatribes. According to modern psychiatry, many 'inner voices' and 'authentic wishes' are nothing more than the product of biochemical imbalances and neurological diseases. People suffering from clinical depression repeatedly walk out on promising careers and healthy relationships because some biochemical glitch makes them see everything through dark-coloured lenses. Instead of listening to such destructive inner voices, it might be a good idea to shut them up. When Sally Adee used the attention helmet to silence the voices in her head, she not only became an expert markswoman, but she also felt much better about herself.

Personally, each of us may have a different view about these issues. Yet from a historical perspective it is clear that something momentous is happening. The number one humanist command- ment – listen to yourself! – is no longer self-evident. As we learn to turn our inner volume up and down, we give up our belief in authenticity, because it is no longer clear whose hand is on the switch. Silencing annoying noises inside my head seems like a wonderful idea, provided it enables me to finally hear my deep authentic self. But if there is no authentic self, how do I decide which voices to silence and which to amplify?

Let's assume, just for the sake of argument, that within a few decades brain scientists will grant us easy and accurate control over many inner voices. Imagine a young gay man from a devout Mormon family, who, after years of living in the closet, has finally accumulated enough money to finance a passion operation. He goes to the clinic armed with $100,000, determined to walk out as straight as Joseph Smith. Standing in front of the clinic's door he mentally repeats what he intends to say to the doctor: 'Doc, here's $100,000. Please fix me so that I will never want men again.' He then rings the bell, and the door is opened by a real-life George Clooney. 'Doc,' mumbles the overwhelmed lad, 'here's $100,000. Please fix me so that I will never want to be straight again.'

Did the young man's authentic self win over the religious brain- washing he underwent? Or did a moment's temptation cause him to betray himself? Or perhaps there is simply no such thing as an authentic self that we can follow or betray? Once we can design and redesign our will, we could no longer see it as the ultimate source of all meaning and authority. For no matter what our will says, we can always make it say something else.

According to humanism, only human desires imbue the world with meaning. Yet if we could choose our desires, on what basis could we possibly make such choices? Suppose *Romeo and Juliet* opened with Romeo having to decide with whom to fall in love. And suppose even after making a decision, Romeo could always retract and make a different choice instead. What kind of play would it have been? Well, that's the play technological progress is trying to produce for us. When our desires make us uncomfortable, technology promises to bail us out. When the nail on which the entire universe hangs is pegged in a problematic spot, technology will pull it out and stick it somewhere else. But where exactly? If I could peg that nail anywhere in the cosmos, where should I peg it, and why there of all places?

Humanist dramas unfold when people have uncomfortable desires. For example, it is extremely uncomfortable when Romeo of the house of Montague falls in love with Juliet of the house of Capulet, because the Montagues and Capulets are bitter enemies. The technological solution to such dramas is to ensure we never have uncomfortable desires. How much pain and sorrow would have been avoided if, instead of drinking poison, Romeo and Juliet could just take a pill or wear a helmet that would have redirected their star-crossed love towards other people.

Techno-humanism faces an impossible dilemma here. It considers the human will to be the most important thing in the universe, hence it pushes humankind to develop technologies that can control and redesign the will. After all, it's tempting to gain control over the most important thing in the world. Yet should we ever achieve such control, techno-humanism would not know what to do with it, because the sacred human would then become just another designer product. We can never deal with such technologies as long as we believe that the human will and the human experience are the supreme source of authority and meaning.

Hence a bolder techno-religion seeks to sever the humanist umbilical cord altogether. It foresees a world that does not revolve around the desires and experiences of any humanlike beings. What might replace desires and experiences as the source of all meaning and authority? As of 2016, there is one candidate sitting in history's reception room waiting for the job interview. This candidate is information. The most interesting emerging religion is Dataism, which venerates neither gods nor man – it worships data.

11

The Data Religion

Dataism declares that the universe consists of data flows, and the value of any phenomenon or entity is determined by its contribution to data processing.[1] This may strike you as some eccentric fringe notion, but in fact it has already conquered most of the scientific establishment. Dataism was born from the explosive confluence of two scientific tidal waves. In the 150 years since Charles Darwin published *On the Origin of Species*, the life sciences have come to see organisms as biochemical algorithms. Simultaneously, in the eight decades since Alan Turing formulated the idea of a Turing Machine, computer scientists have learned to engineer increasingly sophisticated electronic algorithms. Dataism puts the two together, pointing out that exactly the same mathematical laws apply to both biochemical and electronic algorithms. Dataism thereby collapses the barrier between animals and machines, and expects electronic algorithms to eventually decipher and outperform biochemical algorithms.

For politicians, business people and ordinary consumers, Dataism offers groundbreaking technologies and immense new powers. For scholars and intellectuals it also promises to provide the scientific holy grail that has eluded us for centuries: a single overarching theory that unifies all the scientific disciplines from musicology through economics to biology. According to Dataism,

Beethoven's Fifth Symphony, a stock-exchange bubble and the flu virus are just three patterns of data flow that can be analysed using the same basic concepts and tools. This idea is extremely attractive. It gives all scientists a common language, builds bridges over academic rifts and easily exports insights across disciplinary borders. Musicologists, economists and cell biologists can finally understand each other.

In the process Dataism inverts the traditional pyramid of learning. Hitherto, data was seen as only the first step in a long chain of intellectual activity. Humans were supposed to distil data into information, information into knowledge, and knowledge into wisdom. However, Dataists believe that humans can no longer cope with the immense flows of data, hence they cannot distil data into information, let alone into knowledge or wisdom. The work of processing data should therefore be entrusted to electronic algorithms, whose capacity far exceeds that of the human brain. In practice, this means that Dataists are sceptical about human knowledge and wisdom, and prefer to put their trust in Big Data and computer algorithms.

Dataism is most firmly entrenched in its two mother disciplines: computer science and biology. Of the two biology is the more important. It was biology's embrace of Dataism that turned a limited breakthrough in computer science into a world-shattering cataclysm that may completely transform the very nature of life. You may not agree with the idea that organisms are algorithms, and that giraffes, tomatoes and human beings are just different methods for processing data. But you should know that this is current scientific dogma, and it is changing our world beyond recognition.

Not only individual organisms are seen today as data-processing systems, but also entire societies such as beehives, bacteria colonies, forests and human cities. Economists increasingly interpret the economy too as a data-processing system. Laypeople believe that the economy consists of peasants growing wheat, workers manufacturing clothes, and customers buying bread and underpants. Yet experts see the economy as a mechanism for gathering data about desires and abilities, and turning this data into decisions.

According to this view, free-market capitalism and state-controlled communism aren't competing ideologies, ethical creeds or political institutions. They are, in essence, competing data-processing systems. Capitalism uses distributed processing, whereas communism relies on centralised processing. Capitalism processes data by directly connecting all producers and consumers to one another and allowing them to exchange information freely and make decisions independently. How do you determine the price of bread in a free market? Well, every bakery may produce as much bread as it likes, and charge for it as much as it wants. The customers are equally free to buy as much bread as they can afford, or take their business to a competitor. It isn't illegal to charge $1,000 for a baguette, but nobody is likely to buy it.

On a much grander scale, if investors predict increased demand for bread, they will buy shares of biotech firms that genetically engineer more prolific wheat strains. The influx of capital will enable the firms to speed up their research, thereby providing more wheat faster, and averting bread shortages. Even if one biotech giant adopts a flawed theory and reaches an impasse, its more successful competitors will likely achieve the hoped-for breakthrough. Free-market capitalism thus distributes the work of analysing data and making decisions between many independent but interconnected processors. As the Austrian economics guru Friedrich Hayek explained, 'In a system in which the knowledge of the relevant facts is dispersed among many people, prices can act to coordinate the separate actions of different people.'[2]

According to this view the stock exchange is the fastest and most efficient data-processing system humankind has so far created. Everyone is welcome to join, if not directly then through their banks or pension funds. The stock exchange runs the global economy, and takes into account everything that happens all over the planet – and even beyond it. Prices are influenced by successful scientific experiments, by political scandals in Japan, by volcanic eruptions in Iceland and even by irregular activities on the surface of the sun. In order for the system to run smoothly, as much information as possible needs to flow as freely as possible. When millions of people throughout the world have access to all the relevant information, they determine the most accurate price of oil, of Hyundai shares and of Swedish government bonds by buying and selling them. It has been estimated that the stock exchange needs just fifteen minutes of trade to determine the influence of a *New York Times* headline on the prices of most shares.[3]

Data-processing considerations also explain why capitalists favour lower taxes. Heavy taxation means that a large part of all available capital accumulates in one place – the state coffers – and consequently more and more decisions have to be made by a single processor, namely the government. This creates an overly centralised data-processing system. In extreme cases, when taxes are exceedingly high, almost all capital ends up in the government's hands, and so the government alone calls the shots. It dictates the price of bread, the location of bakeries, and the research-and-development budget. In a free market, if one processor makes a wrong decision, others will be quick to capitalise on its mistake. However, when a single processor makes almost all the decisions, mistakes can be catastrophic.

This extreme situation, in which all data is processed and all decisions are made by a single central processor, is called

49. The Soviet leadership in Moscow, 1963: centralised data processing.

communism. In a communist economy people allegedly work according to their abilities and receive according to their needs. In other words, the government takes 100 per cent of your profits, decides what you need and then supplies these needs. Though no country ever realised this scheme in its extreme form, the Soviet Union and its satellites came as close as they could. They abandoned the principle of distributed data processing and switched to a model of centralised data processing. All information from throughout the Soviet Union flowed to a single location in Moscow where all the important decisions were made. Producers and consumers could not communicate directly and had to obey government orders.

For instance, the Soviet economics ministry might decide that the price of bread in all shops should be exactly two roubles and four kopeks, that a particular kolkhoz in the Odessa oblast should switch from growing wheat to raising chickens, and that the Red October bakery in Moscow should produce 3.5 million loaves of bread per day and not a single loaf more. Meanwhile the Soviet science ministry forced all Soviet biotech laboratories to adopt the theories of Trofim Lysenko – the infamous head of the Lenin Academy for Agricultural Sciences. Lysenko rejected the dominant genetic theories of his day. He insisted that if an organism acquired some new trait during its lifetime, this quality could pass

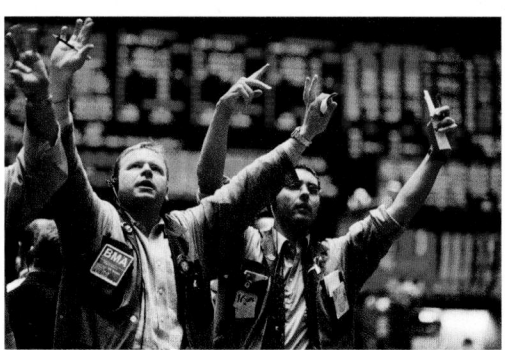

50. Commotion on the floor of the Chicago
Board of Trade: distributed data processing.

directly to its descendants. This idea flew in the face of Darwinian orthodoxy, but it dovetailed nicely with communist educational principles. It implied that if you could train wheat plants to withstand cold weather, their progenies will also be cold-resistant. Lysenko accordingly sent billions of counter-revolutionary wheat plants to be re-educated in Siberia – and the Soviet Union was soon forced to import more and more flour from the United States.[4]

Capitalism did not defeat communism because capitalism was more ethical, because individual liberties are sacred or because God was angry with the heathen communists. Rather, capitalism won the Cold War because distributed data processing works better than centralised data processing, at least in periods of accelerating technological change. The central committee of the Communist Party just could not deal with the rapidly changing world of the late twentieth century. When all data is accumulated in one secret bunker, and all important decisions are taken by a group of elderly apparatchiks, they can produce nuclear bombs by the cartload, but not an Apple or a Wikipedia.

There is a story (probably apocryphal, like most good stories) that when Mikhail Gorbachev tried to resuscitate the moribund Soviet economy, he sent one of his chief aides to London to find out what Thatcherism was all about, and how a capitalist system actually functioned. The hosts took their Soviet visitor on a tour of the City, of the London stock exchange and of the London School of Economics, where he had lengthy talks with bank managers, entrepreneurs and professors. After many long hours the Soviet expert burst out: 'Just one moment, please. Forget about all these complicated economic theories. We have been going back and forth across London for a whole day now, and there's one thing I cannot understand. Back in Moscow our finest minds are working on the bread supply system, and yet there are such long queues in every bakery and grocery store. Here in London live millions of people, and we have passed today in front of many shops and supermarkets, yet I haven't seen a single bread queue. Please take me to meet the person in charge of supplying bread to London. I must learn his secret.' The hosts scratched their heads, thought for a moment, and said: 'Nobody is in charge of supplying bread to London.'

That's the capitalist secret of success. No central processing unit monopolises all the data on the London bread supply. The information flows freely among millions of consumers and producers, bakers and tycoons, farmers and scientists. Market forces determine the price of bread, the number of loaves baked each day and the research-and-development priorities. If market forces make a bad decision, they soon correct themselves, or so capitalists believe. For our current purposes it doesn't matter whether this capitalist theory is correct. The crucial thing is that the theory understands economics in terms of data processing.

Where Has All the Power Gone?

Political scientists also increasingly interpret human political structures as data-processing systems. Like capitalism and communism, so democracies and dictatorships are in essence competing mechanisms for gathering and analysing information. Dictatorships use centralised processing methods, whereas democracies prefer distributed processing. Over the past decades democracy gained the upper hand because under the unique conditions of the late twentieth century, distributed processing worked better. Under alternative conditions – those prevailing in the ancient Roman Empire, for instance – centralised processing had an edge, which is why the Roman Republic fell and power shifted from the Senate and popular assemblies into the hands of a single autocratic emperor.

This implies that as data-processing conditions change again in the twenty-first century, democracy might decline and even disappear. As both the volume and speed of data increase, venerable institutions like elections, political parties and parliaments might become obsolete – not because they are unethical, but because they can't process data efficiently enough. These institutions evolved in an era when politics moved faster than technology. In the nineteenth and twentieth centuries the Industrial Revolution unfolded slowly enough for politicians and voters to remain one step ahead of it and regulate and manipulate its course. Yet whereas the rhythm of politics has not changed much since the days of steam, technology has switched from first gear to fourth. Technological revolutions now outpace political processes, causing MPs and voters alike to lose control.

The rise of the Internet gives us a taste of things to come. Cyberspace is now crucial to our daily lives, our economy and our security. Yet the critical choices between alternative web designs weren't taken through a democratic political process, even though they involved traditional political issues such as sovereignty, borders, privacy and security. Did you ever vote about the shape of cyberspace? Decisions made by web designers far from the public limelight mean that today the Internet is a free and lawless zone that erodes state sovereignty, ignores borders, abolishes privacy and poses perhaps the most formidable global security risk. Whereas a decade ago it hardly registered on the radar, today hysterical officials are predicting an imminent cyber 9/11.

Governments and NGOs consequently conduct intense debates about restructuring the Internet, but it is much harder to change an existing system than to intervene at its inception. Besides, by the time the cumbersome government bureaucracy makes up its mind about cyber regulation, the Internet will have morphed ten times. The governmental tortoise cannot keep up with the technological hare. It is overwhelmed by data. The NSA may be spying on our every word, but to judge by the repeated failures of American foreign policy, nobody in Washington knows what to do with all the data. Never in history did a government know so much about what's going on in the world – yet few empires have botched things up as clumsily as the contemporary United States. It's like a poker player who knows what cards his opponents hold, yet somehow still manages to lose round after round.

In the coming decades it is likely that we will see more Internet-like revolutions, in which technology steals a march on politics.

Artificial intelligence and biotechnology might soon overhaul our societies and economies – and our bodies and minds too – but they are hardly a blip on the current political radar. Present-day democratic structures just cannot collect and process the relevant data fast enough, and most voters don't understand biology and cybernetics well enough to form any pertinent opinions. Hence traditional democratic politics is losing control of events, and is failing to present us with meaningful visions of the future.

Ordinary voters are beginning to sense that the democratic mechanism no longer empowers them. The world is changing all around, and they don't understand how or why. Power is shifting away from them, but they are unsure where it has gone. In Britain voters imagine that power might have shifted to the EU, so they vote for Brexit. In the USA voters imagine that 'the establishment' monopolizes all the power, so they support anti-establishment candidates such as Bernie Sanders and Donald Trump. The sad truth is that nobody knows where all the power has gone. Power will definitely not shift back to ordinary voters if Britain leaves the EU nor if Trump takes over the White House.

That doesn't mean we will go back to twentieth-century-style dictatorships. Authoritarian regimes seem to be equally overwhelmed by the pace of technological development and the speed and volume of the data flow. In the twentieth century dictators had grand visions for the future. Communists and fascists alike sought to completely destroy the old world and build a new world in its place. Whatever you think about Lenin, Hitler or Mao, you cannot accuse them of lacking vision. Today it seems that leaders have an opportunity to pursue even grander visions. While communists and Nazis tried to create a new society and a new human with the help of steam engines and typewriters, today's prophets could rely on biotechnology and super-computers.

In science-fiction films ruthless Hitler-like politicians are quick to pounce on such new technologies, putting them in the service of this or that megalomaniacal political ideal. Yet flesh-and-blood politicians in the early twenty-first century, even in authoritarian countries such as Russia, Iran or North Korea, are nothing like their Hollywood counterparts. They don't seem to be plotting any Brave New World. The wildest dreams of Kim Jong-un and Ali Khamenei don't extend much beyond atom bombs and ballistic missiles: that is so 1945. Putin's aspirations seem confined to rebuilding the old Soviet bloc, or the even older tsarist empire. Meanwhile in the USA paranoid Republicans have accused Barack Obama of being a ruthless despot hatching conspiracies to destroy the foundations of American society – yet in eight years of his presidency he barely managed to pass a minor health-care reform. Creating new worlds and new humans was far beyond his agenda.

Precisely because technology is now moving so fast, and parliaments and dictators alike are overwhelmed by data they cannot process quickly enough, present-day politicians are thinking on a far smaller scale than their predecessors a century ago. Consequently, in the early twenty-first century politics is bereft of grand visions. Government has become mere administration. It manages the country, but it no longer leads it. Government ensures that teachers are paid on time and sewage systems don't overflow, but it has no idea where the country will be in twenty years.

To a certain extent, this is a very good thing. Given that some of the big political visions of the twentieth century led us to Auschwitz, Hiroshima and the Great Leap Forward, maybe we are better off in the hands of petty-minded bureaucrats. Mixing godlike technology with megalomaniacal politics is a recipe for disaster. Many neo-liberal economists and political scientists argue that it is best to leave all the important decisions in the hands of the free market. They thereby give politicians the perfect excuse for inaction and ignorance, which are reinterpreted as profound wisdom. Politicians find it convenient to believe that the reason they don't understand the world is that they don't need to understand it.

Yet mixing godlike technology with myopic politics also has its downside. Lack of vision isn't always a blessing, and not all visions are necessarily bad. In the twentieth century the dystopian Nazi vision did not fall apart spontaneously. It was defeated by the equally grand visions of socialism and liberalism. It is dangerous to trust our future to market forces, because these forces do what's good for the market rather than what's good for humankind or for the world. The hand of the market is blind as well as invisible, and left to its own devices it may fail to do anything at all about the threat of global warming or the dangerous potential of artificial intelligence.

Some people believe that there is somebody in charge after all. Not democratic politicians or autocratic despots, but rather a small coterie of billionaires who secretly run the world. But such conspiracy theories never work, because they underestimate the complexity of the system. A few billionaires smoking cigars and drinking Scotch in some back room cannot possibly understand everything happening on the globe, let alone control it. Ruthless billionaires and small interest groups flourish in today's chaotic world not because they read the map better than anyone else, but because they have very narrow aims. In a chaotic system, tunnel vision has its advantages, and the billionaires' power is strictly proportional to their goals. When the world's richest tycoons want to make another billion dollars, they can easily game the system in order to do so. In contrast, if they felt inclined to reduce global inequality or stop global warming, even they wouldn't be able to do it, because the system is far too complex.

Yet power vacuums seldom last long. If in the twenty-first century traditional political structures can no longer process the data fast enough to produce meaningful visions, then new and more efficient structures will evolve to take their place. These new structures may be very different from any previous political institutions, whether democratic or authoritarian. The only question is who will build and control these structures. If humankind is no longer up to the task, perhaps it might give somebody else a try.

History in a Nutshell

From a Dataist perspective, we may interpret the entire human species as a single data-processing system, with individual humans serving as its chips. If so, we can also understand the whole of history as a process of improving the efficiency of this system through four basic methods:

1. **Increasing the number of processors.** A city of 100,000 people has more computing power than a village of 1,000 people.
2. **Increasing the variety of processors.** Different processors may use diverse ways to calculate and analyse data. Using several kinds of processors in a single system may therefore increase its dynamism and creativity. A conversation between a peasant, a priest and a physician may produce novel ideas that would never emerge from a conversation between three hunter-gatherers.
3. **Increasing the number of connections between processors.** There is little point in increasing the mere number and variety of processors if they are poorly connected to each other. A trade network linking ten cities is likely to result in many more economic, technological and social innovations than ten isolated cities.
4. **Increasing the freedom of movement along existing connections.** Connecting processors is hardly useful if data cannot flow freely. Just building roads between ten cities won't be very useful if they are plagued by robbers, or if some paranoid despot doesn't allow merchants and travellers to move as they wish.

These four methods often contradict one another. The greater the number and variety of processors, the harder it is to freely connect them. The construction of the Sapiens data-processing system accordingly passed through four main stages, each characterised by an emphasis on a different method.

The first stage began with the Cognitive Revolution, which made it possible to connect vast numbers of Sapiens into a single data-processing network. This gave Sapiens a crucial advantage over all other human and animal species. While there is a strict limit to the number of Neanderthals, chimpanzees or elephants you can connect to the same net, there is no limit to the number of Sapiens.

Sapiens used their advantage in data processing to overrun the entire world. However, as they spread into different lands and climates they lost touch with one another and underwent diverse cultural transformations. The result was an immense variety of human cultures, each with its own lifestyle, behaviour patterns and world view. Hence the first phase of history involved an increase in the number and variety of human processors, at the expense of connectivity: 20,000 years ago there were many more Sapiens than 70,000 years ago, and Sapiens in Europe processed information differently from Sapiens in China. However, there were no connections between people in Europe and China, and it would have seemed utterly impossible that all Sapiens could one day be part of a single data-processing web.

The second stage began with the Agricultural Revolution and continued until the invention of writing and money about 5,000 years ago. Agriculture accelerated demographic growth so the number of human processors rose sharply. Simultaneously, agriculture enabled many more people to live together in close proximity, thereby generating dense local networks that contained unprecedented numbers of processors. In addition, agriculture created new incentives and opportunities for different networks to trade and communicate with one another. Nevertheless, during the second phase centrifugal forces remained predominant. In the absence of writing and money humans could not establish cities, kingdoms or empires. Humankind was still divided into innumerable little tribes, each with its own lifestyle and world view. Uniting the whole of humankind was not even a fantasy.

The third stage kicked off with the invention of writing and money about 5,000 years ago, and lasted until the beginning of the Scientific Revolution. Thanks to writing and money the gravitational field of human cooperation finally overpowered the centrifugal forces. Human groups bonded and merged to form cities and kingdoms. Political and commercial links between different cities and kingdoms also tightened. At least since the first millennium BC – when coinage, empires and universal religions appeared – humans began to consciously dream about forging a single network that would encompass the entire globe.

This dream became a reality during the fourth and last stage of history, which began around 1492. Early modern explorers, conquerors and traders wove the first thin threads that encompassed the whole world. In the late modern period these threads were made stronger and denser, so that the spider's web of Columbus's days became the steel and asphalt grid of the twenty-first century. Even more importantly, information was allowed to flow increasingly freely throughout this global grid. When Columbus first hooked up the Eurasian net to the American net, only a few bits of data managed to cross the ocean each year, running a gauntlet of cultural prejudices, strict censorship and political repression. But as the years went by the free market, the scientific community, the rule of law and the spread of democracy all helped to dissolve the barriers. We often imagine that democracy and the free market won because they were 'good'. In truth, they won because they improved the global data-processing system.

So, over the last 70,000 years humankind first spread out, then separated into distinct groups, and finally merged again. Yet the process of unification did not take us back to the beginning. When the diverse human groups fused into the global village of today, each brought along its unique legacy of thoughts, tools and behaviours that it had collected and developed along the way. Our modern larders are now stuffed with Middle Eastern wheat,

Andean potatoes, New Guinean sugar and Ethiopian coffee. Similarly, our language, religion, music and politics are replete with heirlooms from across the planet.[5]

If humankind is indeed a single data-processing system, what is its output? Dataists would say that its output will be the creation of a new and even more efficient data-processing system, called the Internet-of-All-Things. Once this mission is accomplished, *Homo sapiens* will vanish.

Information Wants to Be Free

Like capitalism, Dataism too began as a neutral scientific theory, but is now mutating into a religion that claims to determine right and wrong. The supreme value of this new religion is 'information flow'. If life is the movement of information, and if we think that life is good, it follows that we should deepen and broaden the flow of information in the universe. According to Dataism, human experiences are not sacred and *Homo sapiens* isn't the apex of creation or a precursor of some future *Homo deus*. Humans are merely tools for creating the Internet-of-All-Things, which may eventually spread out from planet Earth to pervade the whole galaxy and even the whole universe. This cosmic data-processing system would be like God. It will be everywhere and will control everything, and humans are destined to merge into it.

This conception is reminiscent of some traditional religious visions. Thus Hindus believe that humans can and should merge into the universal soul of the cosmos – the atman. Christians believe that after death saints are infused with the infinite grace of God, whereas sinners cut themselves off from His presence. Indeed, in Silicon Valley the Dataist prophets consciously use traditional messianic language. For example, Ray Kurzweil's book of prophecies is called *The Singularity is Near*, echoing John the Baptist's cry: 'the kingdom of heaven is near' (Matthew 3:2).

Dataists explain to those who still worship flesh-and-blood mortals that they are overly attached to outdated technology. *Homo sapiens* is an obsolete algorithm. After all, what's the advantage of humans over chickens? Only that in humans information flows in much more complex patterns. Humans absorb more data, and process it using better algorithms than do chickens. (In day-to-day language this means that humans allegedly have deeper emotions and superior intellectual abilities. But remember that according to current biological dogma, emotions and intelligence are just algorithms.) Well then, if we could create a data-processing system that can assimilate even more data than a human being, and process it even more efficiently, wouldn't that system be superior to a human in exactly the same way that a human is superior to a chicken?

Dataism isn't limited to idle prophecies. Like every religion, it has its practical commandments. First and foremost a Dataist ought to maximise data flow by connecting to more and more media, and producing and consuming more and more information. Like other successful religions, Dataism is also missionary. Its second commandment is to link everything to the system, including heretics who don't want to be plugged in. And 'everything' means more than just humans. It means every *thing*. Our bodies, of course, but also cars in the street, refrigerators in kitchens, chickens in their coops and trees in the jungle – all should be connected to the Internet-of-All-Things. The refrigerator will monitor the number of eggs in the drawer, and inform the chicken coop when a new shipment is needed. Cars will talk with one another, and the trees in the jungle will report on the weather and on carbon dioxide levels. We mustn't leave any part of the universe disconnected from the great web of life. Conversely, the greatest sin would be to block the data flow. What is death, if not a condition in which information doesn't flow? Hence Dataism upholds the freedom of information as the greatest good of all.

People rarely manage to come up with a completely new value. The last time this happened was in the eighteenth century, when the humanist revolution began preaching the stirring ideals of human liberty, human equality and human fraternity. Since 1789, despite numerous wars, revolutions and upheavals, humans have not managed to conceive of any new value. All subsequent conflicts and struggles have been conducted either in the name of the three humanist values, or in the name of even older ones such as obeying God or serving the nation. Dataism is the first movement since 1789 that created a genuinely novel value: freedom of information.

We mustn't confuse freedom of information with the old liberal value of freedom of expression. Freedom of expression was given to humans, and protected their right to think and say what they wished – including their right to keep their mouths shut and their thoughts to themselves. Freedom of information, in contrast, is not given to humans. It is given to *information*. Moreover, this novel value may impinge on humans' traditional freedom of expression, by privileging the right of information to circulate freely over the right of humans to own data and to restrict its movement.

On 11 January 2013, Dataism got its first martyr when Aaron Swartz, a twenty-six-year-old American hacker, committed suicide in his apartment. Swartz was a rare genius. At fourteen, he helped develop the crucial RSS protocol. Swartz was also a firm believer in the freedom of information. In 2008 he published the 'Guerilla Open Access Manifesto', which demanded a free and unlimited flow of information. Swartz said that 'We need to take information, wherever it is stored, make our copies and share them with the world. We need to take stuff that's out of copyright and add it to the archive. We need to buy secret databases and put them on the Web. We need to download scientific journals and upload them to file sharing networks. We need to fight for Guerilla Open Access.'

Swartz was as good as his word. He became annoyed with the JSTOR digital library for charging its customers. JSTOR holds millions of scientific papers and studies, and believes in the freedom of expression of scientists and journal editors, which includes the freedom to charge a fee for reading their articles. According to JSTOR, if I want to get paid for the ideas I created, it's my right to do so. Swartz thought otherwise. He believed that information wants to be free, that ideas don't belong to the people who created them, and that it is wrong to lock data behind walls and charge an entrance fee. He used the MIT computer network to access JSTOR, and downloaded hundreds of thousands of scientific papers, which he intended to release onto the Internet, so that everybody could read them freely.

Swartz was arrested and put on trial. When he realised that he would probably be convicted and sent to jail, he hanged himself. Hackers reacted with petitions and attacks directed at the academic and government institutions that persecuted Swartz and that infringe on the freedom of information. Under pressure, JSTOR apologised for its part in the tragedy and today allows free access to much, though not all, of its data.[6]

To convince sceptics Dataist missionaries repeatedly explain the immense benefits of the freedom of information. Just as capitalists believe that all good things depend on economic growth, so Dataists believe all good things – including economic growth – depend on the freedom of information. Why did the USA grow faster than the USSR? Because information flowed more freely in the USA. Why are Americans healthier, wealthier and happier than Iranians or Nigerians? Thanks to the freedom of information. So if we want to create a better world, the key is to set the data free.

We have already seen that Google can detect new epidemics faster than traditional health organisations, but only if we allow it free access to the information we are producing. Free-flowing data can similarly reduce pollution and waste, for example by rationalising the transportation system. In 2010 the number of private cars in the world exceeded 1 billion, and has since kept growing.[7] These cars pollute the planet and waste enormous resources, not least by necessitating ever wider roads and more parking spaces. People have become so used to the convenience of private transport that they are unlikely to settle for buses and trains. However, Dataists point out that what people really want is mobility rather than a private car, and a good data-processing system can provide this mobility far more cheaply and efficiently.

I have a private car, but most of the time it sits idly in the parking lot. On a typical day, I enter my car at 8:04, and drive for half an hour to the university, where I park my car for the day. At 18:11 I come back to the car, drive half an hour back home, and that's it. So I am using my car for just an hour a day. Why do I need to keep it for the other twenty-three hours? Why not create a smart car-pool system, run by computer algorithms? The computer would know that I need to leave home at 8:04 and would route the nearest autonomous car to pick me up at that precise moment. After dropping me off on campus it would be available for other purposes instead of waiting in the parking lot. At 18:11 sharp, as I leave the university gate, another communal car would stop right next to me, and take me home. In this way 50 million communal autonomous cars could replace 1 billion private cars, and we would also need far fewer roads, bridges, tunnels and parking spaces. Provided, of course, that I renounce my privacy and allow the algorithms always to know where I am and where I want to go.

Record, Upload, Share!

But maybe you don't need convincing, especially if you are under the age of twenty. People just want to be part of the data flow, even if that means giving up their privacy, their autonomy and their individuality. Humanist art sanctifies the individual genius, so a Picasso doodle on a napkin nets millions at Sotheby's. Humanist science glorifies the individual researcher, and every scholar dreams of putting his or her name at the top of a *Science* or *Nature* paper. But a growing number of artistic and scientific creations are nowadays produced by the ceaseless collaboration of 'everyone'. Who writes Wikipedia? All of us.

The individual is becoming a tiny chip inside a giant system that nobody really understands. Every day I absorb countless data bits through emails, phone calls and articles; process the data; and transmit back new bits through more emails, phone calls and articles. I don't really know where I fit into the greater scheme of things, or how my bits of data connect with the bits produced by billions of other humans and computers. I don't have time to find out, because I am too busy answering all the emails. And as I process more data more efficiently – answering more emails, making more phone calls and writing more articles – so I flood the people around me with even more data.

This relentless flow of data sparks new inventions and disruptions that nobody plans, controls or comprehends. No one understands how the global economy functions or where global politics is heading. But no one needs to understand. All you need to do is answer your emails faster – and allow the system to read them. Just as free-market capitalists believe in the invisible hand of the market, so Dataists believe in the invisible hand of the data flow.

As the global data-processing system becomes all-knowing and all-powerful, so connecting to the system becomes the source of all meaning. Humans want to merge into the data flow because when you are part of the data flow you are part of something much bigger than yourself. Traditional religions assured you that your every word and action was part of some great cosmic plan, and that God watched you every minute and cared about all your thoughts and feelings. Data religion now says that your every word and action is part of the great data flow, that the algorithms are constantly watching you and that they care about everything you do and feel. Most people like this very much. For true-believers, to be disconnected from the data flow risks losing the very meaning of life. What's the point of doing or experiencing anything if nobody knows about it, and if it doesn't contribute something to the global exchange of information?

Humanism holds that experiences occur inside us, and that we ought to find within ourselves the meaning of all that happens, thereby infusing the universe with meaning. Dataists believe that experiences are valueless if they are not shared, and that we need not – indeed *cannot* – find meaning within ourselves. We need only record and connect our experiences to the great data flow, and the algorithms will discover their meaning and tell us what to do. Twenty years ago Japanese tourists were a universal laughing stock because they always carried cameras and took pictures of everything in sight. Now everyone is doing it. If you go to India and see an elephant, you don't look at the elephant and ask yourself, 'What do I feel?' – you are too busy looking for your smartphone, taking a picture of the elephant, posting it on Facebook and then checking your account every two minutes to see how many Likes you got. Writing a private diary – a common humanist practice in previous generations – sounds to many present-day youngsters utterly pointless. Why write anything if nobody else can read it? The new motto says: 'If you experience something – record it. If you record something – upload it. If you upload something – share it.'

Throughout this book we have repeatedly asked what makes humans superior to other animals. Dataism has a new and simple answer. In themselves human experiences are not superior at all to the experiences of wolves or elephants. One bit of data is as good as another. However, humans can write poems and blogs about their experiences and post them online, thereby enriching the global data-processing system. That makes their bits count. Wolves cannot do this. Hence all the experiences of wolves – as deep and complex as they may be – are worthless. No wonder we are so busy converting our experiences into data. It isn't a question of trendiness. It is a question of survival. We must prove to ourselves and to the system that we still have value. And value lies not in having experiences, but in turning these experiences into free-flowing data.

(By the way, wolves – or at least their dog cousins – aren't a hopeless case. A company called 'No More Woof' is developing a helmet for reading canine experiences. The helmet monitors the dog's brain waves, and uses computer algorithms to translate simple sentiments such as 'I am angry' into human language.[8] Your dog may soon have a Facebook or Twitter account of his own – perhaps with more Likes and followers than you.)

Know Thyself

Dataism is neither liberal nor humanist. It should be emphasised, however, that Dataism isn't anti-humanist. It has nothing against human experiences. It just doesn't think they are intrinsically valuable. When we surveyed the three main humanist sects, we asked which experience is the most valuable: listening to Beethoven's Fifth Symphony, to Chuck Berry, to a pygmy initiation song or to the howl of a wolf in heat. A Dataist would argue that the entire exercise is misguided, because music should be evaluated according to the data it carries rather than according to the experience it creates. A Dataist might explain, for example, that the Fifth Symphony carries far more data than the pygmy initiation song, because it uses more chords and scales and creates dialogues with many more musical styles. Consequently, you need far more computational power to decipher the Fifth Symphony, and you gain far more knowledge from doing so.

Music, according to this view, is mathematical patterns. Mathematics can describe every musical piece, as well as the relations between any two pieces. Hence you can measure the precise data value of every symphony, song and howl, and determine which is the richest. The experiences they create in humans or wolves don't really matter. True, for the last 70,000 years or so, human experiences have been the most efficient data-processing algorithms in the universe, hence there was good reason to sanctify them. However, we may soon reach a point when these algorithms will be superseded, and even become a burden.

Sapiens evolved in the African savannah tens of thousands of years ago, and their algorithms are just not built to handle twenty-first-century data flows. We might try to upgrade the human data-processing system, but this may not be enough. The

Internet-of-All-Things may soon create such huge and rapid data flows that even upgraded human algorithms would not be able to handle them. When cars replaced horse-drawn carriages, we didn't upgrade the horses – we retired them. Perhaps it is time to do the same with *Homo sapiens*.

Dataism adopts a strictly functional approach to humanity, appraising the value of human experiences according to their function in data-processing mechanisms. If we develop an algorithm that fulfils the same function better, human experiences will lose their value. Thus if we can replace not just taxi drivers and doctors but also lawyers, poets and musicians with superior computer programs, why should we care if these programs have no consciousness and no subjective experiences? If some humanist starts adulating the sacredness of human experience, Dataists would dismiss such sentimental humbug. 'The experience you are praising is just an outdated biochemical algorithm. In the African savannah 70,000 years ago, that algorithm was state-of-the-art. Even in the twentieth century it was vital for the army and for the economy. But soon we will have much better algorithms.'

In the climactic scene of many Hollywood science-fiction movies, humans face an alien invasion fleet, an army of rebellious robots or an all-knowing super-computer that intends to obliterate them. Humanity seems doomed. But at the very last moment, against all odds, humanity triumphs thanks to something that the aliens, the robots and the super-computers didn't suspect and cannot fathom: love. The hero, who up till now has been easily manipulated by the super-computer and riddled with bullets by the evil robots, is inspired by his sweetheart to make a completely unexpected move that turns the tables on the thunderstruck Matrix. Dataism finds such scenarios utterly ridiculous. 'Come on,' it admonishes the Hollywood screenwriters, 'is that all you could come up with? Love? And not even some platonic cosmic love, but the carnal attraction between two mammals? Do you really think that an all-knowing super-computer or aliens who contrived to conquer the entire galaxy would be dumbfounded by a hormonal rush?'

By equating the human experience with data patterns, Dataism undermines our primary source of authority and meaning and heralds a tremendous religious revolution, the like of which has not been seen since the eighteenth century. In the days of Locke, Hume and Voltaire humanists argued that 'God is a product of the human imagination'. Dataism now gives humanists a taste of their own medicine, and tells them: 'Yes, God is a product of the human imagination, but human imagination in turn is just the product of biochemical algorithms.' In the eighteenth century, humanism sidelined God by shifting from a deo-centric to a homo-centric world view. In the twenty-first century, Dataism may sideline humans by shifting from a homo-centric to a data-centric view.

The Dataist revolution will probably take a few decades, if not a century or two. But then the humanist revolution too did not happen overnight. At first humans kept on believing in God, arguing that humans are sacred because they were created by God for some divine purpose. Only much later did some people dare say that humans are sacred in their own right, and that God doesn't exist at all. Similarly, today most Dataists claim that the Internet-of-All-Things is sacred because humans are creating it to serve human needs. But eventually the Internet-of-All-Things may become sacred in its own right.

The shift from a homo-centric to a data-centric world view won't be merely a philosophical revolution. It will be a practical revolution. All truly important revolutions are practical. The humanist idea that 'humans invented God' was significant because it had far-reaching practical implications. Similarly, the Dataist idea that 'organisms are algorithms' is significant due to its day-to-day practical consequences. Ideas change the world only when they change our behaviour.

In ancient Babylon, when people faced a difficult dilemma they climbed in the darkness of night to the top of the local temple and observed the sky. The Babylonians believed that the stars controlled their fate and predicted their future. By watching the stars the Babylonians decided whether to get married, plough the fields and go to war. Their philosophical beliefs were translated into very practical procedures.

Scriptural religions such as Judaism and Christianity told a different story: 'The stars are lying. God, who created the stars, revealed the entire truth in the Bible. So stop observing the stars – read the Bible instead!' This too was a practical recommendation. When people didn't know whom to marry, what career to choose or whether to start a war, they read the Bible and followed its counsel.

Next came the humanists with an altogether new story: 'Humans invented God, wrote the Bible and then interpreted it in a thousand different ways. So humans themselves are the source of all truth. You may read the Bible as an inspiring human creation, but you don't really need to. If you are facing any dilemma, just listen to yourself and follow your inner voice.' Humanism then gave detailed practical instructions on how to listen to yourself, recommending techniques such as watching sunsets, reading Goethe, keeping a private diary, having heart-to-heart talks with a good friend and holding democratic elections.

For centuries scientists too accepted these humanist guidelines. When physicists wondered whether or not to get married, they too watched sunsets and tried to get in touch with themselves. When chemists contemplated whether to accept a problematic job offer, they too wrote diaries and had heart-to-heart talks with a good friend. When biologists debated whether to wage war or sign a peace treaty, they too voted in democratic elections. When brain scientists wrote books about their startling discoveries, they often put an inspiring Goethe quote on the first page. This was the basis for the modern alliance between science and humanism, which kept the delicate balance between the modern yang and the modern yin – between reason and emotion, between the laboratory and the museum, between the production line and the supermarket.

The scientists not only sanctified human feelings, but also found an excellent evolutionary reason to do so. After Darwin, biologists began explaining that feelings are complex algorithms honed by evolution to help animals make correct decisions. Our love, our fear and our passion aren't some nebulous spiritual phenomena good only for composing poetry. Rather, they encapsulate millions of years of practical wisdom. When you read the Bible you are getting advice from a few priests and rabbis who lived in ancient Jerusalem. In contrast, when you listen to your feelings, you follow an algorithm that evolution has developed for millions of years, and that withstood the harshest quality-control tests of natural selection. Your feelings are the voice of millions of ancestors, each of whom managed to survive and reproduce in an unforgiving environment. Your feelings are not infallible, of course, but they are better than most other sources of guidance. For millions upon millions of years, feelings were the best algorithms in the world. Hence in the days of Confucius, of Muhammad or of Stalin, people should have listened to their feelings rather than to the teachings of Confucianism, Islam or communism.

Yet in the twenty-first century, feelings are no longer the best algorithms in the world. We are developing superior algorithms that utilise unprecedented computing power and giant databases. The Google and Facebook algorithms not only know exactly how you feel, they also know myriad other things about you that you hardly suspect. Consequently you should stop listening to your feelings and start listening to these external algorithms instead. What's the point of having democratic elections when the algorithms know not only how each person is going to vote, but also the underlying neurological reasons why one person votes Democrat while another votes Republican? Whereas humanism commanded: 'Listen to your feelings!' Dataism now commands: 'Listen to the algorithms! They know how you feel.'

When you contemplate whom to marry, which career to pursue and whether to start a war, Dataism tells you that it would be a complete waste of time to climb a high mountain and watch the sun setting into the waves. It would be equally futile to visit a museum, write a private diary or have a heart-to-heart talk with a friend. Yes, in order to make the right decisions you must get to know yourself better. But if you want to know yourself in the twenty-first century, there are much better methods than climbing mountains, going to museums or writing diaries. Here are some practical Dataist guidelines for you:

'You want to know who you really are?' asks Dataism. 'Then forget about mountains and museums. Have you had your DNA sequenced? No?! What are you waiting for? Go and do it today. And convince your grandparents, parents and siblings to have their DNA sequenced too – their data is very valuable for you. And have you heard about these wearable biometric devices that measure your blood pressure and heart rate twenty-four hours a day? Good – so buy one of those, put it on and connect it to your smartphone. And while you are shopping, buy a mobile camera and microphone, record everything you do, and put in online. And allow Google and Facebook to read all your emails, monitor all your chats and messages, and keep a record of all your Likes and clicks. If you do all that, then the great algorithms of the Internet-of-All-Things will tell you whom to marry, which career to pursue and whether to start a war.'

But where do these great algorithms come from? This is the mystery of Dataism. Just as according to Christianity we humans cannot understand God and His plan, so Dataism declares that the human brain cannot fathom the new master algorithms. At present, of course, the algorithms are mostly written by human hackers. Yet the really important algorithms – such as the Google search algorithm – are developed by huge teams. Each member understands just one part of the puzzle, and nobody really understands the algorithm as a whole. Moreover, with the rise of machine learning and artificial neural networks, more and more algorithms evolve independently, improving themselves and learning from their own mistakes. They analyse astronomical amounts of data that no human can possibly encompass, and learn to recognise patterns and adopt strategies that escape the human mind. The seed algorithm may initially be developed by humans, but as it grows it follows its own path, going where no human has gone before – and where no human can follow.

A Ripple in the Data Flow

Dataism naturally has its critics and heretics. As we saw in Chapter 3, it's doubtful whether life can really be reduced to data flows. In particular, at present we have no idea how or why data flows could produce consciousness and subjective experiences. Maybe we'll have a good explanation in twenty years. But maybe we'll discover that organisms aren't algorithms after all.

It is equally doubtful whether life boils down to mere decision-making. Under Dataist influence both the life sciences and the social sciences have become obsessed with decision-making processes, as if that's all there is to life. But is it so? Sensations, emotions and thoughts certainly play an important part in making decisions, but is that their sole meaning? Dataism is gaining a better and better understanding of decision-making processes, but it might be adopting an increasingly skewed view of life.

A critical examination of Dataist dogma is likely to be not only the greatest scientific challenge of the twenty-first century, but also the most urgent political and economic project. Scholars in the life sciences and social sciences should ask themselves whether we miss anything when we understand life as data processing and decision-making. Is there perhaps something in the universe that cannot be reduced to data? Suppose non-conscious algorithms could eventually outperform conscious intelligence in all known data-processing tasks – what, if anything, would be lost by replacing conscious intelligence with superior non-conscious algorithms?

Of course, even if Dataism is wrong and organisms aren't just algorithms, it won't necessarily prevent Dataism from taking over the world. Many previous religions gained enormous popularity and power despite their factual inaccuracies. If Christianity and communism could do it, why not Dataism? Dataism has especially good prospects, because it is currently spreading across all scientific disciplines. A unified scientific paradigm may easily become an unassailable dogma. It is very difficult to contest a scientific paradigm, but up till now, no single paradigm has been adopted by the entire scientific establishment. Hence scholars in one field could always import heretical views from outside. But if everyone from musicologists to biologists uses the same Dataist paradigm, interdisciplinary excursions will serve only to strengthen the paradigm further. Consequently even if the paradigm is flawed, it would be extremely difficult to resist.

If Dataism succeeds in conquering the world, what will happen to us humans? Initially, Dataism will probably accelerate the humanist pursuit of health, happiness and power. Dataism spreads itself by promising to fulfil these humanist aspirations. In order to achieve immortality, bliss and divine powers of creation, we need to process immense amounts of data, far beyond the capacity of the human brain. So the algorithms will do it for us. Yet once authority shifts from humans to algorithms, the humanist projects may become irrelevant. Once we abandon the homo-centric world view in favour of a data-centric world view, human health and happiness may seem far less important. Why bother so much about obsolete data-processing machines when far superior models are already in existence? We are striving to engineer the Internet-of-All-Things in the hope that it will make us healthy, happy and powerful. Yet once the Internet-of-All-Things is up and running, humans might be reduced from engineers to chips, then to data, and eventually we might dissolve within the torrent of data like a clump of earth within a gushing river.

Dataism thereby threatens to do to *Homo sapiens* what *Homo sapiens* has done to all other animals. Over the course of history humans created a global network and evaluated everything according to its function within that network. For thousands of years this inflated human pride and prejudices. Since humans fulfilled the most important functions in the network, it was easy for us to take credit for the network's achievements, and to see ourselves as the apex of creation. The lives and experiences of all other animals were undervalued because they fulfilled far less important functions, and whenever an animal ceased to fulfil any function at all, it went extinct. However, once we humans lose our functional importance to the network, we will discover that we are not the apex of creation after all. The yardsticks that we ourselves have enshrined will condemn us to join the mammoths and Chinese river dolphins in oblivion. Looking back, humanity will turn out to have been just a ripple within the cosmic data flow.

We cannot really predict the future, because technology is not deterministic. The same technology could create very different kinds of societies. For example, the technology of the Industrial Revolution – trains, electricity, radio, telephone – could be used to establish communist dictatorships, fascist regimes or liberal democracies. Consider South Korea and North Korea: They have had access to exactly the same technology, but they have chosen to employ it in very different ways.

The rise of AI and biotechnology will certainly transform the world, but it does not mandate a single deterministic outcome. All the scenarios outlined in this book should be understood as possibilities rather than prophecies. If you don't like some of these possibilities you are welcome to think and behave in new ways that will prevent these particular possibilities from materializing.

However, it is not easy to think and behave in new ways, because our thoughts and actions are usually constrained by present-day ideologies and social systems. This book traces the origins of our present-day conditioning in order to loosen its grip and enable us to think in far more imaginative ways about our future. Instead of narrowing our horizons by forecasting a single definitive scenario, the book aims to broaden our horizons and make us aware of a much wider spectrum of options. As I have repeatedly emphasised, nobody really knows what the job market, the family or the ecology will look like in 2050, or which religions, economic systems and political structures will dominate the world.

Yet broadening our horizons can backfire by making us more confused and inactive than before. With so many scenarios and possibilities, what should we pay attention to? The world is changing faster than ever before, and we are flooded by impossible amounts of data, of ideas, of promises and of threats. Humans are relinquishing authority to the free market, to crowd wisdom and to external algorithms partly because we cannot deal with the deluge of data. In the past, censorship worked by blocking the flow of information. In the twenty-first century censorship works by flooding people with irrelevant information. We just don't know what to pay attention to, and often spend our time investigating and debating side issues. In ancient times having power meant having access to data. Today having power means knowing what to ignore. So considering everything that is happening in our chaotic world, what should we focus on?

If we think in term of months, we had probably focus on immediate problems such as the turmoil in the Middle East, the refugee crisis in Europe and the slowing of the Chinese economy. If we think in terms of decades, then global warming, growing inequality and the disruption of the job market loom large. Yet if we take the really grand view of life, all other problems and developments are overshadowed by three interlinked processes:

1. Science is converging on an all-encompassing dogma, which says that organisms are algorithms and life is data processing.
2. Intelligence is decoupling from consciousness.
3. Non-conscious but highly intelligent algorithms may soon know us better than we know ourselves.

These three processes raise three key questions, which I hope will stick in your mind long after you have finished this book:

1. Are organisms really just algorithms, and is life really just data processing?
2. What's more valuable – intelligence or consciousness?
3. What will happen to society, politics and daily life when non-conscious but highly intelligent algorithms know us better than we know ourselves?

Notes

1 The New Human Agenda

1. Tim Blanning, *The Pursuit of Glory* (New York: Penguin Books, 2008), 52.
2. Ibid., 53. See also: J. Neumann and S. Lindgrén, 'Great Historical Events That Were Significantly Affected by the Weather: 4, The Great Famines in Finland and Estonia, 1695–97', *Bulletin of the American Meteorological Society* 60 (1979), 775–87; Andrew B. Appleby, 'Epidemics and Famine in the Little Ice Age', *Journal of Interdisciplinary History* 10:4 (1980), 643–63; Cormac Ó Gráda and Jean-Michel Chevet, 'Famine and Market in *Ancien Régime* France', *Journal of Economic History* 62:3 (2002), 706–73.
3. Nicole Darmon et al., 'L'insécurité alimentaire pour raisons financières en France', *Observatoire National de la Pauvreté et de l'Exclusion Sociale*, https://www.onpes.gouv.fr/IMG/pdf/Darmon.pdf, accessed 3 March 2015; Rapport Annuel 2013, *Banques Alimetaires*, http://en.calameo.com/read/001358178ec47d2018425, accessed 4 March 2015.
4. Richard Dobbs et al., 'How the World Could Better Fight Obesity', McKinseys & Company, November 2014, accessed 11 December 2014, http://www.mckinsey.com/insights/economic_studies/how_the_world_could_better_fight_obesity.
5. 'Global Burden of Disease, Injuries and Risk Factors Study 2013', *Lancet*, 18 December 2014, accessed 18 December 2014, http://www.thelancet.com/themed/global-burden-of-disease; Stephen Adams, 'Obesity Killing Three Times As Many As Malnutrition', *Telegraph*, 13 December 2012, accessed 18 December 2014, http://www.telegraph.co.uk/health/healthnews/9742960/Obesity-killing-three-times-as-many-as-malnutrition.html.
6. Robert S. Lopez, *The Birth of Europe* [in Hebrew] (Tel Aviv: Dvir, 1990), 427.
7. Alfred W. Crosby, *The Columbian Exchange: Biological and Cultural Consequences of 1492* (Westport: Greenwood Press, 1972); William H. McNeill, *Plagues and Peoples* (Oxford: Basil Blackwell, 1977).
8. Hugh Thomas, *Conquest: Cortes, Montezuma and the Fall of Old Mexico* (New York: Simon & Schuster, 1993), 443–6; Rodolfo Acuna-Soto et al., 'Megadrought and Megadeath in 16th Century Mexico', *Historical Review*

8:4 (2002), 360–2; Sherburne F. Cook and Lesley Byrd Simpson, *The Population of Central Mexico in the Sixteenth Century* (Berkeley: University of California Press, 1948).

9. Jared Diamond, *Guns, Germs and Steel: The Fates of Human Societies* [in Hebrew] (Tel Aviv: Am Oved, 2002), 167.

10. Jeffery K. Taubenberger and David M. Morens, '1918 Influenza: The Mother of All Pandemics', *Emerging Infectious Diseases* 12:1 (2006), 15–22; Niall P. A. S. Johnson and Juergen Mueller, 'Updating the Accounts: Global Mortality of the 1918–1920 "Spanish" Influenza Pandemic', *Bulletin of the History of Medicine* 76:1 (2002), 105–15; Stacey L. Knobler, Alison Mack, Adel Mahmoud et al., (eds), *The Threat of Pandemic Influenza: Are We Ready? Workshop Summary* (Washington DC: National Academies Press, 2005), 57–110; David van Reybrouck, *Congo: The Epic History of a People* (New York: HarperCollins, 2014), 164; Siddharth Chandra, Goran Kuljanin and Jennifer Wray, 'Mortality from the Influenza Pandemic of 1918–1919: The Case of India', *Demography* 49:3 (2012), 857–65; George C. Kohn, *Encyclopedia of Plague and Pestilence: From Ancient Times to the Present*, 3rd edn (New York: Facts on File, 2008), 363.

11. The averages between 2005 and 2010 were 4.6 per cent globally, 7.9 per cent in Africa and 0.7 per cent in Europe and North America. See: 'Infant Mortality Rate (Both Sexes Combined) by Major Area, Region and Country, 1950–2010 (Infant Deaths for 1000 Live Births), Estimates', *World Population Prospects: the 2010 Revision*, UN Department of Economic and Social Affairs, April 2011, accessed 26 May 2012, http://esa.un.org/unpd/wpp/Excel-Data/mortality.htm. See also Alain Bideau, Bertrand Desjardins and Hector Perez-Brignoli (eds), *Infant and Child Mortality in the Past* (Oxford: Clarendon Press, 1997); Edward Anthony Wrigley et al., *English Population History from Family Reconstitution, 1580–1837* (Cambridge: Cambridge University Press, 1997), 295–6, 303.

12. David A. Koplow, *Smallpox: The Fight to Eradicate a Global Scourge* (Berkeley: University of California Press, 2004); Abdel R. Omran, 'The Epidemiological Transition: A Theory of Population Change', *Milbank Memorial Fund Quarterly* 83:4 (2005), 731–57; Thomas McKeown, *The Modern Rise of Populations* (New York: Academic Press, 1976); Simon Szreter, *Health and Wealth: Studies in History and Policy* (Rochester: University of Rochester Press, 2005); Roderick Floud, Robert W. Fogel, Bernard Harris and Sok Chul Hong, *The Changing Body: Health, Nutrition and Human Development in the Western World Since 1700* (New York: Cambridge University Press, 2011); James C. Riley, *Rising Life Expectancy: A Global History* (New York: Cambridge University Press, 2001).

13. 'Cholera', World Health Organization, February 2014, accessed 18 December 2014, http://www.who.int/mediacentre/factsheets/fs107/en/index.html.

14. 'Experimental Therapies: Growing Interest in the Use of Whole Blood or Plasma from Recovered Ebola Patients', World Health Organization, 26 September 2014, accessed 23 April 2015, http://www.who.int/mediacentre/news/ebola/26-september-2014/en/.

15. Hung Y. Fan, Ross F. Conner and Luis P. Villarreal, *AIDS: Science and Society*, 6th edn (Sudbury: Jones and Bartlett Publishers, 2011).

16. Peter Piot and Thomas C. Quinn, 'Response to the AIDS Pandemic – A Global Health Model', *New England Journal of Medicine* 368:23 (2013), 2210–18.

17. 'Old age' is never listed as a cause of death in official statistics. Instead, when a frail old woman eventually succumbs to this or that infection, the particular infection will be listed as the cause of death. Hence, officially, infectious diseases still account for more than 20 per cent of deaths. But this is a fundamentally different situation than in past centuries, when large numbers of children and fit adults died from infectious diseases.

18. David M. Livermore, 'Bacterial Resistance: Origins, Epidemiology, and Impact', *Clinical Infectious Diseases* 36:s1 (2005), s11–23; Richards G. Wax et al. (eds), *Bacterial Resistance to Antimicrobials*, 2nd edn (Boca Raton: CRC Press, 2008); Maja Babic and Robert A. Bonomo, 'Mutations as a Basis of Antimicrobial Resistance', in *Antimicrobial Drug Resistance: Mechanisms of Drug Resistance*, ed. Douglas Mayers, vol. 1 (New York: Humana Press, 2009), 65–74; Julian Davies and Dorothy Davies, 'Origins and Evolution of Antibiotic Resistance', *Microbiology and Molecular Biology Reviews* 74:3 (2010), 417–33; Richard J. Fair and Yitzhak Tor, 'Antibiotics and Bacterial Resistance in the 21st Century', *Perspectives in Medicinal Chemistry* 6 (2014), 25–64.

19. Alfonso J. Alanis, 'Resistance to Antibiotics: Are We in the Post-Antibiotic Era?', *Archives of Medical Research* 36:6 (2005), 697–705; Stephan Harbarth and Matthew H. Samore, 'Antimicrobial Resistance Determinants and Future Control', *Emerging Infectious Diseases* 11:6 (2005), 794–801; Hiroshi Yoneyama and Ryoichi Katsumata, 'Antibiotic Resistance in Bacteria and Its Future for Novel Antibiotic Development', *Bioscience, Biotechnology and Biochemistry* 70:5 (2006), 1060–75; Cesar A. Arias and Barbara E. Murray, 'Antibiotic-Resistant Bugs in the 21st Century – A Clinical Super-Challenge', *New England Journal of Medicine* 360 (2009), 439–43; Brad Spellberg, John G. Bartlett and David N. Gilbert, 'The Future of Antibiotics and Resistance', *New England Journal of Medicine* 368 (2013), 299–302.

20. Losee L. Ling et al., 'A New Antibiotic Kills Pathogens without Detectable Resistance', *Nature* 517 (2015), 455–9; Gerard Wright, 'Antibiotics: An Irresistible Newcomer', *Nature* 517 (2015), 442–4.

21. Roey Tzezana, *The Guide to the Future* [in Hebrew] (Haifa: Roey Tzezana, 2013), 209–33.

22. Azar Gat, *War in Human Civilization* (Oxford: Oxford University Press, 2006), 130–1; Steven Pinker, *The Better Angels of Our Nature: Why Violence*

Has Declined (New York: Viking, 2011); Joshua S. Goldstein, *Winning the War on War: The Decline of Armed Conflict Worldwide* (New York: Dutton, 2011); Robert S. Walker and Drew H. Bailey, 'Body Counts in Lowland South American Violence', *Evolution and Human Behavior* 34:1 (2013), 29–34; I. J. N. Thorpe, 'Anthropology, Archaeology, and the Origin of Warfare', *World Archaeology* 35:1 (2003), 145–65; Raymond C. Kelly, *Warless Societies and the Origin of War* (Ann Arbor: University of Michigan Press, 2000); Lawrence H. Keeley, *War Before Civilization: The Myth of the Peaceful Savage* (Oxford: Oxford University Press, 1996); Slavomil Vencl, 'Stone Age Warfare', in *Ancient Warfare: Archaeological Perspectives*, ed. John Carman and Anthony Harding (Stroud: Sutton Publishing, 1999), 57–73.

23. 'Global Health Observatory Data Repository, 2012', World Health Organization, accessed 16 August 2015, http://apps.who.int/gho/data/node.main.RCODWORLD?lang=en; 'Global Study on Homicide, 2013', UNDOC, accessed 16 August 2015, http://www.unodc.org/documents/gsh/pdfs/2014_GLOBAL_HOMICIDE_BOOK_web.pdf; http://www.who.int/healthinfo/global_burden_disease/estimates/en/index1.html.

24. Van Reybrouck, *Congo*, 456–7.

25. Deaths from obesity: 'Global Burden of Disease, Injuries and Risk Factors Study 2013', *Lancet*, 18 December 2014, accessed 18 December 2014, http://www.thelancet.com/themed/global-burden-of-disease; Stephen Adams, 'Obesity Killing Three Times as Many as Malnutrition', *Telegraph*, 13 December 2012, accessed 18 December 2014, http://www.telegraph.co.uk/health/healthnews/9742960/Obesity-killing-three-times-as-many-as-malnutrition.html. Deaths from terrorism: Global Terrorism Database, http://www.start.umd.edu/gtd/, accessed 16 January 2016.

26. Arion McNicoll, 'How Google's Calico Aims to Fight Aging and "Solve Death"', CNN, 3 October 2013, accessed 19 December 2014, http://edition.cnn.com/2013/10/03/tech/innovation/google-calico-aging-death/.

27. Katrina Brooker, 'Google Ventures and the Search for Immortality', *Bloomberg*, 9 March 2015, accessed 15 April 2015, http://www.bloomberg.com/news/articles/2015-03-09/google-ventures-bill-maris-investing-in-idea-of-living-to-500.

28. Mick Brown, 'Peter Thiel: The Billionaire Tech Entrepreneur on a Mission to Cheat Death', *Telegraph*, 19 September 2014, accessed 19 December 2014, http://www.telegraph.co.uk/technology/11098971/Peter-Thiel-the-billionaire-tech-entrepreneur-on-a-mission-to-cheat-death.html.

29. Kim Hill et al., 'Mortality Rates Among Wild Chimpanzees', *Journal of Human Evolution* 40:5 (2001), 437–50; James G. Herndon, 'Brain Weight Throughout the Life Span of the Chimpanzee', *Journal of Comparative Neurology* 409 (1999), 567–72.

30. Beatrice Scheubel, *Bismarck's Institutions: A Historical Perspective on the Social Security Hypothesis* (Tubingen: Mohr Siebeck, 2013); E. P. Hannock, *The Origin of the Welfare State in England and Germany, 1850–1914* (Cambridge: Cambridge University Press, 2007).

31. 'Mental Health: Age Standardized Suicide Rates (per 100,000 population), 2012', World Health Organization, accessed 28 December 2014, http://gamapserver.who.int/gho/interactive_charts/mental_health/suicide_rates/atlas.html.

32. Ian Morris, *Why the West Rules – For Now* (Toronto: McClelland & Stewart, 2010), 626–9.

33. David G. Myers, 'The Funds, Friends, and Faith of Happy People', *American Psychologist* 55:1 (2000), 61; Ronald Inglehart et al., 'Development, Freedom, and Rising Happiness: A Global Perspective (1981–2007)', *Perspectives on Psychological Science* 3:4 (2008), 264–85. See also Mihaly Csikszentmihalyi, 'If We Are So Rich, Why Aren't We Happy?', *American Psychologist* 54:10 (1999), 821–7; Gregg Easterbrook, *The Progress Paradox: How Life Gets Better While People Feel Worse* (New York: Random House, 2003).

34. Kenji Suzuki, 'Are They Frigid to the Economic Development? Reconsideration of the Economic Effect on Subjective Well-being in Japan', *Social Indicators Research* 92:1 (2009), 81–9; Richard A. Easterlin, 'Will Raising the Incomes of all Increase the Happiness of All?', *Journal of Economic Behavior and Organization* 27:1 (1995), 35–47; Richard A. Easterlin, 'Diminishing Marginal Utility of Income? Caveat Emptor', *Social Indicators Research* 70:3 (2005), 243–55.

35. Linda C. Raeder, *John Stuart Mill and the Religion of Humanity* (Columbia: University of Missouri Press, 2002).

36. Oliver Turnbull and Mark Solms, *The Brain and the Inner World* [in Hebrew] (Tel Aviv: Hakibbutz Hameuchad, 2005), 92–6; Kent C. Berridge and Morten L. Kringelbach, 'Affective Neuroscience of Pleasure: Reward in Humans and Animals', *Psychopharmacology* 199 (2008), 457–80; Morten L. Kringelbach, *The Pleasure Center: Trust Your Animal Instincts* (Oxford: Oxford University Press, 2009).

37. M. Csikszentmihalyi, *Finding Flow: The Psychology of Engagement with Everyday Life* (New York: Basic Books, 1997).

38. Centers for Disease Control and Prevention, Attention-Deficit/Hyperactivity Disorder (ADHD), http://www.cdc.gov/ncbddd/adhd/data.html, accessed 4 January 2016; Sarah Harris, 'Number of Children Given Drugs for ADHD Up Ninefold with Patients As Young As Three Being Prescribed Ritalin', *Daily Mail*, 28 June 2013, http://www.dailymail.co.uk/health/article-2351427/Number-children-given-drugs-ADHD-ninefold-patients-young-THREE-prescribed-Ritalin.html, accessed 4 January 2016; International Narcotics Control Board (UN), *Psychotropic Substances, Statistics for 2013, Assessments of Annual Medical and Scientific Requirements 2014*, 39–40.

39. There is insufficient evidence regarding the abuse of such stimulants by schoolchildren, but a 2013 study has found that between 5 and 15 per cent of US college students illegally used some kind of stimulant at least once: C. Ian Ragan, Imre Bard and Ilina Singh, 'What Should We Do about Student Use of Cognitive Enhancers? An Analysis of Current Evidence', *Neuropharmacology* 64 (2013), 589.

40. Bradley J. Partridge, 'Smart Drugs "As Common as Coffee": Media Hype about Neuroenhancement', *PLoS One* 6:11 (2011), e28416.

41. Office of the Chief of Public Affairs Press Release, 'Army, Health Promotion Risk Reduction Suicide Prevention Report, 2010', accessed 23 December 2014, http://csf2.army.mil/downloads/HP-RR-SPReport2010.pdf; Mark Thompson, 'America's Medicated Army', *Time*, 5 June 2008, accessed 19 December 2014, http://content.time.com/time/magazine/article/0,9171,1812055,00.html; Office of the Surgeon Multi-National Force–Iraq and Office of the Command Surgeon, 'Mental Health Advisory Team (MHAT) V Operation Iraqi Freedom 06–08: Iraq Operation Enduring Freedom 8: Afghanistan', 14 February 2008, accessed 23 December 2014, http://www.careforthetroops.org/reports/Report-MHATV-4-FEB-2008-Overview.pdf.

42. Tina L. Dorsey, 'Drugs and Crime Facts', US Department of Justice, accessed 20 February 2015, http://www.bjs.gov/content/pub/pdf/dcf.pdf; H. C. West, W. J. Sabol and S. J. Greenman, 'Prisoners in 2009', US Department of Justice, Bureau of Justice Statistics Bulletin (December 2010), 1–38; 'Drugs and Crime Facts: Drug Use and Crime', US Department of Justice, accessed 19 December 2014, http://www.bjs.gov/content/dcf/duc.cfm; 'Offender Management Statistics Bulletin, July to September 2014', UK Ministry of Justice, 29 January 2015, accessed 20 February 2015, https://www.gov.uk/government/statistics/offender-management-statistics-quarterly-july-to-september-2014.; Mirian Lights et al., 'Gender Differences in Substance Misuse and Mental Health amongst Prisoners', UK Ministry of Justice, 2013, accessed 20 February 2015, https://www.gov.uk/government/uploads/system/uploads/attachment_data/file/220060/gender-substance-misuse-mental-health-prisoners.pdf; Jason Payne and Antonette Gaffney, 'How Much Crime is Drug or Alcohol Related? Self-Reported Attributions of Police Detainees', *Trends and Issues in Crime and Criminal Justice* 439 (2012), http://www.aic.gov.au/media_library/publications/tandi_pdf/tandi439.pdf, accessed 11 March 2015; Philippe Robert, 'The French Criminal Justice System', in *Punishment in Europe: A Critical Anatomy of Penal Systems*, ed. Vincenzo Ruggiero and Mick Ryan (Houndmills: Palgrave Macmillan, 2013), 116.

43. Betsy Isaacson, 'Mind Control: How EEG Devices Will Read Your Brain Waves and Change Your World', *Huffington Post*, 20 November 2014, accessed 20 December 2014, http://www.huffingtonpost.com/2012/11/20/mind-control-how-eeg-devices-read-brainwaves_n_2001431.html; 'EPOC

Headset', *Emotiv*, http://emotiv.com/store/epoc-detail/; 'Biosensor Innovation to Power Breakthrough Wearable Technologies Today and Tomorrow', *NeuroSky*, http://neurosky.com/.

44. Samantha Payne, 'Stockholm: Members of Epicenter Workspace Are Using Microchip Implants to Open Doors', *International Business Times*, 31 January 2015, accessed 9 August 2015, http://www.ibtimes.co.uk/stockholm-office-workers-epicenter-implanted-microchips-pay-their-lunch-1486045.

45. Meika Loe, *The Rise of Viagra: How the Little Blue Pill Changed Sex in America* (New York: New York University Press, 2004).

46. Brian Morgan, 'Saints and Sinners: Sir Harold Gillies', *Bulletin of the Royal College of Surgeons of England* 95:6 (2013), 204–5; Donald W. Buck II, 'A Link to Gillies: One Surgeon's Quest to Uncover His Surgical Roots', *Annals of Plastic Surgery* 68:1 (2012), 1–4.

47. Paolo Santoni-Rugio, *A History of Plastic Surgery* (Berlin, Heidelberg: Springer, 2007); P. Niclas Broer, Steven M. Levine and Sabrina Juran, 'Plastic Surgery: Quo Vadis? Current Trends and Future Projections of Aesthetic Plastic Surgical Procedures in the United States', *Plastic and Reconstructive Surgery* 133:3 (2014), 293e–302e.

48. Holly Firfer, 'How Far Will Couples Go to Conceive?', CNN, 17 June 2004, accessed 3 May 2015, http://edition.cnn.com/2004/HEALTH/03/12/infertility.treatment/index.html?iref=allsearch.

49. Rowena Mason and Hannah Devlin, 'MPs Vote in Favour of "Three-Person Embryo" Law', *Guardian*, 3 February 2015, accessed 3 May 2015, http://www.theguardian.com/science/2015/feb/03/mps-vote-favour-three-person-embryo-law.

50. Lionel S. Smith and Mark D. E. Fellowes, 'Towards a Lawn without Grass: The Journey of the Imperfect Lawn and Its Analogues', *Studies in the History of Gardens & Designed Landscape* 33:3 (2013), 158–9; John Dixon Hunt and Peter Willis (eds), *The Genius of the Place: The English Landscape Garden 1620–1820*, 5th edn (Cambridge, MA: MIT Press, 2000), 1–45; Anne Helmrieh, *The English Garden and National Identity: The Competing Styles of Garden Design 1870–1914* (Cambridge: Cambridge University Press, 2002), 1–6.

51. Robert J. Lake, 'Social Class, Etiquette and Behavioral Restraint in British Lawn Tennis', *International Journal of the History of Sport* 28:6 (2011), 876–94; Beatriz Colomina, 'The Lawn at War: 1941–1961', in *The American Lawn*, ed. Georges Teyssot (New York: Princeton Architectural Press, 1999), 135–53; Virginia Scott Jenkins, *The Lawn: History of an American Obsession* (Washington: Smithsonian Institution, 1994).

2 The Anthropocene

1. 'Canis lupus', IUCN Red List of Threatened Species, accessed 20 December 2014, http://www.iucnredlist.org/details/3746/1; 'Fact Sheet: Gray Wolf', Defenders of Wildlife, accessed 20 December 2014, http://www.

defenders.org/gray-wolf/basic-facts; 'Companion Animals', IFAH, accessed 20 December 2014, http://www.ifaheurope.org/companion-animals/about-pets.html; 'Global Review 2013', World Animal Protection, accessed 20 December 2014, https://www.worldanimalprotection.us.org/sites/default/files/us_files/global_review_2013_0.pdf.

2. Anthony D. Barnosky, 'Megafauna Biomass Tradeoff as a Driver of Quaternary and Future Extinctions', *PNAS* 105:1 (2008), 11543–8; for wolves and lions: William J. Ripple et al., 'Status and Ecological Effects of the World's Largest Carnivores', *Science* 343:6167 (2014), 151; according to Dr Stanley Coren there are about 500 million dogs in the world: Stanley Coren, 'How Many Dogs Are There in the World?', *Psychology Today*, 19 September 2012, accessed 20 December 2014, http://www.psychologytoday.com/blog/canine-corner/201209/how-many-dogs-are-there-in-the-world; for the number of cats, see: Nicholas Wade, 'DNA Traces 5 Matriarchs of 600 Million Domestic Cats', *New York Times*, 29 June 2007, accessed 20 December 2014, http://www.nytimes.com/2007/06/29/health/29iht-cats.1.6406020.html; for the African buffalo, see: '*Syncerus caffer*', IUCN Red List of Threatened Species, accessed 20 December 2014, http://www.iucnredlist.org/details/21251/0; for cattle population, see: David Cottle and Lewis Kahn (eds), *Beef Cattle Production and Trade* (Collingwood: Csiro, 2014), 66; for the number of chickens, see: 'Live Animals', Food and Agriculture Organization of the United Nations: Statistical Division, accessed 20 December 2014, http://faostat3.fao.org/browse/Q/QA/E; for the number of chimpanzees, see: '*Pan troglodytes*', IUCN Red List of Threatened Species, accessed 20 December 2014, http://www.iucnredlist.org/details/15933/0.

3. 'Living Planet Report 2014', WWF Global, accessed 20 December 2014, http://wwf.panda.org/about_our_earth/all_publications/living_planet_report/.

4. Richard Inger et al., 'Common European Birds Are Declining Rapidly While Less Abundant Species' Numbers Are Rising', *Ecology Letters* 18:1 (2014), 28–36; 'Live Animals', Food and Agriculture Organization of the United Nations, accessed 20 December 2014, http://faostat.fao.org/site/573/default.aspx#ancor.

5. Simon L. Lewis and Mark A. Maslin, 'Defining the Anthropocene', *Nature* 519 (2015), 171–80.

6. Timothy F. Flannery, *The Future Eaters: An Ecological History of the Australasian Lands and Peoples* (Port Melbourne: Reed Books Australia, 1994); Anthony D. Barnosky et al., 'Assessing the Causes of Late Pleistocene Extinctions on the Continents', *Science* 306:5693 (2004), 70–5; Barry W. Brook and David M. J. S. Bowman, 'The Uncertain Blitzkrieg of Pleistocene Megafauna', *Journal of Biogeography* 31:4 (2004), 517–23; Gifford H. Miller et al., 'Ecosystem Collapse in Pleistocene Australia

and a Human Role in Megafaunal Extinction', *Science* 309:5732 (2005), 287–90; Richard G. Roberts et al., 'New Ages for the Last Australian Megafauna: Continent Wide Extinction about 46,000 Years Ago', *Science* 292:5523 (2001), 1888–92; Stephen Wroe and Judith Field, 'A Review of the Evidence for a Human Role in the Extinction of Australian Megafauna and an Alternative Explanation', *Quaternary Science Reviews* 25:21–2 (2006), 2692–703; Barry W. Brooks et al., 'Would the Australian Megafauna Have Become Extinct if Humans Had Never Colonised the Continent? Comments on "A Review of the Evidence for a Human Role in the Extinction of Australian Megafauna and an Alternative Explanation" by S. Wroe and J. Field', *Quaternary Science Reviews* 26:3–4 (2007), 560–4; Chris S. M. Turney et al., 'Late-Surviving Megafauna in Tasmania, Australia, Implicate Human Involvement in their Extinction', *PNAS* 105:34 (2008), 12150–3; John Alroy, 'A Multispecies Overkill Simulation of the End-Pleistocene Megafaunal Mass Extinction', *Science* 292:5523 (2001), 1893–6; J. F. O'Connell and J. Allen, 'Pre-LGM Sahul (Australia–New Guinea) and the Archaeology of Early Modern Humans', in *Rethinking the Human Evolution: New Behavioral and Biological Perspectives on the Origin and Dispersal of Modern Humans*, ed. Paul Mellars (Cambridge: McDonald Institute for Archaeological Research, 2007), 400–1.

7. Graham Harvey, *Animism: Respecting the Living World* (Kent Town: Wakefield Press, 2005); Rane Willerslev, *Soul Hunters: Hunting, Animism and Personhood Among the Siberian Yukaghirs* (Berkeley: University of California Press, 2007); Elina Helander-Renvall, 'Animism, Personhood and the Nature of Reality: Sami Perspectives', *Polar Record* 46:1 (2010), 44–56; Istvan Praet, 'Animal Conceptions in Animism and Conservation', in *Routledge Handbook of Human–Animal Studies*, ed. Susan McHaugh and Garry Marvin (New York: Routledge, 2014), 154–67; Nurit Bird-David, 'Animism Revisited: Personhood, Environment, and Relational Epistemology', *Current Anthropology* 40 (1999), s67–91; N. Bird-David, 'Animistic Epistemology: Why Some Hunter-Gatherers Do Not Depict Animals', *Ethnos* 71:1 (2006), 33–50.

8. Danny Naveh, 'Changes in the Perception of Animals and Plants with the Shift to Agricultural Life: What Can Be Learnt from the Nayaka Case, a Hunter-Gatherer Society from the Rain Forests of Southern India?' [in Hebrew], *Animals and Society*, 52 (2015), 7–8.

9. Howard N. Wallace, 'The Eden Narrative', *Harvard Semitic Monographs* 32 (1985), 147–81.

10. David Adams Leeming and Margaret Adams Leeming, *Encyclopedia of Creation Myths* (Santa Barbara: ABC-CLIO, 1994), 18; Sam D. Gill, *Storytracking: Texts, Stories, and Histories in Central Australia* (Oxford: Oxford University Press, 1998); Emily Miller Bonney, 'Disarming the Snake Goddess: A Reconsideration of the Faience Figures from the

Temple Repositories at Knossos', *Journal of Mediterranean Archaeology* 24:2 (2011), 171–90; David Leeming, *The Oxford Companion to World Mythology* (Oxford and New York: Oxford University Press, 2005), 350.

11. Jerome H. Barkow, Leda Cosmides and John Tooby (eds), *The Adapted Mind: Evolutionary Psychology and the Generation of Culture* (Oxford: Oxford University Press, 1992); Richard W. Bloom and Nancy Dess (eds), *Evolutionary Psychology and Violence: A Primer for Policymakers and Public Policy Advocates* (Westport: Praeger, 2003); Charles Crawford and Catherine Salmon (eds), *Evolutionary Psychology, Public Policy and Personal Decisions* (New Jersey: Lawrence Erlbaum Associates, 2008); Patrick McNamara and David Trumbull, *An Evolutionary Psychology of Leader–Follower Relations* (New York: Nova Science, 2007); Joseph P. Forgas, Martie G. Haselton and William von Hippel (eds), *Evolution and the Social Mind: Evolutionary Psychology and Social Cognition* (New York: Psychology Press, 2011).

12. S. Held, M. Mendl, C. Devereux and R. W. Byrne, 'Social Tactics of Pigs in a Competitive Foraging Task: the "Informed Forager" Paradigm', *Animal Behaviour* 59:3 (2000), 569–76; S. Held, M. Mendl, C. Devereux and R. W. Byrne, 'Studies in Social Cognition: from Primates to Pigs', *Animal Welfare* 10 (2001), s209–17; H. B. Graves, 'Behavior and Ecology of Wild and Feral Swine (*Sus scrofa*)', *Journal of Animal Science* 58:2 (1984), 482–92; A. Stolba and D. G. M. Wood-Gush, 'The Behaviour of Pigs in a Semi-Natural Environment', *Animal Production* 48:2 (1989), 419–25; M. Spinka, 'Behaviour in Pigs', in *The Ethology of Domestic Animals*, 2nd edn, ed. P. Jensen, (Wallingford, UK: CAB International, 2009), 177–91; P. Jensen and D. G. M. Wood-Gush, 'Social Interactions in a Group of Free-Ranging Sows', *Applied Animal Behaviour Science* 12 (1984), 327–37; E. T. Gieling, R. E. Nordquist and F. J. van der Staay, 'Assessing Learning and Memory in Pigs', *Animal Cognition* 14 (2011), 151–73.

13. I. Horrell and J. Hodgson, 'The Bases of Sow–Piglet Identification. 2. Cues Used by Piglets to Identify their Dam and Home Pen', *Applied Animal Behavior Science*, 33 (1992), 329–43; D. M. Weary and D. Fraser, 'Calling by Domestic Piglets: Reliable Signals of Need?', *Animal Behaviour* 50:4 (1995), 1047–55; H. H. Kristensen et al., 'The Use of Olfactory and Other Cues for Social Recognition by Juvenile Pigs', *Applied Animal Behaviour Science* 72 (2001), 321–33.

14. M. Helft, 'Pig Video Arcades Critique Life in the Pen', *Wired*, 6 June 1997, http://archive.wired.com/science/discoveries/news/1997/06/4302, retrieved 27 January 2016.

15. Humane Society of the United States, 'An HSUS Report: Welfare Issues with Gestation Crates for Pregnant Sows', February 2013, http://www.humanesociety.org/assets/pdfs/farm/HSUS-Report-on-Gestation-Crates-for-Pregnant-Sows.pdf, retrieved 27 January 2016.

16. Turnbull and Solms, *Brain and the Inner World*, 90–2.

17. David Harel, *Algorithmics: The Spirit of Computers*, 3rd edn [in Hebrew] (Tel Aviv: Open University of Israel, 2001), 4–6; David Berlinski, *The Advent of the Algorithm: The 300-Year Journey from an Idea to the Computer* (San Diego: Harcourt, 2000); Hartley Rogers Jr, *Theory of Recursive Functions and Effective Computability*, 3rd edn (Cambridge, MA, and London: MIT Press, 1992), 1–5; Andreas Blass and Yuri Gurevich, 'Algorithms: A Quest for Absolute Definitions', *Bulletin of European Association for Theoretical Computer Science* 81 (2003), 195–225.

18. Daniel Kahneman, *Thinking, Fast and Slow* (New York: Farrar, Straus & Giroux, 2011); Dan Ariely, *Predictably Irrational* (New York: Harper, 2009).

19. Justin Gregg, *Are Dolphins Really Smart? The Mammal Behind the Myth* (Oxford: Oxford University Press, 2013), 81–7; Jaak Panksepp, 'Affective Consciousness: Core Emotional Feelings in Animals and Humans', *Consciousness and Cognition* 14:1 (2005), 30–80.

20. A. S. Fleming, D. H. O'Day and G. W. Kraemer, 'Neurobiology of Mother–Infant Interactions: Experience and Central Nervous System Plasticity Across Development and Generations', *Neuroscience and Biobehavioral Reviews* 23:5 (1999), 673–85; K. D. Broad, J. P. Curley and E. B. Keverne, 'Mother–Infant Bonding and the Evolution of Mammalian Relationship', *Philosophical Transactions of the Royal Society B* 361:1476 (2006), 2199–214; Kazutaka Mogi, Miho Nagasawa and Takefumi Kikusui, 'Developmental Consequences and Biological Significance of Mother–Infant Bonding', *Progress in Neuro-Psychopharmacology and Biological Psychiatry* 35:5 (2011), 1232–41; Shota Okabe et al., 'The Importance of Mother–Infant Communication for Social Bond Formation in Mammals', *Animal Science Journal* 83:6 (2012), 446–52.

21. Jean O'Malley Halley, *Boundaries of Touch: Parenting and Adult–Child Intimacy* (Urbana: University of Illinois Press, 2007), 50–1; Ann Taylor Allen, *Feminism and Motherhood in Western Europe, 1890–1970: The Maternal Dilemma* (New York: Palgrave Macmillan, 2005), 190.

22. Lucille C. Birnbaum, 'Behaviorism in the 1920s', *American Quarterly* 7:1 (1955), 18.

23. US Department of Labor (1929), 'Infant Care', Washington: United States Government Printing Office, http://www.mchlibrary.info/history/chbu/3121-1929.pdf.

24. Harry Harlow and Robert Zimmermann, 'Affectional Responses in the Infant Monkey', *Science* 130:3373 (1959), 421–32; Harry Harlow, 'The Nature of Love', *American Psychologist* 13 (1958), 673–85; Laurens D. Young et al., 'Early Stress and Later Response to Separation in Rhesus Monkeys', *American Journal of Psychiatry* 130:4 (1973), 400–5; K. D. Broad, J. P. Curley and E. B. Keverne, 'Mother–Infant Bonding and the Evolution of Mammalian Social Relationships', *Philosophical Transactions of the Royal Society B* 361:1476 (2006), 2199–214; Florent Pittet et al.,

'Effects of Maternal Experience on Fearfulness and Maternal Behavior in a Precocial Bird', *Animal Behavior* 85:4 (2013), 797–805.

25. Jacques Cauvin, *The Birth of the Gods and the Origins of Agriculture* (Cambridge: Cambridge University Press, 2000); Tim Ingord, 'From Trust to Domination: An Alternative History of Human–Animal Relations', in *Animals and Human Society: Changing Perspectives*, ed. Aubrey Manning and James Serpell (New York: Routledge, 2002), 1–22; Roberta Kalechofsky, 'Hierarchy, Kinship and Responsibility', in *A Communion of Subjects: Animals in Religion, Science and Ethics*, ed. Kimberley Patton and Paul Waldau (New York: Columbia University Press, 2006), 91–102; Nerissa Russell, *Social Zooarchaeology: Humans and Animals in Prehistory* (Cambridge: Cambridge University Press, 2012), 207–58; Margo DeMello, *Animals and Society: An Introduction to Human–Animal Studies* (New York: University of Columbia Press, 2012).

26. Olivia Lang, 'Hindu Sacrifice of 250,000 Animals Begins', *Guardian*, 24 November 2009, accessed 21 December 2014, http://www.theguardian.com/world/2009/nov/24/hindu-sacrifice-gadhimai-festival-nepal.

27. Benjamin R. Foster (ed.), *The Epic of Gilgamesh* (New York, London: W. W. Norton, 2001), 90.

28. Noah J. Cohen, *Tsa'ar Ba'ale Hayim: Prevention of Cruelty to Animals: Its Bases, Development and Legislation in Hebrew Literature* (Jerusalem and New York: Feldheim Publishers, 1976); Roberta Kalechofsky, *Judaism and Animal Rights: Classical and Contemporary Responses* (Marblehead: Micah Publications, 1992); Dan Cohen-Sherbok, 'Hope for the Animal Kingdom: A Jewish Vision', in *A Communion of Subjects: Animals in Religion, Science and Ethics*, ed. Kimberley Patton and Paul Waldau (New York: Columbia University Press, 2006), 81–90; Ze'ev Levi, 'Ethical Issues of Animal Welfare in Jewish Thought', in *Judaism and Environmental Ethics: A Reader*, ed. Martin D. Yaffe (Plymouth: Lexington, 2001), 321–32; Norm Phelps, *The Dominion of Love: Animal Rights According to the Bible* (New York: Lantern Books, 2002); David Sears, *The Vision of Eden: Animal Welfare and Vegetarianism in Jewish Law Mysticism* (Spring Valley: Orot, 2003); Nosson Slifkin, *Man and Beast: Our Relationships with Animals in Jewish Law and Thought* (New York: Lambda, 2006).

29. Talmud Bavli, Bava Metzia, 85:71.

30. Christopher Chapple, *Nonviolence to Animals, Earth and Self in Asian Traditions* (New York: State University of New York Press, 1993); Panchor Prime, *Hinduism and Ecology: Seeds of Truth* (London: Cassell, 1992); Christopher Key Chapple, 'The Living Cosmos of Jainism: A Traditional Science Grounded in Environmental Ethics', *Daedalus* 130:4 (2001), 207–24; Norm Phelps, *The Great Compassion: Buddhism and Animal Rights* (New York: Lantern Books, 2004); Damien Keown, *Buddhist Ethics: A Very Short Introduction* (Oxford: Oxford University Press, 2005), ch. 3; Kimberley Patton and Paul Waldau (eds),

A Communion of Subjects: Animals in Religion, Science and Ethics (New York: Columbia University Press, 2006), esp. 179–250; Pragati Sahni, *Environmental Ethics in Buddhism: A Virtues Approach* (New York: Routledge, 2008); Lisa Kemmerer and Anthony J. Nocella II (eds), *Call to Compassion: Reflections on Animal Advocacy from the World's Religions* (New York: Lantern, 2011), esp. 15–103; Lisa Kemmerer, *Animals and World Religions* (Oxford: Oxford University Press, 2012), esp. 56–126; Irina Aristarkhova, 'Thou Shall Not Harm All Living Beings: Feminism, Jainism and Animals', *Hypatia* 27:3 (2012), 636–50; Eva de Clercq, 'Karman and Compassion: Animals in the Jain Universal History', *Religions of South Asia* 7 (2013), 141–57.

31. Naveh, 'Changes in the Perception of Animals and Plants', 11.

3 The Human Spark

1. 'Evolution, Creationism, Intelligent Design', Gallup, accessed 20 December 2014, http://www.gallup.com/poll/21814/evolution-creationism-intelligent-design.aspx; Frank Newport, 'In US, 46 per cent Hold Creationist View of Human Origins', Gallup, 1 June 2012, accessed 21 December 2014, http://www.gallup.com/poll/155003/hold-creationist-view-human-origins.aspx.

2. Gregg, *Are Dolphins Really Smart?*, 82–3.

3. Stanislas Dehaene, *Consciousness and the Brain: Deciphering How the Brain Codes Our Thoughts* (New York: Viking, 2014); Steven Pinker, *How the Mind Works* (New York: W. W. Norton, 1997).

4. Dehaene, *Consciousness and the Brain*.

5. Pundits may point to Gödel's incompleteness theorem, according to which no system of mathematical axioms can prove all arithmetic truths. There will always be some true statements that cannot be proven within the system. In popular literature this theorem is sometimes hijacked to account for the existence of mind. Allegedly, minds are needed to deal with such unprovable truths. However, it is far from obvious why living beings need to engage with such arcane mathematical truths in order to survive and reproduce. In fact, the vast majority of our conscious decisions do not involve such issues at all.

6. Christopher Steiner, *Automate This: How Algorithms Came to Rule Our World* (New York: Penguin, 2012), 215; Tom Vanderbilt, 'Let the Robot Drive: The Autonomous Car of the Future Is Here', *Wired*, 20 January 2012, accessed 21 December 2014, http://www.wired.com/2012/01/ff_autonomouscars/all/; Chris Urmson, 'The Self-Driving Car Logs More Miles on New Wheels', Google Official Blog, 7 August 2012, accessed 23 December 2014, http://googleblog.blogspot.hu/2012/08/the-self-driving-car-logs-more-miles-on.html; Matt Richtel and Conor Dougherty, 'Google's Driverless Cars Run into Problem: Cars with Drivers', *New York Times*, 1 September 2015, accessed 2 September 2015,

http://www.nytimes.com/2015/09/02/technology/personaltech/
google-says-its-not-the-driverless-cars-fault-its-other-drivers.html?_r=1.

7. Dehaene, *Consciousness and the Brain*.

8. Ibid., ch. 7.

9. 'The Cambridge Declaration on Consciousness', 7 July 2012, accessed 21 December 2014, https://web.archive.org/web/20131109230457/http://fcmconference.org/img/CambridgeDeclarationOnConsciousness.pdf.

10. John F. Cyran, Rita J. Valentino and Irwin Lucki, 'Assessing Substrates Underlying the Behavioral Effects of Antidepressants Using the Modified Rat Forced Swimming Test', *Neuroscience and Behavioral Reviews* 29:4–5 (2005), 569–74; Benoit Petit-Demoulière, Frank Chenu and Michel Bourin, 'Forced Swimming Test in Mice: A Review of Antidepressant Activity', *Psychopharmacology* 177:3 (2005), 245–55; Leda S. B. Garcia et al., 'Acute Administration of Ketamine Induces Antidepressant-like Effects in the Forced Swimming Test and Increases BDNF Levels in the Rat Hippocampus', *Progress in Neuro-Psychopharmacology and Biological Psychiatry* 32:1 (2008), 140–4; John F. Cryan, Cedric Mombereau and Annick Vassout, 'The Tail Suspension Test as a Model for Assessing Antidepressant Activity: Review of Pharmacological and Genetic Studies in Mice', *Neuroscience and Behavioral Reviews* 29:4–5 (2005), 571–625; James J. Crowley, Julie A. Blendy and Irwin Lucki, 'Strain-dependent Antidepressant-like Effects of Citalopram in the Mouse Tail Suspension Test', *Psychopharmacology* 183:2 (2005), 257–64; Juan C. Brenes, Michael Padilla and Jaime Fornaguera, 'A Detailed Analysis of Open-Field Habituation and Behavioral and Neurochemical Antidepressant-like Effects in Postweaning Enriched Rats', *Behavioral Brain Research* 197:1 (2009), 125–37; Juan Carlos Brenes Sáenz, Odir Rodríguez Villagra and Jaime Fornaguera Trías, 'Factor Analysis of Forced Swimming Test, Sucrose Preference Test and Open Field Test on Enriched, Social and Isolated Reared Rats', *Behavioral Brain Research* 169:1 (2006), 57–65.

11. Marc Bekoff, 'Observations of Scent-Marking and Discriminating Self from Others by a Domestic Dog (*Canis familiaris*): Tales of Displaced Yellow Snow', *Behavioral Processes* 55:2 (2011), 75–9.

12. For different levels of self-consciousness, see: Gregg, *Are Dolphins Really Smart?*, 59–66.

13. Carolyn R. Raby et al., 'Planning for the Future by Western Scrub Jays', *Nature* 445:7130 (2007), 919–21.

14. Michael Balter, 'Stone-Throwing Chimp is Back – and This Time It's Personal', *Science*, 9 May 2012, accessed 21 December 2014, http://news.sciencemag.org/2012/05/stone-throwing-chimp-back-and-time-its-personal; Sara J. Shettleworth, 'Clever Animals and Killjoy Explanations in Comparative Psychology', *Trends in Cognitive Sciences* 14:11 (2010), 477–81.

15. Gregg, *Are Dolphins Really Smart?*; Nicola S. Clayton, Timothy J. Bussey and Anthony Dickinson, 'Can Animals Recall the Past and Plan for the Future?', *Nature Reviews Neuroscience* 4:8 (2003), 685–91; William A. Roberts, 'Are Animals Stuck in Time?', *Psychological Bulletin* 128:3 (2002), 473–89; Endel Tulving, 'Episodic Memory and Autonoesis: Uniquely Human?', in *The Missing Link in Cognition: Evolution of Self-Knowing Consciousness*, ed. Herbert S. Terrace and Janet Metcalfe (Oxford: Oxford University Press), 3–56; Mariam Naqshbandi and William A. Roberts, 'Anticipation of Future Events in Squirrel Monkeys (*Saimiri sciureus*) and Rats (*Rattus norvegicus*): Tests of the Bischof–Kohler Hypothesis', *Journal of Comparative Psychology* 120:4 (2006), 345–57.

16. I. B. A. Bartal, J. Decety and P. Mason, 'Empathy and Pro-Social Behavior in Rats', *Science* 334:6061 (2011), 1427–30; Gregg, *Are Dolphins Really Smart?*, 89.

17. Christopher B. Ruff, Erik Trinkaus and Trenton W. Holliday, 'Body Mass and Encephalization in Pleistocene *Homo*', *Nature* 387:6629 (1997), 173–6; Maciej Henneberg and Maryna Steyn, 'Trends in Cranial Capacity and Cranial Index in Subsaharan Africa During the Holocene', *American Journal of Human Biology* 5:4 (1993), 473–9; Drew H. Bailey and David C. Geary, 'Hominid Brain Evolution: Testing Climatic, Ecological, and Social Competition Models', *Human Nature* 20:1 (2009), 67–79; Daniel J. Wescott and Richard L. Jantz, 'Assessing Craniofacial Secular Change in American Blacks and Whites Using Geometric Morphometry', in *Modern Morphometrics in Physical Anthropology: Developments in Primatology: Progress and Prospects*, ed. Dennis E. Slice (New York: Plenum Publishers, 2005), 231–45.

18. See also Edward O. Wilson, *The Social Conquest of the Earth* (New York: Liveright, 2012).

19. Cyril Edwin Black (ed.), *The Transformation of Russian Society: Aspects of Social Change since 1861* (Cambridge, MA: Harvard University Press, 1970), 279.

20. NAEMI09, 'Nicolae Ceauşescu LAST SPEECH (english subtitles) part 1 of 2', 22 April 2010, accessed 21 December 2014, http://www.youtube.com/watch?v=wWIbCtz_Xwk.

21. Tom Gallagher, *Theft of a Nation: Romania since Communism* (London: Hurst, 2005).

22. Robin Dunbar, *Grooming, Gossip, and the Evolution of Language* (Cambridge, MA: Harvard University Press, 1998).

23. TVP University, 'Capuchin Monkeys Reject Unequal Pay', 15 December 2012, accessed 21 December 2014, http://www.youtube.com/watch?v=lKhAdoTynyo.

24. Quoted in Christopher Duffy, *Military Experience in the Age of Reason* (London: Routledge, 2005), 98–9.

25. Serhii Ploghy, *The Last Empire: The Final Days of the Soviet Union* (London: Oneworld, 2014), 309.

4 The Storytellers

1. Fekri A. Hassan, 'Holocene Lakes and Prehistoric Settlements of the Western Fayum, Egypt', *Journal of Archaeological Science* 13:5 (1986), 393–504; Gunther Garbrecht, 'Water Storage (Lake Moeris) in the Fayum Depression, Legend or Reality?', *Irrigation and Drainage Systems* 1:3 (1987), 143–57; Gunther Garbrecht, 'Historical Water Storage for Irrigation in the Fayum Depression (Egypt)', *Irrigation and Drainage Systems* 10:1 (1996), 47–76.
2. Yehuda Bauer, *A History of the Holocaust* (Danbur: Franklin Watts, 2001), 249.
3. Jean C. Oi, *State and Peasant in Contemporary China: The Political Economy of Village Government* (Berkeley: University of California Press, 1989), 91; Jasper Becker, *Hungry Ghosts: China's Secret Famine* (London: John Murray, 1996); Frank Dikkoter, *Mao's Great Famine: The History of China's Most Devastating Catastrophe, 1958–62* (London: Bloomsbury, 2010).
4. Martin Meredith, *The Fate of Africa: From the Hopes of Freedom to the Heart of Despair: A History of Fifty Years of Independence* (New York: Public Affairs, 2006); Sven Rydenfelt, 'Lessons from Socialist Tanzania', *The Freeman* 36:9 (1986); David Blair, 'Africa in a Nutshell', *Telegraph*, 10 May 2006, accessed 22 December 2014, http://blogs.telegraph.co.uk/news/davidblair/3631941/Africa_in_a_nutshell/.
5. Roland Anthony Oliver, *Africa since 1800*, 5th edn (Cambridge: Cambridge University Press, 2005), 100–23; David van Reybrouck, *Congo: The Epic History of a People* (New York: HarperCollins, 2014), 58–9.
6. Ben Wilbrink, 'Assessment in Historical Perspective', *Studies in Educational Evaluation* 23:1 (1997), 31–48.
7. M. C. Lemon, *Philosophy of History* (London and New York: Routledge, 2003), 28–44; Siep Stuurman, 'Herodotus and Sima Qian: History and the Anthropological Turn in Ancient Greece and Han China', *Journal of World History* 19:1 (2008), 1–40.
8. William Kelly Simpson, *The Literature of Ancient Egypt* (Yale: Yale University Press, 1973), 332–3.

5 The Odd Couple

1. C. Scott Dixon, *Protestants: A History from Wittenberg to Pennsylvania, 1517–1740* (Chichester, UK: Wiley-Blackwell, 2010), 15; Peter W. Williams, *America's Religions: From Their Origins to the Twenty-First Century* (Urbana: University of Illinois Press, 2008), 82.
2. Glenn Hausfater and Sarah Blaffer (eds), *Infanticide: Comparative and Evolutionary Perspectives* (New York: Aldine, 1984), 449; Valeria Alia,

Names and Nunavut: Culture and Identity in the Inuit Homeland (New York: Berghahn Books, 2007), 23; Lewis Petrinovich, *Human Evolution, Reproduction and Morality* (Cambridge, MA: MIT Press, 1998), 256; Richard A. Posner, *Sex and Reason* (Cambridge, MA: Harvard University Press, 1992), 289.

3. Ronald K. Delph, 'Valla Grammaticus, Agostino Steuco, and the Donation of Constantine', *Journal of the History of Ideas* 57:1 (1996), 55–77; Joseph M. Levine, 'Reginald Pecock and Lorenzo Valla on the Donation of Constantine', *Studies in the Renaissance* 20 (1973), 118–43.

4. Gabriele Boccaccini, *Roots of Rabbinic Judaism* (Cambridge: Eerdmans, 2002); Shaye J. D. Cohen, *From the Maccabees to the Mishnah*, 2nd edn (Louisville: Westminster John Knox Press, 2006), 153–7; Lee M. McDonald and James A. Sanders (eds), *The Canon Debate* (Peabody: Hendrickson, 2002), 4.

5. Sam Harris, *The Moral Landscape: How Science Can Determine Human Values* (New York: Free Press, 2010).

6 The Modern Covenant

1. Gerald S. Wilkinson, 'The Social Organization of the Common Vampire Bat II', *Behavioral Ecology and Sociobiology* 17:2 (1985), 123–34; Gerald S. Wilkinson, 'Reciprocal Food Sharing in the Vampire Bat', *Nature* 308:5955 (1984), 181–4; Raul Flores Crespo et al., 'Foraging Behavior of the Common Vampire Bat Related to Moonlight', *Journal of Mammalogy* 53:2 (1972), 366–8.

2. Goh Chin Lian, 'Admin Service Pay: Pensions Removed, National Bonus to Replace GDP Bonus', *Straits Times*, 8 April 2013, retrieved 9 February 2016, http://www.straitstimes.com/singapore/admin-service-pay-pensions-removed-national-bonus-to-replace-gdp-bonus.

3. Edward Wong, 'In China, Breathing Becomes a Childhood Risk', *New York Times*, 22 April 2013, accessed 22 December 2014, http://www.nytimes.com/2013/04/23/world/asia/pollution-is-radically-changing-childhood-in-chinas-cities.html?pagewanted=all&_r=0; Barbara Demick, 'China Entrepreneurs Cash in on Air Pollution', *Los Angeles Times*, 2 February 2013, accessed 22 December 2014, http://articles.latimes.com/2013/feb/02/world/la-fg-china-pollution-20130203.

4. IPCC, *Climate Change 2014: Mitigation of Climate Change – Summary for Policymakers*, ed. Ottmar Edenhofer et al. (Cambridge and New York: Cambridge University Press, 2014), 6.

5. UNEP, *The Emissions Gap Report 2012* (Nairobi: UNEP, 2012); IEA, *Energy Policies of IEA Countries: The United States* (Paris: IEA, 2008).

6. For a detailed discussion see Ha-Joon Chang, *23 Things They Don't Tell You About Capitalism* (New York: Bloomsbury Press, 2010).

7 The Humanist Revolution

1. Jean-Jacques Rousseau, *Émile, ou de l'éducation* (Paris, 1967), 348.

2. 'Journalists Syndicate Says Charlie Hebdo Cartoons "Hurt Feelings"', *Washington Okays*', *Egypt Independent*, 14 January 2015, accessed 12 August 2015, http://www.egyptindependent.com/news/journalists-syndicate-says-charlie-hebdo-cartoons-percentE2percent80percent98hurt-feelings-washington-okays.

3. Naomi Darom, 'Evolution on Steroids', *Haaretz*, 13 June 2014.

4. Walter Horace Bruford, *The German Tradition of Self-Cultivation: 'Bildung' from Humboldt to Thomas Mann* (London and New York: Cambridge University Press, 1975), 24, 25.

5. 'All-Time 100 TV Shows: *Survivor*', *Time*, 6 September 2007, retrieved 12 August 2015, http://time.com/3103831/survivor/.

6. Phil Klay, *Redeployment* (London: Canongate, 2015), 170.

7. Yuval Noah Harari, *The Ultimate Experience: Battlefield Revelations and the Making of Modern War Culture, 1450–2000* (Houndmills: Palgrave Macmillan, 2008); Yuval Noah Harari, 'Armchairs, Coffee and Authority: Eye-witnesses and Flesh-witnesses Speak about War, 1100–2000', *Journal of Military History* 74:1 (January 2010), 53–78.

8. 'Angela Merkel Attacked over Crying Refugee Girl', BBC, 17 July 2015, accessed 12 August 2015, http://www.bbc.com/news/world-europe-33555619.

9. Laurence Housman, *War Letters of Fallen Englishmen* (Philadelphia: University of Pennsylvania State, 2002), 159.

10. Mark Bowden, *Black Hawk Down: The Story of Modern Warfare* (New York: New American Library, 2001), 301–2.

11. Adolf Hitler, *Mein Kampf*, trans. Ralph Manheim (Boston: Houghton Mifflin, 1943), 165.

12. Evan Osnos, *Age of Ambition: Chasing Fortune, Truth and Faith in the New China* (London: Vintage, 2014), 95.

13. Mark Harrison (ed), *The Economics of World War II: Six Great Powers in International Comparison* (Cambridge: Cambridge University Press, 1998), 3–10; John Ellis, *World War II: A Statistical Survey* (New York: Facts on File, 1993); I. C. B. Dear (ed.) *The Oxford Companion to the Second World War* (Oxford: Oxford University Press, 1995).

14. Donna Haraway, 'A Cyborg Manifesto: Science, Technology, and Socialist-Feminism in the Late Twentieth Century', in *Simians, Cyborgs and Women: The Reinvention of Nature*, ed. Donna Haraway (New York: Routledge, 1991), 149–81.

8 The Time Bomb in the Laboratory

1. For a detailed discussion see Michael S. Gazzaniga, *Who's in Charge?: Free Will and the Science of the Brain* (New York: Ecco, 2011).

2. Chun Siong Soon et al., 'Unconscious Determinants of Free Decisions in the Human Brain', *Nature Neuroscience* 11:5 (2008), 543–5. See also Daniel Wegner, *The Illusion of Conscious Will* (Cambridge, MA: MIT Press, 2002); Benjamin Libet, 'Unconscious Cerebral Initiative and the Role of Conscious Will in Voluntary Action', *Behavioral and Brain Sciences* 8 (1985), 529–66.

3. Sanjiv K. Talwar et al., 'Rat Navigation Guided by Remote Control', *Nature* 417:6884 (2002), 37–8; Ben Harder, 'Scientists "Drive" Rats by Remote Control', *National Geographic*, 1 May 2012, accessed 22 December 2014, http://news.nationalgeographic.com/news/2002/05/0501_020501_roborats.html; Tom Clarke, 'Here Come the Ratbots: Desire Drives Remote-Controlled Rodents', *Nature*, 2 May 2002, accessed 22 December 2014, http://www.nature.com/news/1998/020429/full/news020429-9.html; Duncan Graham-Rowe, '"Robo-rat" Controlled by Brain Electrodes', *New Scientist*, 1 May 2002, accessed 22 December 2014, http://www.newscientist.com/article/dn2237-roborat-controlled-by-brain-electrodes.html#.UwOPiNrNtkQ.

4. http://fusion.net/story/204316/darpa-is-implanting-chips-in-soldiers-brains/; http://www.theverge.com/2014/5/28/5758018/darpa-teams-begin-work-on-tiny-brain-implant-to-treat-ptsd.

5. Smadar Reisfeld, 'Outside of the Cuckoo's Nest', *Haaretz*, 6 March 2015.

6. Dan Hurley, 'US Military Leads Quest for Futuristic Ways to Boost IQ', *Newsweek*, 5 March 2014, http://www.newsweek.com/2014/03/14/us-military-leads-quest-futuristic-ways-boost-iq-247945.html, accessed 9 January 2015; Human Effectiveness Directorate, http://www.wpafb.af.mil/afrl/rh/index.asp; R. Andy McKinley et al., 'Acceleration of Image Analyst Training with Transcranial Direct Current Stimulation', *Behavioral Neuroscience* 127:6 (2013), 936–46; Jeremy T. Nelson et al., 'Enhancing Vigilance in Operators with Prefrontal Cortex Transcranial Direct Current Stimulation (TDCS)', *NeuroImage* 85 (2014), 909–17; Melissa Scheldrup et al., 'Transcranial Direct Current Stimulation Facilitates Cognitive Multi-Task Performance Differentially Depending on Anode Location and Subtask', *Frontiers in Human Neuroscience* 8 (2014); Oliver Burkeman, 'Can I Increase my Brain Power?', *Guardian*, 4 January 2014, http://www.theguardian.com/science/2014/jan/04/can-i-increase-my-brain-power, accessed 9 January 2016; Heather Kelly, 'Wearable Tech to Hack Your Brain', CNN, 23 October 2014, http://www.cnn.com/2014/10/22/tech/innovation/brain-stimulation-tech/, accessed 9 January 2016.

7. Sally Adee, 'Zap Your Brain into the Zone: Fast Track to Pure Focus', *New Scientist*, 6 February 2012, accessed 22 December 2014, http://www.newscientist.com/article/mg21328501.600-zap-your-brain-into-the-zone-fast-track-to-pure-focus.html. See also: R. Douglas Fields, 'Amping Up Brain Function: Transcranial Stimulation Shows Promise in Speeding Up Learning', *Scientific American*, 25 November 2011, accessed 22 December 2014, http://www.scientificamerican.com/article/amping-up-brain-function.

8. Sally Adee, 'How Electrical Brain Stimulation Can Change the Way We Think', *The Week*, 30 March 2012, accessed 22 December 2014, http://theweek.com/article/index/226196/how-electrical-brain-stimulation-can-change-the-way-we-think/2.

9. E. Bianconi et al., 'An Estimation of the Number of Cells in the Human Body', *Annals of Human Biology* 40:6 (2013), 463–71.

10. Oliver Sacks, *The Man Who Mistook His Wife for a Hat* (London: Picador, 1985), 73–5.

11. Joseph E. LeDoux, Donald H. Wilson and Michael S. Gazzaniga, 'A Divided Mind: Observations on the Conscious Properties of the Separated Hemispheres', *Annals of Neurology* 2:5 (1977), 417–21. See also: D. Galin, 'Implications for Psychiatry of Left and Right Cerebral Specialization: A Neurophysiological Context for Unconscious Processes', *Archives of General Psychiatry* 31:4 (1974), 572–83; R. W. Sperry, M. S. Gazzaniga and J. E. Bogen, 'Interhemispheric Relationships: The Neocortical Commisures: Syndromes of Hemisphere Disconnection', in *Handbook of Clinical Neurology*, ed. P. J. Vinken and G. W. Bruyn (Amsterdam: North Holland Publishing Co., 1969), vol. 4.

12. Michael S. Gazzaniga, *The Bisected Brain* (New York: Appleton-Century-Crofts, 1970); Gazzaniga, *Who's in Charge?*; Carl Senior, Tamara Russell and Michael S. Gazzaniga, *Methods in Mind* (Cambridge, MA: MIT Press, 2006); David Wolman, 'The Split Brain: A Tale of Two Halves', *Nature* 483 (14 March 2012), 260–3.

13. Galin, 'Implications for Psychiatry of Left and Right Cerebral Specialization', 573–4.

14. Sally P. Springer and Georg Deutsch, *Left Brain, Right Brain,* 3rd edn (New York: W. H. Freeman, 1989), 32–6.

15. Kahneman, *Thinking, Fast and Slow*, 377–410. See also Gazzaniga, *Who's in Charge?*, ch. 3.

16. Eran Chajut et al., 'In Pain Thou Shalt Bring Forth Children: The Peak-and-End Rule in Recall of Labor Pain', *Psychological Science* 25:12 (2014), 2266–71.

17. Ulla Waldenström, 'Women's Memory of Childbirth at Two Months and One Year after the Birth', *Birth* 30:4 (2003), 248–54; Ulla Waldenström, 'Why Do Some Women Change Their Opinion about Childbirth over Time?', *Birth* 31:2 (2004), 102–7.

18. Gazzaniga, *Who's in Charge?*, ch. 3.

19. Jorge Luis Borges, *Collected Fictions*, trans. Andrew Hurley (New York: Penguin Books, 1999), 308–9. For a Spanish version see: Jorge Luis Borges, 'Un problema', in *Obras completas*, vol. 3 (Buenos Aires: Emece Editores, 1968–9), 29–30.

20. Mark Thompson, *The White War: Life and Death on the Italian Front, 1915–1919* (New York: Basic Books, 2009).

9 The Great Decoupling

1. F. M. Anderson (ed.), *The Constitutions and Other Select Documents Illustrative of the History of France: 1789–1907*, 2nd edn (Minneapolis: H. W. Wilson, 1908), 184–5; Alan Forrest, 'L'armée de l'an II: la levée en masse et la création d'un mythe républicain', *Annales historiques de la Révolution française* 335 (2004), 111–30.

2. Morris Edmund Spears (ed.), *World War Issues and Ideals: Readings in Contemporary History and Literature* (Boston and New York: Ginn and Company, 1918), 242. The most significant recent study, widely quoted by both proponents and opponents, attempts to prove that soldiers of democracy fight better: Dan Reiter and Allan C. Stam, *Democracies at War* (Princeton: Princeton University Press, 2002).

3. Doris Stevens, *Jailed for Freedom* (New York: Boni and Liveright, 1920), 290. See also Susan R. Grayzel, *Women and the First World War* (Harlow: Longman, 2002), 101–6; Christine Bolt, *The Women's Movements in the United States and Britain from the 1790s to the 1920s* (Amherst: University of Massachusetts Press, 1993), 236–76; Birgitta Bader-Zaar, 'Women's Suffrage and War: World War I and Political Reform in a Comparative Perspective', in *Suffrage, Gender and Citizenship: International Perspectives on Parliamentary Reforms*, ed. Irma Sulkunen, Seija-Leena Nevala-Nurmi and Pirjo Markkola (Newcastle upon Tyne: Cambridge Scholars Publishing, 2009), 193–218.

4. Matt Richtel and Conor Dougherty, 'Google's Driverless Cars Run into Problem: Cars with Drivers', *New York Times*, 1 September 2015, accessed 2 September 2015, http://www.nytimes.com/2015/09/02/technology/personaltech/google-says-its-not-the-driverless-cars-fault-its-other-drivers.html?_r=1; Shawn DuBravac, *Digital Destiny: How the New Age of Data Will Transform the Way We Work, Live and Communicate* (Washington DC: Regnery Publishing, 2015), 127–56.

5. Bradley Hope, 'Lawsuit Against Exchanges Over "Unfair Advantage" for High-Frequency Traders Dismissed', *Wall Street Journal*, 29 April 2015, accessed 6 October 2015, http://www.wsj.com/articles/lawsuit-against-exchanges-over-unfair-advantage-for-high-frequency-traders-dismissed-1430326045; David Levine, 'High-Frequency Trading Machines Favored Over Humans by CME Group, Lawsuit Claims', *Huffington Post*, 26 June 2012, accessed 6 October 2015, http://www.huffingtonpost.

com/2012/06/26/high-frequency-trading-lawsuit_n_1625648.html; Lu Wang, Whitney Kisling and Eric Lam, 'Fake Post Erasing $136 Billion Shows Markets Need Humans', Bloomberg, 23 April 2013, accessed 22 December 2014, http://www.bloomberg.com/news/2013-04-23/ fake-report-erasing-136-billion-shows-market-s-fragility.html; Matthew Philips, 'How the Robots Lost: High-Frequency Trading's Rise and Fall', *Bloomberg Businessweek*, 6 June 2013, accessed 22 December 2014, http:// www.businessweek.com/printer/articles/123468-how-the-robots-lost-high-frequency-tradings-rise-and-fall; Steiner, *Automate This*, 2–5, 11–52; Luke Dormehl, *The Formula: How Algorithms Solve All Our Problems – And Create More* (London: Penguin, 2014), 223.

6.   Jordan Weissmann, 'iLawyer: What Happens when Computers Replace Attorneys?', *Atlantic*, 19 June 2012, accessed 22 December 2014, http:// www.theatlantic.com/business/archive/2012/06/ilawyer-what-happens-when-computers-replace-attorneys/258688; John Markoff, 'Armies of Expensive Lawyers, Replaced by Cheaper Software', *New York Times*, 4 March 2011, accessed 22 December 2014, http://www. nytimes.com/2011/03/05/science/05legal.html?pagewanted=all&_r=0; Adi Narayan, 'The fMRI Brain Scan: A Better Lie Detector?', *Time*, 20 July 2009, accessed 22 December 2014, http://content.time.com/time/ health/article/0,8599,1911546-2,00.html; Elena Rusconi and Timothy Mitchener-Nissen, 'Prospects of Functional Magnetic Resonance Imaging as Lie Detector', *Frontiers in Human Neuroscience* 7:54 (2013); Steiner, *Automate This*, 217; Dormehl, *The Formula*, 229.

7.   B. P. Woolf, *Building Intelligent Interactive Tutors: Student-centered Strategies for Revolutionizing E-learning* (Burlington: Morgan Kaufmann, 2010); Annie Murphy Paul, 'The Machines Are Taking Over', *New York Times*, 14 September 2012, accessed 22 December 2014, http://www. nytimes.com/2012/09/16/magazine/how-computerized-tutors-are-learning-to-teach-humans.html?_r=0; P. J. Munoz-Merino, C. D. Kloos and M. Munoz-Organero, 'Enhancement of Student Learning Through the Use of a Hinting Computer e-Learning System and Comparison With Human Teachers', *IEEE Transactions on Education* 54:1 (2011), 164–7; Mindojo, accessed 14 July 2015, http://mindojo.com/.

8.   Steiner, *Automate This*, 146–62; Ian Steadman, 'IBM's Watson Is Better at Diagnosing Cancer than Human Doctors', *Wired*, 11 February 2013, accessed 22 December 2014, http://www.wired.co.uk/news/archive/2013-02/11/ ibm-watson-medical-doctor; 'Watson Is Helping Doctors Fight Cancer', IBM, accessed 22 December 2014, http://www-03.ibm.com/innovation/ us/watson/watson_in_healthcare.shtml; Vinod Khosla, 'Technology Will Replace 80 per cent of What Doctors Do', *Fortune*, 4 December 2012, accessed 22 December 2014, http://tech.fortune.cnn.com/2012/12/04/ technology-doctors-khosla; Ezra Klein, 'How Robots Will Replace Doctors', *Washington Post*, 10 January 2011, accessed 22 December 2014,

http://www.washingtonpost.com/blogs/wonkblog/post/how-robots-will-replace-doctors/2011/08/25/gIQASA17AL_blog.html.

9. Tzezana, *The Guide to the Future*, 62–4.

10. Steiner, *Automate This*, 155.

11. http://www.mattersight.com.

12. Steiner, *Automate This*, 178–82; Dormehl, *The Formula*, 21–4; Shana Lebowitz, 'Every Time You Dial into These Call Centers, Your Personality Is Being Silently Assessed', *Business Insider*, 3 September 2015, retrieved 31 January 2016, http://www.businessinsider.com/how-mattersight-uses-personality-science-2015-9.

13. Rebecca Morelle, 'Google Machine Learns to Master Video Games', BBC, 25 February 2015, accessed 12 August 2015, http://www.bbc.com/news/science-environment-31623427; Elizabeth Lopatto, 'Google's AI Can Learn to Play Video Games', *The Verge*, 25 February 2015, accessed 12 August 2015, http://www.theverge.com/2015/2/25/8108399/google-ai-deepmind-video-games; Volodymyr Mnih et al., 'Human-Level Control through Deep Reinforcement Learning', *Nature*, 26 February 2015, accessed 12 August 2015, http://www.nature.com/nature/journal/v518/n7540/full/nature14236.html.

14. Michael Lewis, *Moneyball: The Art of Winning an Unfair Game* (New York: W. W. Norton, 2003). Also see the 2011 film *Moneyball*, directed by Bennett Miller and starring Brad Pitt as Billy Beane.

15. Frank Levy and Richard Murnane, *The New Division of Labor: How Computers Are Creating the Next Job Market* (Princeton: Princeton University Press, 2004); Dormehl, *The Formula*, 225–6.

16. Tom Simonite, 'When Your Boss Is an Uber Algorithm', *MIT Technology Review*, 1 December 2015, retrieved 4 February 2016, https://www.technologyreview.com/s/543946/when-your-boss-is-an-uber-algorithm/.

17. Simon Sharwood, 'Software "Appointed to Board" of Venture Capital Firm', *The Register*, 18 May 2014, accessed 12 August 2015, http://www.theregister.co.uk/2014/05/18/software_appointed_to_board_of_venture_capital_firm/; John Bates, 'I'm the Chairman of the Board', *Huffington Post*, 6 April 2014, accessed 12 August 2015, http://www.huffingtonpost.com/john-bates/im-the-chairman-of-the-bo_b_5440591.html; Colm Gorey, 'I'm Afraid I Can't Invest in That, Dave: AI Appointed to VC Funding Board', *Silicon Republic*, 15 May 2014, accessed 12 August 2015, https://www.siliconrepublic.com/discovery/2014/05/15/im-afraid-i-cant-invest-in-that-dave-ai-appointed-to-vc-funding-board.

18. Steiner, *Automate This*, 89–101; D. H. Cope, *Comes the Fiery Night: 2,000 Haiku by Man and Machine* (Santa Cruz: Create Space, 2011). See also: Dormehl, *The Formula*, 174–80, 195–8, 200–2, 216–20; Steiner, *Automate This*, 75–89.

19. Carl Benedikt Frey and Michael A. Osborne, 'The Future of Employment: How Susceptible Are Jobs to Computerisation?', 17 September 2013, accessed 12 August 2015, http://www.oxfordmartin.ox.ac.uk/downloads/academic/The_Future_of_Employment.pdf.

20. E. Brynjolfsson and A. McAffee, *Race Against the Machine: How the Digital Revolution Is Accelerating Innovation, Driving Productivity, and Irreversibly Transforming Employment and the Economy* (Lexington: Digital Frontier Press, 2011).

21. Nick Bostrom, *Superintelligence: Paths, Dangers, Strategies* (Oxford: Oxford University Press, 2014).

22. Ido Efrati, 'Researchers Conducted a Successful Experiment with an "Artificial Pancreas" Connected to an iPhone' [in Hebrew], *Haaretz*, 17 June 2014, accessed 23 December 2014, http://www.haaretz.co.il/news/health/1.2350956. Moshe Phillip et al., 'Nocturnal Glucose Control with an Artificial Pancreas at a Diabetes Camp', *New England Journal of Medicine* 368:9 (2013), 824–33; 'Artificial Pancreas Controlled by iPhone Shows Promise in Diabetes Trial', *Today*, 17 June 2014, accessed 22 December 2014, http://www.todayonline.com/world/artificial-pancreas-controlled-iphone-shows-promise-diabetes-trial?singlepage=true.

23. Dormehl, *The Formula*, 7–16.

24. Martha Mendoza, 'Google Develops Contact Lens Glucose Monitor', Yahoo News, 17 January 2014, accessed 12 August 2015, http://news.yahoo.com/google-develops-contact-lens-glucose-monitor-000147894.html; Mark Scott, 'Novartis Joins with Google to Develop Contact Lens That Monitors Blood Sugar', *New York Times*, 15 July 2014, accessed 12 August 2015, http://www.nytimes.com/2014/07/16/business/international/novartis-joins-with-google-to-develop-contact-lens-to-monitor-blood-sugar.html?_r=0; Rachel Barclay, 'Google Scientists Create Contact Lens to Measure Blood Sugar Level in Tears', Healthline, 23 January 2014, accessed 12 August 2015, http://www.healthline.com/health-news/diabetes-google-develops-glucose-monitoring-contact-lens-012314.

25. Quantified Self, http://quantifiedself.com/; Dormehl, *The Formula*, 11–16.

26. Dormehl, *The Formula*, 91–5; Bedpost, http://bedposted.com.

27. Dormehl, *The Formula*, 53–9.

28. Angelina Jolie, 'My Medical Choice', *New York Times*, 14 May 2013, accessed 22 December 2014, http://www.nytimes.com/2013/05/14/opinion/my-medical-choice.html.

29. 'Google Flu Trends', http://www.google.org/flutrends/about/how.html; Jeremy Ginsberg et al., 'Detecting Influenza Epidemics Using Search Engine Query Data', *Nature* 457:7232 (2008), 1012–14; Declan Butler, 'When Google Got Flu Wrong', *Nature*, 13 February 2013, accessed 22 December 2014, http://www.nature.com/news/

when-google-got-flu-wrong-1.12413; Miguel Helft, 'Google Uses Searches to Track Flu's Spread', *New York Times*, 11 November 2008, accessed 22 December 2014, http://msl1.mit.edu/furdlog/docs/ nytimes/2008-11-11_nytimes_google_influenza.pdf; Samantha Cook et al., 'Assessing Google Flu Trends Performance in the United States during the 2009 Influenza Virus A (H1N1) Pandemic', *PLOS ONE*, 19 August 2011, accessed 22 December 2014, http://www.plosone.org/ article/info%3Adoi%2F10.1371%2Fjournal.pone.0023610; Jeffrey Shaman et al., 'Real-Time Influenza Forecasts during the 2012–2013 Season', *Nature*, 23 April 2013, accessed 24 December 2014, http://www.nature. com/ncomms/2013/131203/ncomms3837/full/ncomms3837.html.

30. Alistair Barr, 'Google's New Moonshot Project: The Human Body', *Wall Street Journal*, 24 July 2014, accessed 22 December 2014, http:// www.wsj.com/articles/google-to-collect-data-to-define-healthy- human-1406246214; Nick Summers, 'Google Announces Google Fit Platform Preview for Developers', Next Web, 25 June 2014, accessed 22 December 2014, http://thenextweb.com/insider/2014/06/25/ google-launches-google-fit-platform-preview-developers/.

31. Dormehl, *The Formula*, 72–80.

32. Wu Youyou, Michal Kosinski and David Stillwell, 'Computer-Based Personality Judgements Are More Accurate Than Those Made by Humans', *PNAS* 112:4 (2015), 1036–40.

33. For oracles, agents and sovereigns see: Bostrom, *Superintelligence*.

34. https://www.waze.com/.

35. Dormehl, *The Formula*, 206.

36. World Bank, *World Development Indicators 2012* (Washington DC: World Bank, 2012), 72, http://data.worldbank.org/sites/default/files/wdi-2012- ebook.pdf.

37. Larry Elliott, 'Richest 62 People as Wealthy as Half of World's Population, Says Oxfam', *Guardian*, 18 January 2016, retrieved 9 February 2016, http:// www.theguardian.com/business/2016/jan/18/richest-62-billionaires- wealthy-half-world-population-combined; Tami Luhby, 'The 62 Richest People Have as Much Wealth as Half the World', *CNN Money*, 18 January 2016, retrieved 9 February 2016, http://money.cnn.com/2016/01/17/ news/economy/oxfam-wealth/.

10 The Ocean of Consciousness

1. Joseph Henrich, Steven J. Heine and Ara Norenzayan, 'The Weirdest People in the World', *Behavioral and Brain Sciences* 33 (2010), 61–135.

2. Benny Shanon, *Antipodes of the Mind: Charting the Phenomenology of the Ayahuasca Experience* (Oxford: Oxford University Press, 2002).

3. Thomas Nagel, 'What Is It Like to Be a Bat?', *Philosophical Review* 83:4 (1974), 435–50.

4. Michael J. Noad et al., 'Cultural Revolution in Whale Songs', *Nature* 408:6812 (2000), 537; Nina Eriksen et al., 'Cultural Change in the Songs of Humpback Whales (*Megaptera novaeangliae*) from Tonga', *Behavior* 142:3 (2005), 305–28; E. C. M. Parsons, A. J. Wright and M. A. Gore, 'The Nature of Humpback Whale (*Megaptera novaeangliae*) Song', *Journal of Marine Animals and Their Ecology* 1:1 (2008), 22–31.

5. C. Bushdid et al., 'Human Can Discriminate More Than 1 Trillion Olfactory Stimuli', *Science* 343:6177 (2014), 1370–2; Peter A. Brennan and Frank Zufall, 'Pheromonal Communication in Vertebrates', *Nature* 444:7117 (2006), 308–15; Jianzhi Zhang and David M. Webb, 'Evolutionary Deterioration of the Vomeronasal Pheromone Transduction Pathway in Catarrhine Primates', *Proceedings of the National Academy of Sciences* 100:14 (2003), 8337–41; Bettina Beer, 'Smell, Person, Space and Memory', in *Experiencing New Worlds,* ed. Jurg Wassmann and Katharina Stockhaus (New York: Berghahn Books, 2007), 187–200; Niclas Burenhult and Majid Asifa, 'Olfaction in Asian Ideology and Language', *Sense and Society* 6:1 (2011), 19–29; Constance Classen, David Howes and Anthony Synnott, *Aroma: The Cultural History of Smell* (London: Routledge, 1994); Amy Pei-jung Lee, 'Reduplication and Odor in Four Formosan Languages', *Language and Linguistics* 11:1 (2010), 99–126; Walter E. A. van Beek, 'The Dirty Smith: Smell as a Social Frontier among the Kapsiki/Higi of North Cameroon and North-Eastern Nigeria', *Africa* 62:1 (1992), 38–58; Ewelina Wnuk and Asifa Majid, 'Revisiting the Limits of Language: The Odor Lexicon of Maniq', *Cognition* 131 (2014), 125–38. Yet some scholars connect the decline of human olfactory powers to much more ancient evolutionary processes. See: Yoav Gilad et al., 'Human Specific Loss of Olfactory Receptor Genes', *Proceedings of the National Academy of Sciences* 100:6 (2003), 3324–7; Atushi Matsui, Yasuhiro Go and Yoshihito Niimura, 'Degeneration of Olfactory Receptor Gene Repertories in Primates: No Direct Link to Full Trichromatic Vision', *Molecular Biology and Evolution* 27:5 (2010), 1192–200; Graham M. Hughes, Emma C. Teeling and Desmond G. Higgins, 'Loss of Olfactory Receptor Function in Hominid Evolution', *PLOS One* 9:1 (2014), e84714.

6. Matthew Crawford, *The World Beyond Your Head: How to Flourish in an Age of Distraction* (London: Viking, 2015).

7. Turnbull and Solms, *The Brain and the Inner World*, 136–59; Kelly Bulkeley, *Visions of the Night: Dreams, Religion and Psychology* (New York: State University of New York Press, 1999); Andreas Mavrematis, *Hypnogogia: The Unique State of Consciousness Between Wakefulness and Sleep* (London: Routledge, 1987); Brigitte Holzinger, Stephen LaBerge and Lynn Levitan, 'Psychophysiological Correlates of Lucid Dreaming', *American Psychological Association* 16:2 (2006), 88–95; Watanabe Tsuneo, 'Lucid Dreaming: Its Experimental Proof and Psychological Conditions',

Journal of International Society of Life Information Science 21:1 (2003), 159–62; Victor I. Spoormaker and Jan van den Bout, 'Lucid Dreaming Treatment for Nightmares: A Pilot Study', *Psychotherapy and Psychosomatics* 75:6 (2006), 389–94.

11 The Data Religion

1. See, for example, Kevin Kelly, *What Technology Wants* (New York: Viking Press, 2010); César Hidalgo, *Why Information Grows: The Evolution of Order, from Atoms to Economies* (New York: Basic Books, 2015); Howard Bloom, *Global Brain: The Evolution of Mass Mind from the Big Bang to the 21st Century* (Hoboken: Wiley, 2001); DuBravac, *Digital Destiny*.

2. Friedrich Hayek, 'The Use of Knowledge in Society', *American Economic Review* 35:4 (1945), 519–30.

3. Kiyohiko G. Nishimura, *Imperfect Competition Differential Information and the Macro-foundations of Macro-economy* (Oxford: Oxford University Press, 1992); Frank M. Machovec, *Perfect Competition and the Transformation of Economics* (London: Routledge, 2002); Frank V. Mastrianna, *Basic Economics*, 16th edn (Mason: South-Western, 2010), 78–89; Zhiwu Chen, 'Freedom of Information and the Economic Future of Hong Kong', *HKCER Letters* 74 (2003), http://www.hkrec.hku.hk/Letters/v74/zchen.htm; Randall Morck, Bernard Yeung and Wayne Yu, 'The Information Content of Stock Markets: Why Do Emerging Markets Have Synchronous Stock Price Movements?', *Journal of Financial Economics* 58:1 (2000), 215–60; Louis H. Ederington and Jae Ha Lee, 'How Markets Process Information: News Releases and Volatility', *Journal of Finance* 48:4 (1993), 1161–91; Mark L. Mitchell and J. Harold Mulherin, 'The Impact of Public Information on the Stock Market', *Journal of Finance* 49:3 (1994), 923–50; Jean-Jacques Laffont and Eric S. Maskin, 'The Efficient Market Hypothesis and Insider Trading on the Stock Market', *Journal of Political Economy* 98:1 (1990), 70–93; Steven R. Salbu, 'Differentiated Perspectives on Insider Trading: The Effect of Paradigm Selection on Policy', *St John's Law Review* 66:2 (1992), 373–405.

4. Valery N. Soyfer, 'New Light on the Lysenko Era', *Nature* 339:6224 (1989), 415–20; Nils Roll-Hansen, 'Wishful Science: The Persistence of T. D. Lysenko's Agrobiology in the Politics of Science', *Osiris* 23:1 (2008), 166–88.

5. William H. McNeill and J. R. McNeill, *The Human Web: A Bird's-Eye View of World History* (New York: W. W. Norton, 2003).

6. Aaron Swartz, 'Guerilla Open Access Manifesto', July 2008, accessed 22 December 2014, https://ia700808.us.archive.org/17/items/GuerillaOpenAccessManifesto/Goamjuly2008.pdf; Sam Gustin, 'Aaron Swartz, Tech Prodigy and Internet Activist, Is Dead at 26', *Time*, 13 January 2013, accessed 22 December 2014, http://

business.time.com/2013/01/13/tech-prodigy-and-internet-activist-aaron-swartz-commits-suicide; Todd Leopold, 'How Aaron Swartz Helped Build the Internet', CNN, 15 January 2013, 22 December 2014, http://edition.cnn.com/2013/01/15/tech/web/aaron-swartz-internet/; Declan McCullagh, 'Swartz Didn't Face Prison until Feds Took Over Case, Report Says', CNET, 25 January 2013, accessed 22 December 2014, http://news.cnet.com/8301-13578_3-57565927-38/swartz-didnt-face-prison-until-feds-took-over-case-report-says/.

7. John Sousanis, 'World Vehicle Population Tops 1 Billion Units', *Wardsauto*, 15 August 2011, accessed 3 December 2015, http://wardsauto.com/news-analysis/world-vehicle-population-tops-1-billion-units.

8. 'No More Woof', https://www.indiegogo.com/projects/no-more-woof.

Acknowledgements

I would like to express my gratitude to the following humans, animals and institutions:

To my teacher, Satya Narayan Goenka (1924–2013), who taught me the technique of Vipassana meditation, which has helped me to observe reality as it is, and to know the mind and the world better. I could not have written this book without the focus, peace and insight gained from practising Vipassana for the last fifteen years.

To the Israel Science Foundation, which helped fund this research project (grant number 26/09).

To the Hebrew University, and in particular to its department of history, my academic home; and to all my students over the years, who taught me so much through their questions, their answers and their silences.

To my research assistant, Idan Sherer, who devotedly handled whatever I threw his way, be it chimpanzees, Neanderthals or cyborgs. And to my other assistants, Ram Liran, Eyal Miller and Omri Shefer Raviv, who pitched in from time to time.

To Michal Shavit, my publisher at Penguin Random House in the UK, for taking a gamble, and for her unfailing commitment and support over many years; and to Ellie Steel, Suzanne Dean, Bethan Jones, Maria Garbutt-Lucero and their colleagues at Penguin Random House, for all their help.

To David Milner, who did a superb job editing the manuscript, saved me from many an embarrassing mistake, and reminded me that 'delete' is probably the most important key on the keyboard.

To Preena Gadher and Lija Kresowaty of Riot Communications, for helping to spread the word so efficiently.

To Jonathan Jao, my publisher at HarperCollins in New York, and to Claire Wachtel, my former publisher there, for their faith, encouragement and insight.

To Shmuel Rosner and Eran Zmora, for seeing the potential, and for their valuable feedback and advice.

To Deborah Harris, for helping with the vital breakthrough.

To Amos Avisar, Shilo de Ber, Tirza Eisenberg, Luke Matthews, Rami Rotholz and Oren Shriki, who read the manuscript carefully, and devoted much time and effort to correcting my mistakes and enabling me to see things from other perspectives.

To Yigal Borochovsky, who convinced me to go easy on God.

To Yoram Yovell, for his insights and for our walks together in the Eshta'ol forest.

To Ori Katz and Jay Pomeranz, who helped me get a better understanding of the capitalist system.

To Carmel Weismann, Joaquín Keller and Antoine Mazieres, for their thoughts about brains and minds.

To Benjamin Z. Keder, for planting and watering the seeds.

To Diego Olstein, for many years of warm friendship and calm guidance.

To Ehud Amir, Shuki Bruck, Miri Worzel, Guy Zaslavaki, Michal Cohen, Yossi Maurey, Amir Sumakai-Fink, Sarai Aharoni and Adi Ezra, who read selected parts of the manuscript and shared their ideas.

To Eilona Ariel, for being a gushing fountain of enthusiasm and a firm rock of refuge.

To my mother-in-law and accountant, Hannah Yahav, for keeping all the money balls in the air.

To my grandmother Fanny, my mother, Pnina, my sisters Liat and Einat, and to all my other family members and friends for their support and companionship.

To Chamba, Pengo and Chili, who offered a canine perspective on some of the main ideas and theories of this book.

And to my spouse and manager, Itzik, who already today functions as my Internet-of-All-Things.

Image credits

1. Computer artwork © KTSDESIGN/Science Photo Library.
2. *The Triumph of Death*, *c.*1562, Bruegel, Pieter the Elder © The Art Archive/Alamy Stock Photo.
3. © NIAID/CDC/Science Photo Library.
4. Moscow, 1968 © Sovfoto/UIG via Getty Images.
5. 'Death and dying' from 14th-century French manuscript: *Pilgrimage of the Human Life*, Bodleian Library, Oxford © Art Media/Print Collector/Getty Images.
6. © CHICUREL Arnaud/Getty Images.
7. © American Spirit/Shutterstock.com.
8. © Imagebank/Chris Brunskill/Getty Images/Bridgeman Images.
9. © H. Armstrong Roberts/ClassicStock/Getty Images.
10. © De Agostini Picture Library/G. Nimatallah/Bridgeman Images.
11. Illustration: pie chart of global biomass of large animals.
12. Detail from Michelangelo Buonarroti (1475–1564), the Sistine Chapel, Vatican City © Lessing Images.
13. © Balint Porneczi/Bloomberg via Getty Images.
14. Left: © Bergserg/Shutterstock.com. Right: © s_bukley/Shutterstock.com.
15. © Karl Mondon/ZUMA Press/Corbis.
16. Adapted from Weiss, J.M., Cierpial, M.A. & West, C.H., 'Selective breeding of rats for high and low motor activity in a swim test: toward a new animal model of depression', *Pharmacology, Biochemistry and Behavior* 61:49–66 (1998).
17. © 2004 TopFoto.
18. Film still taken from www.youtube.com/watch?v=wWIbCtz_Xwk © TVR.
19. © NOVOSTI/AFP/Getty Images.
20. Rudy Burckhardt, photographer. Jackson Pollock and Lee Krasner papers, *c.*1905–1984. Archives of American Art, Smithsonian Institution. © The Pollock–Krasner Foundation ARS, NY and DACS, London, 2016.

21. Left: © Richard Nowitz/Getty Images. Right: © Archive Photos/Stringer/Getty Images.

22. Courtesy of the Sousa Mendes Foundation.

23. Courtesy of the Sousa Mendes Foundation.

24. © Antiqua Print Gallery/Alamy Stock Photo.

25. Woodcut from 'Passional Christi und Antichristi' by Philipp Melanchthon, published in 1521, Cranach, Lucas (1472–1553) (studio of) © Private Collection/Bridgeman Images.

26. Source: Emission Database for Global Atmospheric Research (EDGAR), European Commission.

27. © Bibliothèque nationale de France, RC-A-02764, *Grandes Chroniques de France* de Charles V, folio 12v.

28. Manuscript: *Registrum Gregorii*, *c.*983 © Archiv Gerstenberg/ullstein bild via Getty Images.

29. © Sadik Gulec/Shutterstock.com.

30. © CAMERIQUE/ClassicStock/Corbis.

31. © Jeff J Mitchell/Getty Images.

32. © Molly Landreth/Getty Images.

33. *The Thinker*, 1880–81 (bronze), Rodin, Auguste, Burrell Collection, Glasgow © Culture and Sport Glasgow (Museums)/Bridgeman Images.

34. © DeAgostini Picture Library/Scala, Florence.

35. © Bpk/Bayerische Staatsgemäldesammlungen.

36. Staatliche Kunstsammlungen, Neue Meister, Dresden, Germany © Lessing Images.

37. Tom Lea, *That 2,000 Yard Stare*, 1944. Oil on canvas, 36"x28". LIFE Collection of Art WWII, U.S. Army Center of Military History, Ft. Belvoir, Virginia. © Courtesy of the Tom Lea Institute, El Paso, Texas.

38. © Bettmann/Corbis.

39. © VLADGRIN/Shutterstock.com.

40. *Virgin and Child*, Sassoferrato, Il (Giovanni Battista Salvi) (1609–85), Musee Bonnat, Bayonne, France © Bridgeman Images.

41. © Bettmann/Corbis.

42. © Jeremy Sutton-Hibbert/Getty Images.

43. Left: © Fototeca Gilardi/Getty Images. Right: © alxpin/Getty Images.

44. © Sony Pictures Television.

45. © STAN HONDA/AFP/Getty Images.

46. 'EM spectrum'. Licensed under CC BY-SA 3.0 via Commons, https://commons.wikimedia.org/wiki/File:EM_spectrum.svg#/media/File:EM_spectrum.svg.
47. © Cornell Bioacoustics Research Program at the Lab of Ornithology.
48. Illustration: the spectrum of conciousness.
49. © ITAR-TASS Photo Agency/Alamy Stock Photo.
50. © Jonathan Kirn/Getty Images.

Index

Entries in *italics* indicate photographs and illustrations.

I can use this time to write → take that class @ arab area.

pg. 7 Walking in Love = should be main goal of every Christian

Thank you Lord for helping me through Deg Dis Disease + healing me - taking away the pain, I love you so. Thank you for healing my heart.

pg. 10   or   1 mile + swim

- Go to India

pg. 15 Egggle → Dad   → Learn how to fly plane there
→ Equitorial New Guinea →
→ Joel   whatever (for dreates (me too)

- pg. 71 > Joel
pg. 72 > me

pg. 54   My Biggest Fear   Stepping out of God's will/ not knowing of MBSM, God

To share w/ Joel pg. 71 Don't You Dare pg. 88

Settle.

pg. 72 hmmm...

let Joel know it calms your heart when he holds you - you just love it.

Amanda 91? / Psalm poem pg. 94

me gol vengy I came de une famulvo Royel, y la gente mudume world before me. Vete e aquellos paises. A Istanbul Iras.

You will be bold & break barriers never broken before. Though love, gratitude & meekness & humility, You are from Royal Blood?

We all are some form of royalty, I do suppose... after all we are all Abraham's descendents. wouldn't you agree (yet all/any kind of Royalty).

NEVER
GIVE UP!

NEVER

GIVE UP!

Relentless Determination to Overcome Life's Challenges

JOYCE MEYER

New York   Boston   Nashville

FaithWords
Hachette Book Group USA
237 Park Avenue
New York, NY 10017

Visit our Web site at www.faithwords.com.

Printed in the United States of America

First Edition: October 2008

10 9 8 7 6 5 4 3 2 1

FaithWords is a division of Hachette Book Group USA, Inc.
The FaithWords name and logo are trademarks
of Hachette Book Group USA, Inc.

Library of Congress Cataloging-in-Publication Data

Meyer, Joyce
Never give up : relentless determination to overcome life's
challenges / Joyce Meyer. — 1st ed.
p. cm.
ISBN-13: 978-0-446-58035-9
ISBN-10: 0-446-58035-X
1. Success—Religious aspects—Christianity.   2. Christian life.
3. Determination (Personality trait)   I. Title.
BV4598.3.M47 2008
248.4—dc22
2008016200

CONTENTS

INTRODUCTION

More than anything else, I want you to know you can have a deep, intimate, personal relationship with God through Jesus Christ and the very best life has to offer. God is no respecter of persons (see Acts 10:34) and His promises apply equally to everyone. Yes, you can have the very best God offers, but you will have to be determined to never give up until you have succeeded in every area of life.

I believe God has a great purpose for you, and I urge you to not settle for anything less. He wants to bless you and give you a life that will not only thrill you, fulfill you, and bring you deep joy and sweet satisfaction but also challenge you, stretch you, frustrate you at times, and even lead you to moments when you are tempted to give up.

Think of this book as a manual to use as you pursue the best in every area of your life. It will give you the inspiration you need to keep putting one foot in front of the other when you grow weary on your journey and remind you over and over again, in a variety of ways: *You can do it if you never give up.* It will give you strength to soar above life's tests and trials like a majestic eagle, increase your determination, and teach you how to turn adversities into opportunities. It will empower you to persevere, and build your confidence in God's ability to see you through to the achievement of the goals He has given you. It will strengthen your faith in His desire to fulfill the dreams He has put in your heart.

As you walk through the pages of this book, you will become acquainted with amazing people, people who refused to settle for less than the best and whose stories will inspire you and astound you. With every word you read, I will be cheering you on, reminding you God is on your side, and urging you to never give up.

One of the primary reasons people give up is that they try things, don't succeed, and feel like "a failure." The truth is that we are never a failure unless we give up. When we don't succeed at something, many times we do not have the courage to try again, and we settle for less than we could achieve or enjoy if we would simply keep trying. The fact is, we all have times when things just don't work out the way we hope they will, even though we do our best. We may fail at one thing, or even a few things, but that certainly does not make us a failure in life. I believe these temporary setbacks are part of life and we must experience them in order for us to ever be truly successful. Failing at some things on our way to success humbles us and teaches us lessons we need to learn. For people who never give up, failure is simply the fuel for greater determination and success in the future.

Some of the most successful people in history failed and, instead of being discouraged, refused to give up. For example, consider the following:

- Henry Ford, who invented the automobile, went broke five times before he succeeded in business.
- The great dancer and movie star Fred Astaire took a screen test at MGM studios in 1933. A studio memo reported he was slightly bald, could not act, and could dance a little.
- The family of Louisa May Alcott, the great author who wrote the popular book *Little Women,* thought she should abandon the idea of being a writer and become a seamstress instead.
- A newspaper fired Walt Disney for lack of ideas, and he went bankrupt several times before building Disneyland.
- Enrico Caruso's parents believed a voice teacher who said he had no future in music—he simply could not sing at all. He did not believe the teacher and became one of the most famous opera singers in the world.

- Theodore Roosevelt suffered the deaths of both his mother and his wife on the same day in 1884 before he became a war hero and a very effective president of the United States.
- John Wesley, the founder of Methodism, was often asked to not return to churches after preaching in them once. When he preached on the streets, townspeople kicked him out. When he preached in a meadow, people turned a bull loose on him. But later, because he refused to give up, he preached in a pasture and ten thousand people came to hear him.

The story of Abraham Lincoln also amazes me. In the face of many defeats, he had reason to believe there was no way he could succeed in life or be president of the United States. At twenty-two years old, he failed in business. One year later, he ran for the legislature and lost. When he was twenty-four, he experienced a second business failure. At twenty-six, the woman he loved passed away, and he suffered a nervous breakdown the next year. When he was twenty-nine, he lost another political race, and at thirty-four he made an unsuccessful run for Congress. At thirty-seven, he did get elected to Congress, only to be defeated again two years later. At forty-six, he lost his bid for the Senate, and the next year, he failed in his attempt to become vice president. When he was forty-nine, he was defeated for the Senate again. He had four sons, but only one lived to adulthood. But, at fifty-one years of age, Abraham Lincoln was elected president of the United States, and successfully led the country through one of its most difficult periods. Many people would have said, "No way," but not Lincoln. He never gave up.

In addition to the well-known people who had to persevere, there have been countless others whose names we do not know, but whose failures or mistakes became some of the best-known, best-selling products in the world today.

For example, Ivory soap was never intended to float. It floats because of a manufacturing error—and its buoyancy is the quality that distinguishes it from every other soap on the market.

Similarly, the material used to manufacture Kleenex tissues was originally intended to be used to make filters in gas masks during World War I, but it did not work. It also failed as a cold cream remover. But when someone decided to package and market it in the form of disposable handkerchiefs...well, you know the rest of the story.

I believe you are destined to do great things. God created you for a purpose. He has opportunities He wants to give you and assignments with which He wants to entrust you. I'm sure you have realized by this point in your life that you will face opposition as you follow God. People who are called to greatness meet great challenges. He never promised us it would be easy. In fact, He guarantees us adversities in His Word. He also promises to be with us through difficulties, to fight on our behalf, to strengthen us to overcome any obstacle we confront, and to give us the ability to overcome them under one condition—that we never give up.

No matter what comes your way, *refuse to quit.* Before you get started, I want to remind you of a biblical truth to carry with you through the pages of this book and long after you finish it: "God, who got you started in this spiritual adventure, shares with us the life of his Son and our Master Jesus. *He will never give up on you.* Never forget that" (1 Corinthians 1:9 The Message, emphasis mine).

NEVER
GIVE UP!

CHAPTER 1

NEVER SAY "NO WAY"

"Without the way, there is no going; without the truth, there is no knowing; without the life, there is no living."
THOMAS À KEMPIS

Have you ever faced a situation and said, "There is no way this can ever be"? Maybe some of these thoughts weigh on your mind:

- There is no way I can handle the pressure at work.
- There is no way I can pay my bills at the end of the month.
- There is no way to save my marriage.
- There is no way my children will ever grow up to be responsible adults.
- There is no way I can keep my house clean and straight.
- There is no way for me to open my own business.
- There is no way I can go back to college at my age.
- There is no way I can lose the weight I need to lose.

I want you to know there is always a way. It may not be easy; it may not be convenient; it may not come quickly. You may have to go over, under, around, or through. But if you will simply keep on keeping on and refuse to give up, you *will* find a way.

I want you to make yourself a promise. Promise yourself that you will never again say, "There is no way." The truth is, even when there seems to be no way, with God there is always a way. If you are a born-again

Christian, the Spirit of God lives inside of you. All the creativity in the world resides in Him, and because He lives in you, you have access to all that creativity. The Holy Spirit can give you ideas that never occurred to you and show you ways to do things you never thought you could do.

Instead of dwelling on our difficulties, we need to focus more on the fact that God is for us and His power is at work in us. Often, we give up too easily, saying, "This is too hard" or "This is taking too long." We must stop looking at situations in our lives and thinking, *I really cannot handle this; it's too much for me. I've already tried too many times. I have to accept there is just no way it will ever happen.* Instead, we need to say, "I don't care whether there seems to be a way or not. Jesus is the Way; His Spirit lives in me; and I will find a way!" God promised He would make a way in the wilderness and rivers in the desert (see Isaiah 43:19).

Jesus said in John 14:6, "I am the Way and the Truth and the Life." He is the way, and He will help you find a way where there doesn't seem to be one. First Corinthians 10:13 says, "For no temptation . . . has overtaken you and laid hold on you that is not common to man . . . but with the temptation He will [always] also provide the way out (the means of escape to a landing place), that you may be capable and strong and powerful to bear up under it patiently." In other words, God always has a way for us if we will look for it, wait for it, and refuse to give up on it.

> *I wonder how many times people give up just before a breakthrough, on the very brink of success.*

I wonder how many times people give up just before a breakthrough, on the very brink of success. You can feel the same way for ten years and then suddenly, one day you wake up and everything will change. You don't feel any different than you have felt any other day. Nothing looks any different than it ever has; nothing appears to be happening, but something does happen; and when you go to bed that night, your dream has finally been fulfilled, the situation you lived in for so long is finally over, or you finally achieve the accomplishment for which you labored for years.

I know a young woman who was once working at a job she did not particularly like and was single but wanting to be married. Within one month she became engaged and was hired for her dream job. She waited for what seemed an eternity, but at the right time God made a way. God's ways are not our ways, but His way is always best!

God has a plan for you and He has heard your prayers; you may not realize how close you are to your breakthrough. Even if you have to wait three, four, or five more years, if you will keep pressing on, you *will* have the victory you need. Whatever you do, do not give up on the brink of your breakthrough. Do not stop hoping, believing, and trying. Instead, say, "I will never quit; I will never give up; I will never say, 'No way.'"

THREE KEY ISSUES

I have encountered thousands upon thousands of people over the past thirty years in ministry and I have observed that most people consider three aspects of their lives most challenging and most worth fighting for: health, finances, and family. I have dealt with each of these areas on a personal level over the years and have seen great improvements and victories. If you are willing to persevere through the struggles you face in your health, your family, and your finances, I know you will emerge victorious too.

Never Give Up on Your Health

I'm amazed by the number of people who simply do not feel strong, vibrant, and healthy. I often overhear people telling each other, "Oh, I am so tired"; or, "I would do some of the things I enjoy if I had the energy." This lethargic approach to life is not God's best for us! God wants us to feel good and have the passion and energy we need to enjoy our lives and do everything He calls us to do. He does not want us to be too exhausted or depleted to do the things that give us joy or move us forward in His purposes for our lives.

Though some people do suffer with various conditions that must be treated with medication or therapy, many health problems people struggle with are stress related. Other health conditions develop because people do not take care of themselves; they do not eat healthily, drink enough water, exercise regularly, or get enough rest. If they would simply make some lifestyle changes, their health and quality of life would improve dramatically.

In chapter 8, I discuss in further detail the ten-year period during which I struggled with various health problems. I saw many doctors during that time, and all of them told me my problems were stress related. As I learned about nutrition, exercise, living a balanced life, and reducing stress, my health improved remarkably. I feel better today than I felt thirty years ago.

My daughter Sandra has a similar testimony. She felt badly and suffered with health problems for years. Many of her problems resulted from the stress that came with being a perfectionist's perfectionist. After she gave birth to her twins, she had terrible digestive problems, back pain, and other physical challenges to the point that she had to be taken to the emergency room several times. Sandra finally reached a point where she knew in her heart she could be healthy and feel well. She determined to find the causes of her problems and to solve them. Though she had to be diligent and disciplined, she refused to give up on the prospect of good health. As she applied the lessons she learned about nutrition and exercise, and learned how to keep the stress of perfectionism (worry, fear, etc.) from affecting her physically, she began to feel better and gain strength. Now, she feels terrific and is in good health.

I urge you to do everything you can do to improve your physical health. Make the lifestyle changes you need to make to feel better and be stronger and more energetic. If you need to change your eating habits to include more fruits and vegetables and less sugar and fat, change them. If you need to drink more water and fewer soft drinks, do so. If you need to discipline yourself to sleep a certain number of hours each night, adjust your schedule to accommodate the sleep you need. If you need to exercise more, start exercising more.

In addition, deal with the stress and emotional issues that affect you physically because your mental and emotional state certainly impacts your body. Being a perfectionist may be affecting you, as it did Sandra. Maybe worry is causing your blood pressure to be higher than it should, or perhaps fear or nervousness is affecting your digestive system. Maybe stress and tension are giving you headaches or causing your muscles to stay tight instead of limber. No matter what is affecting you, make sure you seek help for it so it does not make you feel bad or do permanent damage to your body.

If you are tired, lethargic, or simply not feeling well, see your doctor. Find out why you do not feel well and learn what to do about it. Do not settle for poor health when good health is just a few lifestyle changes away. Do everything you can do to feel well and be energetic.

Never Give Up on Your Finances

So many people are trapped in debt these days, struggling to pay their bills at the end of each month, wondering how to build a savings account, and concerned about how to finance their retirement or their children's education.

I remember when Dave and I had to buy clothes for our children at garage sales and drive cars so old we never knew if they would run or not. In the early days of our ministry, we could not afford to stay in hotels when I preached meetings at night, so no matter where we were or how tired we were, we had to drive home. Sometimes we were so exhausted we had to pull over on the side of the road to sleep for several hours before we could continue. I can remember buying canned goods with no labels because they were on sale at the grocery store. I never knew whether I would open those cans to find peaches, green beans, alphabet soup, or cat food, but the "mystery cans" were so inexpensive I had to try them.

I share these memories to let you know I understand financial hardship. I also know from personal experience, and from seeing God turn around the financial situations of countless others, that you should never give up on your finances. Never allow yourself to believe you will always be in debt or never be able to save money.

Bookstores and libraries are full of resources designed to help you break free from the bondage of debt and become a wise steward of your money. With discipline, determination, good counsel, and enough time, you can eliminate your debt, pay cash for your purchases, make investments, and save for the future. Do not think financial freedom is not possible for you, because it is. It may not be easy, but you can achieve it.

Never Give Up on Those You Love

Before Dave and I married, he asked God to send him someone he could help. When God sent him me, he got more than he bargained for! Because of the abuse in my background, I had serious, serious problems. Trying to be in an intimate relationship with me would have been extremely difficult for anyone, and I am sure many men would have given up. But Dave continued to pray and seek God for ways to help me, even when I acted as though I did not want help. At times, he even wept because he did not know what to do. He has even shared that he often went for a drive in the car to pray and cry, and returned home trusting that God would change me. After a few days, he noticed I had suddenly changed. I had not totally changed, but at least he saw progress and that let him know God was working.

I am so thankful today that God was faithful to Dave and to me. He had to lead me along an arduous path of healing, deliverance, and wholeness. He had to teach me to think differently, to trust people, to allow myself to be loved, and how to love others. He even had to teach me to be nice because I grew up thinking I had to protect myself, which meant I was often defensive and harsh.

Our journey was not easy, but Dave refused to give up on me, and we both refused to give up on God. Now we have had more than forty years of marriage, and I can honestly say our relationship is better than ever. Though our early years were difficult, we have had many years of happiness and we look forward to many more.

Like Dave, I also had the opportunity to refuse to give up on someone I loved. My oldest son, David, and I were so much alike we could hardly stand each other at times. At one point, he worked in the ministry and the two of us clashed so intensely I finally decided to tell him to find another job. I did not want to fire him, but I did not believe I could endure the conflict that characterized our relationship. I planned to talk to him and let him know his presence in the ministry simply was not working, but God spoke to my heart, *Don't give up on David.*

Over time, David and I learned to get along well. Now he runs our world missions department, has opened eighteen foreign offices for us, and oversees numerous international outreaches. I am very thankful for his good work and glad God told me to not give up on him.

When you are tempted to give up on your loved ones, remember David and me. Dave refused to give up on me, and I refused to give up on David. Whether you are believing for someone you love to become a Christian, change their behavior, leave a bad relationship, stop using drugs, go back to school, come home, or get a job, keep believing change is possible. Do not give up on the ones you love; your patient love and faithfulness may be exactly what they need to make a complete turnaround.

Love never fails. In other words, it never gives up on people. The apostle Paul describes what love is in 1 Corinthians 13 and mentions that love always believes the best; it is positive and filled with faith and hope. While Jesus was on earth, He gave a new commandment to His followers: that we love one another (see John 13:34). I believe walking in love should be the main goal of every Christian.

God is love (see 1 John 4:8) and He never gives up on us. Let's choose to live with that same attitude. Believe in the power of love to change and transform anything and anyone.

AGAINST ALL ODDS

When I think of people who never gave up, I think of my friend Pennie Sheperd. Her story is the remarkable account of a woman who experienced a tremendous miracle of God's grace and healing in her life because she was determined to reach her goal.

When Pennie Sheperd was thirteen years old, she fell and broke her tailbone. For years after that accident, she lived in chronic pain. When she got out of bed each morning, she felt as though someone had jammed a knife into her back. After she married, there were times her husband had to help her out of bed. She could not stand or sit for long periods of time or bend over to wash her hair in a sink. She slept with an ice pack, or a heating pad, or lots of pillows, but nothing seemed to help. She exhausted every possible avenue of escape from her agony. At times, she was almost completely overwhelmed; at other times, she was depressed by it and thought, *How can I live through one more day?*

Pennie's doctor, Caroline Rogers, diagnosed her with post-trauma arthritis and degenerative disc disease. This caused muscle imbalances, which led to problems in parts of her body where she had not suffered previously. She was trapped in a cycle of pain and suffering that was irreversible—unless, of course, a miracle happened.

Pennie loved God and truly believed He could heal her. She believed every day held the potential for a miraculous healing in her life. So day after day, year after year, she asked God for her miracle.

As Pennie prayed one day, she heard these words in her heart: *Run to your miracle.* Those words did not make sense to Pennie, because she was not a runner and had never been one. She didn't like to run, and she lived in so much pain she did not even want to try. But when God spoke to her heart, *Run to your miracle,* she committed to do so.

The greatest distance she could think of running was a full-length marathon—26.2 miles! So with her family's support and her doctor's permission and offer to act as her trainer, she began the grueling physical and mental training process.

For the next four months, Pennie pushed through the pain to train for the marathon. She kept a strict diet and a rigorous schedule seven days a week, eventually getting up by 3:00 a.m. to get in a long run before a full day of work and family responsibilities. After several months of training, Pennie had no relief from her pain, so she continued to pray about her participation in the marathon to make sure the pain was not an indicator she should stop training. She was determined to pay the price of progress and to not give up. She wanted to be able to stand at the starting line and know she had done everything God asked her to do.

The day of the marathon drew near, but just two weeks before the race, her knee went out. The night before the race, she could hardly walk around her hotel room. She still believed God had spoken to her and felt she had to press through the agony. She was determined to show up at that starting line. She said, "I am going to the starting line even if someone has to carry me to get me there."

Her husband and daughters, her strongest supporters, wrote scriptures and inspiring messages all over her hands and arms. Before sunrise on marathon day, she literally hobbled up to the starting line, with Dr. Rogers beside her, knowing she was exactly where she was supposed to be. She was going to run to her miracle!

As she stood at the line barely able to walk, she turned to Dr. Rogers and cried, "My back doesn't hurt! My back doesn't hurt!" She had not said those words for twenty-eight years. Her knee hurt like crazy, but her back didn't hurt!

Dr. Rogers suggested they begin the race by walking to warm up and see how the knee responded. About a mile and a half into the race, Pennie told Dr. Rogers, "I think I'm going to be able to run."

Throughout the whole race, she had much encouragement. A group of friends prayed for her during the seven hours of the race. Her husband and daughters were at the starting line and at regular checkpoints along the way to encourage and support her.

At the fourteen-mile marker, Dr. Rogers and Pennie were thirty-six seconds off the seven-hour pace, which meant a "sweeper" would pick them up and drive them to the finish line if they did not make up the time. But Pennie gave it her all.

At the twenty-three-mile marker, she hurt so badly her daughters began to run with her—one under each arm to help support her. Pennie crossed the finish line in about seven hours.

On our *Enjoying Everyday Life* broadcast, she said, "Before the race, I had always envisioned in my mind that I would cross the finish line and get my miracle. The truth is, I was not healed at the finish line; I did not receive my miracle at the finish line. I got my miracle at the *starting* line. All I had to do was show up. It doesn't take running a marathon to get a miracle. In my case, it took obedience and commitment."

Pennie Sheperd is a woman who could have said, "No way." Instead, with God's help, the support of family and friends, and a "never say no way" attitude, she never gave up. Pennie's miracle took place in January 2004 and she is still healed today.

GET A GOAL

Right now, I want you to think of an area in your life in which you need to refuse to give up. Come up with a goal—one that will require you to be disciplined and to overcome some obstacles, but one that promises great reward. It may be as basic as making your bed each morning, or as ambitious as running a marathon or climbing Mt. Everest. It may be to break free from a fear of flying or a fear of public

> *Get with God and decide what your goal needs to be and what is worth putting your energy into.*

speaking, or it may be to overcome a physical handicap or a learning disability. It may be cleaning your house or getting out of debt.

If other people think your objective is too easy, then that is too bad. If it is a legitimate goal for you, then stick with it. If other people think it is impossible, do not allow them to discourage you. Just make sure you and God are in agreement and then go after your goal with everything in you.

You will see as you read this book that rewards await those who overcome. I am praying that God will help you be full of "holy determination"—not some kind of fleshly determination or willpower—but true God-given determination. I encourage you to pray and be determined to be disciplined and diligent in every area of your life, because that is the way you will overcome your obstacles and enjoy your hard-won successes. Take life one day at a time and remember that God has rewards in store for you. Don't you dare miss them!

BE SMART ABOUT IT

Many people never accomplish their goals because they do not know how to set them. A popular and easy-to-remember acronym that has been successful in helping countless people reach their goals is the word *smart*:

Specific
Measurable
Attainable
Realistic
Timely

Let me elaborate.

Specific: Make sure your goal is as specific and accurate as possible. For instance, don't simply say, "I want to lose weight"; say, "I want to lose ten pounds in the next three months. Every time you hear yourself speaking of your goals in vague terms, ask yourself, "What does that mean?" You'll soon find yourself making comments such as, "I am going to stop watching television at nine-thirty every night and be in bed by ten" instead of, "I need to watch less television and get more sleep."

Measurable: Goals that are hard to measure are goals that are hard to meet. Before you commit to a goal, decide how you will monitor your progress. For goals involving debt reduction, you can measure your progress with monthly statements. For goals that involve exercise, you can keep track of how many pounds you can lift or how many miles you can run. Whatever your goals are, find ways to measure them, whether that includes a journal, a checklist, a chart, a graph, a spreadsheet, or some other creative way to see how you are doing.

Attainable: Make sure the goal itself is reachable. Don't have a goal of losing thirty pounds in one week or paying off all your debt in one year—when your debt is more than your annual salary. Choose goals that lie just beyond your reach—not so easy that they don't challenge you, but not so difficult that they stretch you to the breaking point.

Realistic: I believe in dreaming big dreams and aiming high, but don't set yourself up for disappointment by trying to reach an unrealistic goal. Assess all the factors that will affect your ability to reach your goals and work within those parameters. Make sure it is realistic, given your health and physical condition, finances, schedule, personal capacity, and other priorities.

Timely: People who set goals without target completion dates rarely accomplish their objectives. Give yourself deadlines by which to accomplish your goals. If you are working toward a goal over a long period of time, consider setting intermediate goals each week or month, to keep you on track.

He Coined the Phrase

You may know of him as one of the most effective political leaders ever to stand on the world stage or as one of the most highly regarded statesmen of all time, but do you know what former British prime minister Winston Churchill suffered before he ever achieved greatness?

Winston Leonard Spencer Churchill's premature birth occurred on November 30, 1874—two months before he was due. He was born into a prominent English family, but his parents had no time for him. His mother did not feed him, but left him in the care of a wet nurse when he was an infant, and with a nanny as he grew older, while she pursued social activities. His father, a busy political leader, never demonstrated much interest in him.

Churchill had a difficult life, certainly one full of challenges, tumult, opposition, and near misses. His parents sent the neglected boy to boarding school as a child. He did not excel academically (except in English and history) or socially. He had a speech impediment (which he never completely lost) and failed to make friends or even get along well with others. Later in life, he told stories about having to dodge cricket balls his fellow students threw at him. After that, at age nineteen, he nearly drowned in an accident in Lake Lausanne, and much later at age fifty-seven, he was hit by a car in New York City.

A graduate of the Royal Military College, the short, stocky, shy Churchill served in the British army as a young man. During that time, he saw combat; participated in a cavalry charge; traveled Europe, Africa, Cuba, and India; and wrote newspaper reports and books.

At age twenty-four, Churchill resigned his military commission to pursue journalism and politics. He sought a seat in

(continued)

Parliament, but lost the election. He then traveled to South Africa as a journalist covering the Boer War, but was captured and thrown into prison. However, he managed to escape, which positioned him as a military hero upon his return to England in 1900. That same year, he ran for Parliament again, and won his seat by a narrow margin.

In 1940, Churchill became prime minister, and in that position, exercised brilliant, courageous, strategic leadership that rescued Great Britain from the edge of seemingly certain defeat to victory in World War II.

In 1941, while visiting the school where he had studied as a youngster, the then–prime minister Churchill is erroneously reported to have given a three-word speech: "Never give in" and then promptly returned to his seat. In reality, Churchill gave a longer speech which included these words: "This is the lesson: never give in, never give in, never, never, never, never—in nothing, great or small, large or petty—never give in except to convictions of honor and good sense. Never yield to force; never yield to the apparently overwhelming might of the enemy."

I encourage you today as Churchill exhorted the schoolboys in 1941: "Whatever you do, *never* give in—and never, never, never give up."

CHAPTER 2

NEVER GIVE UP ON YOURSELF

"By perseverance the snail reached the ark."
CHARLES HADDON SPURGEON

A farmer once took an egg from an eagle's nest. He took it home and placed it under one of his hens and it hatched with a little brood of chickens. The farmer raised the bird with great patience and attempted to tame him. The eagle never really seemed to fit in with the chickens. It always walked alone; it could not seem to relate to or interact with the chickens.

As the eagle grew, he realized something seemed to be wrong deep inside of him. Even though he had never known any other existence but life in the chicken yard, it just did not feel like home to him. He wanted to leave the chicken yard and take to the skies. He even tried to do so, and the farmer finally had to clip the eagle's wings to keep him from flying away.

Since the eagle could not fly, he just sat in the chicken yard looking up at the sky. One day a storm began to brew, the sky grew dark, and all the barnyard animals scurried around for cover; the chickens were terribly frightened, as chickens are prone to be. The eagle sat watching the scene in front of him, realizing the storm did not frighten him in the least.

At that moment, he could not help but stretch out his wings, and as he did, he noticed that the farmer had failed to keep them clipped.

Suddenly, his eye caught sight of a great eagle riding the wind above him, his wings outstretched in majestic form.

The eagle that was raised as a chicken looked again at the chickens scurrying around frantically, then returned his gaze to the eagle soaring peacefully above him, then looked back at the chickens and then up at the eagle again. He heard the eagle let out an awesome, piercing cry. In that instant, he knew he had to get out of that chicken yard! A mighty gust of wind swept beneath his outstretched wings and lifted him into the air. With a shrill scream of victory and freedom, he left the barnyard forever.

YOU HAVE THE HEART OF AN EAGLE

I wanted to share the story of the eagle in the chicken yard with you to fill you with a fresh sense of encouragement to never give up on yourself. I hope it stirs something deep inside of you—something that makes you want to press through every obstacle you face, to be who you are, and to make the courageous choices needed to break free from everything holding you back from the greatness for which you were created.

> *Do you ever feel you are like an eagle in a chicken yard? You know there is much more within you than you are experiencing and expressing in your life right now.*

Do you ever feel you are like an eagle in a chicken yard? You know there is much more within you than you are experiencing and expressing in your life right now. You know God has a great purpose for your life—and you cannot escape or ignore the inner urge to "go for it." But do you also know you will have to work hard; take risks; endure loneliness; leave some things behind; make some difficult decisions; or perhaps be misunderstood, judged, or even criticized in order to achieve and enjoy the fullness of God's destiny for your life?

Know this: All eagles are uncomfortable in a barnyard. They all long for the clear, blue, open skies. When you are living in a place that keeps you from being who you were made to be and doing what you are meant to do, you will be uncomfortable too. When the thought of moving beyond where you are begins to take root in your heart and mind, when a seed of greatness begins to grow, when you have a burning desire to step out of where you are or a desire to be adventurous and do something new or different, pay attention to it. Begin to act on it. But also realize that people around you may not understand your desire to break out of the box. They may want to clip your wings. They may even say, "Now just settle down and be like all the other chickens. Here you have this nice chicken yard and these nice little worms and grubs. Why should you ever want any more than that?"

When you hear such comments and questions, something inside of you may ask, *What is wrong with me? Why do I think as I think? Why do I feel this way? Why can't I just settle down and live a normal life like everybody else?* The reason you cannot just settle down is that you are not a chicken; you are an eagle! You will *never* feel at home in that chicken yard because you were made for something bigger, more beautiful, and more fulfilling.

I encourage you today to fan the flame inside of you. Fan it until it burns brightly. Never give up on the greatness for which you were created, never try to hide your uniqueness, and never feel you cannot do what you believe you were made to do. Realize your hunger for adventure is God-given; wanting to try something new is a wonderful desire; and embracing life and aiming high is what you were made for. You are an eagle!

LIKE AN EAGLE

The eagle is one of nature's best examples of strength, perseverance, and determination. It is a bird who refuses to be denied its destiny, one who never gives up. As you and I continue on our journeys through life and stay committed to never giving up, we can learn some important lessons from this powerful, majestic bird.

Eagles and chickens have one characteristic in common: they are both birds. Aside from that, they could not be more different! I have spoken many times about the fact that different kinds of birds often remind me of the different types of Christians I have observed in the body of Christ. Certain characteristics of these birds are similar to characteristics of some people, and I believe, by the time you read about them, you will see what I mean.

The Chicken

First, I think about chickens. Chickens are generally skittish and afraid of life; they are lazy; and they rarely reach their potential. Chickens simply scratch around barnyards and cluck. We do not see chickens flying, because chickens flap; they do not fly. They need to live inside the confines of a fence and are satisfied to do so. They cannot handle freedom; they must be kept in a chicken yard. As you read in the story at the beginning of this chapter, when a storm comes, the chickens' first response is to flap around in the chicken yard, stirring up dirt, and to run to the chicken house to huddle in fear with all the other chickens, trying to find a little grub worm on the way.

Christians who "flap around" and "run for the chicken house" when difficulties arise are not living the victorious life Jesus died to give us. We have to stop running from things, especially the storms in life. We are more than conquerors (see Romans 8:37). We can be confident in God at all times, and we can stay peaceful in the midst of storms. God has called and equipped us to *overcome,* not to be fearful or intimidated.

The Magpie

Magpies are overly aggressive bullies who abuse other birds. "Magpie Christians" generally drive others away from God and are selfish, arrogant, and self-centered to the point of rudeness. These harsh, pushy Christians cause much disrespect toward the kingdom of God and give fellow believers a bad name.

The Kookaburra

To the kookaburra, life is one big party, one big joke. "Kookaburra Christians" are those who take nothing seriously and laugh at everything (even when laughter is totally inappropriate). They are not sensitive to anything around them and they often wound, hurt, and offend others.

The Bible says we need to be sober-minded (see 1 Peter 1:13). That does not mean we can't have fun, but it does mean we need to pay attention to what is going on around us; be sensitive; and understand there are times when certain stories, jokes, or types of behavior are appropriate and times when they are not.

The Vulture

We know vultures, or buzzards, as the garbage collectors of the bird kingdom. They are drawn to death and uncleanness. These birds enjoy destruction and the filth of life—everything corrupt, rotten, and stinking.

"Vulture Christians" are attracted to people with problems. They love to see people suffer and fail, and they use their words to destroy people's lives. They spread rumors and seem to secretly enjoy ruining the reputation of a person or ministry.

Parrots and Cockatoos

Parrots and cockatoos are the "talkers" in the bird kingdom. "Parrot Christians" are those who "talk the talk," but do not "walk the walk." They appear to know much more than they do because they have learned to repeat what they hear and to mimic what others say. They may have some intellectual knowledge of God, but no personal relationship with Him. They are all noise and no action, all talk and no depth of experience.

The Cuckoo

Cuckoos do not like to work and are happy to sponge off other people. These birds will not even build their own nests; they look for nests other birds have already built, lay their eggs there, and then leave their babies for other birds to raise!

Regrettably, there are cuckoos in the church today. They exist on handouts; they do not want to work. Sometimes they are not even born again, but they operate under a religious spirit. They want to ride on the coattails of other people's faith and enjoy the benefits of the Christian life without investing in it.

The Peacock

If you have ever seen a peacock, you know they strut around as though they own the entire world. "Peacock Christians" are flashy, showy, overly ambitious, worldly, full of pride, and very much in love with and impressed with themselves. They have large egos and feel superior to everyone else. They love to "strut their stuff"—clothes, cars, jewelry, and other things—but are selfish and have little or nothing for people in need.

The Pelican

The big, jolly pelican has a big mouth and is only interested in eating. He seeks whatever he can find to fill his belly.

"Pelican Christians" do the same. They have ravenous appetites and hunger for the Word of God. They could become spiritual giants, but are content to sit on the couch eating junk food, watching television, and ignoring their spiritual lives. They are always happy to let others take positions of spiritual leadership, even though they could do so if they would simply put forth a little effort.

The Canary

Canaries and other caged birds are usually beautiful birds with tremendous potential who spend their lives locked in cages. They do not seem to realize the bondage in which they live; they even sing in the midst of it!

"Canary Christians" are those who go merrily along, either oblivious to or content with the "cages" they live in. Perhaps they are trapped in religious tradition, a lifeless church, a powerless prayer life, inferiority or insecurity, or a host of other problems that keep them from enjoying the freedom and joy available to them in God. They refuse to break free; therefore they never embrace life in all its fullness.

The Crow

If you have ever visited a farm or read a story about one, you know that some fields have scarecrows. The crows must be scared away because they damage the crops. These dirty, noisy animals eat what does not belong to them, reducing the value of a farmer's livelihood and spoiling the fruit of his labors. Crows also enjoy destroying the young of other species. They are cunning, dangerous, and only out for themselves. Nothing matters except their own selfish desires, and they seek their own fulfillment ravenously.

"Crow Christians" also enjoy hurting others and causing destruction. They think nothing of ruining a pastor's reputation or dividing a church. They are some of the most cunning and dangerous people in the church.

One interesting fact about crows is they spend a great deal of time pestering eagles. If you want to be an "eagle Christian," which I will write about in the next section, get ready to be pestered by the crows in the kingdom.

The Eagle

In the midst of all other birds, we find the eagle. Bold, strong, courageous, fiercely devoted the eagle has many wonderful characteristics. Charles Prestwich Scott rightly observed, "Eagles come in all shapes and sizes, but you will recognize them chiefly by their attitudes." Among its other admirable traits, an eagle is loyal and committed, a reliable mate, and a devoted parent. An eagle has sharp, keen eyesight and knows how to fly efficiently by riding the thermal currents of the earth. This bird dwells on the rocks in high places and is confident enough to fly alone and to stand alone in life when necessary and to be strong against the storms, spreading its wings and soaring above the clouds and storms of destruction.

EAGLES OR CHICKENS?

Over the years, I have encountered many people who are eagles but think they are chickens. That is a real problem because eagles who think they are chickens will act like chickens. They will scratch and flap around the barnyard making noise instead of soaring confidently and smoothly above the storms beneath them. The Bible tells us a powerful truth about our thoughts: "As he thinks in his heart, so is he" (Proverbs 23:7). In other words, we become what we think. The things we think about, focus on, and surround ourselves with will shape the people we become. If a bird thinks he is a chicken, he will be fearful and go into hysterics when a storm comes, but if he thinks he is an eagle, he will be strong and courageous.

Let me ask you some important questions, and I really want you to

[handwritten margin notes at top: "Not so good lately in worthiness" / "Not like anything is possible" / "Loving, giving"]

ponder them and answer them honestly: What do you think of your-self? How do you feel about yourself? What is your attitude toward yourself? Do you respect yourself? Do you value yourself?

You have a relationship with yourself. You are with yourself more than you are with anybody else; you can never get away from your-self. You cannot ever go anywhere in the world without yourself, no matter how hard you try. This is why valuing yourself, liking your-self, and feeling good about who you are is so important.

Many cultures in the world suffer from a self-respect crisis, and I think one reason for it is that people don't take care of themselves because they don't value themselves enough. Some even believe put-ting themselves at the bottom of their priority lists is holy! They think God wants them to neglect their own well-being and just sac-rifice...sacrifice...sacrifice and serve everyone else.

Yes, we are called to sacrifice, to serve others, to give, and to not live greedy, self-centered lives. At the same time, we are also called to realize and embrace the truth that we are the people in whom God dwells. We belong to Him; He wants to use us; He wants us to be the kind of people who cause others to want to know Him—and to do these things, we must value and take care of ourselves.

You are God's house (see 1 Corinthians 6:19). You are His build-ing, His temple, and He lives in you! You will destroy the temple by being overcommitted, overworked, living on junk food, not drinking enough water, not getting enough sleep and rest, and failing to exer-cise. In that condition, you cannot be much good to yourself, to the people around you, or to God because you will not be healthy and energized. You will not be able to persevere through difficult times in life or to fully enjoy the good times.

I urge you today: start respecting yourself and taking care of your-self mentally, physically, emotionally, and spiritually. Discipline yourself to get enough rest, to eat healthily, to stay hydrated, and to exercise. You will be surprised how your mind changes, your emo-tions settle, and your body feels strong and healthy. To get started on this healthy lifestyle, let me recommend my book *Look Great, Feel Great*.

I believe you are created to be an eagle. God wants you to be an "eagle Christian," one who can fly high, be bold, live with power, keep circumstances and relationships in perspective, live at peace, stay strong, and soar above the storms of life. Begin to see yourself that way, because that's the way God sees you. Begin to value yourself, because He values you. Walk out of any "chicken mentality" you may have, and live like the eagle you were created to be.

DON'T GIVE UP WHEN YOU ARE ALONE

Most of us, when we think of an eagle soaring, envision a lone bird flying majestically against a bright blue sky. We don't think of eagles in pairs or in flocks. This is because eagles soar alone. They do not fly in flocks, as geese do, or in a covey, as quail do. In the United States, you can watch a congregation of swallows leave San Juan Capistrano, California, every October and then watch them return in March. These birds migrate south for the winter together and make a "group trip" back to California each spring. You might happen upon a bevy of larks, chirping or singing together, but you will not come across a bevy of eagles. They simply do not travel in groups. Something about the character, the strength, and the nature of the eagle gives it the courage to soar by itself.

The first lesson we need to learn if we want to be "eagle Christians" is that we will have to fly alone at times. This does not mean we cannot have relationships or be friends with people. It simply means we will often have to make and act upon decisions that separate us from the pack and enable us to reach higher heights than the people around us may want to go.

One time, I was complaining about being lonely: "I don't have any friends. I seem to spend all of my time alone. Everybody thinks I'm strange and different; they call me a fanatic. Some people in my family don't even like me anymore..." The Spirit of God spoke to my heart: *Just remember this: other birds fly in flocks, but eagles fly alone. What do you want to be, Joyce?*

If you are going to be an "eagle Christian," you may go through some lonely seasons in life. There may be times when you have to go it alone—perhaps as a parent or at work—and follow your convictions in spite of peer pressure. Maybe it's as small as being tempted to tell a white lie to your boss or succumbing to the pressure to "fudge the numbers." As parents, we are often called to stand alone in the best interests of our children. All of your son's friends may be allowed to play video games that you believe are too violent or to surf the Internet without supervision. It is not easy to take a stand against the opposition or your colleagues—and it's even harder when it's your unhappy child!

Maybe it's as simple as being the only one who does not attend a function or do something others are doing. For example, you might have to stay home and spend an evening with God while all your friends go to a party.

Going to the party may not be wrong in itself, but if you are going to be an eagle, you may have to skip the party because you feel you are tired and you know you need to rest. You may be a bit grouchy and realize that you need to spend some quality time with God so you can get your attitude renewed.

In all of these cases, you may have to take a stand and not be overly concerned about what people think or say. Setting limits with your children or with colleagues may be the harder choice. Spending the evening alone praying and studying God's Word may not seem exciting to your emotions. But in both cases, you and those you love will reap the benefits of wisdom and stability. You will also have the peace of knowing you obeyed God instead of merely doing what you felt like doing.

I need to spend the first couple of hours every day with God, because if I don't, I could be grumpy and short with people. But I want to be an eagle for God, so I take time with Him at the beginning of every day. This is what Isaiah 40:31 means when it encourages us to wait for God.

Being an eagle doesn't mean you will never be weary or feel stressed; it means you run to God and allow Him to renew your strength when you do feel weary or pressured. Many of us could avoid losing our tempers or saying words we later regret if we would simply spend some extra time with God when we realize weariness is setting in. This will help you keep going strong and not want to give up on yourself.

Don't stop short of knowing who you are, valuing who you are, and encouraging yourself to be and do everything God intends for you. Even when you feel like an eagle surrounded by chickens, don't quit believing in God's purpose for your life, and never give up on yourself. You may be well aware of your weaknesses, but don't be discouraged by them. Instead of allowing them to separate you from God, run to Him with them. He is the only One who can help you overcome them. We all have faults and flaws, so don't give up because of yours. Remember, God is not surprised by your faults. He knew about them before you did—and He loves you anyway!

Noble and Nobel

Marie Curie, also known as "Madame Curie," was the first woman to ever win a Nobel Prize—and she won two of them, a truly remarkable achievement.

Born Maria Sklodowska in Poland in 1867 to a piano player and a mathematics and physics professor, Marie demonstrated an interest in education at an early age and won academic awards for her school achievements and acclaim for her remarkable memory. She hoped to continue her education following her completion of secondary school at age sixteen, but her father lost much of his money through an unwise investment; so Marie had to go to work as a teacher, dashing her hopes for further study—at least temporarily. When she was eighteen, she took a job as a governess, during which time she endured a sad love affair.

Marie and her sister Bronia, both affected by their father's loss of money, agreed that Marie would use the earnings from her governess job to finance Bronia's education; and when her education was complete, Bronia would fund Marie's studies. Both sisters kept their promise to each other.

Marie traveled to Paris in 1891 to further her education in mathematics, physics, and chemistry at the world-famous university, the Sorbonne. With amazing dedication, Marie lived in a small, cramped student-housing facility, where she survived primarily on bread, butter, and tea.

At the Sorbonne, Marie met physics professor Pierre Curie, and the two married in Paris in 1895. Marie and Pierre committed their lives to science, particularly to the study of radium and polonium, which they isolated in 1898 and Marie named in honor of her native country. Though their research was brilliant,

(continued)

they had to conduct it in its early years under very poor laboratory conditions and were unable to support themselves with their work, so Pierre and Marie both had to spend hours teaching in order to make a living and provide for their two daughters.

Marie's findings earned her a doctorate in 1903, the same year she, Pierre, and Antoine Henri Becquerel received the Nobel Prize for physics. In 1904, she was able to leave her job as a physics teacher at a nearby girls' school and become the chief assistant in Pierre's laboratory. The couple worked diligently and intensely, and among other significant achievements, discovered radioactivity.

The couple's successful scientific partnership came to a tragic end in 1906, when Pierre was hit by a car and killed. Though his death had a profound impact on Marie, it also intensified her resolve to carry on the work the couple had begun. She was soon appointed to fill the teaching position her husband had held, professor of general physics in the faculty of sciences, and thereby became the first woman to teach in the Sorbonne.

In 1911, Marie Curie won a second Nobel Prize, this time for chemistry, and continued in her dedication to her work, particularly in finding therapeutic uses for radium in an attempt to apply it in medical situations to alleviate human suffering.

Marie Curie could have given up on her desire to pursue her education, decided not to suffer through the difficult laboratory arrangements under which she worked, and given up on her work in the wake of her husband's death. But she pressed through her challenges, delays, and difficulties—and contributed greatly to scientific advances that continue to affect and save lives today.

CHAPTER 3

REFUSE TO LIVE IN FEAR

"The only thing we have to fear is fear itself—nameless, unreasoning, unjustified terror which paralyzes needed efforts to convert retreat into advance."

FRANKLIN D. ROOSEVELT

The first African American licensed aviator was a beautiful, vivacious, and courageous young woman named Bessie Coleman, who came to be known as "Queen Bess." She was also the first American woman to earn an international pilot's license. Before she accomplished these remarkable achievements, Coleman dreamed of adventure in her hometown of Atlanta, Texas. When she was twenty-three years old, she finally ventured outside the confines of small-town life and moved to the big city of Chicago to be near several of her brothers and to pursue a life that had been beyond her reach in Atlanta, Texas.

In Chicago, Coleman worked as a manicurist in a barber shop, where she listened to stories about World War I pilots and their adventures in the sky. She dreamed of becoming a pilot herself, but could not afford to do so. At the barbershop, Coleman met two influential businessmen who wanted to help fund her training. One was a newspaper man who thought her story would be good for business. Because flight schools in America did not train black women, Coleman studied diligently to learn French and moved to Paris to attend flight school in 1920. Upon her return to America, she gained much media attention, and people of all races embraced her as they learned about her in the newspapers. Throughout her career, she participated in air shows and was invited to attend important events.

Bessie Coleman died in a plane crash, probably resulting from a wrench in one of the gears, on April 30, 1926. More than ten thousand people attended the funeral of this courageous young woman who refused to give up.

Bessie Coleman faced much discouragement and many obstacles because American flight schools refused to accept her and she had to put her dreams on hold while making the effort involved in learning a foreign language. She had to move alone to a new country, which took great boldness for the determined young woman. Even though her path to success was not as easy as those of her white counterparts, she found a way to do what she really wanted. It required extra time and effort, but she did not let fear or the uncertainties of moving to a foreign country keep her from pursuing her dream—because she was determined to not give up.

NO NEED TO FEAR

For many years, well-known advice columnist Ann Landers typically received approximately ten thousand letters per month, most of which had to do with people's problems and struggles. When asked what she thought was the biggest problem in people's lives, she responded with one word: fear.

Indeed, many of us struggle with fear. Let me begin by saying those of us who are believers in Jesus Christ have no need to fear. He is always with us. He loves us with a perfect love, and as 1 John 4:18 says: "There is no fear in love [dread does not exist], but full-grown (complete, perfect) love turns fear out of doors and expels every trace of terror!"

God knows the minute He calls us to do anything big or small that will take us to a new level—change careers, get married, have children, leave behind all that's familiar for a life on the mission field, go back to work after twenty years as a homemaker, get serious about diet and exercise—the first emotion that typically hits us is fear. Whether it's low-level nervousness or full-blown panic, most of us experience some degree of fear when we think of doing something new. Our minds are filled with excitement, yet we think, *What if this? What if that? What if? What if? What if?*

NEW LEVEL, NEW DEVIL

Any time we step out to do something new, especially something new for God, we will almost always—and almost immediately—face a negative circumstance that will try to discourage us or convince us we cannot do what God asks of us. Something will come up to tell us it is too hard; it does not make sense; it will not work; or we are not qualified to do it. The enemy uses such words and thoughts to discourage us, hoping fervently we will never accomplish what we set out to do.

Revelation 12:4 gives insight on how the devil tries to plant fear in us before we even get started. It says, "And the dragon stationed himself in front of the woman who was about to be delivered, so that he might devour her child as soon as she brought it forth." The woman mentioned in this verse was about to give birth, but the dragon (who represents Satan) had stationed himself in front of her so he could devour her newborn child.

This scene represents a dynamic that often happens in our lives as believers. Every time God puts a fresh, new idea in our hearts or gives us a dream, a vision, or a new challenge for our lives, the enemy will be there to oppose us. I am not referring simply to the times we want to do "big" things for God; the enemy is not selective. We face new levels often. When we receive Jesus as our Savior, that's a new level. When we begin to move into a deeper walk with

God, or begin to pray bolder prayers, those are new levels. When we begin to give time, money, and energy into God's kingdom, we are moving to a new level. God constantly calls us to new levels; some seem big and important, while others may seem relatively small or insignificant. Whatever the case, when we reach a new level, we face a new level of opposition from our enemy, the devil. The Bible tells us that opposition comes with opportunity (see 1 Corinthians 16:9), but God is always with us and we have no need to fear. Some things may seem too great for us, but nothing is impossible with God. He is not surprised or frightened by anything.

If we are going to do anything great for God and if we are determined never to give up on our dreams, we have to take chances; we have to be courageous. When we face situations that threaten or intimidate, we do not need to pray so much for the fear to go away as we need to pray for boldness and a courageous spirit. I can promise you, fear will not go away. It is not something we can get rid of, so we must learn to overcome it.

The spirit of fear will always try to keep us from going forward. For centuries, the enemy has used fear to try to stop people, and he is not going to change his strategy now. But we can defeat fear; we are more than conquerors through Him who loves us (see Romans 8:37). Courage is not the absence of fear; it is pressing forward while the feeling of fear is present.

THE RIGHT PERSPECTIVE ON FEAR

I teach often on Isaiah 41:10 because it encourages us and helps us know what to do when we are afraid. "Fear not [there is nothing to fear], for I am with you; do not look around you in terror and be dismayed, for I am your God. I will strengthen and harden you to difficulties, yes, I will help you; yes, I will hold you up and retain you with My [victorious] right hand of rightness and justice."

What does this scripture mean when it says, "Fear not...for I am with you;...I will strengthen and harden you to difficulties"? It means God makes us stronger and stronger as we go through things. It also means that over time, we become less affected by the difficulties and challenges we face. It is like exercise. When we first do it we get sore, but as we press through the soreness we build muscle and gain strength. We must go through the pain to get the gain.

Consider your life. Are there situations you now handle well that would have previously made you feel fearful and anxious? Of course there are. As you have been walking with God, He has been strengthening you and hardening you to difficulties. In the same way, I can also assure you and encourage you that some of the things bothering you right now will not affect you the same way in five years. I often feel fear when I do certain things for the first time. But after gaining some experience, that feeling is no longer present. We must press through feelings and never allow them to control us.

> *If God removed all fear, we would never grow and overcome obstacles.*

If God removed all fear, we would never grow and overcome obstacles. He often permits difficulty in our lives because He is trying to reveal something that needs to be strengthened or changed in us. Our weaknesses are never revealed in good times, but they quickly show up in times of trial and tribulation. Sometimes He shows us what we are afraid of because He wants to deliver us from that fear and strengthen us for things that will come in the future. In those times, we need to say, "Thank You, God, for allowing me to see that fear in my life. It reveals an area that needs to be dealt with in me." Once that particular area of fear is dealt with, then the enemy will have a very hard time bothering you—and succeeding— in that area again. This is one way God hardens us to difficulties and teaches us to not be afraid.

No matter what we may go through in life, we must always remember that God only does us good as long as we live (see Deuteronomy

34 NEVER GIVE UP!

8:16). Some things may not feel good initially, but they will work out for our good if we keep going forward and trust God each step of the way.

GOD SAYS, "I WILL BE WITH YOU"

The presence of God in our lives helps us overcome fear. If we know by faith that God is with us, we can take on any challenge with confidence and courage. We may not always feel God's presence, but we can trust His Word and remember that He said He would never leave us or forsake us (see Deuteronomy 31:6).

In Joshua 1:1–3, God called Joshua to a great challenge of leadership—taking the children of Israel into the Promised Land: "After the death of Moses the servant of the Lord, the Lord said to Joshua son of Nun, Moses' minister, Moses My servant is dead. So now arise [take his place], go over this Jordan, you and all this people, into the land which I am giving to them, the Israelites. Every place upon which the sole of your foot shall tread, that have I given to you, as I promised Moses."

The Bible simply tells us in this passage Moses had died and Joshua was going to take his place as the leader of God's people. As soon as God gave this news to Joshua, He immediately assured him: "No man shall be able to stand before you all the days of your life. As I was with Moses, so I will be with you; I will not fail you or forsake you. Be strong (confident) and of good courage" (vv. 5–6). Later in this same scene, God encouraged Joshua again, saying, "Be strong, vigorous, and very courageous. Be not afraid, neither be dismayed, for the Lord your God is with you wherever you go" (v. 9). Basically, God was saying to Joshua, "You have a big job to do, but don't let it intimidate you. Fear not. Do not be afraid because I will be with you."

In the Bible, the basis for not fearing is simply this: God is with us. And if we know God's character and nature, we know He is trustworthy. We do not have to know what He is going to do; simply knowing He is with us is more than enough.

Fear is a spirit that produces feelings. When God told Joshua to not be afraid, He was not commanding him to not "feel" fear; He was commanding Him to not *give in* to the fear he was facing. Perhaps you have heard my teaching, "Do it afraid." That basically means, when fear attacks you, you need to go ahead and do whatever God is telling you to do anyway. You may do it with your knees knocking or your palms sweating, but do it anyway. That's what it means to "fear not."

When I feel afraid I often meditate on scripture because it strengthens me to keep pressing forward, no matter how I feel. The Word of God has power inherent in it and will actually strengthen you on the inside. I have personally been encouraged quite often by Genesis 28:15: "And behold, I am with you and will keep (watch over you with care, take notice of) you wherever you may go, and I will bring you back to this land; for I will not leave you until I have done all of which I have told you."

Every man or woman who has ever been given the opportunity to do something great has had to face fear. What will you do when you are tempted to be afraid? Will you run, or will you stand firm, knowing God is with you?

FACING FEAR

One meaning of the word *fear* is "to take flight," so when we use the phrase "fear not," in a very real sense we are saying, "Don't run away from what frightens you. Face the situation; don't run from it. Don't try to hide from it; just meet it head on, even when you'd rather not."

One of the women who works at Joyce Meyer Ministries has a remarkable testimony. When faced with a life-threatening illness, she decided to not run from her fear. She was very fearful when she received her diagnosis, but she did not give up—she pressed through. It is my hope you will find encouragement to face your fears, whatever they may be, with God's empowerment until you see victory, just like Peggy did. Here's how she tells it.

After forty-eight years of living my life "my way," I found myself in the emergency room at a local hospital. My sisters took me there because they were concerned about the way I looked. I was pale; I had trouble sitting up; and I had shortness of breath because my abdomen was enlarged. On March 2, 1996, I was diagnosed with ovarian cancer.

I went through the surgery, a complete hysterectomy (during which surgeons removed a twelve-pound tumor), and then after a four-week recovery from the surgery I took chemotherapy treatments for six months.

When I first heard my diagnosis, I was naturally very scared; I thought I was going to die. The cancer was in late stage three, and I felt I had been handed a death sentence. My doctors were very concerned, and one of them told me some women live five to ten years after surgery and chemo. This was not all that encouraging to me, and it was then I decided to begin praying and evaluating my life. As is the case for so many people, it took a serious wake-up call to get my attention.

I attended church as a child (mostly Sunday school), and I did believe in God. Not long after I began praying, I noticed I had a sense of peace, and I developed a good attitude about my situation. It was hard to explain to my friends and family, but I knew I would be fine if I did not give up hope.

My chemo treatments ended in October 1996. I was put on disability and informed I could not work for at least five years. During one of my many follow-up visits to doctors, one asked if he could pray with me in his office; I said yes. He suggested a church for me to go to. I went three times and on the third time I received Jesus as my Lord and Savior! That was on March 2, 1997. Exactly one year to the day I had been given the "death sentence," I received eternal life! I have been serving God ever since, and I am completely cancer free!

Peggy learned to face her fear of death by turning her life over to Jesus Christ. As she did, she was healed and she learned a valuable lesson: running away from fear does not work, but courageously facing it with God's help brings hope and victory.

I also had a bout with cancer in 1989. I went to the doctor for a routine checkup and was told two days later that I had breast cancer and needed to have a quick mastectomy because the tumor they found was a very aggressive form of cancer. To say I was shocked and frightened would be an understatement, but God quickly reminded me of His Word and gave me courage to go forward. I actually thought of just ignoring it and going on with my life. (People often do that, you know, because it's a way of running from the problem rather than dealing with it. They just ignore things hoping they will go away. The problem is: those things don't go away.) Had God told me to not have surgery, I wouldn't have, but I actually felt in my heart that I was to have the operation, so I did. Like Peggy, I heard all the negative reports about the possibility the cancer could return. But that was nineteen years ago, and I am still cancer-free.

One young woman had a different kind of brush with death, but like Peggy, she never let the odds against her survival keep her from keeping on. Like most recent graduates, Juliane Koepcke must have been filled with excitement after graduating high school. The German native had finished her studies in Lima, Peru, where her father worked as a biologist. Juliane and her mother were seated side-by-side on a flight that would take them from Lima to join her father in Pucallpa.

When the plane carrying ninety-two people flew into an unexpected storm, Juliane noticed the right wing outside her window seat in flames. Her mother commented, "This is the end of everything." The next thing Juliane knew—and the last thing she remembers—she was tumbling through the air, still strapped to her seat.

Several hours later, Juliane awoke somewhere in the vast Amazon rain forest unable to see out of one eye and having suffered a broken collarbone and a deep cut in her arm. After releasing herself from her seat, she began searching for her mother. She found nothing more than some empty seats and the bodies of three young women. Juliane soon realized she was the only one who survived the crash.

Juliane was determined not to die in the rain forest, but had no idea where she was and no one to help her. In her traumatized state, she somehow remembered hearing her father say that a downhill direction in a jungle always leads to water, and water always leads to civilization. So she struck out alone, moving downhill all the way, bushwhacking her way through the rain forest. She heard planes overhead, but could not signal them from beneath the thick growth of the Amazon.

After ten grueling days, Juliane finally saw a small hut and found salt and kerosene inside—exactly what she needed to get the worms out of her skin. On the eleventh day after the crash, she met a group of hunters from Peru, and they took her to an airport where she was able to catch a plane to meet her father. Juliane refused to give up for eleven agonizing days, never knowing what she would face in the jungle. She went on to become a zoologist based in Germany.

THE FEAR OF FAILURE

The fear of death is just one of many fears we may have. The truth is, we have all kinds of fears—everything from fear of heights to fear of spiders to fear of public speaking. In the remainder of this chapter, I want to talk about a different kind of fear that affects the core of our beings—the fear of failure.

There are many reasons why the fear of failure gets a stranglehold on us. It could be rooted in the fear of what other people think of us.

We do not want others to think negatively of us, or maybe we do not want to live with bad thoughts about ourselves so we don't even try to do things, thinking we are protecting ourselves from feelings of failure. Whatever the case, the fear of failure can be so strong that it can keep us from stepping out and taking on challenges and risks.

If you want to be a courageous person who never gives up, you need to squarely face this kind of fear. Fear of failure will keep you from going to the next level in your life, or it will prevent you from even taking the first step toward your goal. It will hold you back from doing and being everything God has for you if you do not confront and conquer it. Always remember, when you take steps to face your fear of failure and deal with it: God is with you! Also, remember that people never fail unless they stop trying. When we do make mistakes along the way, we can always learn from them and allow them to add wisdom to our lives. We can "fail forward"!

What Is It?

The fear of failure is crippling because it holds people back from acting on their desires, and it will certainly hinder you from fulfilling your destiny. The fear of failure is nothing more than the fear we will be unable to do what needs to be done in a specific situation. In other words, we will not succeed at something we attempt to do. This fear is especially active in us when we think of doing something we have never done before. We are afraid we will fail, so we fail to try. When I look at my own capabilities apart from God, I am afraid of everything, but if I keep my eyes on Him I am assured that *through Him* I can do whatever He asks me to do (see Philippians 4:13).

CREATED FOR ADVENTURE

The truth is, we were never created to do the same things all of our lives. God has put a craving for adventure in us, and adventure means trying something we have never done before. Adventure means stepping out, doing something different, doing something a little bit on the edge, and not always living in a zone that we consider "safe."

I assure you, if you are supposed to do something, God will make you able to do it. You do not have to *feel* able, and you do not have to have experience. All you need is a right motive and a heart full of faith. God is not looking for ability; He is looking for availability. He is looking for somebody to say, "Here I am, God, send me. Here I am, use me. I want to serve You, God. I want to do something for You!"

I remember a particular Sunday in my church many years ago. That day I had to confront a dreadful fear of failure that could have kept me from going deeper with God. At that time of my life, I truly loved God, but there were so many biblical truths and principles I did not know. I knew a little bit about what a person needed to do to be saved, but I did not understand victory, overcoming obstacles, power, authority, or being used by God. I had no real hope that my life would ever be any better than it was at the time.

In that church, we had missions Sunday once a year, and on that day, we always sang the song "Here Am I, Send Me." I remember one specific time, something welled up in me from the bottom of my heart and I sang the words with every fiber of my being: "God, here I am! Send me! Send me!" I do not know where I thought He would send me because I had a husband and three small children; I was abused as a child and spent my time simply trying to survive every day. But in my heart, I sensed I wanted Him to use me. I may not have had a lot of ability, but I was available to God. I was willing to say, "I may fall flat on my face, God, but if You want to use me, I am willing to try."

If you want God to use you, do not let the fear of failure stop you from obeying Him as He leads you. God not only sees where you are, He sees where you can be. He not only sees what you have done, He sees what you will do with His help.

Following God is often like walking in a fog. We can only see one or two steps in front of us, but as we take those steps the next ones become clear. With God, we do not always see clearly ahead into the distant future, but if we are willing to trust Him we will have an exciting journey that will make life adventurous and worth living.

Failure or Stepping-Stone?

No one sets out to fail or wants to fail. But as I wrote in the introduction to this book, I do believe "failure" can be an important stepping-stone on the way to success. Failure certainly teaches us what *not* to do, which is often as important as knowing what we *are* to do! So-called failure is all about how we look at it.

Many stories have circulated about how many times Thomas Edison failed before he invented the incandescent light bulb. I have heard he tried seven hundred times, two thousand times, six thousand times, and ten thousand times. No matter how many attempts he made, the number is staggering. But he never gave up. Edison is reported to have said in all his efforts he never failed—not once; he just had to go through many, many steps to get it right! It takes that kind of determination if you are really going to do anything worthwhile.

The man who led IBM to worldwide prominence, former president Thomas J. Watson, had no fear of failure. In fact, he had this to say about it: "Would you like me to give you a formula for success? It's quite simple, really. Double your rate of failure. You are thinking of failure as the enemy of success. But it isn't at all. You can be discouraged by failure—or you can learn from it. So go ahead and make mistakes. Make all you can. Because, remember, that's where you will find success."

Giving in to the fear of failure will surely keep you from reaching your full potential in life. The good news is, you have no reason to fear failure. First of all, God is with you. And secondly, there is no such thing as failure if you simply refuse to quit.

Every time you are tempted to fear, remember God is with you. He will not fail you or forsake you. He is your God; He will help you and hold you in His hand. He is hardening you to be able to face difficulties. He is building in you the strength, stability, and character you need to press through to the good things He has in store for you, and He is building in you the courage to never give up.

I have often pondered why some people do great things with their lives while others do little or nothing at all. I know that the outcome of our lives is dependent not only upon God but also upon something in us. Each of us must decide if we will reach down deep inside and find the courage to press past fear, mistakes, mistreatment at the hands of others, seeming injustices, and all the challenges life presents. This is not something anyone else can do for us; we must do it ourselves.

I want to encourage you to take responsibility for your life and its outcome. What will you do with what God has given you? Are you going to invest your talent and time or hide it because of fear? I truly believe God gives everyone equal opportunity. He said, "I have set before you life and death...choose life" (Deuteronomy 30:19). *Fear is in the category of death, faith and progress fills us with life.* It is your choice, and I believe you will make the right one!

Wired for Success

Alexander Graham Bell, inventor of the telephone and founder of the Bell Telephone Company was smart, curious, hardworking, and persistent. These qualities contributed greatly to the success he enjoyed as a result of his inventions, but the real impetus behind his great inventions was the simple fact that he wanted to help people who lived with challenges and difficulties.

Bell's mother, Eliza Grace, was hearing impaired, and he wanted to find a more effective way to communicate with her than through her ear tube, as other people did. Instead, Bell spoke to her by placing his mouth close to her forehead, believing she could sense the sound waves from his resonant voice. His mother's deafness and his father's work in speech pathology and lifelong interest in helping the hearing impaired communicate gave Bell an interest in helping people communicate too.

A native of Scotland, Bell traveled to London to pursue university studies. While in England, his two brothers contracted tuberculosis and died. Bell also contracted the dreaded disease, and in an attempt to save his life, his parents moved themselves and their last remaining son to Canada. Even though his time at the University of London was cut short, it was there that he became interested in the work of a German physicist named Hermann von Helmholtz. Bell could not read German, but did not let that stop him from finding ways to learn what Helmholtz taught. One of his assertions, which captured Bell's attention, was that electrical tuning forks and resonators could be used to make vowel sounds. Without sufficient training in German, Bell misinterpreted Helmholtz's theory and thought the physicist believed sound could be transmitted over a wire, which he

(continued)

eventually accomplished. Speaking of this erroneous interpretation, he later said, "It gave me confidence. If I had been able to read German, I might never have begun my experiments in electricity." Those experiments in electricity, of course, led to the invention of the telephone.

Bell is best known as the inventor of the telephone, but he also invented several other important devices including a precursor to the iron lung used to help polio patients in the 1950s. This apparatus, too, came about as a result of personal hardship in Bell's life. In 1881, his newborn son Edward died of respiratory failure. In response, Bell designed a metal jacket that would help those with respiratory problems breathe easier. That same year, United States president James A. Garfield was shot and Bell used his abilities to try to save his life by quickly cobbling together a metal-detecting device to find the location of the bullet. Though the president died of his wounds, Bell did his best to help save him.

Let me encourage you to be like Alexander Graham Bell and let the hardships you experience and observe inspire you to be creative. Never give up as you seek to make the world a better place and to improve the lives of those around you.

CHAPTER 4

CONFRONTING YOUR FEARS

*"Put an eagle in a cage and it will bite the bars,
whether they be of iron or gold."*

IBSEN

Chickens often stay in the same chicken yard for their entire lives, but eagles learn to fly to new heights. As you continue on your journey toward being an eagle Christian, I want to call your attention to seven fears you must deal with in life. If you do not deal with these fears, they can keep you in the chicken yard, but as you confront them and overcome them, you will be well on your way to becoming an eagle. Never give up; face your fears with God's help and leave the "chicken mentality" behind!

SEVEN FEARS YOU MUST CONFRONT

1. *Fear of What People Think*

If you want to break free from where you are now and move to the next level in life, the first thing you must deal with is the fear of what people think. So many of our decisions are made to please people, not to please God. We fail to follow our hearts, but we live to please people around us because we want their acceptance and approval. We want them to like us; we want to be part of the group; we don't want them to judge us, criticize us, or speak negatively about us.

The apostle Paul dealt with this issue in Galatians 1:10: "If I were still seeking popularity with men, I should not be a bond servant of Christ." In other words, had he chosen to care about what people thought of him or sought to be popular, he could not have served the Lord the way he did. The same may be true in your life.

Wanting to please people is a fairly natural desire, but if you are out of balance and have an excessive need for people's approval, the devil can use that to steal your destiny. You will never be all God wants you to be if your goal is to please people.

I discovered that the people in my life who didn't want to make a full commitment to serving God tried to keep me from doing so. People usually try to get us to do what they are doing, rather than giving us the freedom to make our own choices. We must not give in to their pressure because eventually each of us will stand before God and give an account of our lives. We won't answer to people; we will answer to God, so we should live to please Him and Him alone.

I believe you really want to go on with God; you have a hunger to go deeper in Him; you want to be an eagle. But to do that, you will have to take the chance that some of your friends or co-workers may not understand you. Let's face it: reading your Bible during your lunch break or taking a walk and praying instead of gossiping may not be popular behavior in the environment where you live or work. People may ridicule you, but the reason for this is that your good choices cause them to feel convicted. They may call you "holier than thou" or say you have become a religious fanatic who thinks you are better than everybody else. They may reject you or speak unkindly about you, but if you know your heart's desire is simply to love God and serve Him with your whole heart, then keep pressing forward and don't let anything stop you.

As I mentioned, most of the time, the people who are upset with you for wanting to move on with God are those who do not want to go on with Him themselves. As soon as you find one other Christian who shares the same values you have and wants to participate in the same activities in which you participate, you have a new friend. God is faithful. When you want to go on in Him, He will give you new friends and surround you with people who are like-minded. I cannot promise you will not have to go through some lonely times first, but I do know God will provide the friends and companionship you need. I have certainly been through lonely times in my life; if you feel lonely, don't be discouraged because things will eventually change.

John 12:42–43 depicts a very sad situation, one that characterizes so many people's lives.

And yet [in spite of all this] many even of the leading men (the authorities and the nobles) believed and trusted in Him. But because of the Pharisees they did not confess it, for fear that [if they should acknowledge Him] they would be expelled from the synagogue; for they loved the approval and the praise and the glory that come from men [instead of and] more than the glory that comes from God. [They valued their credit with men more than their credit with God.]

The men in this passage missed having a deep relationship with God because they cared more about what people thought than about what God thought. Please don't be this way! If you struggle with the need for approval and acceptance, I encourage you to read my book *Approval Addiction*. Do whatever you must do to reach the point where you care more about what God thinks than about what your family, friends, neighbors, or colleagues think. Seek to please Him and you can soar with the eagles instead of scratching around the chicken yard with the chickens.

2. Fear of Criticism

A companion to the fear of what people think about us is the fear that they will criticize us. We want the people around us to think we are smart, wonderful, and always doing the right thing. We don't like any kind of criticism or judgment, nor do we enjoy knowing someone is speaking negatively about us. We don't have to like these things, but we cannot allow ourselves to fear them.

When we suffer from self-doubt, we rely on the approval of others because that makes us feel that we're doing the right thing. Insecurity and the fear of making mistakes causes many people to make most, if not all, of their decisions based on what others think. But what makes us think that they have the answers and we don't? When we are truly confident and secure, the opinions of others cannot control us.

> *When we are truly confident and secure, the opinions of others cannot control us.*

My daughter Sandra suffered with insecurity and perfectionism for many years. I remember her asking excessively for other people's opinions and becoming very confused when they did not agree with her ideas about what she should do in certain situations. I often saw her change a decision to agree with others even though she still felt in her heart that her original decision was correct.

Thankfully, she has received healing from God in this area. Although she is very open to asking for advice, she is secure enough to ultimately do what she feels is right. She is much happier and has the peace of knowing she is following God to the best of her ability.

Many people suffer from insecurities like Sandra's and from other types of insecurity. If that includes you, I have good news: God will set you free. I encourage you to begin confronting the fear of being wrong and believe the Spirit of God lives in you to give you wisdom and to lead you in every situation.

When I think of people who overcame the fear of criticism, I think of Dr. Martin Luther King Jr. When he began to speak out against segregation and the unfair treatment African Americans received in the United States, critics came from many corners, some of them very vicious. In the face of criticism, cruelty, threats, intimidation, and imprisonment, King stayed true to his cause. When an assassin's bullet killed him on April 4, 1968, it struck a man who was determined to never give up. Because King confronted and conquered the fear of criticism, the social changes for which he labored eventually became reality and resulted in great changes in American society.

I believe anyone who does anything significant must first overcome the fear of criticism. We simply cannot follow God and keep all the people happy all the time. To do great things, we must accept that a certain percentage of people won't agree with us—and we can't let that stop us.

3. Fear of Not Pleasing God

If you want to go higher in God, you must also address the fear that your imperfections will keep you from pleasing God. This is a significant struggle for many people; they suffer terribly with secret worries that they are not pleasing to God, they are not doing enough, or they are just not doing things correctly. They're afraid God is displeased with them because they don't read the Bible enough, they don't pray properly, or they don't have enough faith. They live in bondage to the fear that nothing they do is ever enough for God. This is a lie! The devil wants us to believe "enough is never enough." He wants us to feel no matter what we do, God will always want something more or something different.

This is very sad, because the truth is that God isn't nearly as hard to please as we think He is! Yes, He has high standards and desires excellence in every area, but the truth is that nothing about us surprises Him. He knew what He was getting when He called us into relationship with Himself. When we make mistakes, He doesn't sit in heaven wringing His hands saying, "I am shocked by your behavior. I never expected that!" Psalm 139 tells us plainly that He knows what we are going to do before we even do it, so try to keep in mind that God knows *all* about you and loves you anyway!

Read this next statement closely: *If you could be perfect, you would not need Jesus.* Your willingness to keep pressing toward the mark of perfection is all God requires. If you are really in love with Jesus, your love for Him will keep you pressing toward perfection every day. But I can almost guarantee you will also fall short in some way every day. If you really are a sincere Christian, there is no way you can take a lighthearted, "Oh well, I don't care" attitude toward your shortcomings. Your love for Jesus will compel you to want to do better every day. You will get up the next day and try to do better again; maybe you will do better in some area, but you may fail in something else. The message God is trying to communicate to you through these situations is: "You still need Me."

> *If you could be perfect, you would not need Jesus.*

Jesus says in Matthew 5:48, "You, therefore, must be perfect [growing into complete maturity of godliness in mind and character, having reached the proper height of virtue and integrity], as your heavenly Father is perfect." When we think of being "perfect," we can feel intimidated, but we need to understand this word's meaning in its original New Testament language. The Amplified Bible makes this clear. *Perfect* means simply "growing into maturity." We need to rejoice and be happy when we are growing into maturity and stop criticizing ourselves when we are not perfect by human standards.

I want to encourage you to enjoy your journey. After all, life isn't as much about reaching our destination as it is about how we make the trip. People who struggle and are miserable all their lives but finally reach their destination don't accomplish nearly as much as those who may not arrive at their destinations, but enjoy the process.

You may have heard me say: "I'm not where I need to be, but thank God, I'm not where I used to be. I have not arrived, but I am on my way!" That kind of attitude infuriates the devil. Let me encourage you to not set a standard of worldly perfection for yourself. Instead, start getting excited about every step of progress you take. When you are not exactly where you want to be, remind yourself that neither are you where you used to be! Instead of focusing on your faults, focus on your progress. The very fact that you are reading this book shows you are making progress.

> *When you are not exactly where you want to be, remind yourself that neither are you where you used to be!*

You will have to stay focused as you confront and break the fear of not being pleasing to God. You can't allow your faults to distract you from going on with Him. He is the only One who can help you overcome your faults and weaknesses, so focus on Him instead of on them. Hebrews 12:2 teaches us a powerful lesson about staying focused: "Looking away [from all that will distract] to Jesus, Who is the Leader and the Source of our faith." The devil is the author of distraction, and he intends your distraction for your destruction. The devil wants you to be distracted because he knows if you keep your eyes on Jesus, then He becomes not only the author and source, but the finisher of your dream, your vision, and your faith.

Remember, you are no surprise to God. Your faults and failures do not catch Him off guard. He never looks at me and says, "Oh, Joyce! I didn't know you were going to be like *that*! Had I known you were going to do *that* I would not have allowed you to be a Bible teacher!" Read Psalm 139. Every day of your life was written in His book before you ever showed up on planet earth. God knows every word in your mouth, even the ones you have not spoken yet. Your shortcomings do not shock Him. He covered them at the cross with the blood of His Son.

If you really love God, if you are *in* love with Jesus, you will not want to sin or displease God. You will do everything you can to stay away from sin. But you will sin, and you cannot let that distract you or hinder you from moving on with God, because *when* you sin, you have an advocate with the Father. Jesus Christ, the all-righteous One (see 1 John 2:1), is faithful and just to forgive you for all of your sins and to cleanse you from all unrighteousness (see 1 John 1:9). Do not allow your sins, your faults, or your failures to distract you or cause you to believe you are not pleasing to God. If you are in Christ, you please Father God completely because He sees who you are in Jesus, not who you are on your own merits. Remember, we are saved by grace, not by our own works. Salvation is not the result of anything we could possibly do; it is a gift from God to be received by faith (see Ephesians 2:8–9).

4. Fear of Making the Wrong Decision

In order to move forward as an eagle Christian, the fourth fear you need to deal with is the fear of making wrong decisions, which is otherwise known as self-doubt. I do not believe it is God's will for us to doubt ourselves; He wants us to have assurance and confidence.

I once made a decision a handful of people did not like; they did not think I made a good choice. There were also people who liked my decision and believed it was right. But I could not help thinking about the naysayers, and I began to wonder, *Did I do the right thing? Maybe I didn't do the right thing.* Then I talked to God about it: "Lord, did I do the right thing?" He affirmed me, saying I had made the right decision.

Just as you may have had to do, I have had to learn not to listen to self-doubt. God still reminds me from time to time to be careful in this area because Satan is always attacking us with all kinds of fears and doubts. I have learned to not merely believe my feelings, but instead to check my heart and see what I truly believe God is saying.

It is amazing what we will find in our hearts if we simply take time to listen.

If we question ourselves excessively or carry on internal debates about whether or not we have made good decisions, we can all get lost in self-doubt and quickly become double-minded. James 1:7–8 teaches us that people who are double-minded are unstable in all their ways and should not expect to receive anything from God. Also, when we become extremely double-minded, we become confused. When we are really confused, we become discouraged and depressed. Pretty soon, we find ourselves back in the chicken yard scratching around with the chickens when we should be soaring with the eagles.

Don't allow yourself to wallow in self-doubt and become double-minded. If you think you have made a mistake, go to God about it. Say, "God, if I made a mistake, I trust You to show me. I want to know if I made a mistake or a bad decision. Please show me if I'm wrong." Then take some time, get quiet, and see what you sense in your heart. Remember, you will not hear wise direction by listening to your own mind, your emotions, or other people. *Listen to your heart!* As a Christian, the Spirit of God lives in you—in your heart— and He will lead and guide you.

Don't assume you are wrong simply because a handful of people disagree with your decision. If all the devil needs to do to get you off track is find someone to disagree with you, then he is definitely in the driver's seat where your life is concerned. Make sure you stay focused and single-minded, free from self-doubt as you continue to become an eagle Christian.

I've found that doubt often tries to visit me when I am praying. This vague feeling leaves me wondering if God is hearing me and will answer when I should be trusting that He is listening and will respond. Prayer is our greatest privilege and is necessary for God to act in the earth. The Bible tells us we have not because we ask not (see James 4:2), but it also says we must ask in faith with no wavering, hesitating, or doubting (see James 1:6). God works through

ar. Maintaining an attitude of faith requires con-
our part. We can decide to believe or doubt; the
us.

5. *Fear of Missing God*

The next fear we must confront if we are going to be determined to never give up is the fear of "missing" God, of doing something "wrong" as we try to follow Him. All I have to say is: if you miss God somehow, don't worry; He'll find you.

Part of the fear of missing God (making a wrong decision) includes the fear of deprivation. Often, when God puts a dream in our hearts and calls us to do something, one of our first questions is: "What happens if I step out and try to do this and it wasn't God's will? What if He doesn't provide for me?"

If you are afraid of missing God, all you have to do is remember how much God loves you and look at the testimonies of the people who have gone before you. The Bible is full of miraculous stories of God's guidance and provision. He is the God who caused manna to fall from the sky every day for forty years so the children of Israel would have something to eat! If you want to read more remarkable stories, libraries and bookstores carry all kinds of books that chronicle God's leading and provision for people throughout history.

One misconception Christians often have in this area is the idea that we must take upon ourselves all the responsibility of never making a mistake. In reality, the whole point of God's love and grace is that if our hearts are right toward God and we are doing the best we can, when we take a step of true faith after praying and seeking God, if we *do* make a mistake then God will help us get back on track. God has never left us, and He can make that so-called mistake somehow work out for our good. We cannot be afraid of missing God, or we will never be able to take the first step in following Him.

The truth is, everyone makes mistakes while trying to learn to follow God, but that's not the end of the world! Don't be so afraid of failing that you never try. I would actually rather try and then make a mistake than to live my life without ever attempting anything. God likes an aggressive attitude. The fear of missing God is actually the fear of failure; do not allow that fear to influence your decisions. "Step out and find out" is my motto. You cannot drive a parked car! Get going in some direction; if you make a wrong turn you can always go around the block and then head the other way.

6.   *Fear of Change*

Former U.S. president Woodrow Wilson said: "If you want to make enemies, try to change something." For some reason, many people don't like change, and I believe it is often because they fear it.

Many times, when we grow weary or simply become bored with a situation, we get restless and begin to pray: "Oh, God! Something has to change!" Then, when God tries to bring change into our lives, we say, "Lord, what are You doing? I don't think I can take this change!" We often find ourselves caught in the tension between wanting change and fearing change.

Thank God, He never changes. Because He is always the same, we can trust Him through any changing circumstances or situations. Hebrews 13:8 says: "Jesus Christ (the Messiah) is [always] the same, yesterday, today, [yes] and forever (to the ages)." And God Himself speaks in Malachi 3:6, saying: "For I am the Lord, I do not change." This should give us great courage and comfort when we face changes in our lives. We do not need to fear change; we can handle it, because God remains the same.

What if Abraham Lincoln had been afraid of change? He would not have signed the Emancipation Proclamation and the horror of slavery would have continued in America. What if Orville and Wilbur Wright had been afraid of change? We would all be stuck with traveling on the ground or sailing over the water instead of flying through the skies. What if Thomas Edison, whom I mentioned earlier, had feared change? We would still be reading by candlelight after sundown.

I share these vignettes with you to inspire you to think about the many good changes that have made people's lives better and easier through the years. I'm sure you could make your own list of positive changes that have occurred in your life. Remember, the fear of change could keep something wonderful from taking place for you. I urge you to not let that happen, but to boldly embrace the changes that come your way.

Without change there is no growth, and anything that is not growing is dying. Change is a fact of life and the more we fear and resist it, the more we are resisting God. As I noted earlier, God Himself does not change, but He definitely changes people, circumstances, and things.

> *Change is a fact of life and the more we fear and resist it, the more we are resisting God.*

I began to notice a few years ago that the people attending our conferences were all my age or older. I started asking myself where the younger generation was and finally had to realize that I was not offering them anything to which they could relate. I determined that I would make whatever changes I needed to make in order to be relevant to the younger generation because they are the future. We can never change God's Word because it always remains the same, but we can change the way we package it and that is exactly what we did. I changed my clothing style to be in line with today's fashion. We changed the look of our conferences by adding different music styles, updated media presentation, and different lighting techniques. Although these things don't change the Word of God, which is the main thing we are trying to present, they do draw people in. By simply being willing to change some of these exterior things, we now enjoy a congregation of all ages and are delighted to see that a "generation gap" really does not have to exist if we are willing to change some of our old ways.

We all have a tendency to think according to the generation in which we were raised, but we must realize that things are always changing. If we are not willing to change, God will not stop to wait for us and we will wonder why we missed the things He is doing today.

7. Fear of Sacrifice

When we believe God is asking us to do something, we begin with the questions: *What am I going to have to give up if I do this? If I do this, what will it cost me? If I do this, how uncomfortable am I going to be?*

The truth is, anything we do for God requires an investment. Part of loving Him involves a willingness to lay our lives down for Him. Our willingness to sacrifice is part of the test He often puts us through, not so He will know if we really love Him and are serious about following Him, but to prove His love and commitment to ourselves and get it established in our own hearts.

More than thirty years ago, when God filled me with the Holy Spirit and called me into ministry, I lost almost all of my friends, was asked to leave my church, and fell out of favor with many members of my family. I was embarrassed; I was rejected; and it was difficult to press through those experiences without giving up. But it was all part of the price I had to pay to go on with God.

Do not be afraid of sacrifice when God calls you or puts a dream in your heart. Determine you will pay the price and pass the test. I assure you, it is worth it.

Fear of failure; fear of what people think; fear of criticism; fear of not pleasing God; fear of making wrong decisions; fear of missing God; fear of change; fear of sacrifice. Leave these fears in the chicken yard; they have no place in the sky with an eagle. You may be afraid to move away from your chicken yard because it means you have to leave the security of what is familiar to you. I urge you to never give up as you face your fears; spread your wings and fly with courage and boldness.

The Ultimate Overcomer

One of the best-known "overcomers" in American history would have to be Helen Keller, who refused to allow the blindness of her eyes and the deafness of her ears to keep her from enjoying her life, achieving remarkable goals, and making significant contributions to society.

Until she was nineteen months old, Helen Keller's life was relatively normal. But everything changed when she became very ill with what was then called "brain fever." Many people now believe she suffered from scarlet fever, but regardless of her diagnosis, the disease left her severely handicapped, completely destroying her ability to see and hear.

In an effort to help her, her parents took her to see Alexander Graham Bell, who gained fame as inventor of the telephone, but was also a knowledgeable teacher of the deaf and helpful resource for hearing-impaired individuals. Bell referred the Kellers to Boston's Perkins School for the Blind.

The Perkins School provided Helen with a tutor named Anne Sullivan, whom Helen affectionately called "Teacher" until Sullivan's death. As a child, Helen was extremely frustrated by her handicaps and often became violently angry. But Sullivan remained patient and diligent, and over time taught Helen to read Braille, to write, and to communicate effectively despite the fact that she was blind and deaf. To learn more about this, read the play *The Miracle Worker* by William Gibson or watch the movie by the same name.

As a young child, Helen knew she wanted to go to college and was determined not to let her handicaps deny her that privilege. After studying hard and preparing diligently, she was admitted to Radcliffe College, the women's school associated with Harvard

University, and began her studies there, with Anne Sullivan's help, in 1900. Four years later, she had mastered several languages and graduated with honors.

In college, Helen embarked on a writing career that would last half a century and include books, newspaper articles, and magazine pieces. Her first book, *The Story of My Life*, was published in 1903 and eventually found its way into print in more than fifty languages.

Before the end of her life in 1968, Helen received numerous honorary degrees and awards for her humanitarian work and the inspiration she provided for the blind and the deaf, including the prestigious Presidential Medal of Freedom. Perhaps more than that, though, she enjoyed her many opportunities to meet dignitaries such as Oliver Wendall Holmes and every U.S. president from Grover Cleveland to Lyndon Johnson.

One of the primary reasons Helen Keller never gave up on herself was that Anne Sullivan never gave up on her. Throughout this book, I have urged you to not give up on yourself, and I would encourage you in that way again now. But I also want to make sure you remember to do everything you can to inspire others to persevere through their difficulties and overcome their challenges, just as you do.

CHAPTER 5

IT'LL COST YOU!

"The block of granite which was an obstacle in the pathway of the weak becomes a stepping-stone in the pathway of the strong."
THOMAS CARLYLE

You may have heard the name Wilma Rudolph. Perhaps you know of her as an outstanding female athlete whose story was told in the television movie *Wilma* and whose image appeared on a United States postage stamp. Indeed, this phenomenal runner, coach, and sports commentator gained remarkable recognition and acclaim, but do you have any idea what it cost her to achieve her goals and live her dreams?

She was born prematurely, the twentieth of twenty-two children in a poor family reeling from the impact of the Great Depression. In addition to suffering scarlet fever and double pneumonia as an infant, she was afflicted with polio from birth and had to be driven fifty miles twice per week to a hospital for blacks, where she received treatment. In addition, her mother had to rub her legs four times each day for years. At one point, doctors told her she would never walk, but eventually she proved them wrong and walked with a brace until she was nine years old. By the time she was twelve, she could finally walk normally. Within four short years, at age sixteen, she won her first Olympic medal. Four years later, at the 1960 Olympics in Rome, she earned gold medals in three events and became an international superstar.

The little girl who was told she would never walk ended up learning to run—better than any other woman in the world. But she did not do it without pain, without sacrifice, without discipline, or without paying a price. She wanted to see her desires fulfilled—and she did because she never gave up.

To be people who never give up, we must learn to press past both natural obstacles, as Wilma Rudolph did, and unseen forces such as anger, unforgiveness, guilt, shame, and fear. Over the next several chapters, I will be writing at length about overcoming these issues so you can press through anything that is holding you back and go on to be everything God intends for you to be. Press in, press on, press through, and pay the price now so you reach your goals and your potential.

GOING THROUGH

We will all go through situations in life. Many times, we think the phrase "I'm going through something" is bad news, but if we view it properly, we realize *going* through is good; it means we are not stuck! We may be facing difficulties, but at least we are moving forward.

Isaiah 43:2 says: "When you pass through the waters, I will be with you, and through the rivers, they will not overwhelm you. When you walk through the fire, you will not be burned or scorched, nor will the flame kindle upon you." God's Word here is clear: we *will* go through things. We *will* face adversities in our lives. That's not bad news; that's reality.

Let me repeat: we will go through things in life, but the things we go through are the very circumstances, challenges, and situations that make us people who know how to overcome adversity. We do not grow or become strong during life's good times; we grow when we press through difficulties without giving up.

You may remember the tragic day in April 2007 when a gunman opened fire at Virginia Tech, senselessly killing thirty-three bright, young, promising students and himself. One of the students killed that day was Caitlin Hammaren, a vibrant sophomore and the only child of two doting parents. Caitlin's mother, Marian, had to find the will and determination to go on after such unthinkable loss.

After Caitlin's death, authorities gave her laptop to Marian and her husband Chris. They opened it and saw a note taped above the screen: "God, I know that today nothing can happen that you and I can't handle together." Marian knew Caitlin believed those words wholeheartedly, but she could not believe them for herself—not in the midst of the greatest tragedy of her life.

In the weeks and months that followed Caitlin's death, Marian struggled to make peace with her loss and find the strength to live without Caitlin, who had been the center of Marian's life. She devoured books on spirituality and began reading God's Word after many years without having done so. All along the way, the words from the note on Caitlin's laptop continued to surface in Marian's thoughts. But about four months after the Virginia Tech massacre, they became very personal to her. She heard them as though they were being spoken directly to her and she was able to believe them for herself.

Marian wrote in *Guideposts* magazine: "Caitlin's passing had brought me out of my old world and into a new one. A world where things can go wrong—more wrong than I'd ever imagined they could. But with the Lord's help, there was nothing—absolutely nothing—that could happen to me that I couldn't get through."

Today, Marian and Chris still struggle with the loss of their only child. The first Thanksgiving and Christmas after Caitlin's death was extremely difficult, and they mourn the many milestones that they'll never celebrate with Caitlin. But they also are gaining strength every day, and their relationship with God has taken on dimension and depth. Chris is seriously considering becoming a chaplain in the military, and Marian uses her own pain to give others support and love in the face of their own painful situations.

Suffering and tragedy are part of life. The Bible even tells us we will suffer and encounter trials, so we must learn how to deal with them in ways that help us overcome them instead of letting them overcome us. I like the note on Caitlin's laptop and encourage you to take it personally: "God, I know that today nothing can happen that you and I can't handle together."

We need to be balanced in our approach to growth and adversity because growth is not an automatic result of difficulty. Hardships do not necessarily produce growth or strength in us; it is not that simple. Some people shrink back or fall apart when difficulties arise. Like Marian Hammeran has learned, we grow when we determine in the midst of adversity to think and behave in godly ways. When we do what we know is right when it is difficult, uncomfortable, or inconvenient, we grow spiritually, and we are strengthened. We may have to do what is right for a long time before we feel it is "paying off," but if we stay faithful and refuse to give up, good results will come. Once we get through the adversities and challenges we face, we emerge as better people than we were when we went into them.

Determine that you will go all the way through every difficulty you might ever face in life. Make a decision now to keep going forward, no matter how difficult it is. I guarantee you'll be glad you did.

This verse has given me courage and comfort through many of my own battles in life and I believe it will also encourage you: "The Lord God is my Strength, my personal bravery, and my invincible army; He makes my feet like hinds' feet and will make me to walk [not to stand still in terror, but to walk] and make [spiritual] progress upon my high places [of trouble, suffering, or responsibility]!" (Habakkuk 3:19).

MAKE THE RIGHT CHOICES ANYWAY

An important part of never giving up is to make right choices while you are hurting, discouraged, frustrated, confused, or under pressure. The right choice is often the harder choice. And when we're in the middle of terrible stress, we naturally want to take the path of least resistance. But those are exactly the moments when you need to discipline yourself to make the tougher choice. In order to reap "right" results in life, you have to do right when you do not feel like it. I call this "pressing in and pressing on," and knowing how to do it is one of the most important components of becoming a person who never gives up.

You see, progress has a price; and becoming a person who never gives up will cost you.

> *Progress has a price; and becoming a person who never gives up will cost you.*

You will never get where you want to be in life without being willing to sacrifice and push through the obstacles and adversities that stand in your way. Your obstacle may be an attitude, a set of circumstances, a relationship, an issue from your past, a thought or mind-set, a feeling, or a bad habit. Whatever it is, you are the only one who can press through it; no one else can do your pressing for you. You may have tried to overcome your challenges in the past. Perhaps you have tried to the point that you are weary, exhausted, or discouraged. This is precisely the point where you have to summon fresh strength from God and press in one more time.

We always grow weary and may even falter in our determination if we do not continually lean on God. The Bible says that when we wait on God our strength is renewed as the eagle (see Isaiah 40:31). To wait on God means to expect Him to do what needs to be done, to lean on and rely on Him. We must make personal decisions to press through, but we never experience success in anything unless we are relying on God to help us.

One of the definitions I like for the word *press* is: "to exert steady force or pressure against something." I often say, "You have to press against the pressure that's pressing against you!" When something is pressing against you, you must be determined to press against it with greater force, because very little that is truly worthwhile or worth having in life happens without this kind of effort.

Sometimes, you have to press through natural circumstances. For example, let's say your dream is to open a business in a certain neighborhood in your city and the zoning board repeatedly rejects your request to build the facility you need. Or, maybe your dream is to go to college, but you keep getting turned down every time you apply for a scholarship.

I suffered sexual, mental, and emotional abuse as a child. Because of that, I needed to move away from home and begin supporting myself immediately after high school graduation, and I did not get to go to college. My high school teachers recognized a writing gift in me and strongly urged me to try for a college scholarship, but I was unable to do so.

However, that didn't stop God! I've written almost eighty books and never went to school to learn how to do it. I have also received several honorary degrees and two earned degrees because of my writing. I could not go to college the "normal" way, but God had another way to help me do what He wanted me to do. I am amazed when I think of what He will do if we simply refuse to give up. With Him, all things are possible!

The situations and challenges I have described in this section are real; they must be dealt with. But I am more concerned about your response to these obstacles than about the obstacles themselves, because if you can keep your thoughts and attitudes right—and if you will refuse to give up—you will eventually have the breakthrough you need. I can't promise you will get exactly what you want, but if God does not give you what you are asking for then He will give you something better. His ways are above our ways and His thoughts above our thoughts. As you think about your life and make a fresh determination to pay the price of progress, keep in mind some of the things you may need to push through:

- failure
- fear

- disappointment
- rejection
- betrayal
- jealousy
- frustration
- delay
- offense
- self-pity
- difficulties
- people's opinions
- discouragement
- weariness and fatigue
- depression
- anger
- guilt
- shame
- abuse

Of course, there are other dynamics and emotions you may also need to press through, but this list will help you recognize these things as obstacles to be overcome when you encounter them. As you face them in your life, remember, press in and press on!

GO FOR THE FRUIT

Hebrews 12:11 makes a statement I am sure most of us would agree with: "For the time being no discipline brings joy, but seems grievous and painful; but afterwards it yields a peaceable fruit of righteousness to those who have been trained by it [a harvest of fruit which consists in righteousness]."

I don't know many people who get excited about being disciplined. Disciplining ourselves is something we do because we know it is good for us and we know we need to do it, but it's not something we typically call "joyful." Although it is hard, discipline has its rewards. When we discipline ourselves to eat healthily and exercise, we gain energy and begin to be able to maintain the proper weight for our body frame. When we discipline ourselves to spend less money than we earn, we end up with a nice savings account and no pressure from debt. When we discipline ourselves to fellowship with God, pray, and study His Word, we reap the rewards of intimacy with Him.

I think about a woman who faithfully helped me in my home for several years. She used to say to me: "I wish I were small enough to wear your clothes!" She knew that I give away clothes quite frequently, but at that time my clothes were too small for her. Over and over again I heard her say, "Boy, I wish I could wear your clothes!"

Well, the day came when she finally made up her mind and said to me: "I'm going to wear your clothes. When you give away clothes, I'm going to be the first one in line because I'm right here in your house!"

This woman did not sit around wishing she was smaller, nor did she go on a crash diet. She began to change her eating habits, developed a healthier lifestyle than she had been living, and pressed through the pain of not eating foods that were wrong for her. Soon, I began to notice her getting smaller, then smaller and smaller and smaller and smaller. Over the next year, she lost fifty pounds and was able to wear my clothes. She ended up with a lot of them because she was right—it was easier to give them to her right there in my house than to load them into the car and take them somewhere else to give away.

Losing fifty pounds in twelve months required discipline, and it was not easy or joyful. But her discipline paid off and brought about the joy of a desire fulfilled in her life.

Let me call your attention to Hebrews 12:11 again: "For the time being no discipline brings joy, . . . but afterwards . . ." (emphasis mine). We must not buy into the lie that we should only live for the moment or that the present is all we have. We also have a future to consider, and we need to begin to live with an eye toward "afterwards," toward the "later on" times. We have to begin to care just as much or more about later on than we care about right now.

If you want to be thinner when the time comes to wear your swimsuit in June, you need to start eating healthily and exercising before summer arrives. If you want to be able to afford a new car next year, you need to work toward getting out of debt right now. If you dream of living in a nice, clean, orderly home, you have to clear out the clutter and clean it up!

Discipline may not be pleasant for your flesh while you're doing it, but it will give you a tremendous sense of satisfaction in your soul— the satisfaction that comes from knowing you are making good choices. If you will pay the price to be disciplined now, you will enjoy rewards later. If you don't pay the price to do what is right now, then you'll suffer the consequences of an undisciplined life later. You can pay now or you can pay later, but at some point, we all reap the harvest of the choices we've made. We can't simply *wish* our lives were different; we have to press through laziness, fleshly desires, and bad attitudes and refuse to give up on the discipline that will yield good fruit later on.

BE CONSISTENT

I remember a time when I had gained about ten pounds and had tried everything I could think of to lose it. My clothes were getting tight; I was very uncomfortable; and I wanted to lose the weight quickly. For a few days, I ate no fat at all. When that diet didn't work, I ate nothing but protein. When I had no success with all-protein after a couple of days, I decided to eat only nine hundred calories per day. When nothing was working and I still failed to shed my extra pounds, I tried something else and finally concluded my metabolism had gone haywire and my hormones were out of whack. As you can imagine, I became very frustrated.

One day, God spoke to me and said something very simple, but something I will never forget: *Consistency is the key*. Since that day, I have certainly found those words to be true. If we want to make progress in any area of our lives, we need to keep doing the right thing over a period of time. That is what consistency means.

Where my ten pounds were concerned, I didn't need a fad diet; I needed to eat healthily and eat reasonable amounts of food regularly, as a *lifestyle*. I knew that, but continued to struggle until the day I was reminded of a phrase in the Bible: "necessary food" (Job 23:12). God spoke to me from that scripture: *If it's not necessary, don't eat it*. That did it for me. Over the next few months, I lost those ten pounds and haven't regained them since.

The point of the stories about the lady who worked in my home and about my weight battle is to state clearly: none of us can have everything we want now and everything we want later. Remember, "For the time being no discipline brings joy..."

Let me ask you: If you're not happy with the situation you're in right now, will you make the effort to change it? Do you want to be in the same situation this time next year? Or do you want something different? If you want to have something different, then you'll have to pay the price on this end to have what you want on that end. You will have to spend some of this year moving toward your goals for next year. As you move forward, you'll have to make tough choices and you'll come to some painful crossroads. When you reach those places, you will either run back to where you came from or you will press through. If you press through, without giving up, then you'll have the victory you long for later on.

My husband is extremely healthy, and he sometimes feels better than I do physically because he has more strength and more energy. The reason for Dave's health and strength is that he has been exercising every other day since he was about sixteen years old. Like clockwork, he has made a deliberate investment in his physical well-being every other day for many years. His skin looks like that of someone thirty years younger.

I can remember looking at him wishing I didn't have any cellulite or drooping, bulging places, but wishing did not help me at all. Being jealous of Dave didn't help either. I had to start exercising myself! Now, after being diligent and consistent, I don't have nearly as many sags and bulges. I paid the price and got what I wanted!

For much of my life, I seemed to hurt myself often when I tried to exercise, so that became my excuse for not doing it. I tried to lift weights and messed up my elbow. I tried to do sit-ups and hurt my back. I now work with a strength coach who has taught me how to properly lift weights and exercise. I work out three days a week and I don't hurt myself anymore.

Dave has been paying the price to feel good for a long time. He started years ago, so he hasn't suffered the injuries I have. I'm paying the price now, and even though I'm years behind Dave, I know it's worth it. I already feel better today than I felt thirty years ago, and I expect to keep feeling better and better as I continue to exercise and take care of myself.

Another excuse that I used was that I was just too busy. I've finally realized that if we truly want to do something, we will find a way to do it. If we're too busy to be healthy then our lives are probably out of balance and we need to make some changes. Being a minister, I was busy helping people and felt I needed to sacrifice in order to do that. But I've learned that if I don't take time to take care of myself, eventually I won't be able to help anyone else either. Putting yourself at the end of the list of people you take care of is not spiritual; it is actually very foolish.

Do you need to invest in a relationship that desperately needs improvement? If you simply wait for the other person to change, the change might not ever take place. You may have to be the one to take the first step toward reconciliation. I recall a time when I felt my marital relationship was not meeting my needs. I thought of all the things I felt Dave should be doing to make me happier, but nothing changed. I even prayed about it and nothing changed. Then one day the Lord spoke to my heart, telling me I would see change if I would begin to compliment Dave often. That was difficult at first because I thought he should be doing things for me, but as I obeyed God, I saw amazing changes. It wasn't long before Dave was doing most of the things I had wanted him to do. I had to act first and then I saw results.

No matter what you want to see change, you will either press past your pain now and do what is right or you will be exactly where you are today thirty years from now—or perhaps somewhere even worse. God is a good God; He has awesome plans for your life, but you must pay the price of obedience to His Word. I encourage you to start paying today, and to keep your eye on the long-term benefits you will reap tomorrow.

DON'T YOU DARE SETTLE

I want to share with you a verse we often overlook when we read the Bible. It's Genesis 11:31: "And Terah took Abram his son, Lot the son of Haran, his grandson, and Sarai his daughter-in-law, his son Abram's wife, and they went forth together to go from Ur of the Chaldees into the land of Canaan; *but when they came to Haran, they settled there*" (emphasis mine). You may know about Abram and Sarai and Lot, but you may not have ever heard much about this man named Terah.

I believe Terah, Abram's father, failed to take an opportunity God wanted him to take. I believe God wanted Terah to go all the way to Canaan, the Promised Land, but look at the words emphasized in the scripture: "but when they came to Haran, they settled there." In other words, Terah stopped short. He was supposed to go from Ur all the way to Canaan, *but* he stopped when he reached Haran. He and his family started their journey in one place, headed for another, and before they reached their goal, their travels became so difficult they settled. They settled for much less than God had for them because they did not want to press past through pain and keep going.

I believe Abram's father could have been the one with whom God made covenant. God told him to go, but he stopped; he settled. Abram ended up receiving a phenomenal blessing, but I think God also offered it to his father. His father simply quit before he reached the place where he could receive it.

I urge you today not to settle for less than the best God has for you. Don't allow yourself to get into a position to wonder why someone else ends up with something and realize that you had the same opportunity and passed it up.

If we read on in Genesis 11, we learn that "Terah lived 205 years; and Terah died in Haran" (v. 32). He died where he settled. I think many people just settle somewhere and die in that place. They may not die physically, but their dreams die; their visions die; their passion dies; their zeal dies. Their enthusiasm for life dies. Why? Because they gave up and did not press into the best God had for them. They didn't want the pain they were facing, but they ended up with something worse—the pain of disappointment and of destiny denied.

You will face different types of pain and encounter difficulties as you go through life. You simply have to choose which kind of pain you want—the pain of pressing through or the pain of giving up. I'm convinced there is no worse pain than an unfulfilled, dissatisfied life in which you know deep down that you missed God because you didn't have the courage to press on. Don't be that way. Commit right now to go all the way with Him and to press through the challenging times.

In Genesis 12, we read the rest of the story. "Now [in Haran] the Lord said to Abram, Go for yourself [for your own advantage] away from your country, from your relatives and your father's house, to the land that I will show you" (v. 1). Here, God is asking something extremely difficult of Abram, but He also gives him the promise of a reward.

I think, in today's language, God might have said something

like this to Abram: "I know what I'm asking you to do is hard. I know you're attached to your family. I know you'll be lonely. I know packing up and walking away from everything familiar will be difficult. I know you will have lonely nights. I know people will not understand. I know they will talk about you, judge you, and criticize you. So let Me tell you what I will do for you: 'I will make of you a great nation, and I will bless you [with abundant increase of favors] and make your name famous and distinguished, and you will be a blessing [dispensing good to others]'" (v. 2). God went on to promise Abram He would bless him so much that people who came along and blessed him would be blessed, simply because they blessed him.

Throughout the Bible, we find many places where God gives a hard command, but it always comes with the promise of reward. God is not a taker; He is a giver. He never tells us to do anything unless it is for our ultimate benefit. I assure you: everything God ever asks you to do, even if it is difficult, He asks because He has something great in mind for you—but you cannot have it unless you press through that hard place. Pay the price and never give up. Victory is worth everything it costs.

> *Everything God ever asks you to do, even if it is difficult, He asks because He has something great in mind for you.*

I'm often humbled when I think of the extraordinary sacrifices that people have been called on to make in order to simply survive. One young man's story stands out in my mind. In February 2003, Aron Ralston, an experienced and enthusiastic outdoorsman, narrowly escaped death when caught in a fierce avalanche. In a very short time, the avalanche covered Ralston up to his neck and completely buried his skiing companion. With his considerable skill and strength, Ralston freed himself and his friend and got them both to safety.

But Ralston's alpine rescue was not his only heroic act that year. Three months later, while hiking alone through Utah's Canyonlands

National Park, Ralston faced another life-and-death situation, one that required him to make a desperate decision and take a drastic measure in order to stay alive.

Ralston found himself at the bottom of a hole, one hundred feet beneath the surface of the desert, miles away from a paved road, with his arm pinned to a rock wall by an eight-hundred-pound boulder. No matter how hard he tried, the boulder would not budge. Soon, the crushing pain in his hand and arm gave way to the sensation of feeling nothing at all—not a good sign.

As Ralston thought about what to do, he realized he had broken an important hiker's rule: always tell someone where you are going. No one knew where he was; no one was expecting him to return from his hiking trip for several days. By the time anyone missed him, they would not know where to search for him. By the time anyone found him, he would probably be dead. He was in such a hidden place in such a remote desert that his body might never be found.

Five days after he had been trapped, thinking death was imminent, Ralston managed to use his video camera to record messages he hoped someone would find if he did not survive. In an effort to help those who would recover his body identify him, he used his pocketknife to etch his name; the month and year of his birth, October '75; April '03, the month of his hiking trip; and "RIP," into the rock where he was stuck.

Ralston then proceeded to do the only thing he knew to do to save his life. He used his pocketknife to amputate his arm so he could be free from the boulder. Amazingly, he then applied a tourniquet and gave himself basic first aid before he managed to rappel sixty-five feet and hike seven miles out of the canyon. He was nearing the place he had parked his truck, but had to climb eight hundred feet, straight up, to reach it. Bleeding, dehydrated, and almost in shock, Ralston pushed himself to reach his vehicle. As he neared it, he saw rescue teams convening to begin searching for him.

He first received treatment at the nearest hospital, and was then transferred to another facility for surgery on the remaining portion of his arm.

I hope you will never find yourself in circumstances as drastic as Aron's, but I hope you will develop the courage, resourcefulness, and commitment to life he demonstrated when faced with a seemingly impossible situation.

Don't think or say, "This is just too hard" when you know you need to do something. God never requires us to handle more than we can bear, but with every difficulty He always provides a way for us to overcome. Never say, "There is no way," because He is the way (see John 14:6) and He makes a way for us. You can do whatever you need to do in life! You have what it takes!

The King of Denim

Have you ever put on your favorite pair of blue jeans and wondered who invented them? Maybe not, but I believe the story is worth sharing.

For years, Levi Strauss has been the most famous name in denim, but Strauss never intended to patent the popular pants or to make a fortune selling them. He was simply a smart, hardworking Bavarian immigrant determined to make a successful and prosperous life in America after he moved from what is now Germany to New York with his mother after his father died of tuberculosis in 1845. Two of Levi's older brothers, Jonas and Louis, were settled and established in their own wholesale dry goods company in New York prior to his arrival. Following in his brothers' footsteps, Levi began to learn the dry goods business.

When Levi heard about the Gold Rush in California, he decided to head west. He made his home in San Francisco, where he started a dry goods business of his own and represented his brothers' New York company as well. Through his company, he imported dry goods and sold them to small stores in California and throughout the growing towns in the western United States.

In 1872, opportunity knocked for Levi when he received a letter from a customer in Nevada. This man, a tailor named Jacob Davis, explained to Levi in his letter that he had started making pants for working men—with metal rivets in places subject to stress, such as the corners of each pocket and the base of the fly. Because he could not afford to patent the pants alone, he solicited Levi's help in paying for the necessary paperwork and suggested the two men own the patent together. On May 20, 1873, the two received their patent for what they called "waist overalls," now known as "blue jeans."

Levi soon asked Jacobs to join him in his San Francisco factory to oversee the mass production of their riveted waist overalls. At first, seamstresses sewed the pants in their homes, but by the 1880s, demand was growing, and Levi rented factory space for the production of his popular pants. The first 501® jeans, which are still available today, were produced around this time.

Over the years, Levi's business continued to grow, and he became known not only as a smart, hardworking businessman but also as a kind, friendly, down-to-earth employer. He was also a generous philanthropist who contributed large sums of money to local charities, and in 1897, he funded twenty-eight college scholarships to the University of California at Berkeley.

You must realize you will reach goals you never dreamed were possible if you continue to persevere. When you do, be kind and generous, as Levi Strauss was, and remember to share your success and rewards of your hard work with others.

CHAPTER 6

THE CONQUERER'S SPIRIT

*"Difficulties show men what they are. In case of any
difficulty remember that God has pitted you against
a rough antagonist that you may be a conqueror,
and this cannot be without toil."*

EPICTETUS

In 1867, a creative engineer named John Roebling was inspired by an idea to build a spectacular bridge connecting New York City with Brooklyn. However, bridge-building experts throughout the world thought that this was an impossible feat and told Roebling to forget the idea. It just couldn't be done. It wasn't practical. It had never been done before.

Roebling couldn't ignore the vision he had in his mind of this bridge. He thought about it all the time and knew deep in his heart that it could be done. He felt he just had to share the dream with someone else. After much discussion and persuasion he managed to convince his son Washington, an up-and-coming engineer, that the bridge in fact could be built.

Working together for the first time, the father and son developed concepts of how it could be accomplished and how the obstacles could be overcome. With great excitement and inspiration, and the headiness of a wild challenge before them, they hired their crew and began to build their dream bridge.

The project started out well, but when it was only a few months under way a tragic accident on the work site took the life of John Roebling. Three years later Washington was injured, and he suffered brain damage that left him unable to walk, talk, or even move.

"We told them so."

"Crazy men and their crazy dreams."

"It's foolish to chase wild visions."

Everyone had a negative comment to make and felt the project should be scrapped since the Roeblings were the only ones who knew how the bridge could be built. Unbelievably, in spite of his handicap, Washington was never discouraged and still had a burning desire to complete the bridge. Despite his severe handicaps, his mind was still as sharp as ever.

He tried to inspire and pass on his enthusiasm to some of his friends, but they were too daunted by the task. As he lay on his bed in his hospital room, with the sunlight streaming through the windows, a gentle breeze blew the flimsy white curtains apart and he was able to see the sky and the tops of the trees outside for just a moment.

It seemed that there was a message for him to not give up. Suddenly an idea hit him. All he could do was move one finger and he decided to make the best use of it. By moving that one finger, he slowly developed a code of communication with his wife.

He touched his wife's arm with the finger, indicating that he wanted her to call the engineers again. Then he used the same method of tapping her arm to tell the engineers what to do. It seemed foolish, but the project was under way again.

For eleven years Washington tapped out his instructions with his finger on his wife's arm until the bridge was finally completed. Today the spectacular Brooklyn Bridge stands in all its glory as a tribute to the triumph of one man's indomitable spirit and his determination not to be defeated by circumstances. It is also a tribute to the engineers and their team work, and to their faith in a man who was considered mad by half the world. It stands, too, as a tangible monument to the love and devotion of his wife, who for eleven long years patiently decoded the messages of her husband and told the engineers what to do.

—Author Unknown

http://www.dizzyboy.com/stories/inspirational-stories/ inspirational_story_three.html

THE SPIRIT OF A CONQUEROR

Washington Roebling had what I call "the spirit of a conqueror," and I encourage you to get to the point where you have the spirit of a conqueror too. I want to help and inspire you to see yourself differently than you may right now, to see yourself as one who overcomes adversities, not as someone who shrinks back in fear or feels overwhelmed every time a trial comes along. I want you to be and see yourself as someone who is bold in the face of adversity and fully able to handle difficulties with courage, confidence, wisdom, and faith.

You see, adversities are not optional, and it takes a conqueror to overcome them. Jesus Himself said: "In the world you have tribulation and trials and distress and frustration; but be of good cheer [take courage; be confident, certain, undaunted]! For I have overcome the world. [I have deprived it of power to harm you and have conquered it for you]" (John 16:33).

> *Adversities are not optional, and it takes a conqueror to overcome them.*

Paul understood this, and wrote in Romans 8:35–37: "Who shall ever separate us from Christ's love? Shall suffering and affliction and tribulation? Or calamity and distress? Or persecution or hunger or destitution or peril or sword? Even as it is written, For Thy sake we are put to death all the day long; we are regarded and counted as sheep for the slaughter. Yet amid all these things we are more than conquerors and gain a surpassing victory through Him Who loved us." What does it mean to be more than a conqueror? I believe it means that before you ever face adversity, before the battle against you even begins, you already know you will win in the end if you will simply refuse to give up.

Look also at Romans 8:31: "What then shall we say to [all] this? If God is for us, who [can be] against us? [Who can be our foe, if God is on our side?]" I promise you, if God is for you, you have no reason to have any other attitude than that of a conqueror. He is on your side, and He gives you the spirit of a conqueror.

If you do not feel you have a relationship with God, you may not feel that He is for you. He is, but you need to believe that and put your faith in Him to not only forgive your sins but help you throughout your life. Please see the prayer at the end of this book. When you receive Jesus Christ into your heart, you also receive the spirit of a conqueror.

No matter what you go through in life, if you have the spirit of a conqueror and you really know who you are in Christ and truly believe God is on your side, you do not have to be daunted or overwhelmed by any difficulty you may face. Whether you are trying to pay off a mountain of debt, finish your education, fight an illness, leave an intimidating and abusive relationship, start your own business, go into ministry, or simply clean your house, *you can do it* with God on your side. You just need to develop the spirit of a conqueror.

Remember the verse I wrote about earlier: "For the time being no discipline brings joy, but seems grievous and painful; but afterwards it yields a peaceable fruit of righteousness to those who have been trained by it" (Hebrews 12:11). When confronted with adversity, do not ever say, "I just can't take this anymore." Instead, have the spirit of a conqueror and say, "God is for me. He is on my side and I can do all things through Christ who strengthens me."

RESIST THE DEVIL

Many Christians seem to believe their faith should make them immune to the attacks and harassment of the enemy, and they act surprised when he begins to bother them. That simply is not true; as we have learned in this chapter, we *will* go through adversity. Sometimes, but not all the time, our adversities will be the work of the enemy. What we call "a run of bad luck," "a difficult situation," or "a rough day" may actually be the devil's doing. In the face of such adversities, what do we do?

First we need to remember we *do* have authority over Satan, and the Bible tells us we are to *steadfastly* resist him. James 4:7 instructs us: "So be subject to God. Resist the devil [stand firm against him], and he will flee from you." We must understand that we really do not have any power to resist Satan if we are not first submitting to God. This is so important I want to repeat it: we will not be able to exercise power over Satan if we do not first submit ourselves to God—which simply means we need to be obedient to Him. That wholehearted obedience and submission to the Lord is what gives us the authority and the ability to resist the devil.

Resisting the devil does not mean that we merely want our trials or difficulties to go away, but it means we are determined to remain steadfast and display godly behavior at all times. "He Who lives in you is greater...than he who is in the world" (1 John 4:4); and with God's power working in you, not only can you outlast the enemy, you can continue serving God while the battle rages.

Remember, James 4:7 says we must "stand firm" against the enemy. He will not leave simply because we respond to an altar call or wear jewelry with Christian symbols on it. He will not go away because we go to church every Sunday or have a library full of Christian books and CDs. We cannot defeat him unless we are firm in our decisions and determined to never give up.

START STRONG, FINISH WELL

Everything we undertake in life has a beginning and an end. Typically, we are excited at the beginning of an opportunity, a relationship, or a venture; we're also happy when we can celebrate our achievement and have the satisfaction of a fulfilled desire. But between the beginning and the end, every situation or pursuit has a "middle"—and the middle is where we often face our greatest challenges, hurdles, roadblocks, obstacles, detours, and tests. People who are easily led by their emotions rarely finish what they start. They give up when the project is no longer exciting and all they see in front of them is hard work. God wants us to be people who finish what we begin, and He will help us if we let Him.

You may be in the middle of a lot of things right now. Perhaps you're in the middle of getting out of debt. You have paid off all but one of your charge accounts and you're starting to think, *I've done pretty well. I think I will go shopping today because I read about a huge sale at the mall.* Doing "pretty well" is not good enough because that was not your original goal. Discipline yourself a little while longer; you will be so glad you did when you pay off that last charge account and you are completely out of debt. If you need to do something to have some fun or relieve emotional tension, find something that does not require a charge account. Whatever you find yourself in the middle of, determine to see it all the way through to the finish.

Between our beginnings and our endings we must develop the boldness and determination necessary to overcome the overwhelming circumstances we encounter in the middle. The enemy wants us to settle for less than God's best for our lives and to stop short of receiving and enjoying everything God has for us. The devil hates progress and pushes us to give up. God, on the other hand, wants the very best for us; He wants us to finish the races set before us and finish them with joy (see Hebrews 12:1). Ask yourself if you are willing to pay the price to finish.

People who finish well in life are the ones with strong character. They have "staying power." Jesus did not quit when His circumstances were rough, and He is our example. The Bible actually says we are to look away from all that distracts us unto Jesus who is the Author and Finisher of our faith (see Hebrews 12:2).

Certainly, Washington Roebling was a finisher. Another great finisher was the apostle Paul. According to Acts 20:24, none of the obstacles he faced moved him or had any sort of negative effect on him. In this verse, Paul also said he wanted to finish his course with joy. That does not mean that what he endured wasn't difficult; it simply means it did not stop him.

I think most of us want to do and be everything God intends for us, and to enjoy it along the way. Great joy comes with finishing the race God has called you to run. Enjoy the journey and keep your eyes on the prize. Jesus endured the cross for the joy of obtaining the prize that was set before Him.

Paul had a remarkable ministry and finished many things in his life. As he neared the end of his journey on earth, he wrote: "I have fought the good (worthy, honorable, and noble) fight, I have finished the race, I have kept (firmly held) the faith" (2 Timothy 4:7). In this verse, he was basically saying: "I have been through a lot. But I am still here. The enemy tried to take me out, but he did not succeed!"

I think one of the greatest testimonies we can have is summed up in these three words: "I'm still here." When we speak those words, we are saying, "I did not quit. I did not give up. I am still here."

When you feel you are in the "middle" as you reach for your goal, reread some of the stories in this book. You will be inspired by the lives of people who have done amazing things because they pressed through the middle and crossed the finish line.

DON'T BE LED ASTRAY BY DIFFICULTIES

Paul writes in 1 Thessalonians 3:3–5:

> *That no one [of you] should be disturbed and beguiled and led astray by these afflictions and difficulties [to which I have referred]. For you yourselves know that this is [unavoidable in our position, and must be recognized as] our appointed lot. For even when we were with you, [you know] we warned you plainly beforehand that we were to be pressed with difficulties and made to suffer affliction, just as to your own knowledge it has [since] happened. That is the reason that, when I could bear [the suspense] no longer, I sent that I might learn [how you were standing the strain, and the endurance of] your faith, [for I was fearful] lest somehow the tempter had tempted you and our toil [among you should prove to] be fruitless and to no purpose.*

In this passage, Paul was saying: "I was concerned that you would quit and give up because of all the trials and tribulations and afflictions that were coming against you." He was encouraging the Thessalonians, basically saying: "When difficulties come your way, don't be impressed by them; don't let them move you. Don't let them upset you. Just say: 'This too shall pass.' This is part of life, so don't fold under the pressure."

When Dave and I are challenged by difficulty, one of the first things he says is, "I am not impressed!" He shares that if we refuse to allow our difficulties to *impress* us, then they will not *oppress* us or *depress* us either. We must learn to hold steady in the storm and we will arrive safely at our desired destinations.

Any time we try to step out and do something for God, the enemy will oppose it. This has always been true, and it is unavoidable. Paul certainly experienced it and wrote in 1 Corinthians 16:9: "For a wide door of opportunity for effectual [service] has opened to me [there, a great and promising one], and [there are] many adversaries."

Over the years, many people have come to work at Joyce Meyer Ministries. Some of them relocate to St. Louis from other states and have to deal with the challenges of finding a place to live, moving, putting their children in school, and getting established in a new city. Some of these people have been amazed at the "many adversaries" they face during their first few weeks of employment. Over and over again, I've watched the enemy use these difficulties to make them think, *Well, maybe I didn't make the right decision after all. I thought this move was God's will, but maybe I was wrong. Maybe I should take my family and go back home where things were easier.*

Many times, Dave and I have needed to talk with these people and tell them, "The fact that you are having a difficult period of adjustment doesn't mean you didn't hear from God. You just need to hold steady. Go back to what's in your heart. Do not focus on your circumstances. Look at what's in your heart."

If they tell us that they do believe this is God's will for their lives, then Dave and I encourage them, "Then make up your mind that you're going to stay. Don't be double-minded, second-guessing your decision. Determine to push through this difficult season, and refuse to think about turning back. Since you really believe God has led you here, commit to stay—no matter what."

When we say to the enemy, "Look, devil, I know what's in my heart; I know what God told me to do; and I am going to stay here and see what God has for me, no matter how hard it is or how long it takes," then the enemy realizes he cannot control us with trials and tribulations. This builds a new level of faith and strength in us and demonstrates the spirit of a conqueror.

Paul said difficulties are unavoidable. You will face them as you continue to grow in God and follow Him. Do not let them lead you astray, but be determined to overcome them. I know you can make it. You will have to do your part, and it is going to take work, but you can do it!

DEAL WITH YOUR DIFFICULTIES

A person with the spirit of a conqueror must confront and deal with the adversities we face, not run from them. We simply cannot keep trying to escape or avoid situations that are difficult. Anytime we run from a situation, we can almost be sure we will have to go back and face it, or something very similar, at a later time in our lives.

Think about Moses. He ran from Egypt and spent forty years in the desert, where God prepared him to be a great leader. When God appeared to him in the burning bush, He basically said: "Okay, now I want you to go back to Egypt" (see Exodus 3:2–10). Yes, God sent Moses right back to the place he tried to escape.

The Bible is full of similar stories—enough to convince me that running from adversity doesn't do anyone any good. If you run from one difficult situation, there will always be another waiting for you. You see, God knows you need the skills, strength, and training you will gain from adversity, so even if you manage to get out of one difficult situation, He will lead you into another one.

For example, let's say you live next door to some people you really, really don't like. You dislike them so much that you decide to move to get away from them. Then let's say you move to a new neighborhood and your neighbors on both sides have the same personality as the people you wanted to get away from. It may be hard for you to imagine this, but it is possible that God has put them next to you. Why would He do that? Because there may be something in you that needs to be dealt with and will only surface when you are around that type of person.

I used to have an extremely difficult time with people whose personalities were similar to my father's. Because he abused me, I didn't want to be around people who spoke the way he spoke, acted the way he acted, or reminded me of him in any way.

But over a period of time I began to notice God was surrounding me with people who reminded me of my dad. Every time I encountered someone who reminded me of him, I felt insecure and fearful, and I slipped into old behavior patterns. *my mom*

I didn't like my responses, nor did I understand where they came from for quite some time. But I was crying out for God to change me and show me truth. I now realize He began putting me in situations where my weaknesses would be evident so I could say: "Oh. Okay, God, I see that You want to set me free from that!" He was trying to get me to grow up and be a stronger, more mature Christian. We often ask God to deliver us from the wrong things. We want to be delivered from our trials, but we need to ask Him to deliver us from the things in our hearts that hinder His purposes for our lives.

God uses difficulties to make us people He can trust and people He can use. When we face adversity, we must have the spirit of a conqueror. David had to face Goliath before he could become king. We must not try to escape or avoid difficult situations while we are going through them. Instead, we need to say, "God, if this is something I have to go through, then I am going through it victoriously. I want to go through it with the attitude of a conqueror. Whatever You want to do in my life through this, I want You to do. But one thing I ask, God—that You help me press past my feelings and behave in a godly manner all the way through it." Getting away from trouble never needs to be our goal; our goal needs to be to conquer it with Christlike attitudes and behaviors. God teaches us to bless our enemies and pray for them (see Matthew 5:44), not to run from them. He tells us that we overcome evil with good (see Romans 12:21). No matter what is going on in your circumstances, keep doing what you know is right and you will win!

> *God uses difficulties to make us people He can trust and people He can use.*

DO IT FOR GOD

The goal I want you to have when you finish this chapter is this: the next time you face a challenge, you will discipline your attitude, your emotions, your mouth, and your mind to behave in a godly manner all the way through it. That will be hard. You'll feel your flesh rebelling against it. But keep saying, "God, I am doing this for You. I can't do it on my own and I don't even want to do it for myself. I want to do it for You as an act of obedience." Jesus didn't give up. He went all the way through for us, and we can do the same for Him!

I remember one Sunday years ago when my pastor encouraged the congregation to take a moment to say hello to other people and even give them a hug and to tell them we loved them. I looked down the row where I was sitting and saw a woman who had hurt me in a significant way. I strongly sensed the Spirit of God impressing me to give her a hug and let her know I loved her. That was going to require great surrender for me because I felt she needed to apologize to me and ask me to forgive her. But she had no idea she had hurt me! Walking over to her and saying, "I love you" took everything I had! I can't guarantee I was totally sincere, but I know I was obedient to God.

Several months later, God led me to give one of my favorite possessions to that woman. "Now God," I responded, "I don't mind giving it away. I mean, I really would like to keep it, but if You are going to make me give it away, at least let me give to someone I like so I can enjoy seeing her with it!"

God responded to me: *Joyce, if you can give her that, if you can give your favorite possession to someone who really hurt you and is least deserving of it, you will break the power of the enemy. You will destroy his plan to destroy you.*

We do not take steps of obedience and overcome difficult times because we feel like doing so or think obedience is a good idea. We do it because we love God, we know He loves us, we want to obey Him, and we know His ways are always best for us.

Whatever adversities you are facing right now or will face in the days to come, I urge you to confront them, embrace them, and deal with them. Face them like a conqueror. Remember, they are working for your good and God will use them to strengthen you. Embrace them with a conqueror's attitude, and you will find yourself in a place of greater maturity, wisdom, and ability than you have ever known.

Trouble, pain, suffering, and persecution are not good in and of themselves. They come from the enemy, but God intends to work something good out of everything Satan means for harm. The situation itself may not be good, but God is good; and as we continue loving and trusting Him, He will bring good out of it (see Romans 8:28). Believing this gives us the strength and courage to keep pressing on in difficult times. We must have hope. We must believe we will see the Lord's goodness in our lives (see Psalm 27:13).

REACH YOUR FULL POTENTIAL

I fully believe reaching your potential is linked to the way you handle adversity. Winston Churchill said: "Difficulties mastered are opportunities won," and I wholeheartedly agree. If you allow difficulties and challenges to frustrate, intimidate, or discourage you, you will never overcome them. But if you face them head-on and press past the adversities you encounter, refusing to give up in the midst of them and moving forward with the spirit of a conqueror, you will develop the skills and determination needed to be everything you were created to be and experience everything God intends for you.

From Cotton Field to Corporation

The woman who came to be known as Madam C. J. Walker, the first African American female millionaire in the United States was born Sarah Breedlove on a Louisiana plantation in 1867, to former slaves who both died and left her an orphan by the time she was seven years old. She and her sister survived because they were willing to do the grueling, dirty work of picking cotton in the fields of Louisiana and Mississippi.

When she was fourteen, she married a man who died six years later, leaving her alone with a young daughter. Sarah and her daughter moved to St. Louis, Missouri, where her four brothers had found work as successful barbers, and she took a job as a washer-woman, earning $1.50 per day.

During the 1890s, Sarah found herself with a strange scalp condition that caused her to lose most of her hair. She tried every possible remedy to make her hair grow back, but nothing was effective.

Sarah took a job as a sales representative for a hair products company and moved to Denver, Colorado, where she met and married Charles Joseph Walker, and became known as "Madam C. J. Walker."

Still suffering with hair loss, Sarah finally found a solution—in an unusual way. She dreamed about an old man, who told her which ingredients to combine to make her hair grow back. The next day, she mixed the ingredients, tried the remedy, and it was successful!

Madam C. J. Walker quickly realized she had a product she could market and sell, so she opened her own business, producing and distributing "Madam Walker's Wonderful Hair Grower."

(continued)

To get her business off to a strong start, Madam Walker logged many miles as she personally traveled the United States for more than a year, going door-to-door telling people about her products and giving presentations wherever she could find an audience. With diligence and determination, she continued to develop and perfect strategies for increasingly effective sales and marketing of her products.

In 1908, Madam Walker went to Pittsburgh, Pennsylvania, to open a school to train a team of "Walker hair culturists" to help her in her growing business. Two years later, she had made her home and built a factory for the production and distribution of her products in Indianapolis—the metropolis of manufacturing in her day. Before the end of her first year in Indianapolis, Madam Walker gave one thousand dollars toward the building of a local YMCA for African Americans—a remarkable achievement for a woman who once supported herself by picking cotton, but one made possible because she worked hard and refused to give up.

In 1916, Madam Walker left Indianapolis for New York, though her Indianapolis factory continued to operate under the leadership of the forewoman she had trained. Because she refused to give up in the face of the personal problem of hair loss, Madam C. J. Walker went from picking cotton to being a pioneer in the beauty industry and the first African American woman to become a millionaire in the United States. Let her story inspire you to not give up when you face challenges, but to be creative, work hard, and persevere until you reach your full potential.

CHAPTER 7

NEVER GIVE UP ON WHAT'S IMPORTANT TO YOU

"The fishermen know that the sea is dangerous and the storm terrible, but they have never found these dangers sufficient reason for remaining ashore."

VINCENT VAN GOGH

What's important to you? When you stop and think about it, what really matters to you? Is it your job, your reputation, your health, your family, your relationship with God, your possessions, a project, or a cause?

Different things are important to different people. In Old Testament times, rebuilding the devastated walls of his hometown of Jerusalem was extremely important to a man named Nehemiah. It was so important he asked his employer, the king of Persia, to allow him to return to Jerusalem to oversee this project, and the king granted it.

Nehemiah definitely had the spirit of a conqueror, as you'll see as you read his story. He faced a storm of opposition, ridicule, and intimidation as he tried to finish the task so important to him, the project God called him to complete. Similarly, you will face various kinds of storms as you endeavor to do what is important to you.

Nehemiah's story will teach you how to weather them well and emerge victorious.

TAKE COVER

The best way to be safe through a natural storm is to take cover. If you do not seek shelter, the wind will blow you this way and that, and the rain may keep you from being able to see clearly. Besides that, you might get struck by lightning. If you do not take cover, the storm is in control of you. But if you will position yourself indoors through the storm, you will be able to ride it out and be in good shape to do what you need to do when it passes.

Psalm 91:1 gives us instructions for the spiritual storms of life: "He who dwells in the secret place of the Most High shall remain stable and fixed under the shadow the Almighty [Whose power no foe can withstand]." The first place you need to run when a storm hits in your life is to the secret place of the Most High, the presence of God. Read His Word; pray; worship Him; tell Him you trust Him as the winds of adversity blow. These are the spiritual disciplines no foe can withstand. When you practice these habits, you actually construct spiritual walls of protection around yourself. They will not keep the storms of life away, but they will provide protection and enable you to stand strong against them.

Many of life's storms come directly from the enemy. He is a master at orchestrating circumstances that will devastate a life, a dream, a marriage, a family, a business, a school, or a geographic region. He does not want you to build walls of protection around your life. He wants all your walls torn down so he can keep attacking you successfully.

WALLS KEEP THE ENEMY OUT

Upon his arrival in Jerusalem, Nehemiah saw the enemies of God's people were taking advantage of them because the city had no walls to keep the enemies out. When he first met with the people of Jerusalem, he said: "You see the bad situation we are in—how Jerusalem lies in ruins, and its gates are burned with fire. Come, let us build up the wall of Jerusalem, that we may no longer be a disgrace. Then I told them of the hand of my God which was upon me for good, and also the words that the king had spoken to me. And they said, Let us rise up and build! So they strengthened their hands for the good work" (Nehemiah 2:17–18).

The same principles in this story apply to you when you realize your life is a mess and the enemy is attacking you because of the way you have lived. You haven't kept your walls up; you have no protection. But with God's help you can begin now to build the right kind of life; you can start building walls of protection around yourself so you are protected from the enemy. He may come against you, but his plans will not succeed.

If you feel the storms of life have left you battered and tattered for too long and are ready to build some walls of protection, I'm excited for you. But don't expect to do it without opposition. That's just a reality of life. Any time you begin to move forward and do what is right, the enemy will be unhappy. But you don't need to worry about the opposition, because you have what it takes to win every battle. God is for you, in you, with you, and around you. He also works through you to not only win your own battles but to help other people win theirs.

A STORM IS BREWIN'

No sooner had the Jews "strengthened their hands for the good work," than opposition arose. When people began to hear of the rebuilding effort, they ridiculed Nehemiah and the Jews. Three men particularly—Sanballat, Tobiah, and Geshem—began to mock and make fun of them. "They laughed us to scorn and despised us and said, What is this thing you are doing?" (Nehemiah 2:19).

I can remember a similar experience. I wrote earlier about how people responded to me when I received God's call on my life. Many people laughed at me and said, "What do you think you are doing?" This was an extremely difficult time for me, but I did my best to stand through the storm of disapproval. I did everything I could to stay as close to God as possible through studying His Word, praying, and spending time in His presence. I worked to improve my attitudes and my marriage and my relationships with my children. Even in the midst of being misunderstood and laughed at, I kept working with God to overcome my problems and build a healthy self-image through learning about His love and mercy. By doing this, I was building walls of protection. I was putting bricks in my wall as diligently as I could, but the enemy continued to fight against me.

Have you ever been trying as hard as you could to do what was right but felt that the storm raged even stronger? I remember a specific situation that was especially difficult for me. One day, someone I considered a friend walked up to me and said: "Joyce, someone told me you believe God is going to give you one of the largest ministries run by a woman in the whole country. Did you say that?"

I responded, "Yes, I really believe God spoke that to me."

This woman snickered and said, "Well, to tell you the truth, some of us have been talking about that, and we just think that is impossible with your personality."

Hearing such comments is like being in the middle of a violent storm with no protection. I felt as Nehemiah must have felt when Sanballat, Tobiah, and Geshem said to him, between bursts of laughter: "What on earth do you think you are doing? You can't do that!"

Nehemiah stood firm and demonstrated the spirit of a conqueror when he said, "The God of heaven will prosper us; therefore we His servants will arise and build, but you have no portion or right or memorial in Jerusalem" (Nehemiah 2:20).

When the enemy comes against you, you also have to say, "You have no right here. You can have no portion of me, nor can you

I will unify all religions w/ LOVE.

hinder God's will for my life!" If you have given your life to Jesus Christ, you belong to Him, not to the devil. You simply have to remind him of that fact and resist him. You do not have to raise your voice or be dramatic, but you do need to use your authority in Christ and say: "No way, devil. You are not going to stop me. I will not give up on this. No matter what you throw at me, I will not quit!"

I refused to quit and eventually the woman who told me she and her friends thought it was impossible for me to succeed ended up working for me. God always vindicates us if we just keep doing what He tells us to do. Don't ever give up just because someone tells you that you can't succeed.

> *Don't ever give up just because someone tells you that you can't succeed.*

To stand against the enemy in the storms of life, we really must be determined. Sometimes we simply give up too easily. We get a little bit discouraged, or the "Sanballats" of life hurt our feelings, and we begin to cave under the pressure. Never forget that, according to God's Word, *you are more than a conqueror* (see Romans 8:37).

NOT EVERYONE LIKES PROGRESS

Sanballat was a formidable enemy; he did not give up easily. When he heard the building of the city walls was going forward in spite of his objections, "he was angry and in a great rage, and he ridiculed the Jews" (Nehemiah 4:1). The more they built, the angrier he became.

Maybe you know people who become angry when others try to make progress, better themselves, or do something new and different. Why would people become angry about your progress? Because people who do not want to progress themselves do not want you to progress either. They would rather judge you critically than face the fact that they do not really want to make the effort necessary to improve their lives or follow their dreams. Instead of investing in their futures and pursuing their destinies, they prefer to become angry and act as though something is wrong with you. That way, they do not have to consider the possibility that something may be wrong with them. It is called "blame shifting," and people who do it do not want to be blamed for not making right choices so they shift the focus to you and try to discredit you.

Sanballat and his buddies flew into "a great rage" because Nehemiah was leading an effort to help people and improve the community. In your efforts to help yourself and the people around you, do not be surprised if you meet with someone who becomes angry and critical about it. Just stand through that storm; hold your peace; and keep doing what you believe in your heart to be right.

WHAT A MESS!

Sanballat continued his angry rant in Nehemiah 4:2: "And he said before his brethren and the army of Samaria, What are these feeble Jews doing? Will they restore things [at will and by themselves]? Will they [try to bribe their God] with sacrifices? Will they finish up in a day? Will they revive the stones out of the heaps of rubbish, seeing that they are burned?"

Basically, Sanballat was saying, "This mess is too big to ever rebuild!" Yes, the walls were so destroyed that they were nothing more than heaps of charred stones. They were not good building materials for walls—at least not without a lot of effort. In Sanballat's mind, building the walls was a bad idea. They were simply too far gone.

The enemy may be hurling the same accusations at you: "What do you mean you can start over? Look at what a mess your life is in!" "There is no way you can get a college degree, not with the way your mind has been messed up!" "You think you're going to save enough money to send your children to college? Your finances are in such a mess the bank can't even straighten them out!" Whatever he is saying to you, I want you to know you are never too big a mess for God to redeem and nothing is beyond His ability to restore.

If anyone's life has ever been a mess, mine was. If anyone had reason to believe trying to get better simply wasn't worth the effort, I did. And the enemy knew it too. He tried to convince me many times that I would never have a normal life. He told me I would never be "quite right," that I would never be able to have a healthy self-image because I was an incest victim. He said I wouldn't be able to trust men because of the way men had treated me, that I could never have a relationship of godly submission to my husband; that I could never stop trying to control everything around me or keep from losing my temper because of the repressed anger in me. He told me I could never be sweet and nice and kind, and that I would always be rough, harsh, hard, rude, and crude because of the way I was raised. I desperately wanted to change and be the kind of woman I believe God wanted me to be, but I had to stand strong through the battle raging in my mind and emotions.

The devil is a liar; it's that simple (see John 8:44). He does not tell you the truth; he tells you whatever he wants to tell you in order to destroy you. He relentlessly attacks your mind and emotions. When you are tempted to give up because you think your problems are too big, just remember, the devil is a liar and nothing is impossible with God!

The devil hates all progress. Even if you commit to lose a certain amount of weight and do well on your diet for a week, then lose your willpower one day and eat a hot fudge sundae, the enemy will be right there to say: "See? This is too hard. You cannot do it. You have always been overweight and your eating habits are too ingrained in you for you to try to change them now!"

When he sends a storm of accusations against you, stand up and say: "Listen, *liar!* I had a little setback, but it is not over yet! I will meet my goals. I will do what God is calling me to do. It *is* worth the effort, and you will not stop me!" Fight the good fight of faith. You are a soldier in the army of the God, and He is on your side.

We may all have setbacks when we are learning how to do things correctly, but we can have a fresh start every day. All babies fall down while trying to learn how to walk. When they take a tumble, they do not just lie on the floor and cry for the rest of their lives. They get up, dust themselves off, and try again! No matter how many times they fall, they never give up.

PRAY!

In Nehemiah 4:4, we find three words that are vitally important to remember when we are trying to stand through a storm: "And Nehemiah prayed." How did he respond to all the attacks that came against him—the laughing, the anger, the rage, the judgment, the criticism, being told his desired goal was impossible? He prayed!

Let me ask you: What would happen if you prayed every single time you felt afraid or intimidated? What if you prayed every time you were offended, or every time someone hurt your feelings? What if you prayed immediately every time some kind of judgment or criticism came against you? Would your life be different? Would you be able to withstand those storms better? Of course you would.

We can learn an important lesson from Nehemiah's prayer: "Hear, O our God," he said, "for we are despised. Turn their taunts upon their own heads, and give them for a prey in a land of their captivity" (Nehemiah 4:4). Notice that Nehemiah didn't go after his enemies himself; he asked *God* to deal with them. His attitude was, "I'm doing Your will! You told me to build this wall and I am busy building it. You will have to take care of my enemies!"

Many times God tells us to do something or gives us an assignment and we begin doing it. But then the enemy comes against us, and when we turn to fight him, we turn away from God. Suddenly, the enemy has all of our attention. We spend our time fighting him instead of praying and asking God to intervene.

Nehemiah knew better than to let his enemies command his focus. He was aware of them, but he kept his eyes on God and the job God called him to do. And he simply prayed and asked God to deal with those who were attacking him.

Most of us will never forget September 11, 2001. I recently read an inspiring story about Stanley Praimnath, a banker who worked in Tower Two of the World Trade Center and miraculously survived the attacks. Stanley recalls: "For some particular reason, I gave the Lord a little extra of myself that morning [during prayer]. I said, 'Lord, cover me and all my loved ones under your precious blood.' And even though I said that and believed it, I said it over and over and over."

Soon after he arrived at his office on the eighty-first floor, Stanley glanced out of his window and saw Tower One on fire. He and a co-worker decided to evacuate their building, but returned to his office after a security guard assured them Tower Two was safe and secure.

Stanley reached his office just in time to catch a phone call from someone who asked him if he was watching the news. He responded that everything was fine. Then he looked up and saw an American Airlines jet heading straight for him. All he could think to do was dive under his desk, curl into a fetal position, and pray as the plane crashed about twenty feet from him and exploded. Once he got there, he said, he knew "beyond a shadow of a doubt that the Lord was going to take care of [him]."

Stanley saw a wing of the plane ablaze in the doorway to the offices of his department. Miraculously unhurt, he knew he needed to get out of the building, but he was trapped in rubble and could not move.

Stanley prayed fervently, asking God to spare his life and saying, "Lord, you take control. This is your problem now." He remembers suddenly feeling like "the strongest man alive" as strength surged through his body and enabled him to shake off the debris that held him captive. He climbed over the destruction around him and dodged flickering flames saying, "Lord, I have to go home to my loved ones. I have to make it. You have to help me."

Cut and bruised, Stanley stumbled through the dangerous remains of his office, only to realize all exits were blocked. He was trapped against a wall. He fell to his knees and began to pray again, then asked someone on the other side of the wall, "Do you know Jesus?" When the man said yes, the two of them prayed and asked God to help them break through the wall. They did, and Stanley climbed through a small hole, now able to reach the staircase of the collapsing building.

He and his new friend started down the stairs, stopping at every floor to see if anyone needed help. By the time they reached the concourse, the only people they saw were firefighters yelling, "Run! Run!"

The men wanted to run, but were surrounded by fire. If they didn't run through the flames, they would be burned to death. They doused themselves in water under the building's sprinkler system and darted through the flames to safety at last.

Stanley was determined to survive a life-threatening situation, but he could not do it without God's help. While human determination is vital to never giving up, we need to remember to rely on God in every situation, never trying to "make things happen" in our own strength, but doing our part to persevere, while counting on God to bring the breakthrough we need.

I want you to know this: the enemy is really not your problem; he is God's problem. You will waste your time if you turn your attention away from your God-given assignments and opportunities and begin to focus on the enemy. Satan knows that if he can distract you, he can ultimately defeat you. God is your defender; He promises to fight your battles for you. So when the enemy begins to stir up a storm in your life, be like Nehemiah: pray.

A HEART AND MIND TO WORK

Nehemiah has a good report in chapter 4: "So we built the wall, and all [of it] was joined together to half its height, for the people had a heart and mind to work" (v. 6). Clearly, Nehemiah's people were determined to work, even though their enemies came against them persistently. Even though their enemies were stubborn, they would not keep from building!

The enemy cannot hold you back if you are determined! You will have to be more determined than he is, but you can do it because God is on your side. The enemy may oppose you fiercely, but you can build your walls of protection and achieve your goals in life if you simply refuse to give up.

Not only did the walls begin to go up, but the breaches in the walls were being closed, the cracks were being sealed. This made Nehemiah's enemies really furious.

I believe one of the benefits of your reading this book is that the "cracks" in your life will be closed! Any places that may be weak in you are being strengthened as you learn how to be a person who never gives up. This makes the enemy angry, very angry, because he does not want you to move forward in God's purposes for your life.

Some people become afraid when they think the enemy is angry, but there's no need to fear. God is on your side, and He always leads you in triumph (see 2 Corinthians 2:14). To defeat an angry enemy, seek God, stay in His Word, worship Him, and pray, as Nehemiah did. Do not allow yourself to think you have to try to defeat the devil on your own! Pray—keep on doing what you know God is telling you to do! Pray— and keep on doing what you know the devil does not want you to do!

Nehemiah never stopped praying. I wrote about his prayer earlier, but we see him praying again in Nehemiah 4:9: "But because of them we made our prayer to our God and set a watch against them day and night." In other words, they prayed consistently.

I encourage you to pray at all times, not just when you face a crisis. You do not need God only in the midst of disasters; you need Him all the time. Cultivate a lifestyle of prayer, because that is one way you build a wall of protection around yourself and those you love. To learn more about prayer, I suggest my book *The Power of Simple Prayer*.

FIGHT, DON'T FEAR

You can tell by now Nehemiah was a strong and wise leader. Not only did he seek and rely on God, he also made smart decisions and knew how to keep the people encouraged. Nehemiah 4:13–14 says:

> So I set armed men behind the wall in places where it was least protected; and I even thus used the people as families with their swords, spears, and bows. I looked [them over] and rose up and said to the nobles and officials and the other people, Do not be afraid of the enemy; [earnestly] remember the Lord and imprint Him [on your minds], great and terrible, and [take from Him courage to] fight for your brethren, your sons, your daughters, your wives, and your homes (emphasis mine).

I want to echo Nehemiah's words to you today: Fight for your home! Fight for your children! Fight for your right to live free from guilt and condemnation! Fight for your right to live under the grace of God and not be bound to legalism. Fight for your right to be happy! Fight for the dreams God has put in your heart! Fight for what is important to you! Refuse to settle for anything less than everything God has for you. The enemy will come against you, but do not be afraid of him. Instead, remember the Lord and take your courage from Him.

Refuse to settle for anything less than everything God has for you.

After this point in Nehemiah's story, he had a measure of victory: "And when our enemies heard that their plot was known to us and that God had frustrated their purposes, we all returned to the wall, everyone to his work. And from that time forth, half of my servants worked at the task, and the other half held the spears, shields, bows, and coats of mail; and the leaders stood behind all the house of Judah" (vv. 15–16).

The important lesson we learn here is that we must not let down our guard when we experience some relief or a level of victory. We want to be completely free, not merely to get a little relief. Nehemiah could have said, "Listen, everybody! Our enemies know we found out what they were plotting against us. They know God has frustrated their purposes! We can all relax!" But instead, Nehemiah wisely maintained an army against the enemy, even though he knew the enemy had retreated somewhat. He kept the same attitude he had all along: "We are contending to do what God wants us to do. Therefore, we can't slack off; we can't let up; we have to continue to think like fighters, even though our enemy is not active against us at the moment."

TAKE YOUR WEAPON TO WORK

Nehemiah knew how to get a job done. We can read about his strategy in Nehemiah 4:17–18: "Those who built the wall and those who bore burdens loaded themselves so that everyone worked with one hand and held a weapon with the other hand, and every builder had his sword girded by his side, and so worked. And he who sounded the trumpet was at my side."

One of the ways we work with one hand and hold a weapon in the other is to praise God while we work. No matter what you're trying to build—your home, your marriage, your business, financial security, an exercise plan, or an intimate relationship with God—do not forget to worship as you work. Remember to praise God and thank Him for even small steps of progress. You don't have to make a production out of your praise; just keep a thankful heart and an attitude that says, "I love You, Lord. I worship You. I can't do this without You. I need Your help today. Thank You for giving me a goal to work toward and for helping me to accomplish it."

I believe the sword that Nehemiah's builders kept by their sides represents our need to keep God's Word with us all the time. God's Word is a sword for us, and we have to wield it against the enemy. Our swords will not do any good if we keep them in their sheaths, just as a Bible won't help us if it just sits on a shelf gathering dust. They will not be effective against the enemy if we do not use them. To use our swords is to know, believe, and speak the Word of God.

If you wake up one morning and feel you want to give up, use your sword by saying: "I will not give up! God has plans to give me a future and a hope and I am going to keep working so I can experience them!" (see Jeremiah 29:11). God gives us weapons of warfare so we can use them. If you want to win, you will have to remain active. Passivity and wishing never win the battle.

THE REST OF THE STORY

I know you want to know what happened to Nehemiah and the wall. You can read the entire story in the book of Nehemiah, but let me say the project was a success. Nehemiah and the people had to fight their enemies until the walls were finished. He had to stand up to government officials who were not treating the people fairly, and Sanballat continued to pester him and even made one last effort to stop him through intimidation.

When the wall was finished, how do you think the enemies of God's people responded? "When all our enemies heard of it, all the nations around us feared and fell far in their own esteem, for they saw that this work was done by our God" (Nehemiah 6:16).

Even after the building of the wall was complete, Tobiah continued to send threatening letters to Nehemiah. The enemy did not go away, but he didn't stop Nehemiah from completing his task either. Nehemiah learned to stand through the storms of the enemies' assaults, to contend for what was important to him, and to emerge victorious. That is exactly what God wants for you.

The Integrity of a Penney

A poor minister's son from Missouri, James Cash Penney founded one of America's most successful retail businesses on the simple, timeless values of honesty, integrity, respect, and hard work.

In addition to his work as a minister, Penney's father was a farmer, and Penney learned to work on the family farm near Hamilton, Missouri. His father wanted to make sure his children understood the value of money, so he made James work to buy his own clothes, starting at eight years old.

In 1893, Penney graduated from high school, hoping to become a lawyer. Instead, he became a salesman at a local dry goods store, J. M. Hale and Brothers. Penney soon learned that everyone did not live by the same values he held dear. Other clerks took business away from him and he had to learn to be more confident, strengthen his sales skills, and stand up for the sales opportunities that were rightfully his.

In the midst of this time of personal and professional growth, Penney's health began to decline. In danger of contracting tuberculosis, his doctor urged him to relocate from Missouri to a drier climate. He ended up in Denver, Colorado, where he found employment in the dry goods trade he had learned from Mr. Hale.

The industrious and ambitious Penney soon had enough money saved to start his own business, so he opened a butcher shop. It failed, but not because Penney was a bad businessman; it failed because of his integrity. Penney believed in treating all customers with kindness and respect, and when he refused to grant special treatment to a certain influential customer, that customer used his influence to force Penney out of business.

Following the closing of the butcher shop, Penney went to

(continued)

work for the chain of Golden Rule dry goods stores. He enjoyed success there, and eventually became a partner in the company. While working for Golden Rule and living in Wyoming, Penney fell in love with Berta Alva Hess, and married her.

Based on his philosophy of honesty, integrity, and respect, Penney's business continued to grow and he soon opened his own Golden Rule store. He provided quality merchandise at reasonable prices for working-class families and developed a reputation for excellent service. He refused to take credit, requiring customers to pay cash for their purchases.

The owners of Golden Rule eventually sold out to Penney and he changed the chain's name to J. C. Penney Company. Penney moved the company's headquarters to New York in 1914 and opened the first official J. C. Penney store in Missouri in 1918.

Penney had several personal tragedies to overcome as he continued to build his retail empire. In 1910, he suffered the devastating loss of his wife, Berta, who succumbed to pneumonia, leaving him to raise two sons alone. He wrote that her death caused his world to "crash" around him. In 1919, Penney married again. His wife, Mary, bore him a son, but died in 1923, leaving him a widower again. In 1926, he married again. He and his wife, Caroline, had two daughters and remained married until Penney's death.

J. C. Penney built a thriving business and a life upon good, godly principles. He never gave up on his integrity. No matter what you try to accomplish in life, hold fast to the principles in God's Word and you will *always* be successful.

CHAPTER 8

THE KEYS TO SUCCESS

"Our greatest weakness lies in giving up. The most certain
way to succeed is always to try just one more time."
THOMAS EDISON

There are four keys to success in any endeavor you undertake: commitment, determination, waiting on the Lord, and taking time to be refreshed and renewed. If these character traits and habits will become part of the routine of your life, they'll keep you from giving up and move you toward the success you long for.

COMMITMENT

If you want to summarize in one word the idea of "never give up", you can use the word *commitment*. Without commitment, people give up easily; they have no staying power at all. If you want to be one who never gives up, commitment is the key.

Radically Committed

When an eaglet is about one year old, he begins to develop some independence. His eyesight is good at this point, and his talons are sharp. He can find his own food, fly, and even soar a little bit. He lives his life as a young eagle for about three more years, continuing to grow and gain strength, improving in all his abilities, and just getting established in life as an eagle.

But by the time the eagle is about four years old, he begins to change—not deliberately, but instinctively. He feels uncomfortable inside, strange, and uneasy. He probably does not understand what is happening to him. Eventually, in the midst of this strange season of growth, he begins to realize he was not created only to live for his own pleasure. He wants his life to be about more than himself.

At that point, the male eagle leaves to find a female. When he finds one, they begin a game of tag, which is actually a type of courtship. His uneasiness is a longing for love! It is time for him to grow up. The female eagle soars high in the sky in a figure-eight pattern and makes the male eagle chase her. Now he is no longer flying his own course; he is following someone else—someone who appears to be going in a strange direction.

After a little while she dives to the ground, picks up a twig, flies up to about ten thousand feet, and drops the twig. He dives at approximately two hundred miles per hour to catch the twig in midair and takes it back to her. What is her response? She ignores him.

At this point, even though he probably feels very insulted, he has a decision to make: "Am I going to really get committed and see this through to the finish? Is this really what I want? Or should I just forget about this and find something a little easier to do with my time?"

The female eagle repeats this process, and she makes the male's job increasingly difficult. Each time she flies, the twig gets larger and she flies at a lower altitude. That means the twig is going to hit the ground faster, and the male will have to work harder if he wants to win her over.

This game can literally go on for days. Finally, the female gets a branch that is heavier than the male eagle. I can imagine he would want to say, "This is not fair! Now, God, You *know* I can't do that!

Come on, God, You *know* that is too much for me!" This time the female flies only five hundred feet above the ground and drops the branch. If he catches it, they go on together. If not, she flies off and leaves him; she has decided to wait for a male eagle who has the tenacity to be her man.

For the male eagle to continue this process, he must be extremely committed. Once he passes the final twig test, both eagles move from courtship to the final commitment test. She flies high into the sky; he chases her; and suddenly she makes an odd move. In midair, she flips over on her back and sticks her talons up. The male moves over her and locks his talons with hers while they fall toward the earth. At this point he has made up his mind. He is committed, and he would die rather than let her go. Now they begin to sing a love song. They mate for life. Neither of them ever has another mate unless one of them dies. If the female dies, the male raises the young.

Even after the mating process is finished and she is his and they are in their nest, the male eagle continues to court the female for the rest of their lives. Male eagles have been seen to stroke the feathers of their female companions and to bring green twigs home to them long after the courtship is complete. If you are a woman reading this, you are probably wishing every man on earth would learn a few things from the eagle!

This is the kind of commitment that stays steady and strong throughout the course of a lifetime, and I hope it is the kind of commitment you will develop in your life. You may not have it down perfectly, but if you want to move on with God and be blessed in your life, you need this kind of radical commitment—the kind that follows through even when times are tough, when you feel alone or rejected, or when circumstances look bad. Think of the mighty male eagle who stays committed, even though his commitment could cost him his life. And learn to say from the core of your being, "Nothing is going to shake my commitment in this situation. I am going to see it through to the finish. *I will not give up,* no matter what."

God, the Eagle, and You

Many times we find ourselves feeling uncomfortable for the same reason the four-year-old eagle does: it's time for us to grow up. We often feel a vague sort of discontentment and struggle with a sense of restlessness when God wants to take us to a new level. The way to respond is to keep our thoughts clear and our emotions calm because chances are, nothing is seriously wrong at all. The time has simply come to grow stronger and become more mature, and we need to commit to the process.

Is commitment easy? No, but it is worth all it requires. I have fought many battles in life, but I'm now very happy, fulfilled, and satisfied. No, it hasn't been easy, and yes, I've wanted to give up many, many times. But I am committed to God and to the work of ministry to which He has called me. That doesn't mean I'm never tempted to give up; it just means that the strength of my commitment enables me to resist that temptation.

> Is commitment easy? No, but it is worth all it requires.

God wants to take you to a new level of commitment. He wants you, like the eagle, to be fiercely committed to His purposes for your life and to dedicate yourself to Him. He wants you to be in a committed relationship with Him—for life. I can't imagine anything more satisfying, more rewarding, or more adventuresome. He has more in store for you than you have ever asked or imagined, but in order to see His plans become a reality in your life you will need to be very committed!

DETERMINATION

When we receive Christ as our Savior we receive God's Spirit; we receive a new "want-to," a determination that enables us to achieve goals and pursue dreams that seem impossible. In fact, your want-to is one of the strongest forces on earth. If it is turned in the right direction, you will be able to accomplish amazing things in life. I have discovered in my own life that I should not make excuses for anything because the truth is simply this: if I really want to do something, I will do it!

Born in inner-city Detroit to a woman who dropped out of school in the third grade and married when she was only thirteen, renowned pediatric neurosurgeon Ben Carson had a rough start. His parents divorced when he was eight years old, and as a result, his mother worked two or three jobs at a time simply to provide Ben and his brother, Curtis, with the basic needs of life. His "never give up" story is as much a testimony to her determination and perseverance as it is to his.

Neither Ben nor Curtis performed well in school. By the time Ben was in fifth grade, he had the worst grades in his class and other students referred to him as "dummy." During this time, he demonstrated uncontrollable anger and a volatile temper.

Ben's problems concerned his mother greatly. She knew she had to do something. First, she began to pray, and she prayed fervently that she would have the wisdom to know what to do to help Ben and Curtis learn the lessons they needed to know and do well in school. She then instituted strict discipline at home, limiting the time the boys could spend watching television, prohibiting them from playing until they finished their homework assignments each day, and requiring them to read and write reports on two books per week. As a result of the focused attention Ben gave to his schoolwork and his voracious reading habits, Ben not only improved his grades dramatically but also became convinced he was not "dumb" and could learn as well as anyone else. Over time, he learned to control his temper and get along well with others.

By the time he finished high school, Ben was an outstanding student. He graduated with honors and went on to Yale University, where he received a degree in psychology and had plans to continue his education by preparing to be a psychiatrist. Once in medical school, at the University of Michigan, Ben decided to pursue neurosurgery instead of psychiatry. Ben completed his residency in neurosurgery at one of the most respected hospitals in the world, Johns Hopkins in Baltimore, Maryland, and by the time he was in his early thirties, he was that hospital's director of pediatric neurosurgery— the youngest person to ever hold such a position at Johns Hopkins. As of this writing, he remains in this position; serves as a professor of neurological surgery, oncology, plastic surgery, and pediatrics at Johns Hopkins medical school; and maintains a busy schedule speaking to and encouraging young people to maximize their intellectual potential.

If you dream of reaching a goal that seems impossible, remember Ben Carson, who went from being the lowest achiever in his class to being one of the brightest medical minds in the world today. It was possible for him because his mother refused to give up on him and he never gave up on himself.

If you are the parent of a child who struggles in school or in any area of life, let the story of Ben Carson inspire you. Remember what a difference his mother made in his life, and do as she did: pray and ask God to give you wisdom to help your children succeed. Then help them apply the discipline they need and don't let them give up!

If you really want to come out of the bondage that's holding you captive, you will! If you really want to break free from your past and move beyond it, you will. If you really want to develop a positive attitude instead of having a negative one, you will. If you really want to wait and marry the right person and not take "second best" because you fear no one else will come along, you will!

If you're determined enough, nobody—no evil force from hell, no person on earth—can stop you from being successful. If you will obey God and do what He tells you to do, and if you will be determined to outlast the devil every single time he comes against you, then nothing will be able to keep you from reaching your goals.

Be Determined

One of the definitions of *determine* is: "to settle a dispute by an authoritative decision or a pronouncement." This definition encourages me, because I make "pronouncements" often as a way of building my determination in certain areas. Sometimes, when I know I have eaten enough, but my flesh still wants to eat more, I sit at the table and announce aloud: "This is my last bite! I am *finished* with this meal!"

When you are determined in a particular area, you may have to talk to yourself in a similar way. Sometimes the best way to overcome the temptation to give up on something is to say to yourself, "Oh no, you don't! Stop your whining and straighten up right this minute!" Over the years, when I have felt like giving up and had no one to encourage me, I have said to myself: "Joyce, you can make it! It may be hard, but you *can* make it and don't you dare think you can't!"

> Sometimes the best way to overcome the temptation to give up on something is to say to yourself, "Oh no, you don't! Stop your whining and straighten up right this minute!"

At one point in my life, I went through a ten-year period of suffering from various physical problems. *Ten years!* I had lots of symptoms, some of which involved low blood sugar, but most of which were rooted in stress caused from working too hard. I was also going through the "change of life," which was not easy for me. Hormone changes in my body caused migraine headaches and at one point in 1989, as I mentioned, I even had breast cancer and needed surgery.

I can't count the number of times during those ten years I walked onto a platform to preach thinking, *I wonder if I can stand here long enough to preach my message.* I prayed fervently for God to heal me. He did not take away my problems completely, but He did give me strength to do what I needed to do *in spite of them*—every single time. But I didn't want the strength to press through; I wanted the problems gone! Trying to keep going through those years was hard, *really* hard, but I was determined to not be defeated and to not give up. I realize now that my faith became strong during those years and I gained a genuine compassion for the sick.

I thank God for giving me a solution to my problems after ten years. I wanted it to be sooner, but our times are in His hands. Now I'm doing great, and it's wonderful to do what I do and feel good! But many times I had to say, "I would rather do my job and feel good, but hear this devil: I will do it, whether I feel good or not."

While I was suffering, I had to speak words of determination out loud: "I would rather feel good while I do this. Doing it this way is hard, but I am going to fulfill God's call on my life no matter how I feel. I'll do everything God wants me to do and be all He wants me to be!" I remember one morning when I felt really bad. I barely had any strength, but I prayed and told the Lord, "I may not have much strength, but whatever I do have I will use it to serve You the rest of my life." Each time I made a decision like that one, the devil was losing and God was winning.

The primary reason I was so determined to obey God's will whether I felt well physically or not was that I love God. I also knew I would be miserable if I tried to do anything less than everything God called me to do. You may not always want to make the effort to be determined, but I guarantee that you do not want to risk being outside God's will simply because you will not commit to doing everything required to obey it.

Settle the Dispute

Remember, the definition of *determine* carries the idea of settling a dispute. This is important because if you are going to commit to anything, you will have to settle the never-ending dispute between your flesh and your spirit. I often say the flesh is a "gambler," but the spirit is an "investor." As you follow the Holy Spirit and His leading in your heart (your spirit), you will invest in tomorrow by making right choices today.

Galatians 5:17 says, "For the desires of the flesh are opposed to the [Holy] Spirit and the [desires of the] Spirit are opposed to the flesh (godless human nature); for these are antagonistic to each other [continually withstanding and in conflict with each other]." This verse is basically telling us a war rages within us all the time. The flesh and the spirit do not ever get along; they always fight against each other.

This happens in all of us. We have a desire, an impulse, or a sense about something, and we know in our hearts it is right. But our minds try to talk us out of acting on it. Let's say you sense you are to give some money to a family in need. Your heart believes it is the right thing to do, and you believe it will please God and is even inspired by His Spirit. Your flesh will say, "You should not give away that money. You know you need it" or "Don't give anything to those people, they have never done anything for you." The flesh rages against the spirit, and you begin trying to figure out which one to listen to.

I am convinced we miss many blessings in our lives because we try to understand so much instead of simply allowing the Holy Spirit to lead us. We need to settle the dispute between the flesh and the spirit and be determined to obey God, no matter what.

WAITING ON THE LORD

God compares His people to eagles: "But those who wait for the Lord [who expect, look for, and hope in Him] shall change and renew their strength and power; they shall lift their wings and mount up [close to God] as eagles [mount up to the sun]; they shall run and not be weary, they shall walk and not faint or become tired" (Isaiah 40:31). I believe God chose to liken us to eagles to motivate us to rise to our potential in life and to encourage us to wait on Him and find our strength in Him. When success does not come easily, when we find ourselves frustrated and weary in our efforts, we need to wait for the Lord.

What does it really mean to wait for the Lord? It simply means spending time with Him, being in His presence, meditating on His Word, worshipping Him, keeping Him at the center of our lives. One meaning of the word *wait* is "to be twisted or braided together." If we think about a braid in someone's hair, we realize that the hair is woven together so that we cannot tell where one strand ends and another begins. That is the way God wants us to be in our union with Him—so intimately intertwined and tightly woven together with Him that we are truly one with Him, that we are direct representatives of His character. As we wait on Him, we become more and more like Him. Waiting is not passive; it is very active spiritually.

> *Waiting is not passive; it is very active spiritually.*

While we wait, we need to aggressively expect God to do great things in us and in our lives.

An intimate relationship with God will strengthen you in the innermost part of your being. It will strengthen your heart; it will carry you through the hard times in your life with a sense of peace and confidence that all is well, no matter what is happening. It will give you the strength to endure tough situations in such a way that many of the people around you may not even be able to detect even the slightest stress in your life. When you wait on the Lord in the way I describe, you draw everything you need from Him. He is your refuge, your enabler, your joy, your peace, your righteousness, your hope. He gives you everything you need to live in victory over any circumstance.

Isaiah promises that when we wait on God and spend time with Him, becoming more like He is, we will renew our strength. That little prefix *re* means "to go back again, to be made new, to go back to the beginning of something." When we wait on God, our strength is made new again; we can fly as eagles do, over the storms of life; we can walk and run and not faint. We can approach a situation in which we have grown weary with fresh energy and passion. We are encouraged, energized, and less likely to want to give up when success eludes us.

One reason eagles symbolize strength is that they know how to make their strength work for them. They don't expend their energy needlessly. Just as these magnificent birds know how to allow the thermal currents of the physical world to carry them, "eagle Christians" understand the currents of the Holy Spirit. They perceive the moving of the Spirit and can flow through life easily, without strain, striving, and undue stress. This comes from waiting on God and enables them to go through life with strength, power, and vitality.

People often ask me, "How can you and Dave do what you do at your age?" The truth is, we probably feel better than many of the thirty-year-olds in our audiences. One reason for this is that we take care of ourselves physically. We eat healthily; we exercise; we drink plenty of water; and we get enough sleep. We strive to keep excess stress out of our lives and we have learned to wait on the Lord. Jesus said that the weary, the worn out, and the exhausted should come to Him (see Matthew 11:28) and He would cause them to rest.

Many times, between speaking sessions at our conferences, I lie on the bed in my hotel room and rest my body. I also use that time just talking to the Lord. I don't say anything fancy, just, "I love You, Lord; I need You. Help me tonight. Thanks for what You did today." In those moments, I am not only resting my physical body, I am also resting spiritually by trusting God to be my helper and my strength.

Sometimes I grow weary—not just physically, but in other ways too, and need more than a quick rest. When that happens, I spend a day or two being alone as much as possible. I just want to go somewhere with my coffee and sit in a chair and be quiet. I want to spend quiet time with God, and I know that time with Him will renew my strength.

No one is exempt from the need for renewal; everyone needs times of rest, refreshing, and restoration. As you go through life, you grow weary. As you experience disappointments, you grow weary. As you work to meet deadlines and budgets and quotas, you grow weary. As you study for exams, you grow weary.

When weariness sets in, bad situations can arise. Weary people often make comments without thinking, take shortcuts they later regret, and settle for less than the best because they are tired of waiting. Weary people are being unwise if they do not allow their strength to be renewed. I believe many people make poor decisions because they become worn and fail to take the time to draw close to God and wait on Him so they can be strengthened and restored.

I'm convinced that people would make better choices in life if they would simply spend time with God on a regular basis. If they will seek God, they will hear from Him. Many times, they may not even realize He is speaking, but they will find themselves knowing what to do and what not to do in certain situations.

Begin to take time to wait on God and allow Him to renew your strength so you can soar. Begin today to take breaks as soon as you feel weary and say, "I love You, Lord. I need You. I feel a little weary, Lord. Strengthen me."

> *Take time to wait on God and allow Him to renew your strength so you can soar.*

You may not be able to go on a two-week vacation right now, but you can start with several five-minute vacations each day. You may not be able to go away for several days, but even if you have to escape to a bathroom stall or play a worship song in your car, get alone with God for a few minutes and say, "Jesus, I need a little renewal. I love You. I worship You. Strengthen me and renew me." He wants to strengthen and restore you so you will be able to persevere and enjoy success. Waiting on the Lord does not have to be complicated. Just put Him at the top of your priority list. He wants you to have and enjoy a quality of life you may be missing because you are not spending time with Him.

RENEWAL

Psalm 103:2–5 says:

> *Bless…the Lord, O my soul, and forget not [one of] all His benefits—Who forgives [every one of] all your iniquities, Who heals [each one of] all your diseases, Who redeems your life from the pit and corruption, Who beautifies, dignifies, and crowns you with loving-kindness and tender mercy;* Who satisfies your mouth [your necessity and desire at your personal age and situation] with good so that your youth, renewed, is like the eagle's [strong, overcoming, soaring]! *(emphasis mine).*

Most of us find the idea of having our youth renewed very appealing. We'd like to look younger and have the kind of energy we had in our earlier days. But I believe that what the psalmist calls "youth" we refer to as "young at heart"—a fresh, vibrant approach to life, rather than a tired, jaded, negative outlook. Youth is not just a matter of chronological age; it's what goes on inside of you. You live in a body, but you are a spirit, and if you keep your spirit strong, it will affect your body, your mind, your emotions, and your decisions in positive ways.

The psalmist would not have mentioned having our youth renewed like the eagle's if the eagle did not teach us something about renewing youth. I find the eagle's renewal process fascinating and awesome, and I hope you will too.

The Process of Renewal

There comes a time in the eagle's life when he is not as quick as he once was. His takeoff is not as fast as it was years ago; he is slower in flight; his previously sharp talons have grown dull; calcifications have formed on his beak; his feathers have worn and now give off a telltale whistle when he dives for his prey. He's still an eagle, but he has lost much of his strength and prowess.

When this happens, the eagle settles onto a high rock as close to the sun as he possibly can and begins to pluck out his feathers one by one—sometimes as many as seven thousand of them. He's not nearly as concerned with the pain as with the progress. He then seeks a cool, refreshing stream in which to clean himself. The water washes away all the caked mud and dirt, parasites, and insects that may have collected on him over time. When he is fresh, clean, and practically naked, the eagle stands before the sun and begins to wait.

The regrowth he needs will take about forty days. During that time, he sharpens his talons and beak by rubbing them back and forth on a rock. He uses that same rock to beat the calcification off of his beak. Other eagles who have already been through this process may drop food to him. He goes through a quiet season of relative weakness, but then his strength is renewed.

I believe God wants us to learn something from this process the eagle endures. There are times we feel dry, weak, defeated, and discouraged. There are times when life deals us bitter blows and our dreams crumble right before our eyes, when someone we love hurts us, when someone we trust betrays us, when we really thought something was finally going to work out and did not. These are the kinds of situations that make us want to quit. During those times, we really need to be renewed. If we don't take time for renewal, we are in danger of reacting to our circumstances emotionally, which always means responding without wisdom and often contributes to making bad situations even worse.

These are the times we really have to get serious with God. We need more than we usually get from our daily devotional routines. Some people may need to take a lesson from the eagle and set aside a forty-day period of prayer and fasting, seeking God, worshipping Him, and pouring out their hearts to Him. To those who have never fasted this may sound extreme, but desperate people do desperate things in desperate times. Other people need to get *radical* about bringing peace into their lives—turning off the television for a while or turning off their cell phones at a certain time each evening. Maybe you need time to examine your life to decide what is not bearing fruit and needs to be cut off.

I'm convinced that we all need extended or intense times of restoration and renewal. Personally, I try to schedule three or four each year to get away by myself for a week. I make the effort to take these times away, because I believe one of the best gifts we can give ourselves is time alone with God.

> *One of the best gifts we can give ourselves is time alone with God.*

We need times of real quiet because we can connect with God in awesome ways in those places of deep peace and quiet. We do not have to feel we have had a definite "word" from God, nor do we need to have a supernatural experience. Honoring God by giving Him special, set-apart time will produce great results in our lives and bring restoration and refreshment. We will find ourselves making better decisions, displaying godly character with ease, and enjoying all of life much more. Time is one of the most important things we can give God. It tells Him He is important to us and that we realize we cannot manage life properly without Him.

In your pursuit of success, remember to be committed and determined, and wait on the Lord and renew your strength. As you do, you will be empowered to keep going without giving up.

A Voice in the Wilderness

No stranger to hardship or persecution, Marian Anderson was born in 1897, in the "Negro Quarter" of Philadelphia, to a loving family with very limited financial resources. When her father died unexpectedly when she was only ten, money was tight, and her mother went to work as a cleaning woman and a laundress to support Marian and her two sisters.

Marian's amazing singing ability made a way for her to enjoy opportunities otherwise unavailable to her. She began singing in the choir of the Union Baptist Church, where people quickly noticed the quality, range, and richness of her remarkable voice. They knew it was something special. They also knew Marian's family could not afford formal vocal training for her, so the church sponsored a benefit concert, with ten-year-old Marian as the featured soloist, to pay for her voice lessons.

Her family couldn't afford piano lessons either, so she taught herself to play. When she wanted to learn to play the violin, she took a job scrubbing steps to make money to buy her own instrument. Obviously, she was fiercely committed to her music. At one point, she went to apply for admission to a music school in Philadelphia and was treated rudely by a young receptionist. When Marian expressed her desire to pursue enrollment, the young woman replied, "We don't take colored."

At the age of nineteen Marian was introduced to well-known voice teacher Giuseppe Boghetti, who was her teacher, coach, and friend for years. As her abilities and exposure grew, she began receiving invitations to sing and even to tour. With her confidence strong and with strong supporters around her, she arranged to sing at New York's town hall in 1924. The concert was so poorly attended and so negatively reviewed that she considered abandoning music altogether.

But Marian soon bounced back. She went on to win a voice competition sponsored by the Philadelphia Philharmonic Society and then to triumph over more than three hundred other contestants in the Lewisohn Stadium competition. She began touring again, and in 1928, she sang a solo recital at Carnegie Hall.

In 1939, despite her remarkable accomplishments, Anderson was still denied opportunities because of racism. That year, the owners of Washington DC's Constitution Hall refused to allow her to sing because of her race. When Eleanor Roosevelt, wife of the president of the United States, heard what had happened, she made arrangements for Marian to sing at the Lincoln Memorial instead. Approximately seventy-five thousand people attended that outdoor concert. That event was a significant moment in the advancement of civil rights in America and gave many other people who suffered from racism and injustice the courage to pursue their dreams.

Anderson went on to become the first African American to appear as a soloist at New York's Metropolitan Opera. She also sang at inauguration ceremonies and received many prestigious awards, including the Presidential Medal of Freedom in 1963 and a Grammy Award for Lifetime Achievement in 1991. As her outstanding career drew to a close, she launched her 1956 farewell tour with a triumphant concert at a place that once refused to even allow her inside its doors—Constitution Hall in Washington DC.

No matter what doors seem closed to you today, keep going and never give up. Persevere with determination, and opportunities that seem impossible today will open for you tomorrow.

CHAPTER 9

OVERCOMING THE OBSTACLES TO SUCCESS

"When you get into a tight place and everything goes against you, till it seems as though you could not hold on a minute longer, never give up then, for that is just the place and time that the tide will turn."

HARRIET BEECHER STOWE

A donkey fell into a deep well, and the farmer who owned him had no idea how to get the donkey out. After much thought, he concluded the best solution would be to call some of his friends to help bury the donkey in the well. After all, he reasoned, the donkey was old and trying to get him out of the well was going to be a lot of trouble.

The neighboring farmers arrived with the shovels and they all began to pitch dirt into the well, on top of the donkey. The donkey began to make horrible noises—for a while. Then the donkey was silent. The men peered down into the well and saw an amazing sight.

Every time they threw a shovel full of dirt into the well, the donkey simply shook it off, so it ended up under his feet instead of on top of him. They kept shoveling, trying to bury the helpless donkey and he kept shaking off the dirt! Before long, the pile of dirt with which they were trying to bury him became tall enough to raise him out of the well, and he simply stepped off of the pile onto level ground.

If the donkey had just stood there and allowed the dirt to bury him, he would not have survived. But he was determined to get out of that pit. Every time a shovel full of dirt landed on him, he shook it off and climbed on top of it.

I want you to be like the donkey. Use the difficulties and obstacles you face as stepping-stones to your goals in life. When a seemingly impossible situation comes your way, do not just allow it to bury you. Be creative about overcoming it and determined to make it work for you, not against you. Turn it to your favor, even if you have to get on top of it one step at a time. Let the circumstances that could suffocate you be the very situations that strengthen you and raise you to a new level. Be a person who is willing to work with God to develop a determined, "can do" attitude; act on the truth that His power is at work within you and that you can do all things through Christ who strengthens you (see Philippians 4:13), no matter what stands in your way.

CERTAIN TO SUCCEED

True success does not come easily or without hurdles for anyone. It is the result of hard work, patience, determination, creativity, sacrifice, and moving beyond mistakes—but it does come. The only way you will ever be a failure is to give up, and you'll be most tempted to do so when you face opposition.

Scripture clearly teaches us we have an enemy who wants to frustrate and defeat us. His primary goals are to "steal and kill and destroy" us (John 10:10), and he spends his time roaming around "like a lion roaring [in fierce hunger], seeking someone to seize upon and devour" (1 Peter 5:8). He will oppose and interfere with our success in every possible way, doing his best to discourage us so we will give up.

I want you to be successful in every area of your life—your work, your relationships, your marriage, your parenting, your finances, your health, your ministry, your creative expressions, your pursuit of God's plans, and everything else in which you are involved. But I know lasting success requires effort, and as you work toward it, the enemy will be there to try to steal, kill, or destroy it. It's crucial to learn to keep moving forward in strength when success does not come as easily as you would like and when you face the obstacles you are sure to encounter along the way.

CLEARING HURDLE #1: THE TEMPTATION TO QUIT

The temptation to quit is part of being human, but giving in to it is equal to giving up. We do not avoid the temptation by sitting still and refusing to deal with it or wishing God would remove it. It's one of the realities of the Christian life and a hindrance to success we must work to overcome. The Bible tells us in Luke 22:40, "Pray that you may not [at all] enter into temptation." Tempting believers is part of the devil's job. As long as we live, we'll be tempted; and as long as we follow Jesus, our job is to resist temptations.

> *Tempting believers is part of the devil's job. As long as we live, we'll be tempted; and as long as we follow Jesus, our job is to resist temptations.*

Because we don't always realize that feelings of discouragement and thoughts of quitting are from the enemy, we don't always resist such temptations as we should. Some thoughts the enemy may plant in your mind to tempt you to give up might sound like this:

- This is too difficult. It will require too much effort.
- I really am not qualified to do this.
- I am facing too many problems and can't possibly solve them all.

- I have no one to help me.
- My friends and family think I'm crazy for pursuing this.
- This will require too much of a sacrifice.
- I don't have the money to do this.
- I'll never be able to finish this.
- What was I thinking when I agreed to this assignment?
- I knew this would never work.
- I need to go back to my old job/city/way of life/ministry. Everything was better there.

I encourage you to begin to recognize temptations to quit as works of the enemy; and I want you to start resisting each temptation with everything in you. Don't flirt with temptation or consider any temptation insignificant. Don't let the devil lure you into passivity or wait until you've been in a depressed, hopeless slump for three days, listening to the enemy list reasons to abandon your cause. Resist the devil at his onset! Declare war against temptation. Show the enemy no mercy. The *instant* you feel tempted to give up, you need to say aloud, "*I will not quit. I refuse to give up. I will* finish what God has called me to do."

In over thirty years of ministry, I've often been tempted to quit, especially during the early years when we traveled extensively. There were many occasions when I woke up in hotel rooms, exhausted from preaching the previous night and weary of hearing about problems at the office—someone quit; an urgent matter needed my attention; we received a bill we weren't expecting; and so on. At times I became extremely discouraged, sometimes to the point of tears. I didn't always want to get dressed and go preach to people about pressing through their difficulties when I felt overwhelmed by the pressures in my own life.

Over the years, I've had to resist many temptations to want to give up, and I've learned the best way to work at resisting temptation is to do your job in prayer. It's far wiser and more effective to pray and ask for God's help as you stand against temptation than to try to exert willpower alone. When you seek His strength, He gives it to you.

Jesus told the disciples twice in one day to pray that they wouldn't enter into temptation. If it was that serious a threat in His eyes, we'd better take it seriously too. Look at Luke 22:40 and the verses following: "And when He came to the place, He said to them, Pray that you may not [at all] enter into temptation....And when He got up from prayer, He came to the disciples and found them sleeping from grief, and He said to them, Why do you sleep? Get up and pray that you may not enter [at all] into temptation" (Luke 22:40, 45–46).

Be sure that you work with God and pray that you won't surrender to the temptation to give up. Ask Him to help you finish what He has called you to do, be all He wants you to be, do all He wants you to do, and have all He wants you to have.

CLEARING HURDLE #2: PERSONAL PROBLEMS

When the enemy sees we are determined to complete an assignment from God, one of his favorite and most effective tools to make us give up is personal pain. I've seen so many people who face personal crises and problems as they walk through life, and I've experienced personal pain myself as I've headed toward the accomplishment of goals or the fulfillment of dreams.

I once had to conduct a seminar during a time when I was really heartbroken over a serious situation involving one of our children. The place where I was speaking had a small room behind the platform, a nice place for speakers to take breaks and be refreshed between speaking sessions. When the time came for me to speak, I would walk onto the platform and conduct the meeting as usual. Then I would go behind the platform and cry. When the next session began, I dried my tears and carried out my responsibilities again; then I went back to that little room and cried some more. I repeated that cycle for the duration of the seminar. I was hurting personally, but I still fulfilled my responsibility.

To be people who remain determined and diligent, we must learn how to make it through times of personal pain. People of character simply don't say, "Well, I'm not going to do my job tonight because I have a problem! I'm going through a personal crisis, so how could people expect me to minister to them?"

That may sound silly to you, but there are people who allow their personal problems to rule their lives and dictate which commitments they keep and which ones they cancel. If they have personal problems, they decide they can't work in the church nursery on their assigned Sunday. They don't want to do their jobs or keep their word. They fail to do what they said they would do because they think their individual crises should excuse them from their commitments.

Helping someone when you're struggling with pain of a personal nature is extremely difficult; it takes commitment and determination. As a minister, I can assure you, ministering to hurting people when you are hurting too is quite a challenge. Listening to other people's problems when you have problems of your own can be extremely hard. But it helps you develop great compassion and it makes you stronger.

When you're suffering, you still need to keep your word if at all possible and do your best to help others. In fact, those are precisely the seasons when you need to be diligent to keep your commitments and actively look for ways to bless the people around you. Times of personal testing are exactly the times you *need* to go all the way through what you promised God and keep serving Him.

We all experience personal problems, and there are times in life when it is impossible for us to keep a commitment, but we should do our best to keep those times to a minimum. Our challenges may pertain to our health, our children, our marriages, aging parents, stress at work, time management—and the list goes on. We cannot allow these difficult circumstances to derail us as we head toward fulfilling God's plan for our lives. The people who refuse to be deterred are the ones who are determined to stay focused on their jobs, who stay devoted to God, who remain committed to their families, who are determined to confront every obstacle that comes their way. They are winners in life.

CLEARING HURDLE #3: REJECTION

When people reject you, you may not want to keep doing what you're supposed to do; you may want to find a place to hide and nurse your wounds, but don't! Keep moving forward, no matter who disapproves or tries to stop you.

If you look back over the course of your life, you'll probably find that every time God has tried to take you to a new level, someone usually disapproved. In some way, that person sent you the message "If you do this, I'm not going to be your friend." Do you know why this happens so often? Probably more than anything, Satan uses the pain of rejection from other people to keep us from pursuing God's will. The pain of being rejected by those we love is so powerful it often causes us to abandon our determination to do what God has called us to do. Rejection causes us to convince ourselves we cannot do anything to please anybody and makes us want to give up. We need to remember that we may not please people all the time, but if we please God, that is what really matters.

Jesus dealt with the issue of rejection in Matthew 10:5–14. In this passage, He warned His disciples that some people would reject them, and He told them how to handle that rejection: "And whoever will not receive and accept and welcome you nor listen to your message, as you leave that house or town, shake the dust [of it] from your feet" (v. 14). In a very real sense, they were to "shake off" the rejection and go on.

When God filled me with the Holy Spirit, He was equipping me for the call He was about to release in my life—to begin teaching His Word.

As soon as that happened, I experienced quite a bit of rejection, which I mentioned earlier. Almost every friend I had thought I was crazy. At that time, in the circles in which I moved, women did not preach and teach the Bible. When everyone around me seemed to reject me, I had to be determined to not give up on what God called me to do.

Many people will want to attach themselves to you when you are going "up" in life—when you are doing something others think is good or when you are gaining visibility, prestige, or respect. But when you do something unpopular or out of the ordinary, you may lose some friends. If you want to do anything great for God, you may have to be willing to endure rejection and loneliness for a while. It is worth it, though, and in the end, you will be glad that you pressed through. Realize that no matter how much rejection you face, God accepts you completely. He wants you to break through the barrier of rejections and keep making progress toward your goals.

CLEARING HURDLE #4: WANTING TOO MUCH TOO SOON

People have asked me through the years, "What has been the most difficult aspect or experience of your ministry?" I always respond, "Not giving up when we were laying the foundation." Though we have certainly faced our share of situations that tempted us to want to give up, nothing has been as challenging as staying focused and diligent during those early years, when so much of the ministry was unseen and unknown. We certainly wanted a fruitful ministry, but we knew we needed a strong foundation before we could build one. That meant everything needed to be done with excellence.

Dave and I were growing personally at the same time we were try-ing to grow the ministry God had called us to. He spoke to our hearts about excellence, integrity, and keeping all strife out of our personal lives and ministry. We were being stretched like never before. Dave and I were learning how to work together, and keeping strife out of our personal relationship was something we both had to work hard to achieve.

We didn't have many people to help us and we had no experience, so everything we did was a huge step of faith. Every decision seemed major to us and we were learning how to hear from God. It was often tiring and discouraging, especially when we did our best and the growth was painfully slow.

I began by teaching a Bible study in my living room for five years. For the last half of that time period, God added a second Bible study in someone else's home, so I taught two Bible studies per week, with approximately twenty-five or thirty people in each one. I also did other types of ministry—counseling, helping people gain freedom from various types of bondage, passing out evangelistic tracts on downtown streets, praying (a lot)—anything I could do to serve God. I had quit my job in order to minister to others and, although we struggled with finances, God gave us the grace to not give up.

After five years, God promoted me. I went to work at a church in St. Louis, and I worked there for five years. My first paycheck was sixty-five dollars and that seemed like a fortune to me. There, I started a women's meeting. God's hand was upon that gathering; it grew very quickly and we soon had about five hundred ladies attending each week.

Over time, I became an associate pastor in that church and then went on to teach at the church's Bible college. Little by little, God trained me and strengthened me, but I often wanted more.

The time came when God spoke to my heart and said, *Now I'm finished with your work in this church. Take your ministry, and go north, south, east, and west.* As I obeyed that word, which took time, I found myself with more meetings than I had held before, but they were smaller in size. I felt as though I was starting all over again! I was ministering to more people, but they were in smaller groups— seventy-five in one place, sixty in another, one hundred over here, eighty-five over there. When we had meetings of one hundred and fifty or two hundred people, we thought we were really doing something! I was working harder than ever and had more responsibility.

Each year, the ministry grew and our meetings became larger and larger. We reached three hundred, then four hundred, then four hundred and fifty. By that time we were becoming known and beginning to travel to hold meetings in other cities. And then the growth stopped—at least for a season.

But one morning while Dave was getting ready for work, the Spirit of God visited him in a powerful way and he began to weep. God showed him the condition in which many people live and said to Dave, *All these years I've prepared you to go on television. Now I want you to go on television. It's time for this ministry to go on television. You have the answers people need.* We had those answers because God had used the times of slow growth to teach us what He wanted us to share with others.

When we went on television, our ministry tripled in three months. Since then, it has grown so quickly we've been running to keep up with God. We now have more people on staff than used to attend my meetings. We have great vision and plans for the future, and I can honestly say God has caused our ministry to grow in exactly the right way and at precisely the right times. I'm glad He has led us as patiently as He has, even though I grew frustrated at times.

Whatever you believe God wants to do in your life, be patient as He brings it to pass. Eagerly embrace the times of teaching and preparation He takes you through, even when it seems painfully slow. Don't long for too much too fast, but be thankful every day that God is bringing growth, expansion, and new opportunities your way in His perfect timing.

CLEARING HURDLE #5: INCONVENIENCE

We live in an "instant" society. We want everything to be easy, comfortable, and convenient. When faced with a little work we say, "Oh, woe is me. That's too hard. I have to unload the dishwasher." I was around before dishwashers existed and all I have to say to those who complain about dishwashers is, be thankful you didn't have to wash the dishes by hand! And you didn't have to dry them!

We have escalators and elevators so we don't have to climb stairs. We have push-button microwaves so we can cook our food in seconds, so we will not miss our favorite television programs. We have instant potato flakes so we do not have to peel, boil, and mash our potatoes. We have drive-through windows so we don't have to get out of our cars for our prescriptions, dry cleaning, or tacos. We have push-button this and push-button that, and instant this and instant that. We are addicted to ease and comfort.

The sad part about all these conveniences is, although we view them as "good," they actually work against us in many ways. They cause us to seek the easy way out when taking the more difficult way just might build something of worth in us. You will never face an obstacle God is not aware of. When your journey becomes difficult, remember that even though the world prefers "the easy road," God did not build you for comfort. He designed and equipped you for adversity, whether you know it or not. He made you in such a way that the best comes out of you when you face hard situations and you experience joy and strength when you overcome them.

CLEARING HURDLE #6: JEALOUSY

Years ago, Dave and I both smoked cigarettes, and God was dealing with us about quitting. It was not good for our health, and it was an expensive, stinky habit. Dave simply said to God, "Well, I know I need to stop smoking, but I am going to continue as I am until You give me the grace to quit. I'm not going to worry about it; I cast my care on You." The next day, he got up not wanting to smoke, and to this day, he has never wanted another cigarette.

Do you think I was able to stop smoking as easily as Dave was? No! Let me repeat: *No!* I had to *suffer!* I tried to quit so many times I lost count. My routine of "walking it out" went something like this: I couldn't stand to be without a cigarette, so I went and bought a pack. Then I reminded myself I was quitting, so I threw it away. Then I reached the point where I had to have a cigarette, so I rummaged through the trash can to find one! Then I felt terrible about smoking, so I put out the cigarette and threw it away again. And in the middle of the night I would dig it out and relight it! There I was, in the wee hours of the morning, puffing on a cigarette under the biggest load of condemnation you can imagine!

At that time, Dave and I attended a Spirit-filled church and after services, I would make a beeline to the parking lot, roll down my car windows, lie down in the car, and smoke a cigarette—and I thought none of our church friends knew. They probably thought our car was on fire because of the smoke rolling out of the windows.

All this time, Dave was going through life smoke-free. Many times I asked, "Dave, isn't this *hard* for you?"

"Oh no," he responded. "God's grace is on me."

Furious and wanting a cigarette, I wanted to scream, "Well, where is His grace for *me?!*" Perhaps I didn't have enough faith at the time to receive His grace for me, but whatever the problem was I kept smoking and resented Dave because quitting had been so easy for him.

Since that time, I've learned that one obstacle to experiencing the breakthroughs we long for is the tendency to look at others and become jealous of the way God brought breakthrough in their lives. We think something is unfair about the fact that they enjoyed quick success, seemingly without effort, while we have to struggle and suffer. Part of overcoming the obstacle of jealousy is trusting God enough to believe that His plan for each of us is perfect. It may not seem fair at times, but God does know what He is doing and He knows how He wants to use us in the future.

In my case, I finally got determined and said with much prayer and leaning on God, "I am going to quit smoking! With God's help I *will* quit smoking! I do not care how hard it is, *I am going to quit!*" After that, I went through a difficult period for about thirty days and then it was over. I finally had a breakthrough and have never smoked or wanted to smoke again.

Why didn't God deliver me the easy way like He did Dave? I don't know. But I do know that I learned an important lesson about trusting His timing for me without being jealous of the way He works for others.

CLEARING HURDLE #7: SELFISHNESS

Another obstacle to the pursuit of our goals is the fact that we can be selfish, especially with our personal space and our freedom. If we're going to be committed to never giving up, we're going to have to sacrifice our selfish wants at times. Persistence and determination are hard on the "self." After all, if I'm committed to something, then what happens when the time comes and I don't feel like doing it? We often think freedom is being able to do whatever we want to, whenever we want to do it. But the truth is, if we ever could live that way, it would just drive us to be more and more selfish and self-centered.

If we aren't determined to never give up, it's easy for the devil to talk us out of doing whatever we should be doing by appealing to our selfish desires. Persistence requires us to discipline our feelings and do what is right, no matter how we feel about it.

We need to surrender our freedom and our personal space once in a while and learn to say no to ourselves. We cannot trust our flesh; it won't get us where we need to be when we need to be there. It will want to stop when we need to keep pressing on. It will tempt us with our personal space and freedom—at the expense of our personal growth.

It says that feeling good right now is more valuable than living a disciplined, committed lifestyle that will result in the achievement of our goals and the fulfillment of our destinies.

We need to surrender our freedom and our personal space once in a while and learn to say no to ourselves.

Not only does God want you to succeed, He *created* you to succeed. And He does not want you to settle for anything less. Don't allow failures or mistakes to cause you to give up; let them teach you what to avoid the next time around. As you move toward success, wait on the Lord and let Him renew your strength. Success will not come easily, but it *will* come, if you persevere.

The Freedom Fighter

British political leader William Wilberforce is one of the world's best-known champions of freedom and opponents of cruelty and injustice. In 1780, this Cambridge-educated son of a wealthy merchant became a member of Parliament when he was twenty-one years old, the youngest age permitted. A charming and articulate speaker, Wilberforce enjoyed representing his constituency in Parliament, but did not align himself with any particular cause or legislative agenda during his early years.

In 1785, Wilberforce had what can only be called a remarkable encounter with God. After that, he wanted to do everything he could do to serve God and thought he had to choose between his political career and full-time ministry. Eventually, he realized he didn't have to be a professional minister to do God's work, but that he could do God's work most effectively outside the walls of a church, in the halls of government.

By 1787, Wilberforce had emerged as Parliament's dominant voice against the slave trade. He had very few supporters, because the wealth of the British Empire depended on slave labor and Britain led the world in taking slaves from Africa and shipping them under horrific conditions to docks where they were sold like livestock to plantation owners and then forced into manual labor and often treated terribly.

Speaking of the slave trade, Wilberforce said: "Let the policy be what it might, let the consequences be what they would, I am from this time determined that I would never rest until I have affected its abolition." Wilberforce certainly did not rest in his efforts. He endured sickness, hardship, controversy, the betrayal of colleagues he believed supported him, and the mockery of others, as he tirelessly and passionately worked to stop human trafficking. He presented his bill for the abolition of the slave trade to Parliament eighteen times before it finally passed.

In 1807, Wilberforce finally saw the goal for which he had labored, sacrificed, and endured ridicule accomplished. The slave trade was officially abolished. However, the end of the slave trade did not liberate people who were already enslaved, so that became Wilberforce's next objective. Wilberforce died in 1833, knowing the end of slavery was imminent. One month after his death, the law abolishing slavery was enacted.

Injustice continues to run rampant in the world today. You may or may not be called to publicly champion a specific cause, as Wilberforce was, but you can treat people well, stand up for what is right in your life, and help those who cannot stand up for themselves. Like Wilberforce, you can serve God wherever you are—as a university student, a businessperson, a stay-at-home mother, a medical or legal professional, or whatever you do. Be a person who knows what is right, according to the Word of God, and who never gives up in your efforts to see right prevail.

CHAPTER 10

TESTIMONY BEGINS WITH T-E-S-T

*"Accept the challenges so that you may feel
the exhilaration of victory."*
GENERAL GEORGE S. PATTON

I'm sure you know people with amazing stories of the way God has worked in their lives. I always love to hear a great testimony, but I also know that behind every extraordinary account of someone's life lies some kind of challenge or difficulty. No one ever has a testimony without a test.

We must pass all kinds of tests as we go through our lives, and passing them is part of never giving up. It's vital for us to understand the important role that tests and trials play in our lives, because understanding them helps us endure them and actually be strengthened by them. Everything God permits us to go through will ultimately be good for us—no matter how much it hurts, how unfair it is, or how difficult it is. When we encounter tests and trials, if we will embrace them and refuse to run from them, we will learn some lessons that will help us in the future and become stronger.

> It's vital for us to understand the important role that tests and trials play in our lives.

In this chapter, I want us to

explore five of the most common, most important tests we must pass in life.

TRIALS AND TESTS

Trials "try" us, and tests "test" us. Most of the time, the purpose of them is to show us who we really are, to reveal character in us. We can think all kinds of good thoughts about ourselves, but until we are put to the test, we don't know whether those things have become realities in us or not. We may consider ourselves generous, honest, or deeply committed to a particular truth or ideal, but the depth of these dynamics only reveals itself when we're under pressure.

We never know what we really believe until our beliefs are tested. We never know how godly we are until our faith is tried. We can't predict how we will behave under pressure until the stress hits an all-time high. We don't know whether we really are nice people until we have to be kind to someone when we don't feel like it. When we go through tests, we learn whether or not we really have the character and commitment we think we have. I believe it's very important for us to really know ourselves; so tests are good for us because they affirm strengths and reveal weaknesses. Don't be afraid to face your weaknesses. God's strength is available specifically for our weaknesses.

> *Don't be afraid to face your weaknesses. God's strength is available specifically for our weaknesses.*

Why did Jesus—our perfect Jesus—have to go out into the wilderness so the devil could tempt Him for forty days and forty nights? God the Father already knew Jesus would not succumb to the temptation. I believe that, in His humanity, Jesus went into the wilderness to build His own confidence and to show the devil He would be faithful to His heavenly Father, no matter what.

Other people cannot always pray all your giants away; they cannot fight all your battles for you. Your pastor can only go so far with you, but then you have to take the ball and make the touchdown. Your best friend or work colleague can only pray you a certain amount of the way through, but sooner or later, you have to learn how to pray. You have to learn how to find your own scriptures. You have to learn how to stand your own ground. You have to get to the point where every single time you have a problem, the first place you run is to God. You get your answers and advice from Him, not from your pastor, your best friend, or the person who sits next to you at work.

First Peter 4:12 gives us great insight into the purpose of trials: "Beloved, do not be amazed and bewildered at the fiery ordeal which is taking place to test your quality, as though something strange (unusual and alien to you and your position) were befalling you."

One reason we must go through trials is to test our quality. Often, we find ourselves wishing we had the faith of Sister so-and-so. I can assure you, if she has a strong and vibrant faith, she did not develop it easily. Just as muscles are built through exercise, firm faith comes from the furnace of affliction.

Sometimes people say to me, "Oh, I wish I had the kind of ministry you have, Joyce." Well, I did not get it by wishing. These people were not around when I was feeling I couldn't hold on one more second, begging God to help me not to quit or give up. They don't know the tests and trials I've faced along the way.

No one who does anything worthwhile for God has traveled an easy road. Doing great things for God requires character, and character is developed by passing life's tests and staying faithful to Him through the trials.

> *No one who does anything worthwhile for God has traveled an easy road.*

After many years of tests and trials, I have become more steadfast and patient than I ever thought possible. I'm still growing in these areas, but circumstances do not move me the way they used to. James 1:2–3 says: "Consider it wholly joyful, my brethren, whenever you are enveloped in or encounter trials of any sort or fall into various temptations. Be assured and understand that the trial and proving of your faith bring out endurance and steadfastness and patience."

I must say that before my trials worked steadfastness and patience in my life, they brought out many other negative qualities, mindsets, and attitudes I didn't know I had. One reason God allows us to go through tests and trials is so the hidden things in our hearts can be exposed. Until they are exposed, we cannot do anything about them. But once we see them, we can begin to face them and ask God to help us. God does not allow us to go through difficult times because He likes to see us suffer; He allows us to walk through them so we will recognize our need for Him. Never be afraid of the truth, because it is the truth that makes us free (see John 8:32).

Look again at James 1:3. The point is that everything you go through ultimately does work out for your good because it makes you stronger and builds your endurance; it develops godly character; it helps you to know yourself and to be able to deal with things on an honest level with God and take care of them so you can go on.

The next time you encounter some sort of test or trial, determine to believe it is for your good. Say to God: "Okay, Lord, I believe this is going to work out well for me. It doesn't feel good. I don't like it at all. I don't understand it! It hurts! It doesn't seem fair! But I believe You will use it for my good." Placing your faith in God opens the door for Him to work miracles out of messes!

THE "GET OVER IT" TEST

Another test we all have to pass in life is the "get over being offended" test, which can also be called the "bitterness-resentment-unforgiveness" test. The only way to get over being offended is to forgive. The sooner you do it, the easier it will be. Don't let offense take root in your heart, because it will be more difficult to deal with if you do.

One of our first responses when someone hurts or offends us should be to pray: "God, I choose to believe the best. My feelings are hurt, but You can heal me. I refuse to be bitter; I refuse to be angry; I refuse to be offended." You need to say with your mouth that you will not be offended because offense is a trap! It's the bait of Satan. He uses offenses to pull us away from God and His principles.

The word *offense* comes from the Greek word *skandalon*. A skandalon was the part of an animal trap that held the bait; its purpose was to lure a victim. Offense is the devil's bait to lure us into a trap of full-blown bitterness, resentment, and unforgiveness, so he can destroy our lives.

We Christians must learn to be good at forgiving people, because we will be doing it all of our lives. When we forgive, we are actually doing ourselves a favor. We are freeing ourselves from the agony of anger and bitter thoughts. As long as we live, we will encounter people who hurt us, reject us, disappoint us, use the wrong tone of voice with us, fail to understand us, or let us down in times of need. Those dynamics are part of human nature and they are part of the territory that comes with relationships. Why should we ruin our lives over other people's bad behavior? We need to take the high road and forgive!

> *We Christians must learn to be good at forgiving people, because we will be doing it all of our lives.*

Jesus knew this truth. That is why He responded as He did when Peter asked Him: "Lord, how many times may my brother sin against me and I forgive him and let it go? [As many as] up to seven times?" Jesus answered: "I tell you, not up to seven times, but seventy times seven!" (Matthew 18:21–22). What He meant was: "Forgive and keep on forgiving. Just keep it up, keep it up, keep it up, keep it up." Every time you decide to forgive, you are maintaining your freedom! You're refusing to be filled with negative thoughts and emotions and your decision enables God to release you from the trap of resentment.

We bless other people when we forgive them, but forgiveness is more for our benefit than for the benefit of those who hurt us. We extend forgiveness for ourselves because holding bitterness, resentment, and unforgiveness only hurts us. Most of the time, those who have offended us don't even realize what they have done and are not even aware we are suffering.

When people hurt or offend you, you may not be able to help the way you feel, but you can help what you do about it. If you are hurt, you are hurt; but God will heal your feelings if you will *by faith* obey His Word, pray for your enemies, and bless those who mistreat you (see Matthew 5:44). Turn the entire situation over to God and refuse to be offended.

I heard a story about a pastor I know. He was hosting a guest speaker in his church and the guest made critical remarks about the way the church handled their services. The guest preacher was young and inexperienced, and the pastor decided not to let his comments offend him. He sat in the front row of the church and repeated in a low whisper, "I will not get offended. I will not get offended." This story has stuck with me for a long time because it is an excellent example of someone who took quick action to make sure no root of offense or bitterness got into his heart.

Jesus tells us to bless those who hurt us (see Matthew 5:44), but how do we do that? One way is by not speaking negatively about them or telling others what they did to us. I believe we invite trouble into our lives when somebody does something wrong or hurtful to us and we tell other people about it for no reason except to gossip. When we hurt, we want sympathy, we tend to want people to know who hurt us and how it happened, but God does not want us to respond that way. He wants us to keep our mouths closed, to refuse to speak ill of others, and to pray for them and bless them. This is one way we overcome evil with good (see Romans 12:21)!

The Bible is full of stories of people who refused to be offended. Job was doubly blessed when he prayed for those who hurt him (see Job 42:10). Moses prayed for Miriam and Aaron right away when they judged him and came against him (see Numbers 12:13). You see, Moses was a godly man. When wronged, he fell on his face before God, saying basically, "Oh, God, oh, God, forgive them, forgive them, forgive them. They don't know what they're doing."

> *The next time someone hurts you, do yourself a favor and forgive quickly, refuse to be offended, and bless and pray for that person.*

The next time someone hurts you, do yourself a favor and forgive quickly, refuse to be offended, and bless and pray for that person. Do this over and over again. You will have many opportunities in life to pass the tests of offense, and I want you to pass with flying colors!

THE "JUDAS KISS" TEST

Judas was the disciple who betrayed Jesus with a kiss. He agreed to reveal who Jesus was for thirty pieces of silver, and said he would do it by kissing him, which was an acceptable greeting in Jesus' day. The "kiss" traditionally is a greeting of love and affection; therefore, the "Judas kiss" inflicts a deeper wound than ordinary offenses do. This kind of attack comes from someone with whom you have a close relationship, someone you love and trust. Many times a person who becomes a "Judas" in your life is someone you have helped and treated well.

There are many sad stories about pastors who have mentored young men in their congregations, taught them everything they know, given them opportunities to grow in their ministry gifts, and helped them earn respect, only to have these men leave their churches to start new ones, taking many of the members of the original church with them. This is an excellent example of a "Judas kiss." I believe such a situation is one of the worst tragedies that can happen to a church and one of the most painful situations a pastor can endure when he or she has invested his life in training an emerging leader.

What a sad turn of events we read about in Luke 22:48, when Judas betrayed Jesus with a kiss—a sign of love. It was not just anybody who betrayed Jesus; it was one of His twelve disciples; one of His close associates; one of His personal friends; someone to whom He gave a position of honor. Judas betrayed Jesus for *money* because he was greedy, so he *sold* Jesus for a few pieces of silver. And he did it with a kiss!

Being betrayed by someone you love is one of the hardest situations you will ever endure, but if it happens, you must let God heal you and move beyond it, or it will destroy you. You will simply be alone with your hurt, your resentment, and your pain. In addition, your ability to enjoy God and His good plan for you will be diminished, if not entirely destroyed. Do yourself a favor and totally forgive. God never asks us to do anything without giving us the ability to do it. You don't have to spend your life feeling hurt, bitter, angry, and offended. You can forgive and be free!

THE "MOTIVE" TEST

I like to define a motive as "the *why* behind the *what*." A motive is the reason we do what we do. We often say we are doing things for God, but sometimes we do not understand why we do them. We only know *what* we are doing, but we have not taken the time to truly understand what motivates us. Impure motives can cause many problems, one of which is being overcommitted, which results in unnecessary stress in our lives. Surely we won't live with extreme stress if we are obeying God and doing only what He wants us to do.

Never agree to do something in order to impress people or because you fear what they may think or say about you if you don't. God wants us to help and bless people, but a "good work" done with a wrong motive is no longer a good work. Don't say yes with your mouth if your heart is screaming no.

Take the motive test as often as you can. Begin to ask yourself questions that will help you assess your motives, such as:

- Why did I agree to serve on that committee? Am I looking for recognition?
- Why did I want to attend a certain function? Am I seeking social status?
- Why did I say I would lead the missions group at church? Do I really have a heart for evangelism and a longing to serve God, or do I want people to talk about what a good church member I am?
- Why do I really want that promotion at work so much? Is it motivated by God or worldly ambition?

As you evaluate your motives, you will begin to see what is in your heart. Pass the test by making sure your motives are pure and right before God. The motive test is a lifelong test. I frequently reevaluate my motives and discontinue things I find I am doing for the wrong reason, and that helps me keep my priorities in order.

THE "LOVING THE UNLOVELY" TEST

We all face the "loving the unlovely" test. When we ask God to teach us about love, we shouldn't expect Him to send us someone who is easy to love; that would not be much of a test. The test of loving those who are unlovely takes place when people come into our lives and affect us like sandpaper on our souls. These are the people who grate on our nerves, people we would rather not interact with. And yet, we know in our hearts God Himself has put these people in our lives.

Galatians 6:1–3 basically tells us how to pass the "loving the unlovely" test:

> *Brethren, if any person is overtaken in misconduct or sin of any sort, you who are spiritual [who are responsive to and controlled by the Spirit] should set him right and restore and reinstate him, without any sense of superiority and with all gentleness, keeping an attentive eye on yourself, lest you should be tempted also. Bear (endure, carry) one another's burdens and troublesome moral faults, and in this way fulfill and observe perfectly the law of Christ (the Messiah) and complete what is lacking [in your obedience to it]. For if any person thinks himself to be somebody [too important to condescend to shoulder another's load] when he is nobody [of superiority except in his own estimation], he deceives and deludes and cheats himself.*

This passage teaches us that when we are around people who are difficult to deal with, we will still treat them well and help restore them if we are truly spiritual. We won't make more out of their faults than necessary; and we will look for the good and not just the bad in them. We'll be concerned more about them than about how they're making us feel. Instead of judging them, we'll examine ourselves. We'll try to understand our weaknesses and ask God to help us rather than focusing on how much we think He needs to help those "unlovely" people. Loving those we consider unlovely is one of the most Christlike qualities we can develop.

THE "TEST OF TIME"

The final test to which I want to call your attention is the "test of time." Will you still be here—following God, obeying Him, living by His Word—five years from now? Will you still be doing everything

He has called you to do ten years from now? Will you walk with Him every day for the rest of your life, even through difficulties, and refuse to give up?

Let me encourage you as you think about passing the test of time: Make every effort to be on fire for God. Don't be passionate for Him for six months and then allow yourself to become lukewarm. Don't be disciplined one day and be lazy the next, or be diligent to live your life by the Word for one week and then ignore what God says a week later. Have staying power; be consistent over the course of time; be committed to God for the long haul. I have been serving God seriously for more than thirty years and I believe my greatest testimony is simply, "I'm still here!"

Whatever you may be facing right now, go ahead and go through it the right way so you won't have to go through it again. Remain stable in your attitudes and actions. Keep your commitments and continue reaching out in love to others who are hurting also.

In God's school, we never fail. We get to keep taking our tests over and over again until we pass them. I want you to pass your tests the first time so you don't have to waste the time I did in my life going around and around the same mountains. God's ways are always best, and the sooner we submit to them, the better our lives will be.

Not Your Average Cosmetic Change

Chances are, anytime you see a Cadillac in a certain shade of pink, you know what it means: the owner is a tremendously effective salesperson for Mary Kay cosmetics.

In 1963, the Mary Kay company had eleven salespeople. In 2007, it had more than 1.7 million sales representatives, called "Independent Beauty Consultants," in more than thirty countries around the world. What was the secret behind such phenomenal growth? A woman who never gave up, Mary Kay Ash.

She was born in 1918, in Hot Wells, Texas. By the time Mary Kay was six years old, she had the great responsibility of caring for a seriously ill relative while her mother worked, often more than fourteen hours per day, to provide for the family. Anytime Mary Kay faced a new or challenging situation, her mother told her "You can do it." Because of her mother's encouragement, she grew up feeling empowered, confident, and truly able to do anything.

As a young girl, Mary Kay wanted to grow up and become a doctor. In 1938, while studying to pursue that dream, she took a part-time sales job selling household goods for Stanley Home Products and quickly realized she was extremely talented as a salesperson. She soon chose business over medicine because she enjoyed such amazing success in sales.

Several years later, Ash went to work for a gift company, where she eventually became national training director. After twenty-five years as an outstanding employee in the area of direct sales, Ash resigned from the company because a man she had trained was promoted over her and received double the salary she was making.

But Mary Kay kept going. She realized she was not the only woman being treated unfairly in corporate environments, so

(continued)

she decided to write a book to help other women enjoy the opportunities that were not available to her. As she penned her thoughts, she realized she was not only offering women advice, she was writing the strategy and philosophy of a new company. She once said, "I envisioned a company in which any woman could become just as successful as she wanted to be. The doors would be wide open to opportunity for women who were willing to pay the price and had the courage to dream."

By 1963, she had a wealth of sales experience, a plan, five thousand dollars, and a twenty-year-old son to help her start her new business, called "Beauty by Mary Kay." According to the company Web site, "It was a first—a company dedicated to making life more beautiful for women. It was founded not on the competitive rule but on the Golden Rule—on praising people to success—and on the principle of placing faith first, family second and career third." Mary Kay often called it "a company with heart."

As a woman, Mary Kay had to struggle to achieve her full potential in the business world. Because she refused to give up, she was able to start an enormously successful company that would empower other women to do great things. I want to encourage you with a statement this innovative and determined businesswoman once made: "With your priorities in order, press on, and never look back."

CHAPTER 11

TRY SOMETHING NEW

"You are never too old to set a new goal or to dream a new dream."

C. S. LEWIS

Someone once said, "Never be afraid to try something new. Remember, amateurs built the ark. Professionals built the *Titanic*." Sometimes we're tempted to give up because whatever we're doing doesn't work. When that happens, we have to try something new. As long as we're making progress in life, we will have new opportunities and face new challenges. We have dreams and goals that we want to pursue, but we also need solutions to the challenges that we experience along the way. If we know what to do, then we need to do it; if we don't, then we have the privilege of trusting God to guide us. The Bible says we are to do what the crisis demands and then stand firmly in our place (see Ephesians 6:13). Do what you can do and God will do what you cannot do.

As we continue learning to be determined and to not give up, we must understand how to embrace new things. Many times we say we want something new and then when new things come, they frighten us. New experiences and opportunities can be daunting, but they don't have to be. If we insist upon clinging to old methods, old ideas, old relationships, or old mind-sets and refuse to accept or embrace the new, we aren't going to progress in our lives. I want you to go forward—as far and as fast as God's will dictates—without giving up when you face a challenge that requires you to think or act in a new way. I believe God is doing a new thing in your life, and I want you to be ready to embrace it with passion.

You remember that God chose Joshua to succeed Moses as the leader of the Israelites and to take them into the Promised Land. God gave him the daunting task of leading an entire nation into a new place, a place where they'd never been before. They knew they were leaving behind a land of terrible oppression and hard labor as slaves in Egypt, and they heard they were headed to a land flowing with milk and honey, but they had never seen that land before. It was going to be new to them. Their story, even though it took place thousands of years ago, teaches us some important lessons about new experiences in our lives today.

In Joshua 3, we read that officers went through the Israelites' camp, "Commanding the people: When you see the ark of the covenant of the Lord your God being borne by the Levitical priests, set out from where you are and follow it" (v. 3).

The ark contained the Ten Commandments, which represent the will of God, and a piece of manna—the same manna with which God fed the Israelites in the wilderness for forty years. The manna represents God's presence and His miraculous provision. On our journey through life, when we are headed into new situations, we need to follow the ark, figuratively speaking, just as the Israelites followed the ark literally as they headed toward a new homeland. To "follow the ark" in our day is to follow the leading of the Holy Spirit, who lives within all believers.

The first question we need to ask when we face a new situation is, "Is this God's will for me?" The second is, "Do I have inner peace about this?" We also need to ask ourselves if we sense His presence in the new situation. If we believe it is God's will, we can trust God to provide everything we need each step of the way.

As a young woman, I didn't know how to seek God as I do now, and I made a decision I don't believe I would have made had I known how to follow the ark.

At eighteen years of age, I married the first young man who paid attention to me because I was afraid no one would ever want me, due to the sexual abuse I experienced as a child. I received Jesus as my Savior at the age of nine, but I never had any spiritual education or training after that point, so I'd never heard the idea of "being led by the Holy Spirit." I didn't know how to seek His guidance or realize that asking Him to lead me was important.

I look back now and see that the Holy Spirit was trying to lead me all along, because deep in my heart, I knew the relationship I was about to enter was not right. My mind told me I shouldn't miss this opportunity, but deep inside I felt uneasy about it. I simply didn't know what to do with the still, small voice telling me, "This will never work." But I was so desperate for someone to love me, I refused to pay attention to what I knew inside and I just followed my head, my flesh, my emotions, my plan, and my "want to."

The situation became five years of a terrible mess. This man and I often traveled to other states, and on two occasions—once in Oakland, California, and once in Albuquerque, New Mexico—he simply left me stranded and I had to leave all my meager possessions behind and return home on a bus. I was eighteen years old, lonely, frightened, and confused.

For most of the time we were married, he refused to work. He ran around with other women, lied, stole from me, and ended up going to prison for writing bad checks. One night I woke up and he was trying to remove my wedding ring from my finger so he could sell it. I could have saved myself five years of heartache and trouble had I simply known how to listen to God and allow Him to lead me into His perfect will for my life.

We need to "follow the ark" instead of following our personal desires, our fears, our emotions, our good ideas, or the advice of others. We follow the ark by taking one step of obedience to God after another.

The ark will lead us to new places, just as the Old Testament ark did the Israelites, because God does not take us backward. As we follow Him, we will find ourselves in unfamiliar places, uncomfortable places, or places in which we may be unsure of ourselves. Our journeys may not be easy, but personally, I would rather follow God into a new and uncomfortable place than remain in a place that is no longer life-giving for me.

> *I would rather follow God into a new and uncomfortable place than remain in a place that is no longer life-giving for me.*

FAILING TO FOLLOW

When we do God's will, His presence is with us and we are sure to succeed. But when we fail to follow Him, we invite all kinds of problems.

Some people have miserable lives for no other reason than the fact that they refuse to follow the leading of the Holy Spirit. Some people keep friends who influence them negatively. They may be fun to be with, or perhaps they've been friends for a long, long time, but they don't want to listen to God and obey His will. It's always best to have close friends who can build your faith in God rather than those who diminish it. Don't just decide who you'll be friends with, but ask God to give you relationships that are right for you.

Sometimes we pray for something but aren't willing to follow the Holy Spirit even when He tries to lead us into the answer to our prayer. For example, I know of people who refuse to attend singles groups at church because they don't want to go by themselves. Even when they believe God wants them to do so, they won't go. They want to date and eventually get married, but they want God to work on their terms. I can understand where it might take some courage to go alone, but doing the difficult thing is often good for us.

It stretches our faith and often leads us into the very thing we are looking for. We must be careful that we are not planning and praying that our plan will work, when we should be praying first and then following God's plan.

Are you willing to say yes to God even if it means saying no to your friends or even to yourself? What if you're invited to go somewhere with friends or a family member, but you don't have peace about going? It might be a party where people will be using drugs, or a movie that contains nudity. You'll have to compromise your conscience in order to go. What will you do? Are you willing to spend the evening at home alone in order to follow God?

In each of these situations, people who go ahead and do what they want to do are following their own desires over God's will. Is it possible, if they were to follow the leadership of the Holy Spirit, something truly wonderful would happen? Maybe a nice evening at home would provide the peace, quiet, rest, and refreshment they need after a hectic work week. Maybe they would make some new friends they could really enjoy if they would be bold enough to attend the singles group alone.

My daughter Sandra met her husband when she attended a church event all by herself. She was determined to follow God's leading, and she received a great blessing. I remember that it was difficult for her to go alone, but she told me that she was not going to sit home, get depressed, and complain about her life when she could make a different choice.

We need to be people who want to obey God and follow Him more than anything else. We need to be strong and courageous and do whatever God leads us to do. Even if you are the only person in the whole world doing the right thing, I encourage you to obey God. Do not follow your flesh, your friends, or your own wishes. Be like the Israelites who followed the ark and, by doing so, followed God into the good things He had for them.

YOU HAVE NOT PASSED THIS WAY BEFORE

I believe one of the most important verses in Joshua 3 is verse 4: "Yet a space must be kept between you and it, about 2,000 cubits by measure; come not near it, that you may [be able to see the ark and] know the way you must go, *for you have not passed this way before*" (emphasis mine).

These words grip my heart, and I want you to pay particular attention to them: "you have not passed this way before." Everything about those words says "new." Whether it is a job, a place to live, a relationship, an addition to your family, a position of influence, an exercise program, a church, a hobby, or a way of serving your community—if you have not passed that way before, it is new to you. You may not know how to get there or what to do when you do. But if you follow the ark (God), you will arrive at your destination and He will direct you in everything you need to do.

One of the biggest obstacles to following God is fear—simply because "you have not passed this way before." Things we have never done before always make us feel a little more timid than usual. People have asked me if I'm ever nervous or fearful, and my answer is yes. When I am doing something for the first time, or something I have no experience with, I am tempted to lose my confidence just like everyone else. My best advice is: press forward into what you believe God's will is for you. Don't let the feeling of fear rule your decisions. Be diligent to apply God's Word to your life and to stay close to Him.

When you follow God into a new place, you may feel stretched. Perhaps you receive a job promotion and you know you don't have all the natural skills and knowledge you need to do the new job well. Then you become worried or fearful because you think you're "in over your head." The job may be over your head, but it's not bigger than God. If He leads you into it, He will help you fulfill the responsibilities that go with it. God's power and presence enable us to do things we could not do on our own.

It's important to remember God is on your side as you go into new situations, because the enemy will always be lying in wait to try to keep you from following Him. Most of the time, he attempts to make us turn around and go back to our old circumstances by using the "what if" strategy. The goal of this scheme is to stir up such fear that you decide to not walk into the new place. In actuality, it looks something like this.

A man gets a new job in a new city. He and his wife have been praying for a new opportunity and a better job for him, and they believe this is it. They are so excited because God has orchestrated every aspect of the job search, the interview, and the offer. He has even blessed the man with a nice raise, and the new company is willing to pay moving expenses.

As the man prepares to write his resignation letter at his current job, a question enters his mind: *What if I end up having to work a lot more overtime than I have to work here?* He brushes that aside, but soon he wonders, *What if the cost of living is so much higher where we're going that my raise doesn't make much difference in our lifestyle?* Then he asks himself, *What if I don't even like the job after I've done it a while?*

That night, his wife says to him, "*I've been thinking: What if the children don't like their new school as much as they love their school here? What if we can't find a church as wonderful as our church here?*"

The man and his wife find themselves being tempted to walk away from the answer to their prayers, because of "what if" questions and lies. They have to stop relying on their minds and follow what God has placed in their hearts. They have to refuse to fear and begin thinking, *What if we are so radically happy in our new city we never want to come back here again? What if this job is the best thing to ever happen to our family?*

The enemy will call to your attention every possible "negative" about a new situation, but he will not bother to mention any of the positives. To focus on the positives, you have to spend time with God and be willing to follow the ark, as the Israelites did.

> *Don't be afraid to step out and do something new. Perhaps you haven't passed this way before, but God is with you.*

Don't be afraid to step out and do something new. Perhaps you haven't passed this way before, but God is with you, so press through the fear and opposition, and refuse to give up.

BE SENSITIVE TO GOD'S TIMING

Once you know something new is God's will for your life, you may want to step into it before God releases you to do so. Or, after you have had to wait for a while, you may decide you do not want to do it after all! We should not run ahead of God or lag behind Him. God has an appointed time for every purpose under heaven (see Ecclesiastes 3:1).

I tried to go on television before God opened the door for me, and my project failed miserably. When His time came, I was thinking about settling into an easier schedule, and I didn't want to start a major new endeavor. My plan at that time was to work less, not more. Following God frequently means that we don't get to do things our own way or according to our own timing.

Moses is a classic example of a man who struggled with God's timing. God spared his life as a baby and then sent him to grow up in Pharoah's palace. God had a great plan for Moses' life—a plan for Moses to lead His people out of the bondage and slavery of Egypt into freedom. But before God told Moses it was time to step out, he took steps of his own. He saw an Egyptian mistreating one of his kinsmen and, acting on his emotions, he killed the man. When he found out two Hebrew men witnessed the murder, he was afraid and he fled to the desert, where he stayed for forty years. The very fact that Moses behaved emotionally and fearfully shows he wasn't yet ready for the task God had in mind for him.

During his time in the desert, God worked with Moses, helped him grow up, and eventually revealed Himself to him in the burning bush (see Exodus 3:2). When God was ready to use him, Moses didn't want to cooperate. He asked God repeatedly to use someone else and gave Him all sorts of reasons he would not be a good deliverer. But Moses was God's choice, and after considerable protest, he did stand before Pharoah and speak on God's behalf. Like many of us, first he tried to step out too soon and then, when the time was right, he didn't want to step out at all.

I want you to know that God has His own timing. He may drop a dream, vision, or idea in your heart at one point in your life and then not ask you to do anything about it for years. In Moses' case, he underwent a forty-year training period in the desert before he was prepared to fulfill God's call on his life. I hope you do not have to wait that long, but if you do, just be patient. Resist the urge to get out in front of Him; and don't allow yourself to lag behind Him. Trying to move outside of His timing—either too quickly or too slowly—will cause you the kind of frustration that makes you want to give up, because His presence will not be there. Whether God asks you to wait two weeks or two decades, be sensitive to His timing in your life and be willing to wait for Him as He leads you into the next new place He has for you.

GOD WILL BE WITH YOU

In chapter 3, I made a brief reference to God's promise to Joshua: "As I was with Moses, so I will be with you" (Joshua 1:5). God repeats this pledge in Joshua 3:7: "This day I will begin to magnify you in the sight of all Israel, so they may know that *as I was with Moses, so I will be with you*" (emphasis mine).

God was calling Joshua into new territory. He was saying: "Come on now, Joshua, you've never done this before. You've never passed this way before. Previously, you worked under Moses. Moses carried all the main responsibility and you just helped him, but now Moses is gone. But you don't have to be concerned, and here's why: because as I was with Moses, so I will be with you."

I believe we allow ourselves to become overly concerned about our weaknesses. When faced with a new situation or opportunity, we often think of all the reasons why we can't do it: *I don't have enough experience. I don't have a college degree. I don't speak well in public. My computer skills are rusty. I don't know how to run a household. I'm not sure I will be a good parent. I've never been successful on a diet in my life. I just don't know if I will be able to do this new assignment.* This kind of self-doubt will sabotage you before you ever get to the new place God has for you. You will paralyze yourself with your own thinking. I call this "the paralysis of analysis."

The truth is, you don't have to be worried about your weaknesses. God is not at all surprised by them because He knows everything there is to know about you. God is not looking for *ability;* He's looking for *availability.* I encourage you to wake up every day and say: "Here I am, God. Is there anything You want me to do? Do You have something new for me? I'm going to be bold and courageous in You, Lord. If You want me to keep doing the same thing for another ten years, I can do it. And if You want me to launch out into something I have never done before, I can do that too because You are with me."

The truth is, you don't have to be worried about your weaknesses. God is not at all surprised by them.

The Germinator

In Europe and the United States during the nineteenth century, many women died after giving birth in hospitals. Up to 25 percent of these new mothers contracted the aggressive infection puerperal sepsis, which was commonly called "childbed fever," and never recovered.

In the late 1840s, Dr. Ignaz Semmelweis, a young Hungarian doctor, worked in the maternity wards of a hospital in Vienna. The hospital had two maternity wards. Male medical students and physicians delivered babies in one, while female midwifery students attended the other. Semmelweis noticed that the rate of death from childbed fever was three to five times higher in the ward staffed by medical students than in the ward where midwives delivered babies.

At that time, no one understood the relationship between germs and disease, so the only way for Semmelweis to figure out why so many more women died at the hands of male physicians and medical students than under the care of midwives was to figure out what the two groups did differently. He soon realized the medical students often went straight from classes during which they performed autopsies and dissected corpses to attend women in labor.

Semmelweis acted on his hunch that the students might somehow be transferring disease from deceased patients to birthing mothers and ordered the men working in delivery rooms to wash their hands with a solution of chlorinated lime water before they examined the women. The results were astounding. With the hand-washing policy in place, the death rate in the ward dropped from more than 18 percent to less than 1.5 percent.

(continued)

We might think Semmelweis would be applauded for his outstanding work, but instead, he was widely and sharply criticized and even met with hostility. The director of the hospital where he worked felt Semmelweis had undermined his leadership, so he refused to grant the doctor any promotions or the teaching position he desired. His colleagues ridiculed him to the point that he eventually resigned and left his position at the hospital and returned to Hungary. In Hungary, Semmelweis took a job at a hospital where his methods reduced mortality rates from childbed fever to less than 1 percent.

Members of the medical community continued to dismiss Semmelweis's assertion that hand washing was a necessary step in reducing infection and death. In the midst of opposition, he held firm to his beliefs and wrote a book about them in the early 1860s. The scientific establishment ridiculed his book and continued to criticize his work.

In 1865, Semmelweis suffered a mental breakdown, which some attribute to the constant hostility and rejection he endured over his discovery of the importance of hand washing. His friends committed him to a mental institution where he died, ironically, of an infection of childbed fever, which he contracted through a cut finger.

We know today that Semmelweis was right. Science has proven the connection between germs and disease, and thorough hand-washing procedures are now mandatory in medical environments. Thankfully, Semmelweis never gave up on his scientific convictions and was willing to suffer for them so countless lives could be saved.

CHAPTER 12

YOUR OWN WORST ENEMY

*"Nothing is more wretched than the mind of
a man conscious of guilt."*

PLAUTUS

Amerian poet and author Maya Angelou is known all over the world. I once read a biographical sketch on her and I want to share a portion of it with you:

> Maya Angelou was born Marguerite Annie Johnson in Saint Louis, Missouri. Her parents divorced when she was only three years old, and she and her brother, who gave her the nickname "Maya," were sent to live with their grandmother in the small town of Stamps, Arkansas. There, the young Maya endured racial discrimination, which was legal in America's southern states at that time. Also in Arkansas, she observed and absorbed powerful faith in God and the gentle manner and courtesy that marked the tradition of her African American people. Angelou says her grandmother and extended family taught her the values that later shaped her life and work.
>
> When Maya was seven years old, she went to visit her mother in Chicago, where she suffered sexual abuse by her

mother's boyfriend. The shame of that incident so para-
lyzed Angelou that she shared it with no one except her
brother, with whom she had a close relationship. Later, she
heard the news that her uncle had killed her attacker. The
little girl believed, in some way, her words had murdered
the man, and she did not speak for five years.

Can you imagine being a child and carrying such guilt you didn't speak for five years? Guilt is *that* powerful; it can silence a voice as eloquent and powerful as Angelou's. Feeling guilty is like being locked in an invisible prison cell. Guilt robs you of the ability to believe you are worthy of success in life, and you are never truly free as long as you carry its burden.

Jesus came to break the bondage of guilt in our lives. He paid with His life for everything we have ever done wrong and everything we will ever do wrong. When we believe in His work on the cross and receive Him as Lord and Savior, He looks at us and says, "Not guilty." We are justified in Christ Jesus, which means that God sees us as if we had never sinned. His sacrifice gave us the right to a brand-new life that includes success and enjoyment.

NO CONDEMNATION

"We all make mistakes." That phrase has been around for years, and it is certainly true. Equally true is the fact that most of us really *hate* making mistakes, doing things wrong, or causing problems. Many times, the guilt we feel over our mistakes is much worse than the mistakes themselves.

> *Many times, the guilt we feel over our mistakes is much worse than the mistakes themselves.*

We should of course be sorry when we do wrong, and we should always repent, which means to be willing to turn away from the sin and begin doing what is right with God's help. We may even need a brief time of grieving over very serious mistakes. But we are not supposed to hate ourselves, reject ourselves, and feel guilty and ashamed for days and weeks on end when we sin or make mistakes. We may feel *convicted* of sin when we do something wrong, but we should not experience condemnation. We are supposed to receive the forgiveness God freely offers us and go on with our lives, constantly seeking to know God better and better.

The apostle Paul had plenty of reasons to feel guilty. He'd persecuted and killed many Christians before he had a personal encounter with Jesus Christ and began to follow Him. In Romans 8:1, this man who knew what it means to be both guilty and free of guilt writes: "Therefore, [there is] now no condemnation (no adjudging guilty of wrong) for those who are in Christ Jesus, who live [and] walk not after the dictates of the flesh, but after the dictates of the Spirit."

Human nature causes us to feel condemned when we make mistakes, and no one likes feelings of condemnation. Many people I know want to do everything right, but that isn't going to happen as long as we're in bodies of flesh. If we could be perfect in our behavior, we wouldn't need God and we would miss out on the enjoyment and blessings of a relationship with Him. Jesus died for us because we could not be good enough on our own to enjoy a relationship with God who is completely holy.

Sometimes, because we feel guilty over our sins or mistakes, we struggle to enjoy God or the lives He has given us. No matter how hard we try, we will never be able to live without making mistakes. If we believe we have to suffer and feel terrible and go on a guilt trip every time we make a mistake, then our lives will be miserable. The way to overcome this is by faith.

We must believe the truth of God's Word more than we believe our emotions or our

> *No matter how hard we try, we will never be able to live without making mistakes.*

thoughts. It really is possible to *know* in the depths of our hearts that we are not guilty, even if we *feel* we are. At such times, we need to confess the truth of God's Word instead of focusing on our feelings.

If you want to honor God, put your faith to work believing God forgives you immediately when you ask Him to do so. Even if you *feel* guilty you can be assured you are not because 1 John 1:9 says: "If we [freely] admit that we have sinned and confess our sins, He is faithful and just (true to His own nature and promises) and will forgive our sins [dismiss our lawlessness] and [continuously] cleanse us from all unrighteousness [everything not in conformity to His will in purpose, thought, and action]." When feelings of guilt and condemnation arise, press through them in faith and say, "I don't care how I feel. I have been forgiven!" People will never enjoy God's plan for their lives unless they stop making decisions based on feelings. Honor God by deciding to trust His Word more than you trust your feelings.

PAID IN FULL

Reportedly, the IRS once received a letter that read:

"Dear IRS, I have cheated on my taxes many, many times. I can't remember ever paying what I owed. I feel so guilty about this that I cannot sleep at night. Please find enclosed a check for $20,000.00. P.S.: If I still have trouble sleeping, I'll send more money."

Guilt will cause people to do unusual things. Most of the time, guilt is nothing more than our pathetic way of sacrificing something to make up for what we've done wrong. The truth is, either Jesus did a complete and thorough job when He paid for our sins at the cross and does not need our help, or He did not finish the work. I assure you, He finished the work; He was the final, the only, the complete sacrifice ever needed for our sin. No additional sacrifice will ever need to be made. In my case, I finally realized that I was punishing myself with feelings of guilt and refusing to do anything I might enjoy as a way of paying for my imperfections. I also realized that my pathetic attempt was not pleasing, nor was it acceptable to God at all.

We humans tend to divide our sins into categories. We determine how big or bad we think a certain sin is and then decide how long we should feel rotten about it in order to pay for it. We also feel we have no right to be blessed, succeed at anything, or enjoy ourselves until we satisfy the debt.

You may be thinking, *Now wait a minute, Joyce, I can't just do something wrong and not care about it.* I have even thought that myself in the past. Let me say this: if you are a true believer in Jesus Christ, there is no way you could ever sin and not care. That is not possible because He has given you a new heart. You know sin offends God, and you love God, so of course you will care. But there is a difference between caring about your sin and feeling condemned because of it. Caring about it simply means the convicting power of the Holy Spirit is at work in a heart that is tender toward God, but feeling guilty and condemned is an indication that the enemy is at work. Always remember to receive conviction, which is from the Holy Spirit, but aggressively reject condemnation, which is from the enemy.

THE GUILT CYCLE

Many years ago, I was on the evangelism board at a church Dave and I attended. Every Wednesday night, we walked around neighborhoods, knocking on doors and telling people about Jesus.

At that time, I really loved God, but I was "doing" witnessing, not *being* a witness. I didn't understand that Jesus promised that the Holy Spirit would give us power to *be* His witnesses, not simply to do the activity of witnessing (see Acts 1:8). In order to *be* a witness, I had to let God teach me the truth of His Word and change my heart and behavior. I was controlling, manipulative, selfish, and angry. I was frustrated, aggravated, upset, and generally hard to get along with.

Over time, I did receive the power to "be." That doesn't mean I never make mistakes or never have a bad day, but it does mean I have more good days than bad ones. My behavior has improved dramatically since those days and continues to get better and better all the time.

I want you to understand that my improvement was very difficult and extremely slow for a long time. I didn't make much significant progress until I learned how to break the cycle of guilt in my life. If I made a mistake on Monday, I spent Tuesday, Wednesday, and part of Thursday feeling guilty about what happened on Monday. So I ruined not only Monday for myself but also Tuesday, Wednesday, and Thursday. By the time Thursday afternoon rolled around, I finally felt a little better because I believed I had suffered long enough to pay for what I did on Monday.

> *I didn't make much significant progress until I learned how to break the cycle of guilt in my life.*

Then what do you think happened on Thursday afternoon, just as I was finally starting to feel good about myself again? I made another mistake on Thursday night. Then Thursday night was ruined, which led to my feeling bad on Friday, Saturday, and Sunday. In the course of an entire week, I had only a few hours of freedom from guilt. What a miserable existence!

We must realize guilt does not change us, but it does trap us in the cycle of sin and more guilt. The truth is, the guilt itself is sin because the Bible says whatever is not of faith is sin. Guilt is certainly not something we have by faith, because it is rooted in fear. A right response to sin is to receive conviction from God, admit the sin, confess it, turn from it, and receive God's complete forgiveness through Jesus Christ.

I have learned to receive forgiveness quickly, because I know God wants me to have a joyful, blessed life. He wants the same for you. Guilt will steal your joy and blessings, but you can enjoy life and have the strength to keep going if you learn to repent quickly, receive forgiveness quickly, and quickly get back to doing what you are supposed to do instead of staying trapped in guilt and condemnation.

> *You can enjoy life and have the strength to keep going if you learn to repent quickly, receive forgiveness quickly, and quickly get back to doing what you are supposed to do.*

DOUBLE PAYMENT

In Philippians 3:8–12, Paul is basically saying he wanted to be perfect, but had not yet attained it; he knew he had not yet "arrived." Read his words in verses 13–14: "I do not consider, brethren, that I have captured and made it my own [yet]; *but one thing I do* [it is my one aspiration]: forgetting what lies behind and straining forward to what lies ahead, I press on toward the goal to win the [supreme and heavenly] prize to which God in Christ Jesus is calling us upward" (emphasis mine).

Notice Paul writes: "but one thing I do." What was his "one thing," the thing he knew would help him make more progress than anything else he could do? It was forgetting what was behind him. How do we forget what lies behind us, the situations in our pasts, especially those things that cause us to feel guilty? We simply stop thinking about them and stop talking about them, and we press on to the good things ahead.

Many times when we teach on Philippians 3, we stop with verse 14, but I want you to see something powerful in verses 15–16: "So let those [of us] who are spiritually mature and full-grown have this mind and hold these convictions; and if in any respect you have a different attitude of mind, God will make that clear to you also. Only let us hold true to what we have already attained and walk and order our lives by that."

Paul is saying people who are spiritually mature know how to let go of past mistakes and go on without suffering for days on end. He is indicating that people who feel guilty and bad about themselves every time they make mistakes still need to grow up in God because they are spiritually immature.

Spiritual maturity says, "I have a job to do. God's called me; He's gifted me; He's put me in this place; I have an assignment. I don't have time to be trapped in guilt over what I did wrong last week; I need to live today! I want to change and grow spiritually, but I know I must let go of past mistakes. Jesus' sacrifice gives me the right to succeed."

We must come to the point in life where we decide whether we really believe what the Bible says or not. Do we believe Jesus took our sins upon Himself and that He took our punishment, rose from the dead, and is interceding for us before the Father right now? Do we believe Jesus is our substitute and that what He did for us we do not have to do again? If we do believe He did these things for us, then we need to act on that belief and refuse to stay trapped in yesterday's mistakes. He always calls us to go forward, so we need to forget what lies behind us and go on.

I remember well when God taught me this lesson in a profound way. I was a committed Christian; I already had a ministry; and I was teaching a Bible study. But I felt guilty almost all the time. I always had a vague feeling on the inside that something simply was not right about me. Over and over again in my mind, I wondered, *What's wrong with me?*

One day I went shopping and parked my car in the back of the store's parking lot. I still chuckle when I think of that because it certainly was not the kind of car anyone would want to steal. Maybe I parked in the back of the lot because I felt so pitiful, as though I didn't deserve to park closer to the store.

I had done something wrong that day, and as I was walking into the store, the Lord posed a question in my heart, asking me how I was going to deal with my sin.

I responded, "Oh, I'll just receive the sacrifice Jesus made when He died on Calvary."

God then continued to place different things in my heart: *I see, and when do you plan to do that?*

I suddenly understood what He was trying to show me, and said, "Oh, probably in about three days."

He continued: *Don't you realize all you are doing is trying to pay for your sin and I have already done that? You can't pay for your sin. I've done it through the sacrifice of My Son.*

The Lord then impressed upon me that I would be doing Him a favor to receive His forgiveness and simply let go of the guilt and get on with His business. He showed me He was trying to use me and I was of no value to Him in the negative, downtrodden position I was in.

None of us can be of value to God while we're trapped in guilt and condemning feelings. Only immature Christians beat themselves up for every mistake they make. But mature believers apply the work of the cross in their lives, accepting by faith that Jesus shed His blood to forgive our sins, receiving His forgiveness, and going on about the things God has called us to do without a trace of guilt. This may be a new way of thinking for you, but it is biblical and it is your right as a believer in Jesus Christ.

THE RIGHT TO FEEL GOOD

You have a blood-purchased right to feel good about yourself and to succeed in life without feeling guilty. I want you to think about this: Jesus paid for your freedom with His own blood. He purchased you and set you free from slavery to sin and guilt. *You have a right to feel good about yourself and to succeed in life without feeling guilty.* You do not even have to be in the neutral zone and feel "okay" about yourself; you can actually feel good. Not only have you been set free from feeling bad about yourself, you can like yourself!

Dave is a very happy, peaceful person. For years, people have asked him, "What's it like for you for Joyce to be the one in front of people? Doesn't that bother you?"

Dave did go through some transition in the early years, but it didn't take him long at all to accept completely that I was called to be more visible than he is. He is the type of person who simply sets his mind to agree with God.

When our ministry first began to take on a public component, Dave complained to God, saying, "Well, this doesn't seem fair. The man is usually the one who is most visible and 'out front.' Why haven't You anointed me to teach, God?"

Dave knew God does not have to explain Himself or justify His actions to anyone. I'm not sure exactly how He answered Dave's question, but I do know Dave soon came to me and said, "It's obvious that God has anointed you to do what you're doing. It's obvious He has called you and I believe He wants me to back you up and be your covering. I just want to tell you, from now on I'm with you; I'm for you. I'll go where you go."

One of the main reasons Dave is such a happy, contented person is that he never walks around feeling belittled because I often receive more public attention than he does. He does not feel guilty about what he is not called or anointed to do. He knows what he is anointed to do; he is great at those things; and he enjoys doing them. He is never in competition with me. He is happy because he knows who he is and he likes himself.

Many people are not like Dave, though. They don't like themselves and they don't feel secure in who they are. In fact, I used to be that way. But now, I have a good relationship with myself. I like the person God created when He made me.

I'll never forget being in a certain city and seeing the newspaper headline that read: "Meyer says she likes herself!" I am not sure why that was newsworthy, but perhaps it was because society views having a low self-image as admirable and has somehow convinced us we are arrogant if we actually like ourselves. I believe something is wrong with that mentality. Self-hatred is often connected with being religious; and it leads to all kinds of bondage. God loves us, so we need to love ourselves, not in a haughty way, but in a healthy, balanced, biblical way because we believe He who knew no sin became sin for us so we might be made the righteousness of God in Christ (see 2 Corinthians 5:21). We have no right to hate what God purchased with the blood of His Son, our Lord and Savior, Jesus Christ. God loves us and we need to agree with Him, not with the devil who tries to make us feel guilty.

Small Change, Big Changes

Some of the greatest blessings of my life have resulted from opportunities to be a blessing to someone else. One of the most powerful stories I know is the remarkable account of Osceola McCarty, a woman who refused to give up on her dream of helping others receive a benefit she never had.

The Mississippi native was born in 1908 to a mother who had to work long, hard hours to raise young Osceola alone. She worked as a cook for a local county clerk and sold candy at the local school to make extra money. Surrounded by need and influenced by her mother's work ethic, Osceola began working at a young age and saved the money she earned by ironing after school.

Osceola had already decided she wanted money to help care for her grandmother in her old age, and when she was in the sixth grade, she also realized one of her aunts needed her help. That aunt had no children, so when she was unable to walk after a hospital stay, Osceola had to help her. By the time her aunt had improved enough for Osceola to return to school, her fellow students had advanced so far beyond her she decided to not go back.

So the kindhearted young girl began a career that would last decades. She simply did what she knew how to do—washing and ironing. And she did it for years. Osceola charged $1.50 to $2.00 per bushel to do laundry in the early days. As the cost of living increased, so did her prices. When she reached the point where she had to charge $10.00 per bushel, the frugal washwoman became more committed than ever to saving money. She put it in her existing savings account, and more remarkably, she refused to take it out. She just left it—and it grew.

As her savings became significant, her bankers encouraged

her to let her money work for her, suggesting ways it would earn more interest than it did in savings accounts. She took their advice, and her balances continued to increase. Even though Osceola, who never married, had money in the bank, she kept washing and ironing. She lived conservatively; she did not have a car; her house was not air-conditioned for many years; and she walked everywhere she went, even pushing her own shopping cart about a mile to buy groceries.

Osceola's hometown of Hattiesburg, Mississippi, is also home to one of the state's universities—the University of Southern Mississippi. At one point during Osceola's lifetime, the school did not admit black students. By the time she was older, they did—the woman with only a sixth-grade education wanted to use her money to give deserving African American students in financial need the opportunity to receive a college degree. So, in 1995, she donated $150,000 to the university and established the Osceola McCarty Scholarship. In an interview after her gift became public knowledge, McCarty said: "I can't do everything, but I can do something to help somebody. And what I can do, I will do."

Let me encourage you to think of ways you can bless others, as Osceola McCarty did. Have some kind of dream that involves more than getting ahead personally, but that also includes helping people. Be as committed to that dream as you are to other goals you set in life.

CHAPTER 13

NEVER BE ASHAMED

"True innocence never feels shame."
JEAN-JACQUES ROUSSEAU

You may know the story of a quiet, courageous woman named Rosa Parks, a hardworking African American seamstress who refused to surrender her bus seat to a white man. At that time, in the mid-1950s, the law required African Americans to give their seats to white people, to use separate restrooms and water fountains, and to submit themselves to all sorts of shameful rules and regulations for no other reason than the fact they were not white. These people suffered terribly because of who they were, not because of anything they did.

Rosa Parks pressed past the shame society put on her because of her race when she kept her seat that December day in 1955. Upon her death in 2005, she was the first woman to lie in state at the United States Capitol. Imagine that! A woman who was once refused a seat on a bus because of who she was decided one day to throw off the cloak of shame she had worn all her life. Her refusal to submit to shame launched a movement that rolled the reproach off an entire race of people, and her courage earned her a place of honor in American history.

IT'S ABOUT WHO YOU ARE

You cannot press through the challenges and difficulties of life if down deep inside of you, in the core of your being, you are ashamed of who you are. I'm not talking about being embarrassed because of something you've done, but about *who you are*. Guilt is a feeling you have about what you do; shame relates to who you are as a person. Shame and guilt are very close relatives, but shame lodges far deeper in a person's heart than guilt does, and I believe it is a more serious issue.

Shame often hides in your very nature, in the depth of your soul. It manifests itself as a constant feeling of uneasiness, a low-level but continual sense of condemnation, and a never-ending internal whisper that tells you, "You're not good enough." It is evident in the way you feel about yourself, relate to others, view relationships, and think about opportunities that present themselves to you.

YOU'RE NOT WHO YOU WERE

A person who is abused almost always develops a shame-based nature. That was certainly true for me. The sexual abuse I suffered as a child produced a root of shame deep inside of me. Early on, I was simply ashamed of what my father was doing to me, but at some point during my childhood, I stopped being ashamed of what he was doing and became terribly ashamed of myself because he was doing it to me. I thought, *Something must be wrong with me if he wants to do this to me.* I now understand that by thinking that way, I poisoned my soul. Everything in me became tainted as the destructive power of shame coursed through my thoughts and personality. I took on a false sense of responsibility; I took on myself the blame and shame for someone else's sin.

Now that shame is gone. I have been set free through God's love and acceptance. I am able to say, "I was abused as a child; I was filled with hatred and bitterness; I was unwilling to forgive my father for abusing me; I was ashamed of who I was; I became a very unpleasant person who tried to control everything and everyone." I can tell you all about what I was because the events that caused such shame are in the past and the characteristics I previously displayed do not describe me anymore. When Jesus died on the cross, I died with Him; when He was raised to a new life, I was raised with Him. I am a new creature in Christ; old things have passed away and all things are made brand new (see 2 Corinthians 5:17). The same thing is true for you if you have accepted Jesus as your Savior.

Whatever is in your past, do not feel ashamed and as if you must hide it. Telling everyone you know for no reason might not be wise, but if you do need to tell someone, don't be ashamed to do it. What God has delivered you from may encourage other people to believe they can receive help too.

When I think about the terrible situations I encountered as a child—some of the things I endured—and especially when I have a reason to talk about them, I honestly feel I am talking about someone I knew long ago, not myself. Freedom is not letting your yesterday affect your today. What you were is not what you are! You are God's child now, and your future is bright.

> *Freedom is not letting your yesterday affect your today.*

CHARACTERISTICS OF A SHAME-BASED NATURE

I believe, more than anything, that all of us simply want to be free to be who we are and to be accepted for who we are—and shame keeps us from being able to do that. The word *shame* means "to be confounded, confused, dry, disappointed, and stopped." To be confounded means "to be defeated, overthrown, and damned." The word *damned* means "doomed to punishment." Therefore, if we have shame-based natures, we are doomed to punishment. We punish ourselves every day by hating ourselves, rejecting ourselves, and being ashamed and embarrassed because of who we are. People in this condition have no idea how wonderful life could be if they would simply stop being ashamed and learn to accept themselves.

People with shame-based natures may exhibit a variety of personality traits or struggle with a number of issues because of the shame and sense of unworthiness in their lives. I want to call some of these issues to your attention so you can overcome them or help someone else break free from shame and live in the wonderful acceptance Jesus died to give us.

Depression

People who have shame-based natures often are depressed. They're emotionally down, discouraged, and hopeless because they feel terrible about themselves. Many people battling depression alienate themselves from others. They live isolated lives full of self-doubt and totally lacking in confidence. If you've had a lifetime struggle with depression, ask yourself, *How do I feel about myself?* No medicine can change the way you feel about yourself. I encourage you to see a doctor if you need to, but more importantly you should saturate yourself with the truth of God's Word and begin to renew your mind on who He says you are, because that will change the way you feel about yourself.

I know many people struggle with anxiety, depression, and other conditions that significantly reduce their ability to enjoy life. With God's help, I have overcome many of these issues, but I have experienced few "instant breakthroughs" on the road to healing. Most of the time, leaving these problems behind has to be done one step at a time.

Perfectionism

Perfectionists receive their self-worth by trying to do everything right so people will applaud them and tell them how perfect they are. Some become addicted to approval from others. Do you know how painful it is to be addicted to approval and the kind of person who becomes angry or unhappy if others offer healthy constructive criticism, correction, or even suggestions? You have to get to the point where somebody can say, "I don't like your hair that way," and you can say, "You know what? I do." Or someone might say, "I don't think that color looks good on you." But you can say, "Okay, but I love it." This does not mean you should be belligerent and never take advice from anybody. Please keep the balance. I'm simply saying that if you are addicted to others' approval, you will never have any freedom. Satan can arrange ten times every day to find somebody who gives you the wrong look, the wrong glance, or the wrong word that will cause you to try to change yourself to suit others; so, therefore, you are never true to yourself.

Because of the abuse I endured as a child and the root of shame it produced, I suffered horrible guilt every time I failed to perform perfectly. That intense guilt and shame rendered me unable to do much that would make a difference or amount to anything for God. It was an underlying pressure that drained me of the ability to enjoy anything I did.

Hard-Heartedness

People who struggle with toxic, poisonous shame are often cold and unfriendly. I was like that—extremely hard-hearted, cold, and unfriendly. I thought I didn't need anyone. I refused to allow anyone to really get close to me because I imagined they wouldn't like me if they really knew me. I was afraid to open my heart to anyone because I thought they would hurt me. So I managed to keep them out. People living with this kind of shame are often quick-tempered. Deep down inside they're angry about their lives and the things that have happened to them, but they express it wrongly. Hard-hearted people usually have an edge of harshness about them and spend their lives being misunderstood. Everyone thinks they are mean, but in reality they're deeply wounded and hurt. Always remember, "hurting people hurt people."

Fear

Shame-based nature produces timidity and fear of all types, and compulsive issues such as: eating disorders, gambling, drugs, alcohol, excessive spending, stealing, lying, and sexual perversions. People who suffer these things may become workaholics. Or they often hide by spending excessive time on the computer or watching television. They can even hide from themselves by sleeping too much. They say they are just tired, but the truth is they are probably tired of themselves.

False or Excessive Responsibility

I never really knew what neurosis was until I studied it—and then I realized it was a big problem for me. People who are neurotic take on too much responsibility, and in conflict they always assume they are at fault.

For many years, I had a false sense of responsibility. I didn't know that was what was wrong with me, but I felt so pressured—so burdened—under the weight of everything I thought people expected me to do and all of my so-called assignments in life. The only problem was: I had assigned them to myself! No one had put those burdens on myself but me.

Once again, this issue was rooted in my childhood. While I was being abused by my father, I was trying to keep it from my mother. My mother was a sweet lady who got hurt all the time; I didn't want her to hurt anymore. I was afraid of my father. He said to me repeatedly, "You'd better not tell." Under those circumstances, I constantly tried to keep everyone "fixed." I didn't want to make Daddy angry, and I didn't want Daddy to become upset with Mom because I knew his anger would cause him to mistreat her. So I had to keep Daddy happy; I had to keep Mother happy. I had one brother, and I tried to keep him from finding out or getting hurt. I thought I had to keep everything fixed. So in an almost unconscious way, I became the peacemaker in the house. The minute anything went wrong, I tried to fix it. I spent my life thinking, *Uh-oh, Dad's getting mad. What can I do to bring peace?* This habit led to a false sense of responsibility that I dealt with for years and years and still have to guard against.

Being responsible is a good quality, but taken too far, even a good thing can become a problem. Do you tend to feel responsible for everything that doesn't work out right? If you have a child who didn't turn out as you hoped, do you think it's your fault? Neurotic people always assume situations and circumstances are their fault. You may be wondering, *What could I have done differently so my child would have turned out better?* You know what? Sometimes you can do everything right, and people still make wrong choices in their lives. We must stop taking the responsibility for everything that goes on in the world, and begin to live our own lives and enjoy them. Don't let someone else's bad choice affect your joy.

NAKED AND UNASHAMED

The Bible says that Adam and Eve were in the garden with no sin; they were naked and unashamed. I believe this lets us know that God wants us to be completely open and honest, first with ourselves and then with others. If we want to live victorious Christian lives, we have to accept the fact that Jesus took our shame on the cross.

There is far too much pretense in the world. People try really hard to cover up their weaknesses, mistakes, and sins. Even though many people hide behind much pretense, they're actually longing to see someone who is genuine and real. As Christians we should be willing to set examples for the world to follow.

In Genesis 3:7, 21 we see Adam and Eve tried to cover themselves after they sinned. Similarly, we try to cover up all kinds of things about ourselves. How do we do this? We make excuses; we blame other people; we pretend things do not bother us when they do.

My daughter Sandra had frequent outbreaks of pimples as a teenager. She put so much makeup on those pimples trying to cover them up that she actually drew attention to them. Similarly, I tried so hard to cover up the abuse in my childhood that I ended up with a rough, tough "I don't need anybody" attitude that actually drew attention to me. People knew something was not right. They didn't know what it was, but I do recall being asked, "What is wrong with you?"

The truth is I was hurting and lonely and afraid, but I didn't know how to tell anyone, so I tried to cover it with a pretend personality I thought would keep others from getting too close to me.

Many people do not know who they are. They have tried to be so many people, they lost themselves long ago. If this describes you, I encourage you to stop trying to be like somebody else. Stop trying to put on an act for others and just be real. You do not need to cover yourself or hide behind anything or anybody. Be brave enough to put your shame aside and let the real you step into the light for the people around you to see and enjoy. Start by being really open with God. Tell Him everything about yourself—your secret thoughts, fears, and desires. He already knows everything before you tell Him, but doing so will be a type of release as you begin being totally honest with someone you can trust, and God is that someone. He knows all about you and loves you unconditionally. The more you are open with yourself and God, the more you will be able to be genuine with other people.

> *The more you are open with yourself and God, the more you will be able to be genuine with other people.*

LET ME INTRODUCE YOU

You are a wonderful person, and if you have suffered under the weight of shame for most of your life, allow me to reintroduce you to yourself. You need to meet the *real* you because you have been deceived into thinking you are someone you are not. You might say you have been the victim of identity theft. You are a child of God. His power is in you and He will enable you to do whatever you need to do in life. You are a new creation, the righteousness of God in Christ (see 2 Corinthians 5:17,21). You have an assignment from God and a great future. Your past has been washed away in the blood of Jesus. You are awesome—totally, absolutely awesome! The Bible says many wonderful things about you, and you need to learn what they are by reading and studying it. Don't ever allow what someone else thinks or says about you to be the factor that determines your value, because what God says is the only things that really matters.

Psalm 34:4–5 says, "I sought (inquired of) the Lord and required Him [of necessity and on the authority of His Word], and He heard me, and delivered me from all my fears. They looked to Him and were radiant; their faces shall never blush for shame or be confused."

Let me urge you today to seek the Lord, require Him in your daily life, and trust the authority of His Word. Look to Him, and He will set you free from shame. As you press past the shame you have carried in your life, you will be able to be and do everything God has called you to be and do. Embrace the unconditional acceptance God offers you through Jesus Christ and you will have the power to never give up.

From Darkness Into the Light

During the fall of Michelle's sophomore year in college, her life became overwhelming to her—even though nothing identifiable was wrong with her. In addition to anxiety, she struggled with everyday activities, she suffered a dramatic loss of energy, and her sense of hope for the future—and the present—took a major nosedive. Without intervention, Michelle's hope, energy, and coping abilities began to return as spring approached. By April of that year, she was her old self again.

Little did she know that depression would become the norm for her for months at a time over the course of the next twenty years. At first she thought she simply struggled adjusting to school each fall, but even after she finished college and began her career, every autumn as the leaves fell to the ground, her moods and outlook on life plummeted too. Her symptoms grew worse with every passing year—to the point that she often wanted to die and be with the Lord and thought she had done so when she awoke in the middle of the night.

Michelle sought all kinds of help—counseling (both psychological and pastoral), doctors, nutritionists—but no one could help her. She eventually enrolled in a clinical trial at the National Institutes of Mental Health in Bethesda, Maryland, and became one of the first people in the world to be diagnosed with seasonal affective disorder.

Thrilled to have a diagnosis, but deeply upset by the knowledge of that ongoing chemical glitch in her brain chemistry, she returned to her home and career in New York and tried to hold on to her life. For the next seven years, she and her doctor worked to find the right balance of extending her day artificially by sitting

(continued)

for four hours a day under a very bright full spectrum light and taking antidepressant medication, which was extremely difficult for her to choose to do. Her Christian training and upbringing had taught her that her daily walk with God would give her strength, and that God could heal her depressions.

Today, Michelle knows the Lord saved her life. "I was so frustrated for so many years," she says. "I couldn't understand why God wouldn't give me relief... it seemed so little to ask. But God's ways aren't our ways, and His timing is often different than ours would be. But He does give us the strength to go through our afflictions and the insight to be able to use our suffering in constructive ways. Now I know that although God didn't heal me, He did save my life. Without His grace, I can say without hesitation that I wouldn't be here today."

CHAPTER 14

YOUR SECRET WEAPON

"How much more grievous are the consequences of anger than the causes of it."
MARCUS AURELIUS

Vincent Newfield worked at our ministry, and I want you to read this story that he shared with us:

Has God ever spoken something very strong in your heart? He did to me in November 1998. While sitting in my closet praying, the Lord spoke to my heart and said, I am moving you and your family to St. Louis to work for Joyce Meyer. Your gift will be a blessing to her, and you and your family will be blessed greatly.

Wow! Was that really God? He had never spoken to me that way before. Repeatedly, His words rang in my ears for weeks. The strange thing was, I had yet to fill out an application or send in my resume. But in January 1999, I did. Within weeks my references were being called. I was in awe. Could God have really spoken to me? I thought. Is He really moving us to St. Louis? A new journey of faith began to unfold.

March came and I was contacted by the ministry. After two interviews by phone, I was sent a sample writing assignment. After sweating it out for three weeks, I received word that they liked it. The next step would be a face-to-face interview in June. I couldn't believe it! What God had said in my heart was actually happening, and rather quickly. Excitement, hope, and joy filled my soul. My faith had risen to an all-time high. Then...the bottom dropped out.

On May 12, 1999, I received one of the most discouraging phone calls of my life. "Vincent, I am sorry to inform you that the interview has been cancelled. The writing position you applied for has been put on hold for six months to a year." My heart sank within me. All hope, happiness, and faith were gone. What happened? I don't understand. Things were moving along, and then suddenly it's over. *Allison, my wife, stood faithfully beside me, encouraging me to not give up hope. "If God spoke to you, it's going to happen," she said. "It's just a matter of time."*

After three days I mustered enough faith to listen to a teaching tape by Charles Stanley fittingly titled Unshakable Faith. *Slowly but surely, God encouraged me as I drove and listened. A ray of hope began breaking forth on the horizon of my heart. Then all of a sudden, WHAM! I found myself in one of the worst accidents ever. My car was totaled, and I was further humiliated to find that the accident was my fault. Feeling totally discouraged, I returned home and discovered that both of our computers broke earlier that day. We published a magazine and this meant it couldn't be finished. Its theme, ironically, was "Encouragement in the Face of Hopelessness."*

Seething anger was now surging. "Where are You, God? What are You doing? I don't understand? Why are You allowing this in my life?" For days on end, disappointment, discouragement, and despair became very close friends. Question after question filled my mind. The biggest one that would haunt me was: Did I really hear from God about moving to St. Louis? *I tried with everything in me to forget about St. Louis and working for Joyce Meyer, but it never went away. It seemed like everywhere I looked, I saw St. Louis, Missouri—license plates, garbage can lids, household cleaners, you name it.*

God made a way for us to finish that issue of the magazine and taught me a tremendous lesson in patience, endurance, and trusting Him. My faith had been put to the test. Each day I had a choice to believe God had spoken to my heart or reject it. I cried many tears and experienced a wide range of emotions, including anger and depression. But deep inside, I wanted to believe, and my willingness to trust in God's track record kept me going.

Over a year had passed; it was now July 2000. Allison and I had published the last issue of our magazine, and I was looking for new employment. One more time, I thought. I'll send Joyce Meyer Ministries some updated sample work and call them one more time. That was nearly seven years ago. Since then I have been given the awesome privilege of writing for Dave and Joyce and overseeing her monthly magazine, which impacts millions every month.

My point: If God has put something in your heart that will not seem to go away, keep on believing. He will bring it to pass. Cry if you have to... work through anger, frustration, and doubt... but don't give up.

PUSH THROUGH ANGER AND UNFORGIVENESS

Vincent admitted that he was angry at times during the long process of making the changes he wanted. Everyone is tempted to be angry occasionally; we may even be tempted to stay angry for days, weeks, or months on end. The Bible acknowledges that we will be angry at times, but it clearly instructs us not to stay angry. We need to be people who are "calm, cool, and collected" because we have learned how to handle our anger and disappointment in a biblical way. Scripture plainly tells us: "When angry, do not sin; do not ever let your wrath (your exasperation, your fury or indignation) last until the sun goes down" (Ephesians 4:26). We can *be* angry, but we must not *stay* angry; we can be disappointed, but we need to get reappointed. God will always give us a new beginning when something did not work out the way we planned. We need to learn to keep our hearts completely cleansed of all negative emotions because they war against our peace.

If we don't deal with our negative emotions, we will soon find ourselves with all kinds of problems, including physical ones. For example, if we are angry with someone and stay angry long enough to let that anger take root in our hearts, we will develop a root of bitterness (see Hebrews 12:15) and become resentful and miserable. The longer we allow ourselves to remain that way, the more negatively it affects our health. Try to remember that being upset takes a lot of energy and unless you have energy to waste I suggest you learn how to stay calm, cool, and collected.

I am continually amazed by the number of people I encounter who love God and are trying to move forward with Him while still harboring anger, unforgiveness, or other negative emotions in their hearts. Sometimes they are angry at someone for a situation or offense that happened many years ago. They often say, "Well, I have really tried to get over this, but I just can't." Yes, they can. God does not ever ask us to do anything impossible. If He tells us to forgive, then we can forgive. If He tells us not to be offended, then there is a way for us to keep from being offended. He tells us to stay calm, cool, and collected; therefore, it is possible to do it.

> *God does not ever ask us to do anything impossible. If He tells us to forgive, then we can forgive.*

WHAT'S THAT SMELL?

Many times, where anger and other negative emotions are concerned, we stuff them deep in our hearts and pretend nothing is wrong. Negative emotions buried alive never die; they have to be thoroughly dealt with or they will keep showing up one way or another. Anger and unforgiveness, for example, have roots, deep underlying feelings that result in angry outbursts and unkind actions. They can even be the root cause of depression. We may develop ways of ignoring our negative feelings and finding temporary relief, but rotten roots always produce rotten fruit; and before long, that root of anger will produce more ugly feelings and behaviors, which may be a little different from our previous feelings, but still come from the same root.

I'm sure you have experienced opening your refrigerator door and noticing a bad odor. Immediately you wonder, *What's that horrible smell?* The only way to eliminate the smell is to find its source. You may cover it up temporarily with some kind of deodorizer, but unless it's completely eliminated, it always comes back.

Most of the time, the odor doesn't come from something in the front of the refrigerator. It is usually something hidden, something you would not even think could cause such a stink. You have to take everything out of the refrigerator, piece by piece, to discover the source of the smell. You have to clear the shelves and go through the drawers until, at last, you finally find an old, moldy, unidentifiable piece of food you had long forgotten!

I've run across many people who say their lives are like their refrigerators—they stink. When this happens, they have to examine every aspect of their hearts to find the cause of the problem. Often, the source is anger or offense they have denied, covered, or tried to hide. It lurks in the hidden places of the heart, causing great damage and breeding rottenness in the core of a person. Just as a little puff of deodorizer will not cover a terrible smell in a refrigerator, trying to camouflage deeply rooted anger and unforgiveness or other negative emotions will not eliminate the cause of the bad attitudes and actions that result.

We must stop trying to spray deodorizer on our lives. We cannot continue to say, "Well, I'll just go to church today and I'll be better" or "I will ask the minister to pray for me and everything will be fine." No, we must begin to ask ourselves what is making us miserable, and we must examine our hearts until we find out what they hold. We need to start asking ourselves, "What is this really all about?"

Many times, people say, "Well, so-and-so made me mad. You wouldn't believe what he did to me!" I need to tell you today that it is never about what other people do to us. We must "own" our responses. Instead of blaming other people for our bad behavior, we need to start asking ourselves: "Why am I so angry and frustrated most of the time? Why do I get upset so easily? What is going on in my heart that causes me to explode so often? Why do I get angrier than seems appropriate in certain situations? Why am I so insecure that I blow up every time someone tries to give me even the tiniest piece of advice? Why do I get upset every time things don't go according to my plan?"

We have to let God deal with other people. We need to leave them to Him and get busy working with Him in our own lives to learn how to live in a world full of difficult people and still have peace and not allow them to control us. Even if the person who hurts us so often were to go away and leave us alone, someone else would show up to take that role! The answer is not to eliminate challenging people or situations from our lives; the answer is for us to change so we can learn to deal with them and stay calm, cool, and collected.

STOP STUFFING YOUR STUFF

One of the primary ways people avoid dealing with pain is to "stuff it." So often, when people hurt us, we stuff it down deep in our hearts instead of dealing with it—and when we stuff pain for too long, it finally explodes in one big ugly fit of anger. It took me years to understand why I would explode over some seemingly minor situation that certainly was not sufficient cause for my ridiculous behavior. I now know that the explosions came from many negative emotions I had stuffed deep inside and was refusing to deal with. The "incident" that seemed to be the problem was just the trigger for the explosion that was hidden and ready to go off at any time.

I was ignoring the real problem and blaming my bad behavior on anything and anyone I could. No matter how spiritual I pretended to be on Sunday at church, my family knew the real me. I managed to ignore the problem for years by making excuses and blaming others, but eventually I had to let God "clean out the refrigerator" so to speak and get to the root of the problem.

God uses the truth to set us free (see John 8:32), but it is not the truth about someone else that sets us free; it is the truth about ourselves that we need! Facing truth about myself was very difficult and emotionally painful, but it was also the beginning of my healing. Whatever the truth is, go ahead and admit it. If you're angry, admit it. If you're afraid, admit it. If you are jealous of someone, admit it. Go to God and say, "You know what, God? I know I have a bad attitude. It really stinks and I can smell it. I want to understand why I have this problem with jealousy. What is in me? Would You show me why I have this problem?"

> *God uses the truth to set us free (see John 8:32), but it is not the truth about someone else that sets us free; it is the truth about ourselves that we need!*

Maybe the Lord will show you immediately what your struggle is. Perhaps you are insecure; maybe you do not know who you are in Christ. You may compare yourself to others too often. Perhaps you suffered a major hurt or disappointment years ago and you have not been willing to forgive or allow God to heal you yet.

Make a commitment to start being honest and owning your feelings; refuse to stuff them and immediately stop making excuses and blaming others for your negative emotions. You will probably have to talk to God a lot, and you may even need to seek help from a trustworthy friend or minister, but whatever you have to do is worth doing if it helps you to be free and enjoy life.

Whatever has hurt, offended, or angered you, determine today that you are going to go through the pain of facing it and dealing with it. A friend of mine who is a pastor told me recently that when he has a day where he feels depressed, impatient, frustrated, or easily upset he asks himself what happened the day before that he has not dealt with. He said God almost always shows him something he did not deal with the previous day and helps him recognize that as the root of his bad behavior.

One night I tossed and turned until almost 5:00 a.m. and finally asked God, "What's wrong? Why can't I sleep?" He immediately brought to my remembrance a situation that occurred the previous day in which I had been unkind to someone and made an excuse to myself rather than apologizing as I should have. As soon as I asked for His forgiveness and determined I would apologize as soon as I could, I was able to go to sleep. Make a habit of asking, "God, what is this mood or attitude really all about?" You may be surprised by some of the things you learn.

In my life, I've had situations and circumstances hurt and anger me, and over a period of time, I learned how to deal with them. I won't tell you that letting myself feel the pain those situations inflicted and making myself confront my hurt and my anger was easy, but it has been the only way for me to overcome and move beyond them. Facing ourselves and taking responsibility for our attitudes and actions is one of the most difficult things we can ever do, but it is also one of the most important.

OWNING MY ANGER

Dave and I married when he was twenty-six years old and I was twenty-three. One day, he began talking about all the fun he enjoyed as a child. Even though his family didn't have much money and his father died as a result of heavy drinking, Dave had happy memories of his childhood. He had a godly mother who prayed for her children and loved them and kept her family together. Her goodness overrode every difficult thing and every bad thing about Dave's younger days. In the course of conversation, Dave made a comment that hit me hard: "I just remember all the fun we had growing up."

When I heard those words, a feeling of deep sadness came over me and I thought, *I cannot ever remember having fun as a child. In fact, I never even got to be a child.* My father always made me feel guilty if I ever tried to play and have fun as a little girl. In our house, laughter was "noise." So when I grew up, all I ever did was work. I actually became a workaholic because I never learned to play and I thought work was the only activity others approved of. As an adult, the more I thought about my past and all the good I was never able to experience in life, the angrier I became without even realizing why.

As I grew older and developed a relationship with God and came to know His Word, I learned Ephesians 6:12: "For we are not wrestling with flesh and blood [contending only with physical opponents], but against the despotisms, against the powers, against [the master spirits who are] the world rulers of this present darkness, against the spirit forces of wickedness in the heavenly (supernatural) sphere." That verse taught me my father was not my enemy. Yes, he was the one who violated me, but he was under the influence of the enemy. He was simply the devil's puppet. I had to stop blaming my father and begin to heal. I had to stop allowing the devil to continue the cycle of pain he had initiated in my life.

When I realized my childhood had been stolen and I could never get it back, I became angry. I was so angry that it came out in ways that made no sense. I finally realized I was not capable of getting my childhood back, but that God's Word promises He would give me double blessing for my former trouble (see Isaiah 61:7). I learned that we don't overcome evil with more evil, but according to Romans 12:21, we overcome evil with good. If I wanted to "get the devil back" for what he stole from me, I needed to spend the rest of my life helping as many people as I could.

I can now say that I am no longer angry and that I have forgiven my father completely. That did not happen overnight; in fact, it was a long, painful process. But I did confront my anger, rather than try to hide it, pretend it did not happen, or act like it was okay. I was able to release my father to let God deal with him and to press through it and get to the place of peace and forgiveness where I am now.

I will never be able to be a child again. My father, who is now deceased, could never repay what he stole from me. He could never compensate for my losses. But God is my Restorer! When I was about fifty years old, I began to learn how to be a child again. I am laughing and smiling and relaxing and having fun. I may not have had a good start in life, but I am going to have a great finish—and that would never be possible without pressing through my anger and learning to forgive.

As believers, we have an enemy. He will oppose us any way he can. My book *Battlefield of the Mind* makes clear that he uses our own thoughts and feelings to come against us. I like to say that the devil sets us up to get upset. The enemy knows we cannot enjoy power if we have no peace. He knows that the love of God cannot flow through us if we are upset. He will always attempt to upset us, but we can learn to stay calm, cool, and collected at all times. The next time something happens that could easily upset you, ask yourself if it is worth it. Will being upset change it? Can you afford to waste your energy being upset? Will it distract you from God's purpose for your life? Every day we are faced with good and evil; we decide which to choose.

USE YOUR SECRET WEAPON

Every time we suffer hurts, injustices, or offenses, we need to remember that people are not our enemies; Satan is our enemy. God has given us a secret weapon, one that is sure to defeat the devil and destroy his strategies and plans. I mentioned it earlier, but I want to stress how truly important and powerful it is. It's found in Romans 12:21: "Do not let yourself be overcome by evil, but overcome (master) evil with good."

You have a secret weapon against the enemy, and he absolutely hates it because he knows he cannot stand against it. I call it a secret weapon because most people miss it. Your weapon is your God-given ability to be good to people who offend you. Your flesh may want revenge, but God says press through your pain by repaying evil with good. This is difficult to do when you are emotionally distraught; it is much easier to do if you can train yourself to remain calm, cool, and collected no matter what the devil is doing.

When I teach on this scripture, someone always says, "But I can't be good to that person who hurt me. I just can't do it!" Yes, you can. As I wrote earlier, God does not ask us to do things without giving us the ability to do them.

When you do not think you can obey Romans 12:21, all you have to do is say, "God, by Your grace and mercy, I am going to be good to that person. I am not going to tell others what he has done to me. I will not speak ill of that person, but I will pray for him, as You want me to. If I see the person who hurt me, I am going to walk right up to him and say hello. I am going to be kind and obey Your Word and overcome evil with good, by Your grace."

You may not have warm, fuzzy feelings toward a person who has hurt you, but as a follower of Jesus Christ, you must deal with your anger in a biblical way. Do that, as an act of your will, and the right feelings will eventually follow.

To me freedom means I am able to make choices about how I will behave and not be a prisoner to negative emotions. I can act according to God's Word instead of reacting to situations. The devil may be alive and well on earth, but he is not going to control me any longer and he has no right to control you either. God is on your side, and that makes you more than a conqueror.

> *Freedom means I am able to make choices about how I will behave and not be a prisoner to negative emotions.*

Strength in Weakness

Sometimes, all it takes is a little teamwork to turn complete failure into phenomenal success. That was the case with Spencer Silver and Art Fry, two chemists who worked at the 3M company during the 1970s.

In an effort to develop a new adhesive stronger than those in use at the time, Silver achieved the opposite result. He created an adhesive that was less sticky than most of 3M's current products and did not dry. It did stick to certain objects but peeled off easily. Silver could not think of a single use for his not-so-sticky adhesive—nor could anyone else. He completely failed to formulate the stronger, stickier glue 3M had hoped to develop.

Four years later, Silver's 3M colleague Art Fry found himself frustrated every Sunday when he sang in his church choir. He liked to place bookmarks in his hymnal to mark the pages of the hymns to be sung each week so he could locate them quickly and without flipping too many pages. Week after week, Fry's bookmarks fell out of his hymnal. He could have taped or glued them—but the adhesives in tape and glue were too strong to keep from damaging his pages when he removed his bookmarks.

One day, Fry remembered that Silver had developed a weak adhesive and applied some of Silver's product to his bookmark. It was exactly what he needed! His bookmarks stuck to his hymnal pages, but lifted off easily and without damaging the paper.

Before long, Fry began using his "bookmarks" for other purposes—attaching notes to files, leaving brief messages for colleagues, or using them to organize papers or notebooks. His co-workers were intrigued and began wanting the little notes with weak adhesive for themselves.

In the late 1970s 3M coined the name Post-it® for Fry's

bookmarks. They tested the product in 1977 and found little interest in the market. But they knew their self-attaching notes represented a stroke of brilliance and they refused to give up. So, in 1979, they launched a massive promotional and consumer sampling effort in an attempt to catch the attention of the American public with their ingenious new product. In 1980, more than ten years after Spencer Silver's apparent failure with weak adhesive, 3M introduced Post-it® notes across America. Today, Spencer Silver's mild adhesive coats little notes of all sizes, shapes, and colors and can be found in homes, offices, and schools around the world.

CHAPTER 15

THE STRENGTH TO SOAR

"Refuse to be average. Let your heart soar as high as it will."
A. W. TOZER

Being a person who never gives up does not mean barely surviving or simply making it through life with no strength or joy at the end of each day. No, there is a quality of strength, a certain kind of victory, a sense of triumph, and a unique passion that accompanies a person who will not quit. Such a determined, persistent, overcoming individual has not only the ability to survive but the power to thrive.

Think again about the eagle. A person who thrives doesn't scratch around the chicken yard with roosters and hens, nor do they behave fearfully, as chickens do. Instead, he or she has the spirit of a conqueror and the strength to rise and soar like a majestic eagle.

LEARNING TO FLY

Deuteronomy 32:11 says: "As an eagle that stirs up her nest, that flutters over her young, He spread abroad His wings and He took them, He bore them on His pinions." I want you to notice the action in this verse: the eagle stirs up her nest, flutters over her young, spreads her wings broadly, and bears her young on her wings.

If we are going to have strength to soar, we need to understand all this verse means. Let me begin by sharing with you about the eagle's nest. The eagle builds its nest high on a rock, with sturdy sections of rock behind and beneath it. The strong foundation of the nest is composed of tree limbs that can be as large as two inches by six feet. An eagle's nest can weigh as much as two tons when complete and can be up to twenty feet deep and eight to ten feet wide.

Once the foundation is laid, the female eagle stops gathering materials for the foundation and stays home to do the "finish work," but the male eagle continues to bring home vines, leaves, animal fur, anything he can find to build and "furnish" the nest. When the nest is nearly complete, the female lays two or three eggs. In preparation for the eaglets to be born, she pulls feathers from her own breast to pad the nest for the babies. Like most mothers, she is willing to suffer herself in order to bring comfort to her babies. While she watches over the eggs, the male eagle brings her all her food and lets her stay in bed. (I am going to suggest that Dave go eagle watching!)

The male eagle continues to collect things for the nest. He may leave the nest one day and return several times with items such as golf balls, tennis balls, old shoes, tin cans, and any kind of shiny metal he can find. If the mother eagle thinks the nest is too crowded with junk, she throws out what she does not want. I probably tell Dave two or three times each week, "Throw that away. We will never use it!" Sometimes I just do it myself when he isn't watching!

Meanwhile, when the time comes for the baby eaglets to be born, they begin to peck on the shells of their eggs with a small, tooth-like appendage on the end of their beaks. The eaglets use this "tooth" to peck and peck and peck and peck and peck until they finally break through the shell.

God puts that kind of determination in the baby eagles from the very beginning. It is necessary because it helps prepare them for the adversities they will face later in their lives. In fact, I have read if a person tries to intervene and help an eaglet break through its tough shell, the baby may die. It needs to have the opportunity to be tenacious and press through the obstacle of its shell in order to live the way God intended. If it doesn't learn to not give up in the face of difficulty, it cannot survive.

The baby eagles spend the first three months of their lives in the comfortable nest their parents have prepared. It's deep; it's wide; it's sturdy; it's cozy. Momma's there; and everything is great. Their lives are easy; they are well cared for.

But the eaglets get a big surprise when they are about twelve weeks old. Their mother suddenly begins to throw all of their toys out of the nest. She gets in the nest and starts beating around with her wings, throwing things overboard and getting rid of them. Next, she begins to pull out all of the comfortable material in the nest—the feathers and the animal fur—and leaves the babies sitting on thorns and sticks. This is what the Bible means when it mentions that the mother eagle "stirs up her nest." The reason she stirs the nest is that she wants her babies to get out and fly. She knows she must teach them to fly or they will not survive.

Before long, the mother eagle becomes more proactive. She actually begins to nudge them out of the nest. The little eaglets, who have no idea how to fly, fall through the sky, probably very frightened. They do not know what is happening to them, and I am sure they do not want to look at the ground that is fast approaching. Soon, though, they hear a *whooooooooosh* as the mother eagle sweeps up under them to catch them.

I can only imagine what a little eagle would say if he could: "Oh! Thank God! Oh, Momma, boy, was I scared. I was wondering if you loved me, but you must love me because you saved my life. I don't know why you did that to me, but can we just go back to the nest, please? I want my toys back."

At that point, the mother eagle takes the babies right back up to the nest and then nudges them out again. She keeps repeating the process, over and over again, until they finally understand that they have no choice but to fly. The mother eagle does this because she loves them and wants them to have the best life they can possibly have. Most baby eagles won't get out of the nest without this push. Similarly, most of us will also choose comfort over challenge unless we have no choice.

Do you feel God is working in your life the same way the mother eagle does with her young? Has He been taking some of your toys away from you lately? Has He been pulling some of the padding out of your nest so you find yourself sitting on prickly branches? Is He making changes in your life that are leaving you uncomfortable? Is He saying to you, "Come on, it is time to fly"?

Sometimes we feel God has dropped us. We do not know what is going on in our lives; we feel as though we are plummeting toward the earth and we dread the crash landing. We must not be tempted to give up in these situations. We cannot allow ourselves to be afraid when we go through situations we do not understand. God is simply trying to teach us how to fly—and before we fall too far, we'll hear it: *whooooooooosh!* He will be there just in time to save us!

PREENING AND PREPARATION

An eagle begins every day sitting on a rock, systematically passing each of his wing feathers (as many as one thousand two hundred on each wing), through his mouth, one by one. He exhales onto each feather and oils it with oil from a special gland located near his tail. This is called preening; it takes about an hour; and it is similar to a steam cleaning. It reconditions, waterproofs, and prepares the eagle's feathers for the flying he will do that day. This preparation is vital to the eagle's survival and ability to function at his optimum capacity.

Some believers never gain the strength to soar because they do not take time to prepare themselves to be their best every day. I don't know about you, but I have to prepare every day before I go out and deal with people or do anything else. The best form of daily preparation for maximum strength is time with God. Nothing will prepare you to face what you have to face at work, at home, in relationships, or in daily life like taking some time to commune with God before your busy day begins.

When I first understood that I needed to prepare myself for every day by spending time with God, some of my children complained. "Every morning you get up and go to your study, Mom. Can't you just act like a normal mother and fix our breakfast once in a while?"

They were teenagers by that time, so they really didn't need help with breakfast. I replied: "Listen, you're old enough to put cereal in a bowl and pour milk on it. And you should be glad I go to my study every morning. You will have a much nicer day if I take this time with God!"

My daughter Sandra, who is thirty-eight years old now, is spending some time with me while I am working on this book. Yesterday morning, I told her I was going to spend my early morning time with the Lord, and she replied, "Are you going to get nice?" We both laughed because we've learned that even something as simple as being nice to others can be impossible without that daily preparation in God's presence and Word.

I know many people feel extremely pressed for time, and the very thought of adding something else to your schedule makes you shudder. I can hear you now: "Well, I have a house full of children and teenagers. How do you expect me to spend time with God?" Or, "I work three jobs. How do you expect me to spend time with God?"

All I have to say is, the more you have to do and the busier you are, the more you really need to spend time with God. I do not know how you need to adjust your schedule, but I know the time you give to God is no different than the money you give to God. If you give Him time, He will give it back to you. He is in charge of time; He knows how much time you need to accomplish the things you really need to do, and He can protect and help you manage it if you will spend time with Him.

> *If you give Him time, He will give it back to you. He is in charge of time.*

Stop trying to fit God into your schedule and make a decision to put Him first; then work the rest of your schedule around Him.

You don't have to live life feeling as though you'll go over the edge any minute or thinking you simply cannot make it one more day. If you're in a difficult marriage, the best thing you can do for yourself, before you do anything else, is spend time with God right away, every day. If your job entails great responsibility and high stress, the best thing you can do for yourself is to take some of your lunch break, get in your car, and spend time with God. If you have a child who is challenging to raise and hard to deal with, spending time with God before you begin your day will be the best investment you can make. Taking a few five-minute "spiritual vacations" throughout the day is also helpful. Pause for even a few minutes and pray or meditate on scriptures that encourage and strengthen you.

Life does not have to overwhelm you or catch you off guard. Prepare yourself for the challenges you will face each day, and prepare yourself to soar with the eagles by taking time to be with God every day.

COME UP HIGHER

In addition to spending time with God in order to stay strong, eagle Christians also have to be careful to allow the right influences in their lives. Who are you spending time with? Who has input in your life? Who are you listening to?

As I wrote earlier, an eagle who hung out with chickens all the time would surely soon begin to take on some qualities of chickens. Proverbs 4:23 instructs us: "Keep and guard your heart with all vigilance and above all that you guard, for out of it flow the springs of life." Notice these words are a command: "guard your heart." It is your job to guard your heart. Guarding your heart is not something God will do for you.

The very issues of life—the most important things—flow out of your heart, so be very diligent to guard it. On a practical level, this means to be very careful what you allow into your life through the "ear gate" (what you listen to), and through "the eye gate" (what you look at and watch).

Be very careful who you spend time with, and remember, an eagle does not fly with all kinds of other birds, so you cannot hang around with just any group of people. If you spend time with negative people, you will develop a negative attitude. On the other hand, if you surround yourself with those who are positive, you will find your attitude becoming more optimistic. If you spend time with stingy people, you are going to get stingy; but if you hang out with givers, their attitudes and actions will challenge you to come up higher and be more generous.

We need to be the kind of people who *want* to be challenged by others, because that helps us maintain our strength to soar. Let me encourage you to pray for God to put people in your life whose behavior will convict you to live at a higher level. God wants all of us to come up higher. I believe we should live with the goal of never giving up or being content where we are, but to always be pressing forward to come up higher and move to the next level in our faith and in our lives.

BUILD YOUR SPIRITUAL MUSCLES

God wants you to be strong enough to soar over the storms of life and to fulfill all the plans He has for you. For that to happen, you have to build spiritual muscles. You cannot simply *wish* you had spiritual muscles; you have to *build* them just as you have to build muscles in your physical body.

It would be great if we could develop muscles by taking a pill or having someone pray for us. Everyone would want those prayers! We all know that is a silly idea because the only way to develop muscles is to work out!

I always said I would *never* be the kind of person to go to a gym three times a week because I hated exercise! I was okay if I was doing an activity I liked such as walking or golfing (and happened to be getting exercise while I was doing it); but I didn't care much for the whole "let's just exercise" routine. As I mentioned earlier, I now go to the gym three times a week to work out. What made me change my mind? I finally wanted to be stronger, be healthier, and have more energy. That was enough to make me do what I needed to obtain my desired goal.

I'll tell you exactly what happened. God spoke to my heart and said, *You need to do something now so you'll be strong for the last third of your journey*. He wants me to have strength to soar.

I believe I am in the "last trimester" of my ministry, and I want to give birth to everything God has planned for me. In order to travel internationally, do conferences, and fulfill my other responsibilities, I have to have a strong body to house my spirit.

I found a coach and started working out. I started with five-pound weights, and at first I thought I would never be able to lift them. Then I went to seven and a half, then to ten, then to twelve and a half, then to fifteen. Now I can lift twenty pounds easily and sometimes even twenty-five, depending on the exercise I am doing.

It only took nine weeks to work up to this level. You will be amazed at the progress you make if you'll simply be willing to start doing whatever God is leading you to do. It may not be lifting weights, but I am sure He is dealing with you about something because God is always urging us upward. Doing the difficult things God asks is what helps us build spiritual muscle.

Difficult situations, ones that are hard to bear, are the ones that give you strength to soar, so do not run away from them. You will reach the point where circumstances that were once very difficult for you become easy. In fact, you rise above the storms of life, just as the eagle does.

> *Difficult situations, ones that are hard to bear, are the ones that give you strength to soar, so do not run away from them.*

The Gilded Cage

Many times when we think of people who refused to give up, we think of those who overcame obstacles such as poverty or economic disadvantage, physical or mental handicaps, or some other kind of tragic circumstance. But this is not always the case. People born to lives of privilege have challenges too, as we learn from the life of Florence Nightingale.

She was named after the city in which she was born—Florence, Italy—in 1820. The daughter of wealthy parents who had enjoyed a two-year honeymoon in Europe after their wedding, Florence Nightingale was expected to pursue the interests of the privileged, marry well, and live the traditional life of a sophisticated upper-class Englishwoman.

But Florence had other plans. In 1837, in the garden of her family's Hampshire home, she received a "call" from God, in which she sensed Him calling her to do His work, but she did not know specifically what He wanted her to do. Eventually, she realized she was supposed to be a nurse, but her parents felt nursing was "beneath" Florence because of her social background, academic ability, vivacious personality, and beauty. In their opinion, a woman of such distinction should not even consider such a "low class" occupation, and they refused to allow Florence to seek the training she needed to follow her God-given desire to become a nurse.

After considerable conflict in the family, Florence's parents eventually but reluctantly agreed to allow her to take a three-month course in nurse's training. This positioned her to become superintendent of a women's hospital in London in 1853.

In March 1854, the Crimean War broke out when Britain, France, and Turkey declared war on Russia. British newspapers soon reported that wounded soldiers were unable to receive proper medical care because of inadequate medical facilities on the front. A British official and acquaintance of Florence's asked her to oversee the process of staffing military hospitals in Turkey with female nurses who could provide proper care for the men injured in battle. She arrived in Turkey with thirty-eight nurses ready for duty in November 1854.

At first, doctors did not want the nurses working with them and refused to ask for their help. But within days, the medical needs of wounded soldiers overwhelmed the doctors and they recognized the importance and value of having the nurses assist them. As a result of Florence's work, military hospital conditions improved dramatically and mortality rates dropped significantly. Florence earned the deep respect of British soldiers not only for her skills as a nurse and administrator but also for her kindness and compassion to them.

Florence won many prestigious awards throughout her lifetime and made invaluable contributions to the modern health care system, and her influence extends throughout the nursing profession to this day. She raised nursing to an admirable occupation for women and wrote classic books on nursing, books that are still in print.

Florence Nightingale had to overcome enormous family opposition and social pressure to do what she believed God called her to do. She never gave up on His plan for her life and stayed faithful to it until her death at age ninety. Like Florence, be willing to risk the disapproval of others to follow God and stay committed to everything He asks you to do.

CHAPTER 16

NEVER GIVE UP ON YOUR FUTURE

"My interest is in the future because I am going to spend the rest of my life there."
CHARLES F. KETTERING

Throughout this book I've been urging you to never give up. Never giving up means marching into your future with boldness and confidence, seeing each new day as an opportunity to move forward in all the best God has for you and taking each new challenge as a mountain to be climbed instead of a boulder that will crush you.

You have a great future ahead of you, but you will not be able to fully enter into it and enjoy it if your past still holds you captive. The past has the potential to keep you from experiencing the joy, freedom, and blessings of the present and the future—*if you let it*. God wants you to break free from your past and set your face toward your future with hope, courage, and expectation. And the best way I know to never give up on your future is to refuse to be trapped in your past.

DON'T LOOK BACK

God is serious about setting you free from your past. Perhaps no story in the Bible better illustrates this than the unfortunate account of Lot's wife. Let me remind you.

A man named Lot and his family lived in a city consumed by wickedness and sin. God was angry at the degradation in that city and one nearby, so He decided to destroy them. He sent two angels to Lot's house to tell him to take his family and flee. The angels warned them: "Escape for your life! *Do not look behind you* or stop anywhere in the whole valley; escape to the mountains [of Moab], lest you be consumed" (Genesis 19:17, emphasis mine). Lot's wife made the mistake of disobeying the command to not look back. Right then and there, she turned into a pillar of salt.

Jesus did not want us to forget what happened to this woman. Luke 17:32 is a very short Bible verse in which Jesus spoke only three words: "Remember Lot's wife." That was His way of saying: "Stop looking back. The past is finished. Don't look behind you; look to the future ahead!"

As long as I thought and talked about my past, I felt I had no future. God was providing one, but I would have totally missed it had I not finally understood I had to "let go in order to go on!"

Sometimes when we focus on our past, we can't even see our future. We become discouraged, hopeless, and depressed. Don't behave as if your past is more important than your future by giving it too much of your time. If you do, it will keep you trapped in days gone by and steal your enjoyment of the present moment and your hope for the future. When Lot's wife looked back, she lost her life. She lost her family, and she lost her future. I want you to know: you can be destroyed by looking back! You probably will not turn into a pillar of salt, but you can become as "dead" as a pile of salt on the inside if you allow excessive focus on the past to steal the life you have today.

NEW EVERY MORNING

Holding on to your past only gives you opportunity to relive it. The Bible says that as a man thinks so is he (see Proverbs 23:7). I like to make this point by saying, "Where the mind goes the man follows." If I put my mind on the past, I will keep repeating it one way or another, but if I put it on the future I will make progress toward God's dream for me.

You don't have to be afraid of repeating the past. If you believe God is greater than your sins, mistakes, and shortcomings, you will have the spiritual energy and the strength and the grace of God to help you do better in the future. The dreams of your future have no room for the disappointments of the past. They will keep you weighted down and stuck.

Many times, before your feet even hit the floor in the mornings, the enemy begins to remind you of everything you did wrong the previous day or everything that did not work out well. In doing so, the enemy's goal is to use yesterday to keep you from living today. He always wants to use your past against you. But God does not want us living under the tyranny of the past. Every day can be a new beginning if we allow it. God's mercy is greater than yesterday's mistakes.

> *The enemy's goal is to use yesterday to keep you from living today.*

God's mercies are new every single day! So every morning when you wake up and Satan starts reading you an inventory of yesterday's failures, you need to say aloud: "God's mercies are new every morning. I receive Your mercy right now, God. I thank You for forgiveness. Thank You for putting my past behind me and calling me forward into a great future." Instead of wondering what terrible thing will happen before the end of the day, start the day by saying, "Something good is going to happen to me today!"

Chances are, the enemy will be there to whisper to you, "Well, you will just be disappointed and you won't do any better today than you did yesterday"; but you can say, "I don't know exactly what the day holds, but I am going to have a positive attitude. I am going to press forward and do my best, and at the end of this day, I will get over the mistakes I make and get ready for the next new day. You might as well leave me alone, devil, I will never give up on the new mercies God gives me every morning and I refuse to be trapped in my past." You will be amazed at how much some good positive self-talk will help and even energize you. Being negative drains us all, but faith, hope, and a positive attitude open the door for God to work miracles.

You are the only one who can silence the enemy when he rehearses the past. No one can do it for you. If you are determined to overcome him and to press into the wonderful future God has for you, you will need to learn to talk back to the devil. You cannot borrow someone else's determination; you have to dig down deep and get a firm grip on your inner strength and say: "I will not give up! The enemy will not use my past to steal my future! I am not going to let yesterday affect today!"

Fulfilling your destiny demands letting go of what lies behind. I had to stop mourning over what I had lost and take an inventory of what I had left. I offered that to God, even though it did not seem like much, and He has done wonderful things in my life. He will do the same thing for you if you will work with Him instead of against Him. God called Himself "I AM" (see Exodus 3:14). He said that was His name. Why? What kind of a name is that? He is making a strong point that He is ever present, ready to work in our lives right now. He is with us to help us right now! He is able to make even past mistakes work to our benefit if we trust Him. Will you let go of the past and live every day fully while you press into your future?

FORGET THE FORMER THINGS

The prophet Isaiah urges us to forget the past: "Do not [earnestly] remember the former things; neither consider the things of old" (Isaiah 43:18). Remembering and considering are processes of the mind, so Isaiah is basically saying: "Get your mind off your *old* sins, your *old* failures, your *old* mistakes, your *old* friends, your *old* life, the *old* nature, your *old* job—everything about your past. Do not let your mind dwell on such "former things."

After I left the job in which I taught well-attended Bible studies at a church and was leading Joyce Meyer Ministries, people who remembered the Bible studies occasionally asked: "Don't you miss the good old days when we were all together at the church for Bible study? Weren't those good times?"

Yes, those were good days. They were so important to the development of the ministry I have today, and we did have wonderful times studying the Word together. But I have to say I really don't miss those times. They were good days, but the days I am living now are better! If we miss the "old days" long enough and keep thinking about them and talking about them, we will completely think ourselves out of the ability to enjoy today or the future.

I don't know exactly what God has in His plan for me, but I do know it will be good. I am excited about walking it out with Him, and in order to do that I have to refuse to allow my mind to "drift" into directions that will not benefit me. Do your own thinking on purpose, and don't just wait to see what happens to fall into your mind.

We see things with our mind. What do you see? After Abraham had endured a great loss, God led him to a mountaintop. He told him to look north, south, east, and west and then said, "You can have whatever you see!" (see Genesis 13:14–17). Wow! That sounds like a good plan to me. Why look back and get more of what we have had when we can look forward and enjoy things too wonderful for us to even imagine?

As we continue reading in Isaiah 43, God says: "Behold, I am doing a new thing! Now it springs forth; do you not perceive and know it and will you not give heed to it? I will even make a way in the wilderness and rivers in the desert" (v. 19).

Notice, God does not say, "I *will* do a new thing"; He says, "I *am doing* a new thing." God is doing a new thing, but to experience it, you have to turn your focus away from what God did in the past, and onto what He is doing now. Failing to do this will result in a life full of regret.

To regret means "to feel sorry or to grieve over; to mourn; a sense of loss and longing for something or someone gone; distress over desire unfulfilled or an action performed or not performed." Regret binds you to your past and keeps you focused on former things. But God is doing a new thing *now*. Be a person who lets go of past regrets, lives in the present, and never gives up on a fabulous future.

I do not believe God releases the blessings He allots for us each day if we live in the past. He knows we will not enjoy

> *Be a person who lets go of past regrets, lives in the present, and never gives up on a fabulous future.*

them if we are stuck in a memory fifteen years old or if we cannot get past a mistake we made in high school. Come into the present; receive the blessings God has for you today, and look forward to the future with great anticipation.

GOD FORGETS!

Do you really believe God forgets your sins once you repent? He does. Elsewhere in this book, I mentioned that Jesus paid in full the price for our sins, that we are forgiven and have no reason to feel guilty or condemned once we have repented. Now I want to go a step further and make sure you understand God not only *forgives* but also *forgets* all your sins. He does not forgive you and then say: "Oh boy, I remember when I had to forgive Johnny for pulling his sister's hair and making her cry. Now he wants me to forgive him for cheating on his algebra test. His list of forgiven sins is getting awfully long!" No, if Johnny were to say to God, "I know You've already forgiven me for pulling my sister's hair a couple of years ago, but now I need Your forgiveness for cheating on a test," God would say, "Your sister's hair? You asked me to forgive you for that? I have absolutely no recollection of that; there's no record of it anywhere. Now, what did you want to tell Me about your algebra test?"

In both the Old and New Testaments, God makes sure we understand how completely He forgets our sins. In Jeremiah 31:34 He says: "For I will forgive their iniquity, and I will [seriously] remember their sin no more." In Hebrews 10:17, the writer references Jeremiah's words: "And their sins and their lawbreaking I will remember no more."

I want us to see this verse in context:

For by a single offering He has forever completely cleansed and perfected those who are consecrated and made holy. And also the Holy Spirit adds His testimony to us [in confirmation of this]. For having said, This is the agreement (testament, covenant) that I will set up and conclude with them after those days, says the Lord: I will imprint My laws upon their hearts, and I will inscribe them on their minds (on their inmost thoughts and understanding), He then goes on to say, And their sins and their lawbreaking I will remember no more (vv. 14–17).

This passage is not talking about a forgiveness that happens the day we receive Christ and takes care of all of our *previous* sins. God's forgiveness is ongoing for the duration of our lives; it is for every day. When Jesus died on the cross two thousand years ago, He not only forgave everything we had done in our pasts, but He also committed Himself to forgive every sin we would commit in the future. He knows our thoughts before we think them; He knows our words before they come out of our mouths; He knows every wrong decision we will ever make—and it's all already covered. All we have to do is stay in relationship with Him. After all, what He wants from us more than anything else is not perfect performance, perfect behavior, or perfect attitudes, but hearts that really love Him. Always remember, God is not surprised by our bad behavior. He knew about it long before we did and He wants us anyway. He is excited about helping us grow into all He knows we can be.

God buries our sins—past, present, and future—in the sea of forgetfulness and remembers them no more. Often, we preach God's forgiveness, but we fail to focus on the fact that He also forgets them. "I will remember your sins no more!" says God. Determine today to stop remembering what God has forgotten.

Throughout this book, I have mentioned many of the devastating and dysfunctional situations of my past, and for years I allowed the impact of those circumstances to keep me from moving forward into my future. I did not progress into the new life God had for me because I kept using my past as an excuse to continue in old thought patterns and behaviors.

Jesus said, "And you will know the Truth, and the Truth will set you free" (John 8:32). When we are willing to face the truth about ourselves, we will be set free. When I accepted the fact that I was blaming other people and the circumstances of my past for my failure to go forward into the future, I was finally able to take responsibility for my life, deal with my past, and become free to pursue my future. I found the strength to believe 2 Corinthians 5:17: "Therefore if any person is [ingrafted] in Christ (the Messiah) he is a new creation (a new creature altogether); the old [previous moral and spiritual condition] has passed away. Behold, the fresh and new has come!"

Being a new creation in Christ does not mean all your problems and weaknesses vanish into thin air the minute you make a commitment to live for God. It means you become brand-new spiritual "clay." Let me explain.

Jesus, who is called "the Seed," comes to live in us as the seed of Almighty God, bringing with Him a seed of everything God is. In the physical world, seeds must be planted; they need time to become rooted and grounded; and they must be nurtured. They need sunlight and water. Someone has to keep the weeds from choking the life out of them and hindering their growth.

A similar process is necessary in our spiritual lives. We do not become "new creations" overnight, but through a process. God's Word needs to be planted in our hearts, and it needs time to take root in us. We need the water of God's Word (see Ephesians 5:26) and the light of His Spirit to strengthen us. It is also extremely helpful to take advantage of every opportunity we have to be nurtured by more mature believers who can guide us.

As we continue and continue and continue in these good, godly habits, those seeds God planted in us begin to grow up like plants and become trees of righteousness (see Isaiah 61:3). Over time, we realize we are changing not because of our own effort but because we are spending time with God and growing in His Word. As we dwell in His presence and live by His truth, His image is re-created in us and we truly become new.

THE PAST IS PAID FOR

The enemy loves to remind you of your past. He takes the mistakes, disappointments, hurts, and offenses from former seasons in your life and replays them in your mind like broken records. In between every verse of those same old songs, he sings this chorus: "Now you have to pay. You have to pay for everything you did wrong. You have to pay! You have to pay! You have to pay!"

As I mentioned earlier, many people believe the lie that we somehow need to compensate for the misjudgments or wrongdoings of our pasts. I do not want you to fall into this trap. Let me urge you not to ever try to "pay" for your past by giving up your future. Do not believe the lie that a bad past disqualifies you for a great future. Do not allow yourself to believe you have made so many mistakes in the past that you have no hope for the future.

In Jeremiah 29:11, God says: "For I know the thoughts and plans that I have for you, says the Lord, thoughts and plans for welfare and peace and not for evil, to give you hope in your final outcome." No matter what your past may hold, determine in your heart to believe that God has good plans for you. Your mistakes cannot change this truth. The only way it would not be true in your life would be for you to reject it. If you are not willing to believe it, you are not likely to see it come to pass for you. But if you will believe it and not stop believing it, you will soon see God's thoughts and plans toward you are for your welfare and peace, not for evil, regardless of your past. He is for you; He has great plans for you; and you have every reason to have hope for your future.

He Stuck with It

You may never have heard of Thomas Adams, but you have probably enjoyed the product developed as a result of his persistence. Let me explain.

During the 1860s, success seemed impossible for Adams. After failing at several jobs and trades, he became a photographer. He was creative and enjoyed his work, but struggled to make a living at it.

About the same time, Mexican general Antonio de Santa Anna went into exile in the United States and stayed at Adams's home in Staten Island, New York. Santa Anna encouraged the innovative Adams to see what he could do with chicle, a substance derived from the sapodilla trees of Mexico. Santa Anna believed the chicle might be used to make synthetic rubber and told Adams he could secure the substance for him at very little cost from friends in Mexico.

So Adams went to work experimenting with chicle in a warehouse in New York City. He tried mixing it with rubber to make bicycle tires, but failed. He also tried to make toys, masks, rain boots, and other items, but he never succeeded. After experimenting diligently for about a year and finding no use for chicle, he decided to throw his remaining chicle stock in the East River and forget about it.

Before he disposed of the chicle, he paid a visit to a local drugstore, where he overheard a little girl asking to buy a piece of chewing gum for a penny. At that time, chewing gum in America was made from paraffin wax. In that moment, Adams realized he could probably make chewing gum from chicle, not realizing this had been taking place in Mexico for years. That night, Adams and his son made pieces of chewing gum from chicle—little pieces without any artificial flavors—and wrapped them in brightly colored tissue papers. They sold each stick of gum for a penny.

Adams finally found success selling his gum, which he called "Adams' New York Gum Number 1." In 1888, his "Tutti-Frutti" gum became the first to be sold in vending machines and by the end of the nineteenth century, he had established the most profitable chewing gum company in America. His company gained a monopoly in the chewing gum business in 1899 when it merged with the six largest chewing gum manufacturers in the United States and Canada. This company developed and sold Chiclets®, with tremendous success, beginning in the early 1900s.

Next time you see a package of chewing gum, remember Thomas Adams and let it remind you to never give up.

CHAPTER 17

THE POWER OF HOPE

"Hope is necessary in every condition."
SAMUEL JOHNSON

On Christmas Eve 1981, a group of people embarked on a tragic journey they would never forget. A woman named Pat, her husband Gary, her two teenage stepsons, and a close friend were flying to Colorado to go skiing. Gary was piloting the small plane that was to get them to their destination. He had his pilot's license but had not yet been instrument certified, so he did not file a flight plan.

Flying over the mountains taxed the small engine, and they began to go down. Gary saw a tiny clearing along the tree line in the middle of the mountain range. He aimed for it and although they crash-landed, he was able to keep much of the plane intact and right side up.

The impact of the crash broke Pat's back, but everyone else's injuries were minimal. Pat was unconscious for a while, but she vaguely remembers her husband saying that he was leaving to get help.

The four stayed in the cramped cabin of the plane and waited overnight for help to arrive. The next day they thought their prayers had been answered. About a hundred yards away a military helicopter landed and a group of soldiers jumped out. The stranded passengers thought their ordeal was over. They assumed Gary had made it to safety and that this was their rescue. A few minutes later the group realized the men were *not* looking for crash victims. Horrified, they screamed for help, but the helicopter engine was so deafening the soldiers could not hear them. The white top of the plane blended into the snow; the snow was too deep for the boys to make it very far away from the plane, and they had nothing to use as a signal. There was nothing else they could do. The helicopter took off. The cold silence of the mountain crushed their spirits.

The victims ate snow for Christmas dinner and tried to build a fire, to no avail. They sat in the freezing temperature and endured the pitch-blackness of night. In the back of Pat's mind, she asked God, *If we are all going to die, can I go first? I can't watch the others die in front of me.*

Thankfully, they had a Bible with them. Through it, the Lord began to encourage them to press on when it seemed there was no hope.

Because no flight plan was filed, no one knew the group was missing. However, the emergency transponder was sending out a signal. Little did the crash victims know a commercial jet flying overhead heard the signal, but since there was no report of a missing plane, it was assumed that the transponder was being used as a drug drop. (In the 1980s, drug traffickers dropped loads of drugs packed with transponders so teams on the ground could locate them.) However, when the same flight heard the transponder a couple of days later, the alert went out and hundreds of people began to search the mountains. It was difficult to pinpoint the location of the transponder because the signal echoed off the mountains.

The weather grew worse the sixth day; a blizzard was setting in. But then the cold and hungry survivors began to hear noises outside. Suddenly out of nowhere a man with a big smile appeared in the window of the plane! Reinforcements were called in. Even against orders a helicopter pilot landed his Huey on a jagged sheet of ice in blizzard conditions. Pat and the boys were loaded on the helicopter, but their friend would have to be skied down the next morning. The snow was chest deep now, so a team of rescuers with emergency supplies stayed with him overnight. The next morning it took seven hours to ski him to a landing zone.

Pat had surgery on her broken back and to this day has a steel rod helping to hold her back together. The boys both lost about half of each foot due to frostbite, and the friend lost both of his legs at the knee.

For months, no one had seen or heard from Gary. Pat knew he was gone.

It took several months, but Pat began to make a strong recovery. She was encouraged to press on. She began to feel God had something for her to do. She realized she could either feel sorry for herself and live as a victim, or go on with life.

Nine months after the accident, Gary's body was discovered. He had apparently fallen into a ravine and frozen to death.

Pat decided to press on and find the work God was calling her to do. She went through a decade of searching and emotional wilderness, trials, and pain—all in preparation for the work she now does with a man she married after Gary's death.

Always remember that you have two choices when you find yourself in difficult circumstances: you can give up or go on! Only you can make the choice, but if you choose to go on you can encourage many others throughout your life who are faced with the same choice you once had to make.

> *You have two choices when you find yourself in difficult circumstances: you can give up or go on!*

HOLD ON TO YOUR HOPE

Pat and the other plane crash survivors learned how to hope in a seemingly hopeless situation. People like Pat, who have seen God's faithfulness in the past, tend to be very hopeful. They know a bad situation can turn into a wonderful testimony in a matter of minutes. They know how to hold on to hope and refuse to give up.

People who have lost hope view life from the perspective of dread.

Dread, which is closely related to fear, steals the ability to enjoy ordinary life and makes people anxious about the future. It keeps them from looking forward to the next hour, the next day, the next month, or the next decade. Their thoughts about the present are negative and their outlook on the future is filled with fear, pessimism, doubt, and worry.

Hope, on the other hand, is the opposite of dread; and it is a close relative of faith. Hebrews 11:1 tells us faith is "the assurance (the confirmation, the title deed) of the things [we] hope for." When we have hope, our outlook on life and the future is positive. We can have hope because we trust in God's love, His power to provide for us, and His ability to lead us in every situation. Hope keeps us from worrying, allows us to leave our unanswered questions in God's hands, empowers us to stay at peace, and enables us to believe the best about the days to come. People with hope are happy, optimistic, and full of strength and courage.

Because hope is such a powerful force, the enemy goes after it with a vengeance. If he can steal your hope, he can set you on the path toward total despair and depression—and that's his intention. He will plant thoughts such as these in your mind:

- You have always been this way. You will never change.
- No one will ever want to marry you.
- You might as well settle into this entry-level job, because you will never be smart enough to get a promotion.
- You might as well go buy some clothes in a larger size because you will never lose this weight.
- Your children will never amount to anything.
- You will not have enough money for retirement.
- No one in your family has lived more than seventy years, so you cannot expect a long life.
- You will never own a new car.
- You will never own your own home.
- You will never get out of debt.

If you read these statements carefully, you will notice they have a common thread running through them: self-pity! The devil puts thoughts in our minds to make us feel sorry for ourselves and resent the people who have what we are convinced we can never have.

Self-pity is a very destructive negative emotion. Self-pity makes us blind to our blessings and the possibilities before us; it steals our hope for today and for tomorrow. People who pity themselves think, *Why should I try to do anything? I'll just fail.*

I used to love to sit and drink my coffee, feeling sorry for myself and thinking about how mistreated I was. But I finally realized self-pity is idolatry because it is self-focus carried to an extreme and it rejects God's love and ability to change things for us.

I encourage you to be determined to not waste one more day of your life in self-pity. When you lose hope and begin to feel sorry for yourself, stop right that minute and say, "I refuse to feel sorry for myself. I may be in a difficult season of life right now, but I will not stop hoping for better things! The Roman orator Cicero said, "While there's life, there's hope." This is true. As long as you are alive, you have the ability to hope.

The enemy wants you consumed with hopelessness and will tell you all sorts of things about yourself, your life, other people, and God if he thinks you will believe them and lose hope. But the devil is a liar; you must not believe anything he says. The psalmist refused to give up when the enemy assaulted him with hopelessness. He knew how to talk to himself to overcome such attacks, and said, "Why are you cast down, O my inner self? And why should you moan over me and be disquieted within me? *Hope in God* and wait expectantly for Him, for I shall yet praise Him, my Help and my God" (Psalm 42:5, emphasis mine).

Remember, God has thoughts and plans for your good, to give you hope for your future (see Jeremiah 29:11). If you will hold on to your hope and fight for it when the enemy tries to take it away, you will see amazing things take place in your life. Being hopeful helps you to press on instead of giving up.

HOPE IN THE MIDST OF DISAPPOINTMENT

A friend recently told me a story that illustrates what happens when a person refuses to give up hope even after significant disappointment.

A pastor met with a member of his church, a businessman who was very excited and full of faith because God was working in his life in significant ways. The man's business was selling large equipment to companies, and he told his pastor he had recently bid on a very lucrative contract. He believed he would get this contract. In fact, the man was absolutely *convinced* the vendor would give him the business.

The pastor believed God showed him the man would not receive the contract and advised the man to reconsider his expectations so he would not be disappointed when it did not materialize. The man refused, saying, "I *know* I am going to get that contract. I'm believing for it. My faith is strong. I am going to get it."

He did not. The contract went to a competitor.

At first the businessman was upset, but then the pastor gave him some wise advice: "I believe the best course of action at this point is to go talk to the people who gave the contract to someone else and let them know you would be glad to help them and be involved in their business any way you can. Tell them you understand why they bought the equipment from someone else, but that you would like to do the installation because you have the best installers anywhere." He encouraged him to remain hopeful and refuse to be defeated by one setback or disappointment.

The businessman took his pastor's advice and decided to press through his disappointment and not give up hope for a relationship with the company that did not buy from him, even though he would have to accept a different type of relationship than he initially wanted.

He did get the contract for installation and while the businessman's employees were installing the equipment, the head of the company noticed what an excellent job they were doing. The next time he purchased a large piece of equipment, he bought from the businessman—and it was a much, much bigger sale than the first contract would have been. So he ended up with "the best of both worlds" because he refused to stop hoping in the face of a major disappointment.

When things don't work out the way we wish they would, we can either be sad about what we did not get, or we can be creative and look to see what we can do with what is left.

The businessman chose to have a positive, hopeful attitude toward a disappointing situation. He pressed through the temptation to be angry with the company that did not award him the first contract or jealous of the people who did get it, and he ended up with the best deal of all. He thought God would bless him one way—with the first contract—but God rewarded him in another way. First, He blessed him with a profitable business arrangement, and second, He taught him the importance and value of staying hopeful in the aftermath of disappointment.

HOPE WHEN THE HEAT IS ON

One way we develop hope is by learning about how God came through for people in seemingly hopeless situations, such as Pat and the businessman who sold large equipment. The Bible is also full of stories that give us hope and build our faith. One of them is the remarkable account of three young Hebrew men—Shadrach, Meshach, and Abednego—who insisted upon worshipping the one true God and refused to worship a golden image (see Daniel 3:1–6). As punishment for their rejection of the false god, the young men were thrown into a fiery furnace—turned up seven times hotter than its usual temperature. They would be incinerated instantly; their lives were finished; the situation was hopeless.

Was the situation really hopeless? Not with God involved. My favorite part of this story takes place when the king who tried to incinerate them looked into the furnace and saw a fourth man in the furnace with them (many scholars believe him to be the pre-incarnate Jesus), and then realized the chains or ropes with which the young men had been bound had become loose. When the three young men came out of the furnace unharmed, they received promotions in the king's government! Everything about a terrible, seemingly hopeless situation turned out well for them.

Like Shadrach, Meshach, and Abednego, you may be facing a situation that seems utterly hopeless. I urge you today to ask and allow the Lord to restore your hope. Hope simply says, "I believe something good can and will happen." Sadness, disappointment, and despair have to flee in the presence of hope. As long as you have hope, you will not be able to give up.

> *Sadness, disappointment, and despair have to flee in the presence of hope.*

On the Right Track

During his early years, Andrew Carnegie was expected to follow in his father's footsteps and become a weaver in his home country of Scotland. Prior to the industrial revolution, which brought steam-powered looms to Scotland in 1847, a weaver could make a decent living working with his hands. But when mechanized looms arrived, companies no longer needed hand loom weavers. When this happened to Carnegie's father, the family's standard of living dropped dramatically and his mother had to go to work. She opened a grocery store and made extra money mending shoes. Reflecting on those days, Carnegie wrote that that experience taught him "what poverty meant" as he watched his father beg for work. As a result, he wrote: "Then and there came the resolve that I would cure that when I got to be a man." He grew up with the determination to do all he could to provide jobs to hardworking people and to ensure he did not live an impoverished life.

In 1848, Carnegie's mother took a drastic step in an effort to avoid complete devastation and financial ruin for her family. She borrowed money to move herself, her husband, and her children to America. Two of her sisters had already settled in Pittsburgh, so the Carnegie family joined them there. Carnegie and his father both found work in a cotton factory.

Carnegie then took a job in Pittsburgh's telegraph office as a messenger boy. He distinguished himself by working hard and eagerly volunteering for new or additional responsibilities. At the same time, he developed a disciplined reading habit and found ways to see plays and experience other cultural and educational activities.

Through his work at the telegraph office, the diligent young

Carnegie caught the attention of Thomas A. Scott of the Pennsylvania Railroad. Scott soon hired Carnegie as his personal secretary. Through dedication and an outstanding work ethic, Carnegie moved through the ranks at the Pennsylvania Railroad and became Scott's successor as superintendent of the Pittsburgh division.

After the Civil War, Carnegie saw great potential in the iron industry, so he resigned from the railroad and set out to make a fortune in iron. And make a fortune he did. With smart decisions, calculated risks, and a keen eye for disciplined spending, he became one of the wealthiest men and greatest philanthropists of his day.

Carnegie believed "the man who dies rich dies disgraced," so he began giving away money as he grew older, focusing on educational and cultural causes. By the time of his death, he had funded the establishment of two thousand five hundred public libraries and given away approximately 350 million dollars.

Hard work and the relentless pursuit of his goals made Andrew Carnegie, once the impoverished son of an unemployed weaver, a man his peers called "the richest man in the world." He set his mind to becoming a wealthy entrepreneur and one who employed others. By working hard, taking advantage of opportunities, and embracing responsibility, he did it. When you set your mind to the accomplishment of your goals and do what you can to reach them, you will achieve them.

CHAPTER 18

LIKE A SUDDEN WIND

*"The highest reward for a man's toil is not what he gets for it
but what he becomes by it."*
JOHN RUSKIN

I once heard a story called "The Obstacle in Our Path" that I want to share with you.

In ancient times, a king placed a boulder on a roadway. Then he hid himself and watched to see if anyone would remove the huge rock. Some of the king's wealthiest merchants and courtiers came by and simply walked around it.

Many loudly blamed the king for not keeping the roads clear, but none did anything about getting the big stone out of the way. Then a peasant came along carrying a load of vegetables. On approaching the boulder, the peasant laid down his burden and tried to move the stone to the side of the road. After much pushing and straining, he finally succeeded. As the peasant picked up his load of vegetables, he noticed a purse on the road where the boulder had been. The purse contained many gold coins and a note from the king saying that the gold was for the person who removed the boulder from the roadway. The peasant learned what many others never understand.

The important lesson to remember is that work has its rewards. This is true in corporate settings, in your home, in your relationships, and in every area of life. If you are willing to pay the price, you can have the prize. The same principle applies in our spiritual lives as we walk with God. He requires us to work to accomplish what He calls us to do, to fulfill His purposes for our lives, and to follow the dreams He plants in our hearts; and He promises that if we are diligent, we will have a reward.

I believe many people are unhappy today because they only want to commit to things that are easy or convenient. This saddens me because these people often cheat themselves out of the rewards God has for them simply because they want to avoid difficulty. If they would be willing to exert effort, they would reap great benefits.

God wants to bless us in many ways. Sometimes He requires us to do something difficult before we receive certain blessings because He wants those blessings to be rewards for our diligence. There are no rewards for starting something; there are no rewards for quitting. But if we never give up, we *will* receive our rewards.

GOD IS A REWARDER

Perhaps the one verse that most clearly presents the promise that God rewards those who never give up is Hebrews 11:6: "He is the rewarder of those who earnestly and diligently seek Him [out]."

We must believe God is a *rewarder* of those who diligently seek Him. If we are diligent, we can expect rewards. We cannot manipulate God; we must believe His Word and do what He asks us to do. He tells us the diligent will be blessed, so we simply keep on keeping on and refuse to give up.

You know by now that doing something the right way one time does not necessarily bring great reward. It is doing what is right over and over and over and over that will bring the good result you desire. You must be diligent when everyone else around you slacks off or gives up. Diligence is not easy, but it has its rewards—and I guarantee you, they are worth every effort and every sacrifice you make.

THE WORK BEHIND THE REWARDS

There is no such thing as a "drive-through" breakthrough! You can get a drive-through hamburger or a drive-through milkshake; you can even drop off and pick up your dry cleaning at a drive-through, but you do not have the option of drive-through convenience if you want a reward from God. There are no shortcuts with Him. If you want the reward, you have to do the work.

At this season of my life, I am living in the rewards that come with years of diligence—not one year, not two years, not even ten years, but *decades* of diligence. Knowing I have obeyed God and followed His call on my life is the most important reward, and being able to live my dream of ministering to others is an added benefit. It thrills me to realize that although I did not have a great beginning in life due to the devil's attack, I am having a great finish and God has been exalted as King.

I want to remind you that I did not see rewards quickly. For many years I actually felt that I was doing all I knew to do and still not making much progress. Actually, I was making progress, but there were so many things in my life that needed fixing I couldn't get past what was still wrong in order to see what had changed for the better. I want you to know this so you will not be discouraged if you struggle with similar feelings. Most of the time, you really are making progress when you do not think you are moving forward even one inch. God is working in you and moving you along, little by little, in such a way that you may not even notice you are no longer where you were last year. But God notices; and He wants you to keep going, even if you only advance one inch at a time. The Bible says God delivers us from our enemies little by little (see Deuteronomy 7:22). Merely refusing to give up is a victory in itself.

> *God is working in you and moving you along, little by little, in such a way that you may not even notice you are no longer where you were last year.*

THERE'S A REASON FOR THOSE BLESSINGS

When we see people who seem to be showered with God's blessings and are prospering in life, people who appear perfectly at peace and are always joyful and seem to have everything work out in their favor, we wonder why they live such blessed lives. The world calls them lucky, but we know better. We know there is reason for the good life they now enjoy.

Most of the time, people we refer to as "blessed" did not start their lives in a blessed state. They have had their share of challenges and adversities; they have gone through difficulties; they have been diligent to live by God's Word; they have stayed faithful to Him; they have given to others when they had little to give; they have prayed when they did not feel like praying; and they have been patient, loving, and kind to people when they did not want to be nice to anyone. They did these things in obedience to God's Word, and because they knew doing them would bring joy to their lives.

When God saw that these people made wise decisions, followed Him wholeheartedly, and refused to give up over a period of time, He blessed them. He is a God who loves to bless us, but He does test us to see if we are determined and willing to live for Him even when we don't get what we would like.

In Matthew 5, Jesus says the pure in heart are blessed (see v. 8), but we must remember that becoming pure in heart is not easy or quick. He also says that the makers and maintainers of peace are blessed (see v. 9); but peacemaking and peacekeeping require humility, sacrifice, and a willingness to adapt and adjust to other people. God's blessings are provided freely by His grace, but we must be willing to obey Him if we want to enjoy them.

I have been through many difficult situations no one knows about. I had a lot of lonely times, and a lot of hard years. I endured many years of rejection and judgment and criticism as a woman in ministry. But now I have the reward. It was not easy, but it is worth it.

Your reward will come too if you will refuse to give up! I shudder to think what my life would be like now had I given up during one of those times I was so tempted to quit. I thank God that He kept pushing me forward and sometimes even dragging me while I was kicking and screaming about every inconvenience. At times, I felt as if I could not go backward; I didn't know how to go forward, but God would not let me quit. He provided people to help, money to pay bills, open doors, strength, wisdom, and everything else I needed. The road was narrow, steep, and difficult to climb, but I eventually realized it was the only road that was truly going anywhere. I am so glad now that I did not give up, and if you refuse to give up, your reward will come also.

TO THOSE WHO OVERCOME...

We have said that even though we like to be thought of as strong, determined, and able to overcome difficulties, we still try to avoid obstacles, opposition, and adversities that will eventually give us what we say we want. We can receive much encouragement from the book of Revelation because it is full of promises of rewards for those who overcome:

> *He who is able to hear, let him listen to and give heed to what the Spirit says to the assemblies (churches). To him who overcomes (is victorious), I will grant to eat [of the fruit] of the tree of life, which is in the paradise of God (Revelation 2:7).*

> *He who overcomes (is victorious), I will make him a pillar in the sanctuary of My God; he shall never be put out of it or go out of it, and I will write on him the name of My God and the name of the city of My God, the new Jerusalem, which descends from My God out of heaven, and My own new name (Revelation 3:12).*

*He who overcomes (is victorious), I will grant him to sit
beside Me on My throne, as I Myself overcame (was victorious) and sat down beside My Father on His throne (Revelation 3:21).*

These scriptures remind us of our heavenly reward when we overcome adversity with faith, perseverance, and character. We gain victory in our lives when we choose to embrace the difficult path of
overcoming.

SQUEEZE THROUGH THE NARROW PLACE

If you want to reach any worthy goal at all or do anything significant for God, you will find you always have to go through a narrow
place. Any time God leads you toward a broader place—a position
of greater influence, greater enjoyment in life, or a fulfilled desire—
you will have to squeeze through a narrow place. Your narrow place
may be a time when you have to walk away from negative relationships, when you have to discipline your mouth to speak positively
instead of complaining, or when you have to put yourself on such
a tight budget you can no longer afford some of the small pleasures
of life, such as a cup of gourmet coffee or a movie ticket. You may
have to choose to work while others are entertaining themselves.
These kinds of disciplines will squeeze you and press you, but they
will also lead you to the broad places and blessings God has for you.
When you are pressed and squeezed to the point you feel you can
hardly breathe, make sure you stay focused on the reward ahead. It
is like working all week and looking forward to payday on Friday.
When the devil tries to discourage you, just say aloud, "Payday is
coming!"

Jesus spoke of the narrow gate in Matthew 7:13–14: "Enter through
the narrow gate; for wide is the gate and spacious and broad is the
way that leads away to destruction, and many are those who are

entering through it. But the gate is narrow (contracted by pressure) and the way is straitened and compressed that leads away to life, and few are those who find it."

Many people never leave the broad road that leads to destruction. Why? Because there's plenty of room on the broad road. You will have plenty of company on the broad road. It is the easy way. Jesus calls it "spacious," and when I think of that, I think of a person who has room to carry all their fleshly baggage with them on that road! The broad road may be easy and wide, but it does not lead to anything good. In fact, it leads to destruction.

Those who really want to find life must follow what Jesus calls the "compressed" way. It's straight; it's narrow; it's not as easy to walk as the broad road. When we decide to walk that narrow path, God begins requiring more of us. He starts taking away some of our fleshly baggage. He pulls the padding out of our nest. He asks us to let go of some of our old ways, to adjust some of our attitudes, to raise the standards in our relationships and conversations, and to make some changes in the ways we spend our time and money. Living on the narrow path requires that we let God out of the "Sunday Morning Box" we often try to keep Him in and invite Him into every day and everything. We cannot have any places in our lives where God is not welcome.

Everything God asks us to do as we walk the narrow way may not be easy, but it will be good for us. It may not feel like a blessing as we go through it, but it leads to blessing in the end. It will require us to overcome some things, but it will lead us to wonderful rewards.

I encourage you to get on the narrow road and stay there. There may not be many people on it with you, but if you look carefully you will see Jesus because the narrow way is always the road He travels. Do not go back to the broad road when the narrow one loses its appeal. The broad road is deceptive. It may be fun for a while, it may be easy, but it leads to trouble. Pay the price to stay with God on the road that leads to life.

PAYDAY IS COMING

The promise of reward is one dynamic that keeps us moving in the right direction and refusing to quit through hard times. When you become discouraged or weary as you press through, do not give up, because the price you pay is worth it in the end. Payday is coming!

God may be giving you some tough assignments right now. He may be asking you to do some things you feel you absolutely cannot do, but I want you to know: if you will press in and press on and press through, God has a reward for you. The promise of reward is awesome. It keeps us motivated and encouraged.

A scripture that immediately comes to my mind when I think about the rewards God promises when we never give up is Hebrews 11:6, which tells us God "is the rewarder of those who earnestly and diligently seek Him [out]." We must always remember that God rewards those who diligently seek Him. His rewards may not come in the form you expect or at the time you expect, but they will come.

His rewards may not come in the form you expect or at the time you expect, but they will come.

One of my favorite passages of scripture is Isaiah 61:7–8: "Instead of your [former] shame you shall have a twofold recompense; instead of dishonor and reproach [your people] shall rejoice in their portion. Therefore in their land they shall possess double [what they had forfeited]; everlasting joy shall be theirs. For I the Lord love justice."

Look with me at the first part of the passage: "Instead of your [former] shame you shall have a twofold recompense." The word *recompense* means "reward," or "payment for past hurts." Recompense reminds me of the word *compensation*. When I think of workman's compensation, I think of payment made to a person who has been injured on a job. Always remember, if we get hurt while working for God, He takes care of us; He "pays" us, so to speak. We are not on the world's payroll; God is watching over us. If someone comes against us, if someone hurts us, if someone rejects us, if someone wounds us, we need to keep serving God and doing right, and He will make sure we get what we deserve in the end.

Realizing that I did not have to try to collect from people who hurt me was life-changing for me. The truth is, they could not pay me. They could not give me back what they took from me, but God can always give you more than people take from you.

We find another promise of reward in Joel: "And I will restore or replace for you the years that the locust has eaten—the hopping locust, the stripping locust, and the crawling locust, My great army which I sent among you. And *you shall eat in plenty and be satisfied* and praise the name of the Lord, your God, Who has dealt wondrously with you. And My people shall never be put to shame" (Joel 2:25–26 emphasis mine).

Note the words I have italicized: "you shall eat in plenty and be satisfied." This part of the promise means so much to me because I spent so many years dissatisfied and discontent. No matter what I had, I wasn't satisfied. No matter what anybody did for me, I wasn't satisfied. No matter what I accomplished, I was not satisfied. Why? Because only God can satisfy.

Whatever you have lost in your life, He will restore. That's a promise. He will repay what has been stolen from you. As you trust Him, He will make sure you "shall eat in plenty and be satisfied and praise the name of the Lord."

YOU'RE ON YOUR WAY

There is a reason you chose to read this book. I suspect something about its title, *Never Give Up!*, caught your attention. Maybe you have lived your entire life with a hope, a dream, or a goal you feel is so important it is worth never giving up on; you will make any sacrifice to see that dream come true or fight your way through any hardship to reach it. Perhaps you are facing a situation right now and you know the only way to survive it is to never give up. Maybe you have given up on something in the past and want to make sure you do not give up again.

Whatever the reason, I want to personally encourage you one more time to *never give up*. I also want to make sure you know God often moves suddenly when people have pressed in and pressed on for years, refusing to give up. Don't despair, because your breakthrough may "suddenly" come today, and if not today, then perhaps tomorrow or the next day. He hears every prayer you have ever prayed about your struggles, and He sees the commitment you demonstrate and watches you as you overcome adversity in pursuit of all He has for you. I believe there will be moments and seasons in which He will suddenly give you the provision you need, open the doors of opportunity you are waiting for, give you the ideas you are missing, align the relationships that need to come together, or do whatever you are longing for Him to do in your life. It will happen, and it may seem to be a "suddenly" for you.

I believe most of the time, only the *results* come suddenly. Behind every seemingly sudden victory, breakthrough, great opportunity, achievement, or blessing lies a degree of faithful perseverance and determination known only to those who have done the pressing.

> *Most of the time, only the* results *come suddenly.*

You may have to wait much longer than you'd like before you finally enjoy the satisfaction of a goal accomplished or a desire fulfilled. Waiting—and waiting patiently and expectantly—are part of never giving up. Whatever you are believing God to bring to pass in your life, be patient, enjoy the journey, and refuse to give up.

You may not be able to change your circumstances, but you can be determined to wait on God and trust Him to do it. You may have to wait longer than you thought or hoped you would, but God will not be late. While you wait, you can be confident God will come through for you. When He does, you will be so glad you persevered.

God has a wonderful, unique plan for you, and I encourage you to be determined to see the fullness of that plan come to pass in your life. When life seems difficult and you are tempted to quit, remember those who have gone before you. Remember those who never gave up and eventually enjoyed the rewards of their faith and hope.

Will you decide right now that no matter what happens in life, you will never give up? The choice is up to you, but I assure you, whatever you have to overcome to reach your goals and enjoy the success God has for you will be worth the effort.

Be determined; be committed; be patient; and be diligent in every area of your life—and whatever you do, never give up!

PRAYER OF SALVATION

The most important relationship of your life is a personal relationship with Jesus Christ. If you would like to receive Him as your Lord and Savior, and enter into the greatest relationship you have ever known, please pray the prayer below:

Father,

You loved the world so much You gave Your only begotten Son to die for our sins so that whoever believes in Him will not perish but have eternal life.

Your Word says we are saved by grace through faith as a gift from You. There is nothing we can do to earn salvation.

I believe and confess with my mouth that Jesus Christ is Your Son, the Savior of the world. I believe He died on the cross for me and bore all of my sins, paying the price for them.

I believe in my heart that You raised Jesus from the dead and that He is alive today.

I am a sinner; I am sorry for my sins; and I ask You to forgive me. By faith I receive Jesus Christ now as my Lord and Savior. I believe that I am saved and will spend eternity with You! Thank You, Father. I am so grateful! In Jesus' name, Amen.

APPENDIX

STORIES OF INDIVIDUALS WHO NEVER GAVE UP

ABOUT THE AUTHOR

JOYCE MEYER is one of the world's leading practical Bible teachers. A #1 *New York Times* best-selling author, she has written more than seventy inspirational books, including *The Confident Woman, I Dare You,* the entire *Battlefield of the Mind* family of books, her first venture into fiction with *The Penny,* and many others. She has also released thousands of audio teachings as well as a complete video library. Joyce's *Enjoying Everyday Life®* radio and television programs are broadcast around the world, and she travels extensively conducting conferences. Joyce and her husband, Dave, are the parents of four grown children and make their home in St. Louis, Missouri.

JOYCE MEYER MINISTRIES
U.S. & FOREIGN OFFICE ADDRESSES

Joyce Meyer Ministries
P.O. Box 655
Fenton, MO 63026
USA
(636) 349-0303
www.joycemeyer.org

Joyce Meyer Ministries—Canada
Lambeth Box 1300
London, ON N6P 1T5
Canada
1-800-727-9673

Joyce Meyer Ministries—Australia
Locked Bag 77
Mansfield Delivery Centre
Queensland 4122
Australia
(07) 3349 1200

Joyce Meyer Ministries—England
P.O. Box 1549
Windsor SL4 1GT
United Kingdom
01753 831102

Joyce Meyer Ministries—South Africa
P.O. Box 5
Cape Town 8000
South Africa
(27) 21-701-1056

OUR GOALS:

Me: Fast 3x's/montr

Joel " 2x's

FITNESS CONTEST: 10K

Me - Fitness Competition
in Brazil

Wedding invite